Research in Mathematics Education

Series editors
Jinfa Cai
James Middleton

More information about this series at http://www.springer.com/series/13030

Florence Mihaela Singer • Nerida F. Ellerton
Jinfa Cai
Editors

Mathematical Problem Posing

From Research to Effective Practice

Springer

Editors
Florence Mihaela Singer
Department of Educational Sciences
University of Ploiesti
Bucharest, Romania

Nerida F. Ellerton
Department of Mathematics
Illinois State University
Normal, IL, USA

Jinfa Cai
Department of Mathematical Sciences
University of Delaware
Newark, DE, USA

Research in Mathematics Education
ISBN 978-1-4614-6257-6 ISBN 978-1-4614-6258-3 (eBook)
DOI 10.1007/978-1-4614-6258-3

Library of Congress Control Number: 2015938695

Springer New York Heidelberg Dordrecht London

Printed on acid-free paper

Springer Science+Business Media LLC New York is part of Springer Science+Business Media (www.springer.com)

Foreword

This is the second book in the *Research in Mathematics Education* series. Since the publication of the first edition in 1983 of *The Art of Problem Posing* by Brown and Walter, there has been increased effort to incorporate problem posing into school mathematics at different educational levels around the world. In the field of mathematics education, problem posing has been viewed not only as a means to understand students' mathematical thinking but also as a means to teach mathematics with understanding. This volume has at least the following three features. First, it presents the state of the art of research in mathematical problem posing. Readers will be well informed about problem-posing research as a line of scientific inquiry. The 52 authors of the 26 chapters pay careful attention to both past accomplishment and future directions of studies. Thus, this book should be useful for graduate courses related to mathematical problem posing and problem solving or as a foundation upon which to propose lines of inquiry into problem posing. Second, this book includes many great ideas to assist those implementing problem-posing tasks into classrooms; many of these ideas have already been tested in classrooms. Thus, this book can be used by mathematics teacher educators for designing and implementing teacher professional development sessions for practicing teachers. Third, this book truly has an international scope. Authors from 16 different countries have not only used diverse conceptualizations of problem posing but also presented a wide range of approaches for investigating issues related to problem posing.

As we indicated in the Foreword of the first book of the series, *Research Trends in Mathematics Teacher Education*, we have designed the solicitation, review, and revision process of volumes in the series to produce thematic volumes, allowing researchers to access numerous studies on a theme in a single, peer-reviewed source. Our intent for this series is to publish the latest research in the field in a timely fashion. This design is particularly geared towards highlighting the work of promising graduate students and junior faculty working in conjunction with senior scholars. The audience for this monograph series consists of those in the intersection between researchers and mathematics education leaders—people who need the highest quality research, methodological rigor, and potentially transformative implications ready at hand to help them make decisions regarding the improvement of teaching, learning, policy, and practice. With this vision, our mission of this book series is:

1. To support the sharing of critical research findings among members of the mathematics education community
2. To support graduate students and junior faculty and induct them into the research community by pairing them with senior faculty in the production of the highest quality peer-reviewed, research papers
3. To support the usefulness, and widespread adoption, of research-based innovation

We are grateful for the support of Melissa James from Springer in developing and publishing this book series, as well as the support for the publication of this volume.

We thank the editors (Singer, Ellerton, and Cai) and all of the authors who have contributed to this comprehensive and insightful book!

Jinfa Cai
James Middleton

Preface

Mathematical Problem Posing Today: A Cross-Cultural View

The era of information and communication technology creates new social environments and needs. Living in a world where interdependency and dynamics become main features of the global society, young generations have to face unpredictable changes they should learn coping with. Consequently, education systems all over the world support (or at least should pay attention to) a very fast process of changing priorities. Inherently, teaching and learning strategies are influenced by this context.

As a practice of learning and thinking, problem posing may play an essential role in this change. Since 1970, when Paulo Freire introduced the term problem-posing education in his book *Pedagogy of the Oppressed* as a metaphor for emphasizing critical thinking, the problem-posing methodology extended to various domains of knowledge. Within learning environments that offer a range of activities, sources for study, opportunities for interaction, and an emphasis on exploration and application, students can actively construct meaning in both the natural and simulated worlds, in the classroom. Teachers and students might create knowledge together in a variety of contexts and generate and address critical questions about the knowledge they produce. In Freire's vision, all these could help to develop more democratic, diverse, critically thinking members of society.

Mathematics as a tool for rational thinking can play an important role in preparing the fluent thinkers needed in the dynamic world of today (and tomorrow). For a long time, both the mathematics community and school practice have ranked *problem solving* as the top component of the mathematical domain. However, arguments in favor of problem posing come from at least two directions: from the past, where history shows that problem posing is the agent of change within scientific paradigms, and from the future, where the knowledge economy and the knowledge society trigger unprecedented demands and put enormous pressure on educational systems all over the world. We started this book envisioning that a fresh look at problem posing is by all means a necessary step nowadays. The aim of *Mathematical Problem Posing: From Research to Effective Practice* is thus threefold: to present an updated overview of contemporary research on problem posing; to draw attention to successfully applied experiences; and to identify main directions for further research and new teaching and learning practices.

In a structured way, the book starts with multiple perspectives for defining the field of problem posing in the context of mathematics education, continues with the place problem posing holds in the school curriculum, and concludes with problem posing in teacher education programs and teacher professional development.

The book is multidimensional from a range of perspectives. From a conceptual view, the papers included in this collection present different epistemological, philosophical, and pedagogical approaches to problem posing. Concerning methodology, the studies of the volume range from qualitative research to quantitative meta-analysis. They range, with respect to the target population of students, from primary graders to intermediate and upper secondary grades. They also range, with respect to the target population of teachers, from preservice teachers (for all grades) to in-service teachers working at various levels of education. However, maybe the most important dimension of the book is its multicultural coverage. The authors come from different geographical areas: 16 countries are listed with the authors' affiliations (Australia, Belgium, Canada, China, Czech Republic, Israel, Italy, Japan, Malaysia, Norway, Romania, Serbia, Singapore, Sweden, the Netherlands, and the United States of America), from 4 continents, to which we can even add the diversity of backgrounds and experiences in a variety of cultural environments of many of the authors. This cultural diversity brings into the book various representations, expressions, knowledge, skills, and attitudes towards approaches to problem posing. The cultural diversity of authors' backgrounds makes the multiplicity of perspectives presented in the book deeply authentic. It also shows that problem posing is becoming more and more a global phenomenon.

The collection of articles in this book covers the way from research to effective practice by offering a large gamut of ideas, critical analyses, and successful experiences. The book starts with defining the field of problem posing in the context of mathematics education. In this first part, Jinfa Cai and his colleagues come up with a vision of problem posing as lenses for understanding and improving students' learning of mathematics. In a more specific approach, Ragnhild Hansen and Gert Hana show how problem posing can be emphasized in a modelling perspective, while Jasmina Milinković conceptualizes problem posing via transformation, and Sergei Abramovich and Eun Kyeong Cho explain how to use digital technology for mathematical problem posing. Further, Cinzia Bonotto and Lisa Dal Santo look at the connection between problem posing and creativity in relation to problem solving, while Vincent Matsko and Jerald Thomas explore ways to foster creativity in mathematics classrooms. Florence Mihaela Singer and Cristian Voica develop a framework for using problem posing as a tool for identifying and developing mathematical creativity.

The second part of the book provides practical examples of using problem posing in school mathematics teaching. Here, Victor Cifarelli and Volkan Sevim relate reformulation and sense-making within the problem-solving process to problem posing; Sharada Gade and Charlotta Blomqvist discuss the role of explicit mediation for developing problem-posing capacity of fourth and fifth graders, while Kees Klaassen and Michiel Doorman find in problem posing good opportunities for providing students with content-specific motives. A content like statistical literacy is seen by Lyn English and Jane Watson as a relevant opportunity for problem posing in the elementary school. Further, Mitsunori Imaoka, Tetsu Shimomura, and Eikoh Kanno describe effective ways of using computers for problem posing in upper grades, a topic that is rarely addressed. From Singapore, Kwek Meek Lin proposes a research experiment in which problem posing is used as an assessment tool in the lower secondary school. In the final two chapters of this part of the book, from a multicultural perspective, Xianwei Van Harpen and Norma Presmeg analyze the outcomes of a comparative investigation of high school students' mathematical problem posing in the United States and China, while Limin Chen, Wim Van Dooren, and Lieven Verschaffel come up with a design experiment for enhancing the development of Chinese fifth graders' problem-posing and problem-solving abilities, beliefs, and attitudes.

From a focus on students who are involved in problem posing in the classroom, as we have seen in second part of the book, the authors move to the (future) teacher who is to orchestrate such activities and discuss, in the third part of the book, mathematics problem posing in teacher education programs and teacher professional development. More specifically, Roslinda Rosli, Mary Margaret Capraro, and their colleagues address the relationship between problem solving and problem posing in a study with middle grade preservice teachers. The same relationship is addressed by Vrunda Prabhu and Bronislaw Czarnocha in the context of an integrated teaching/research methodology that has become known as Teaching-Research/New York City (TR/NYCity) methodology. Further, Rosa Leikin explains how to teach in a dynamic geometry environment and to use it as a tool for mathematical problem posing and geometry investigations by using examples from a course with prospective mathematics teachers. Ilana Lavy describes studies conducted in dynamic geometry environments that adopted "what if not" strategies. Todd Grundmeier provides details of the results of an exploratory study that incorporates problem posing in a mathematics course for prospective elementary and middle school teachers, where the content coverage included problem solving, data analysis and probability, discrete mathematics, and algebraic thinking. From a different perspective, problem posing was used as a motivational tool; this aspect is addressed by Alena Hošpesová and Marie Tichá in a study investigating primary school teacher training. Two other studies also explore ways in which problem posing has been investigated in preservice teacher training. Thus, Michal Klinshtern, Boris Koichu, and Avi Berman find unexpected perceptions of teachers as problem posers, while Helena Osana and Ildiko Pelczer succeed in classifying problem posing in mathematics professional development in a few distinct categories, despite the continuing paucity of empirical studies. Giving prospective elementary teachers the opportunity to pose personally and socially relevant mathematics problems is an important focus for Sandra Crespo's chapter. Finally, in a study with prospective and practicing middle school teachers, Nerida Ellerton shows how problem posing can become an integral component of the mathematics curriculum and introduces the concept of a *Pedagogy of Problem Posing*.

In the final part, we provide an overview of this book's special contributions to the field. We comment there how the book takes into consideration past literature, energizes present practices, and looks towards future learning, teaching, and research endeavors. Beyond the diversity of approaches and cultural spaces reflected into this collection, the book brings together the visions of experienced contemporary personalities who have researched and written on problem posing as well as those of some remarkable young professionals who have embarked on promoting this new and emerging field.

Bucharest, Romania Florence Mihaela Singer
Normal, IL, USA Nerida F. Ellerton
Newark, DE, USA Jinfa Cai
March, 2015

Contents

Contributors

Sergei Abramovich Department of Curriculum and Instruction, State University of New York at Potsdam, Potsdam, NY, USA

Avi Berman Department of Education in Science and Technology, Technion—Israel Institute of Technology, Haifa, Israel

Charlotta Blomqvist Berghemsskolan, Umeå University, Umeå, Sweden

Cinzia Bonotto Department of Mathematics, University of Padova, Padova, Italy

Jinfa Cai Department of Mathematical Sciences, University of Delaware, Newark, DE, USA

Mary Margaret Capraro Texas A&M University, College Station, TX, USA

Robert M. Capraro Texas A&M University, College Station, TX, USA

Limin Chen Liaoning Research and Training Center for Basic Education, Shenyang Normal University, Shenyang, China

Eun Kyeong Cho Department of Education, University of New Hampshire, Durham, NH, USA

Victor V. Cifarelli Department of Mathematics and Statistics, University of North Carolina at Charlotte, Charlotte, NC, USA

Sandra Crespo Michigan State University, East Lansing, MI, USA

Bronislaw Czarnocha Hostos Community College, City University of New York, New York, NY, USA

Lisa Dal Santo Department of Mathematics, Istituto Comprensivo Valdagno 2, Valdagno, Italy

Michiel Doorman Freudenthal Institute for Science and Mathematics Education, Faculty of Science, Utrecht University, Utrecht, The Netherlands

Nerida F. Ellerton Department of Mathematics, Illinois State University, Normal, IL, USA

Lyn D. English School of Curriculum, Queensland University of Technology, Brisbane, Australia

Sharada Gade Department of Science and Mathematics Education, Umeå University, Umeå, Sweden

The Graduate Centre, City University of New York, New York, NY, USA

Dianne Goldsby Texas A&M University, College Station, TX, USA

Elsa Gonzalez y Gonzalez Texas A&M University, College Station, TX, USA

Todd A. Grundmeier Department of Mathematics, California Polytechnic State University, San Luis Obispo, CA, USA

Gert M. Hana Department of Mathematics, Bergen University College, Bergen, Norway

Ragnhild Hansen Department of Mathematics, Bergen University College, Bergen, Norway

Alena Hošpesová University of South Bohemia, České Budějovice, Czech Republic

Stephen Hwang Department of Mathematical Sciences, University of Delaware, Newark, DE, USA

Mitsunori Imaoka Hiroshima University, Hiroshima, Japan

Chunlian Jiang Faculty of Education, University of Macau, Macau, China

Eikoh Kanno Department of Mathematics Education, Aichi University of Education, Kariya, Japan

Kees Klaassen Freudenthal Institute for Science and Mathematics Education, Faculty of Science, Utrecht University, Utrecht, The Netherlands

Michal Klinshtern Department of Education in Science and Technology, Technion—Israel Institute of Technology, Haifa, Israel

Boris Koichu Department of Education in Science and Technology, Technion—Israel Institute of Technology, Haifa, Israel

Meek Lin Kwek Department of Mathematics, Raffles Girls' School (Secondary), Singapore

Ilana Lavy Department Head of Management Information Systems, The Academic College of Yezreel Valley, Yezreel Valley, Israel

Roza Leikin Faculty of Education, University of Haifa, Haifa, Israel

Vincent J. Matsko University of San Francisco, San Francisco, CA, USA

Jasmina Milinković Teacher Education Faculty, University of Belgrade, Belgrade, Serbia

Anthony J. Onwuegbuzie Sam Houston State University, Huntsville, TX, USA

Helena P. Osana Department of Education, Concordia University, Montréal, QC, Canada

Ildiko Pelczer Department of Mathematics and Statistics, Concordia University, Montréal, QC, Canada

Vrunda Prabhu Formerly at Bronx Community College, City University of New York, New York, NY, USA

Norma Presmeg Department of Mathematics, Illinois State University, Normal, IL, USA

Roslinda Rosli The National University of Malaysia, Bangi, Malaysia

Volkan Sevim School of Education, Virginia Commonwealth University, Richmond, VA, USA

Tetsu Shimomura Department of Mathematics Education, Hiroshima University, Hiroshima, Japan

Steven Silber Department of Mathematical Sciences, University of Delaware, Newark, DE, USA

Florence Mihaela Singer Department of Educational Sciences, Faculty of Letters and Science, University of Ploiesti, Bucharest, Romania

Jerald Thomas Graduate School of Education, Aurora University, Aurora, IL, USA

Marie Tichá Institute of Mathematics of the Academy of Sciences of the Czech Republic, University of South Bohemia, Prague, Czech Republic

Wim Van Dooren Faculty of Psychology and Educational Sciences, University of Leuven, Leuven, Belgium

Xianwei Van Harpen Department of Mathematical Sciences, University of Wisconsin-Milwaukee, Milwaukee, WI, USA

Lieven Verschaffel Faculty of Psychology and Educational Sciences, University of Leuven, Leuven, Belgium

Cristian Voica Department of Mathematics, University of Bucharest, Bucharest, Romania

Jane M. Watson Faculty of Education, University of Tasmania, Hobart, Australia

About the Authors

Sergei Abramovich is Professor of Mathematics Education in the Department of Curriculum and Instruction at the State University of New York, Potsdam, United States. He was born in St. Petersburg (then Leningrad), Russia, did all his studies at the university there, and earned Ph.D. in mathematics in 1981. During the last two decades he has been working with preservice and in-service teachers, public school faculty and their pupils, and colleagues in mathematics education and mathematics and has been disseminating results of his research (both individual and collaborative) through presentations at international conferences across 20 countries. During his career he has authored/coauthored 170 publications and is a sole author/editor of 5 books on the use of computers in education. He is the founding editor of the International Mathematical Virtual Institute *Open Mathematical Education Notes, coedits the journal Spreadsheets in Education*, and serves on editorial boards of other six professional journals.

Avi Berman is Professor Emeritus of Mathematics at the Technion—Israel Institute of Technology. His research interests include Matrix Theory: Nonnegative Matrices, Completely Positive Matrices, Combinatorial Matrix Theory, and Stability; Mathematics Education: Teaching Linear Algebra and Giftedness and Creativity. At the Technion he holds the Israel Pollak Chair and served as the head of the Department of Education in Science and Technology, the Center for Pre-University Education, and the Interdepartmental Committee on Applied Mathematics. He was a member of the Senate Steering Committee and the Board of Governors Council. He also was the head of the Israeli Society for the Promotion of and Research on Giftedness and Creativity. He is the author of 6 books, 4 edited books, and more than 100 papers. He has supervised 22 Ph.D. students, 21 M.Sc. students, and 6 postdoctoral fellows.

Charlotta Blomqvist is presently Rektor at Grisbacka Elementary School, Umeå, Sweden. Prior to being Rektor, she has for over a decade taught Mathematics, Swedish, Science, History, and English at Berghem Elementary School, Umeå.

Charlotta's expertise includes working as a Developer of Mathematics at the Umeå Municipality and also as an Expert at Skolverket, the Swedish National Agency for Education, Stockholm.

Cinzia Bonotto is full Professor in the scientific area MAT 04 (Mathematics Education, History and Foundation of Mathematics) in the Department of Mathematics at the University of Padova (IT). She is the scientific director of research projects on the teaching of mathematics, mainly at the pre-university level, and has a long experience in teacher training at all levels of schooling. Presently her research themes regard mathematical modelling, problem solving, and problem posing.

Jinfa Cai is a Professor of Mathematics and Education and the director of Secondary math education at the University of Delaware. He is interested in how students learn mathematics and solve problems and how teachers can provide and create learning environments so that students can make sense of mathematics. He received a number of awards, including a National Academy of Education Spencer Fellowship, an American Council on Education Fellowship, an International Research Award, and a Teaching Excellence Award. He has been serving on the Editorial Boards for several international journals, such as the *Journal for Research in Mathematics Education*. He was a visiting professor in various institutions, including Harvard University, Beijing Normal University, and East China Normal University. He has served as a Program Director at the U.S. National Science Foundation (2010–2011) and a co-chair of American Educational Research Association's Special Interest Group on Research in Mathematics Education (AERA's SIG-RME) (2010–2012). He will be chairing a plenary panel at the ICMI-13 in Germany in 2016.

Mary Margaret Capraro graduated from the University of Southern Mississippi. She joined Texas A&M University in 2000 as a clinical professor in Mathematics Education. She earned a position as an Assistant Professor in the Department of Teaching, Learning, and Culture in 2007 and was promoted to Associate Professor in 2011. Her research interests include teacher knowledge and preparation in mathematics education and student understanding of mathematical concepts. She was previously employed with the Miami Dade County Schools as both a teacher and an assistant principal. She has 75 peer-reviewed articles and 125 national and international presentations. She is currently Co-PI of the Aggie STEM Center and the STEM Collaborative working extensively with public schools and school districts around Texas planning mathematics PDs and designing interdisciplinary **project-based learning activities. She is associate editor of the *Middle Grades Research Journal*.**

Robert M. Capraro is Co-Director of Aggie STEM, Director of STEM Collaborative for Teacher Professional Learning, and Professor of Mathematics Education in the Department of Teaching Learning and Culture at Texas A&M University. Dr. Capraro's expertise is applied research in school settings, program evaluation, the teacher as change agent for STEM school improvement, and STEM student achievement. He recently received the best paper award from the International Conference on Engineering Education. He is currently involved in research in 8 school districts with more than 20,000 students and 80 teachers. His editorial work includes Associate Editor of the *American Educational Research Journal*, *School Science and Mathematics*, and *Middle Grades Research Journal*. He is the author or coauthor of 3 books, several book chapters, and more than 100 articles on mathematics education, quantitative research methods, and teacher education. His total external funding is ~7 million. He co-directs the Aggie STEM Camp each summer.

Limin Chen obtained in 2011 the degree of Doctor in Educational Sciences at the University of Leuven, Belgium. She is a lecturer working in the Liaoning Research and Training Centre for Basic Education of Shenyang Normal University, and in charge of in-service mathematics teachers' training in Liaoning Province. Her major research interest is educational psychology, mathematics education, and teacher education. Limin Chen is a member of the Middle School Mathematics Education Committee of Liaoning Province and Cognitive Science Society of Liaoning Province.

Eun Kyeong Cho is an Associate Professor in the department of education at the University of New Hampshire in the United States. Dr. Cho has earned an Ed.D. in Curriculum and Teaching from Teachers College, Columbia University, in 2005. Her research interests are preservice teacher preparation, professional development, technology integration, and diversity. Regarding the topic of mathematical problem posing, she has presented and published papers on the technology-enabled problem posing since 2006. She currently serves as a consulting editor of a professional journal, *Young Children*, published by the National Association for the Education of Young Children (NAEYC), and a consulting editor of a research journal, *Early Childhood Research Quarterly* (ECRQ).

Victor V. Cifarelli is Professor of Mathematics Education and the Coordinator of the Mathematics Education Program in the Department of Mathematics and Statistics at the University of North Carolina at Charlotte. He received his Ph.D. in Mathematics Education from Purdue University. His doctoral research focused on the role of reflective abstraction as a learning process in mathematical problem solving. Dr. Cifarelli has presented his research at AERA, PME, and PME-NA. His articles have appeared in *Constructivist Foundations, the Journal of Mathematical Behavior*, *Focus on Learning Problems in Mathematics*, and the Proceedings of PME and PME-NA. In addition to his research, Dr. Cifarelli has co-directed numerous projects for mathematics teachers under the Eisenhower Professional Development Program and the North Carolina Department of Public Instruction.

Sandra Crespo is an Associate Professor of mathematics education at Michigan State University. Her research focuses on the curriculum and pedagogy of teacher education. She seeks to understand how to support novices to learn ambitious and equity-oriented mathematics teaching. She is especially interested in the practice of problem posing because it redistributes power dynamics in the classroom with regard to who produces and who consumes mathematical knowledge. Problem posing is associated with the production of disciplinary knowledge and with critical pedagogy. This makes problem posing a key practice for teachers of mathematics to develop as learners of mathematics and of mathematics teaching. She is currently involved in a longitudinal study of prospective teachers documenting how they develop their problem posing practices over time. Her work has been published in *Educational Studies in Mathematics*, *Journal of Mathematics Teacher Education*, *For the Learning of Mathematics*, *and Teaching Children Mathematics*.

Bronislaw Czarnocha a quantum physicist, entered mathematics education through the practice and formulation of mathematics teaching-research in high schools and colleges as the integration of Action Research and teaching experiment methodology of Vygotsky. He had formulated a distinct branch of teaching-research, TR/NYCity model. He has been co-PI on several NSF and European Community grants as well as on several CUNY grants with interest in the development of schema of thinking with the help of the Triad of Piaget and Garcia. More recently he has been investigating rich implication of Koestler's bisociation theory of the Aha! Moment for student learning. Bisociation that is "the spontaneous leap of insight…which connects previously unconnected matrices of experience [frames of reference] and makes us experience reality on several planes at ones" provides an intrinsic characterization and explanation of teaching-research as a uniquely creative bisociative framework, which connects habitually not connected domains of teaching and research.

Lisa Dal Santo is currently a primary teacher in the Istituto Comprensivo di Valdagno, Vicenza (IT). She has graduated in Primary Teacher Education at the University of Padua. Her research themes regard problem posing, problem solving, and creativity in mathematics. She attended a course in Educational Robotics and she has a specialization in Clinical Pedagogy.

Michiel Doorman works as a researcher and teacher trainer at the Freudenthal Institute for Science and Mathematics Education at Utrecht University in the Netherlands. His interests are context-based mathematics education, the role of modelling in learning mathematics, and coherency between mathematics and science teaching. He has been involved in various national projects on curriculum design. Currently, he is involved in EU projects with an emphasis on inquiry-based learning for developing flexible problem-solving skills and improving the relevancy of mathematics and science for students.

Nerida F. Ellerton is Professor of Mathematics at Illinois State University. She holds two doctoral degrees—in Physical Chemistry (Adelaide) and in Mathematics Education (Victoria University, NZ). She was Dean of Education at the University of Southern Queensland (1997–2002) and Professor at Edith Cowan University (1992–1997). She has taught in schools and universities and served as consultant in numerous countries. She has authored more than 150 peer-reviewed articles and 19 books including *Rewriting the History of School Mathematics in North America 1607–1861*, and *Abraham Lincoln's Cyphering Book and Ten other Extraordinary Cyphering Books*, both published by Springer. She was editor of the *Mathematics Education Research Journal* (1993–1997). Since 2011 she has been Associate Editor of the *Journal for Research in Mathematics Education*. Research interests include history of mathematics education, language factors in mathematics learning and teaching, problem posing, algebra education, and mathematics education in Southeast Asia.

Lyn D. English is a Professor of mathematics education at the Queensland University of Technology. Her research has been supported by continuous grants from the Australian Research Council since 1990 and includes statistics education, engineering education, mathematical modelling and reasoning, and mathematical problem solving and posing. Lyn is founding editor of the international journal, *Mathematical Thinking and Learning*, and is editor (jointly with David Kirshner) of the third edition of the *Handbook of International Research in Mathematics Education*. She is a fellow of *The Academy of the Social Sciences in Australia and* is founding editor of the international journal, *Mathematical Thinking and Learning*. Lyn is a recipient of the *Mathematics Education Research Group of Australasia* Career Research Medal.

Sharada Gade is currently Assistant Professor in Mathematics Education, at Umeå University, Sweden. In collaboration with teachers largely at Grades 4–6 and 7–9, she has conducted close-to-practice classroom research since 2008. For such theory/practice manner of practitioner inquiry, she has drawn on narrative as well as cultural historical activity theory and/or CHAT perspectives. Such work falls at the intersection of three domains, namely CHAT perspectives, practitioner inquiry, and mathematics education. Subsequent to her Doctorate from the University of Agder, Norway, Sharada has held Postdoctoral Visiting Fellowships at Homi Bhabha Centre of Science Education (TIFR), Mumbai, India; Umeå Mathematics Education Research Centre, Umeå University; The Graduate Centre, City University of New York (CUNY), USA; and The Oxford Centre for Sociocultural and Activity Theory Research (OSAT), Department of Education, University of Oxford, United Kingdom.

Dianne Goldsby a clinical Full Professor, has been a clinical faculty member of the Teaching, Learning and Culture Department at Texas A&M University for 12 years. She usually teaches four undergraduate/graduate courses each semester. Her primary focus is teaching mathematics education courses, including mathematics methods, a problem-solving course, and integrated mathematics and science. She has authored peer-reviewed journal articles and research summaries and presented over 60 times at international, national, regional, and local conferences. She is currently an SERA division co-chair, reviewer for mathematics teaching journals, associate editor of the newsletter *Focus on Mathematics Pedagogy and Content*, and a former associate editor for *School Science and Mathematics Journal*. She is the coordinator for the department's three online M.Ed. programs. Her research interests center on preservice teacher perceptions of mathematics and mathematics teaching.

Elsa Gonzalez y Gonzalez is Assistant Professor in the Department of Educational Leadership, Curriculum and Instruction since September 2014 at Texas A&M—Corpus Christi; previously she was a Visiting Assistant Professor and Senior Research Associate in the department of EAHR at Texas A&M University and has been working in the College of Education and Human Development since 1998. Her research interests include higher education leadership, methodological issues in cross-language qualitative data analysis, women in higher education, and access and retention of underrepresented women and Latina students and faculty in STEM fields.

Todd A. Grundmeier is a Professor of Mathematics in the Mathematics Department at Cal Poly. Dr. Grundmeier's teaching interests are varied and include all classes in the undergraduate mathematics curriculum with a focus on classes for prospective elementary school teachers and prospective secondary school mathematics teachers. Professionally, Dr. Grundmeier's research focus is undergraduate mathematics education and includes interests in problem posing, geometry, calculus, technology, and self-inquiry. He is also the Co-PI of the Cal Poly Robert Noyce Scholarship program and has done work on many state and federally funded projects in mathematics education.

Gert M. Hana is an Associate Professor at Bergen University College (BUC) in Norway. He received his Ph.D. degree from the University of Bergen, within theoretical mathematics. Hana is engaged in the Department of Mathematics at BUC, where he is involved in teaching and Educational Research. He can be contacted at gmh@hib.com.

Ragnhild Hansen is an Associate Professor at Bergen University College (BUC) in Norway. She received her master and Ph.D. degrees from the University of Bergen, both within applied mathematics. Her bachelor degree includes subjects from mathematics, physics, informatics, and statistics. Hansen has been engaged as a researcher in projects at the Institute of Marine Research and at Nansen Environmental and Remote Sensing Center, both in Bergen. From her education and professional engagements, she has got experience with mathematical models and modelling. Her employment at BUC started in 2006 in the Department of Mathematics. In addition to teaching, she is involved in Educational Research. She can be contacted at rhan@hib.com.

Alena Hošpesová is an Associate Professor of the Faculty of Education at the University of South Bohemia in České Budějovice. Her research interests focus mainly on primary school teacher education (pre- and in-service) and on ways of improving teachers' competencies (via self and joint reflection, problem posing, methods of inquiry-based education). She cooperates on this topic with Marie Tichá.

She is an active member of the Czech and international research communities (e.g., she was a member of the International Committee of The International Group for the Psychology of Mathematics Education (PME); she was one of the organizers of Thematic Working Groups on teacher education on several recent conferences of ERME; she is a member of informal community Learners' Perspective Study). She is a member of several Councils of Ph.D. Studies in didactics of mathematics in the Czech Republic, and she supervises Ph.D. students.

Stephen Hwang is currently a postdoctoral researcher working with Jinfa Cai in the Department of Mathematical Sciences at the University of Delaware. His research interests include the teaching and learning of mathematical justification and proof, the nature of practice in the discipline of mathematics, the development of mathematical habits of mind, and mathematics teacher preparation.

Mitsunori Imaoka was a Professor in the Department of Mathematics Education at Hiroshima University until 2012 and a member of meetings organized by MEXT of Japan to study and discuss improvements to the national curricula for high school mathematics during 1998–1999 and 2004–2009. He is now teaching mathematics in Hiroshima Institute of Technology. In his own teaching in the 1990s, he found that student problem posing produced authentic mathematical thinking. He therefore became motivated to research effective ways of involving upper-grade students in problem-posing. His first report on his problem-posing work was published in the Journal of JASME, 2001. Since then, in collaboration with Tetsu Shimomura and Eikoh Kanno, he has continued practical computer studies on problem posing by prospective teachers, and on various types of problem-posing activities by high school students.

Chunlian Jiang is currently an Assistant Professor in the Faculty of Education, University of Macau (UM), Macao. Before joining UM, she was an Associate Professor in the Department of Mathematics and Statistics, Central China Normal University (CCNU), China. She received a Ph.D. degree from Nanyang Technological University, Singapore, a master's degree in mathematics education, and a bachelor degree in mathematics from CCNU. She has teaching experience in high schools in both China and Singapore. Her research interest includes mathematical problem solving, mathematical problem posing, mathematics Olympiad, education of mathematically gifted students, and use of IT in mathematics teaching and learning. She has published articles in these areas as well as reflective articles on practical teaching and learning of mathematics in classrooms. She served on the editorial board for the *Mathematics Bulletin* in 2004–2008.

Eikoh Kanno taught in public high schools in Aichi Prefecture, Japan, from 1999 to 2012. He is currently a part-time lecturer in Aichi University of Education. When teaching high school mathematics classes, he frequently put into practice teaching approaches that incorporated problem posing. During the last 10 years, he has presented papers at workshops organised by the Japan Society of Mathematical

Education. In the mathematics classes in which he incorporated problem posing, his students learned actively.

Kees Klaassen received an M.Sc. degree in theoretical physics in 1986 and in 1995 completed his Ph.D. in physics education. He is currently a senior researcher in the Freudenthal Institute for Science and Mathematics Education at Utrecht University in the Netherlands. His main research interest concerns the development of coherent learning pathways along which students know all the time what they are doing and why. Further interests are the nature and role of theory in design-based research, educational challenges surrounding cross-cutting scientific subjects such as macro-micro thinking, and the facilitation of science education by tapping core causal knowledge as advance organizer.

Michal Klinshtern got her first and second degrees in Applied Mathematics from the Technion—Israel Institute of Technology and is now a Ph.D. student in Mathematics Education in the Department of Education in Science and Technology at the Technion. Her Ph.D. research project focuses on teachers' use of mathematical problems in their teaching. She serves as a teaching assistant, in several service courses in the Department of Mathematics at the Technion.

Boris Koichu is the head of mathematics education R&D group in the Department of Education in Science and Technology at the Technion—Israel Institute of Technology. His research focuses on learning for and through mathematical problem solving and problem posing, with special focus on the interplay of cognitive, affective, and situational factors involved. He is an author of more than 60 research papers published in peer-reviewed journals, edited books, and peer-reviewed conference proceedings and is a coeditor of a book on mathematical creativity and giftedness. He is a member of the *European Mathematical Society Committee for Education* and a member of the *International Committee of Mathematical Creativity and Giftedness*. He received an M.Sc. in mathematics from Lviv State University, Ukraine, and Ph.D. in mathematics education from the Technion. Following 2-year postdoctoral fellowship at the University of California, San Diego, he joined the faculty of the Technion in 2006.

Meek Lin Kwek is a mathematics educator at Raffles Girls' School (Secondary) in Singapore. In her role as a teacher specialist in the school's Centre for Pedagogical Research and Learning, she played an instrumental role in establishing the Centre's basic research infrastructure for engendering a culture of informed practice in the community. A teacher-researcher herself, Meek Lin collaborated and supported other teachers in practitioner inquiry projects with the objective of developing teacher capacity. In her current position as the school's Lead Teacher, she continues to extend these efforts through her academic leadership in the Professional Learning Academy.

Meek Lin holds a Master of Education in Mathematics Education from the National Institute of Education, Nanyang Technological University. Her research interests include assessment, problem solving, and teacher knowledge. Her research studies often culminated in conference presentations including, for example, "Dual conceptions of Functions? What do you mean?" and "Alternative Assessments for Learning in Mathematics."

Ilana Lavy is an Associate Professor with tenure at the Yezreel Valley College and is the department head of Management Information Systems. Her Ph.D. dissertation (at the Technion, Haifa, Israel) focused on the understanding of basic concepts in elementary number theory. After finishing doctorate, she was a postdoctoral research fellow at the Education faculty of Haifa University. Her research interests are in the field of preservice and mathematics teachers' professional development as well as the acquisition and understanding of mathematical and computer science concepts. She has published over 90 papers and research reports (part of them is in Hebrew).

Roza Leikin is a Professor of Mathematics Education and Education of Gifted students in the Faculty of Education, University of Haifa. All her research, design, and implementation activities embrace three interrelated areas: Mathematical creativity and ability; Mathematics teacher knowledge and professional development; Mathematical challenges in education. Lately she is interested in advancing contribution of neurocognitive research methodologies to the field of mathematics education. In the University of Haifa, she is the Founding Director of the Interdisciplinary Center for the Research and Advancement of Giftedness and Excellence (RANGE), the Head of the M.A. program in Gifted and Talented Education, and the Director of the Edmond J. Safra Brain Research Center for the Study of Learning Disabilities. Roza Leikin is the President of MCG—the International Group for Mathematical Creativity and Giftedness (affiliated with ICMI http://igmcg.org/)—and the Head of the National Advisory Mathematics Education Council of the Israel Ministry of Education. She edited 10 volumes related to research in mathematics education and published about 130 papers in peer-reviewed research journals, books, and conference proceedings. For more details, see http://ps.edu.haifa.ac.il/roza-leikin.html

Vincent J. Matsko studied at Carnegie Mellon University. After teaching at the university for several years, he began teaching at secondary schools for students especially talented in mathematics and science. It was then that he became interested in fostering students' creativity in mathematics, especially when it came to problem posing. He is an avid problem writer himself and enjoys the challenge of creating novel problems. Please also see www.vincematsko.com.

Jasmina Milinković works as an Associate Professor at the Teacher Education Faculty, University of Belgrade, Serbia, where she teaches Methodology of Teaching Mathematics, Mathematical Games, Mathematics Assessment, and related courses. She earned her doctoral degree at the Teacher Education Faculty, University of Belgrade, Serbia, after completing her master's degree at the University of Wisconsin—Madison, Department of Curriculum and Instruction. Earlier, she had graduated from the Mathematics Department at the University of Belgrade, Serbia, and from the Teachers Training Academy for Elementary School Teachers. Her research interests have focused on the implications of theory of representations in different areas related to the learning and teaching of mathematics. Besides writing research papers, as associate of the Mathematics in Context research project, she was coauthor of the Mathematics in Context middle school curriculum. She has also written mathematics textbooks for primary grades in Serbia.

Anthony J. Onwuegbuzie is Professor in the Department of Educational Leadership and Counseling at Sam Houston State University. He teaches doctoral-level courses in qualitative research, quantitative research, and mixed research, including program evaluation. His research areas include disadvantaged and underserved populations such as minorities, children living in war zones, students with special needs, and juvenile delinquents. Additionally, he writes extensively on an array of qualitative, quantitative, and mixed methodological topics. Dr. Onwuegbuzie has secured the publication of more than 400 works, including more than 300 journal articles, 50 book chapters, and 2 books. His current *h-index* is 58, which indicates sustained impact. Dr. Onwuegbuzie has made more than 700 presentations and 100 workshops worldwide that include more than 25 keynote addresses across 6 continents. Dr. Onwuegbuzie is former editor of American Educational Research Association's flagship journal, *Educational Researcher* (*ER*). He is currently a coeditor of *Research in the Schools*.

Helena P. Osana is Associate Professor in the Department of Education at Concordia University in Montreal, Canada. She teaches elementary mathematics methods courses in the teacher education program and courses on cognition, mathematics education, and research methods at the master's and Ph.D. levels. She has received several grants from the Social Sciences and Humanities Research Council of Canada and the Fonds de recherche société et culture Québec for her research on the development of children's thinking in mathematics and the instructional conditions that foster their learning in the classroom. Her particular focus is on the pedagogical factors that positively impact children's meaningful use of symbols and manipulatives. She also investigates the problem solving and algebraic thinking of elementary preservice teachers and is actively involved in professional development initiatives for in-service teachers on children's thinking and inquiry in the mathematics classroom.

Ildiko Pelczer is a Ph.D. student in Mathematics Education at Concordia University in Montreal, Canada. Her research focuses on preservice teachers' mathematical knowledge for teaching with a special interest on their specialized content knowledge. She has collaborated on studies concerning pedagogical interventions to enhance teachers' problem posing abilities and studies on identifying features of mathematical creativity in students' thinking.

Vrunda Prabhu whose Ph.D. in Point Set topology addressed fine points of Čech compactification, was a born educator from her early childhood. She was a Montessori mother who incorporated the philosophy embodied in Maria Montessori's book *The Discovery of the Child* into her son's upbringing. Later, while working with a Dalit community in Tamil Nadu, India, she had formulated the new program "Montessori for Mothers" as the tool to "discover the child" and as the foundations for a new literacy campaign. Vrunda had acquired her initial teaching-research skills in the RUMEC community studying and applying APOS, Piaget-based learning theory into calculus. A co-PI on two NSF grants and several internal CUNY grants, she had focused her attention on the design of creative learning environment in remedial mathematics classes, culminating in the discovery of Koestler's bisociation theory of the Aha! Moment for teaching mathematics. She passed away prematurely in the spring of 2013 leaving behind a wealth of education materials as well as several notebooks of her poetry, both being published posthumously.

Norma Presmeg is Professor Emerita in the Mathematics Department of Illinois State University. Her research interests include a long-standing interest in creativity, going back to her thesis on Albert Einstein's creative thought, which led to her research interest in visualization in mathematics education. Mathematical problem solving, problem posing, and the role of metaphors in mathematics education were themes that emerged from this research, along with the importance of affective issues. Concurrently, the role of culture in learning and teaching mathematics was a theme that led to interest in ethnomathematics. All of these threads are supported by a theoretical interest in semiotics, which is ongoing, and more recently by the potential of transcendental phenomenology to introduce new questions in research in mathematics education.

Roslinda Rosli recently completed her doctoral studies in the Department of Teaching, Learning, and Culture at Texas A&M University with an emphasis area in mathematics education. She is a senior lecturer in the Department of Teaching and Learning Innovation at the National University of Malaysia. She teaches statistics, research methodology, and mathematics education courses. Dr. Roslinda presented at over a dozen local, national, and international conferences and is currently engaged in continuing to grow her resume of published works. Her research areas include mathematical knowledge for teaching, problem posing, problem solving, and teacher education. This book chapter was part of the scholarly work from her Ph.D. dissertation research, and she is looking forward to publishing additional works in the near future.

Volkan Sevim is Assistant Professor of Mathematics Education in the School of Education at Virginia Commonwealth University. He received his Ph.D. in Curriculum and Instruction with a specialization in Mathematics Education from the University of North Carolina at Charlotte. His doctoral research focused on students' understanding of quadratic functions. His current research examines learning processes involved in the coevolution of problem solving and problem posing. Dr. Sevim has presented his research at NCTM, AERA, PME, and PME-NA. Dr. Sevim has taught upper level mathematics courses at a public high school in Charlotte, NC, for 5 years. In the last 4 years, he has been teaching mathematics education methods courses to prospective elementary teachers at university level.

Tetsu Shimomura is a Professor in the Department of Mathematics Education at Hiroshima University. My field of research is analysis, especially Sobolev functions. Since 2002, I have continued the practical study on problem posing by university students of prospective teachers and by high school students.

Steven Silber is currently a doctoral student in the Department of Education at the University of Delaware, working with Jinfa Cai in the Department of Mathematical Sciences. His research interests include students' and teachers' mathematical thinking and reasoning while engaging in mathematical tasks, and first-year undergraduate students' transition to college-level mathematics courses.

Florence Mihaela Singer is a Professor at the University of Ploiesti, Romania. She is a recipient of a Fulbright grant from Harvard University. She has authored/coauthored more than 280 scientific works, including research articles published in *New Ideas in Psychology*; *Journal of Mathematical Behavior*; *Teaching and Teacher Education*; *ZDM*; *Mind, Brain, and Education*; and mathematics textbooks for grades 1–12. She has worked as international consultant in Kuwait, Republic of Moldova, Tajikistan (within the World Bank education programs), and Estonia (under the Council of Europe contract). She had 12 talks as invited professor/plenary lecturer abroad and she served as scientific reviewer for various journals, international projects, and research reports. Her domains of expertise are curriculum design and development, teacher training, textbook development (mathematics), and educational management. Her domains of interest are cognitive science and neuroscience, with a special focus in the psychology of learning.

Jerald (Jay) Thomas is an Associate Professor of education at Aurora University in Aurora, IL (USA). His teaching focuses on research methods in the areas of learning theory, motivation, and assessment. He is a past president of the National Consortium for Secondary STEM Schools (NCSSS) and serves as the editor of the NCSSS Journal. His research explores motivation and creativity among secondary students who are gifted and talented in STEM fields. He has also authored, coauthored, and contributed to numerous books and articles in the area of gifted STEM education.

Marie Tichá is a research worker at the Institute of Mathematics of the Academy of Sciences of the Czech Republic, department of Didactics of Mathematics. She is also an external teacher in the Faculty of Education of Charles University in Prague. She focuses on the study of possibilities of developing mathematical literacy of students aged 6–12 years. In connection with it she studies in cooperation with A. Hošpesová the ways of improvement of both in-service and preservice teachers' professionalism with an emphasis on their subject didactic competence. They concentrate on the benefits of problem posing, especially on the potential of the use of problem posing accompanied by a qualified pedagogical reflection as motivational, educational, as well as diagnostic tool. Recently they strengthened attention to opportunities of inquiry-oriented mathematics education.

Wim Van Dooren obtained in 2005 the degree of Doctor in Educational Sciences at the University of Leuven, Belgium. Since 2014 he is Associate Professor in educational sciences of that same university. His major research interest is educational psychology and (psychology of) mathematics education. Wim Van Dooren is a member of the editorial board of several international journals and Associate Editor of *Educational Studies in Mathematics*.

Xianwei Van Harpen received her Ph.D. in Mathematics Education from Illinois State University. Her earlier research interest includes mathematical problem posing, mathematical creativity, and international comparative studies in mathematics education. After working at Illinois State University for a few years as an Assistant Professor, she is currently serving at the University of Wisconsin-Milwaukee as a course coordinator of developmental mathematics. Her current work is focused on improving undergraduate students' mathematical abilities. Specifically, she is responsible for curriculum design and instructor training in addition to teaching. She also serves as a mentor for graduate student instructors in their development as future mathematics professors.

Lieven Verschaffel obtained in 1984 the degree of Doctor in Educational Sciences at the University of Leuven, Belgium. From 1979 until 2000 he fulfilled several research positions at the Fund for Scientific Research—Flanders. Since 2000 he is a full professor in educational sciences of that same university. His major research interest is educational psychology and (psychology of) mathematics education. Lieven Verschaffel is a member of the editorial board of numerous international journals and Series Editor of the book series *New Directions in Mathematics and Science Education* published by Sense Publishers.

Cristian Voica is an Associate Professor in the Department of Mathematics, University of Bucharest, Romania. He is the author or coauthor for more than 50 research papers on Algebraic Geometry and on Mathematics Education, and for more than 15 books. What distinguishes him from others in his field of activity is that he enjoys teaching: as recognition of that, he received from ANOSR (a national students organization) the title "Bologna Teacher" in 2014. In the last years, Cristian was involved in many projects concerning teacher training and mathematics education in Romania.

Jane M. Watson is Professor Emerita of Mathematics Education in the Faculty of Education at the University of Tasmania, where she has taught in preservice, graduate, and in-service programs in Education since 1985. She has been involved in many state and national professional development programs for teachers of mathematics and has had continuous Australian Research Council funding for research since 1991. Her main research interest is in statistics education, and in 2006 she published a book with Erlbaum in the United States entitled *Statistical Literacy at School: Growth and Goals*, based on the research carried out in Australia. She is a Fellow of the Academy of the Social Sciences in Australia and a recipient of the *Mathematics Education Research Group of Australasia* Career Research Medal.

Part I
Defining the Field: Interpreting Problem Posing in the Context of Mathematics Education

Chapter 1
Problem-Posing Research in Mathematics Education: Some Answered and Unanswered Questions

Jinfa Cai, Stephen Hwang, Chunlian Jiang, and Steven Silber

Abstract This chapter synthesizes the current state of knowledge in problem-posing research and suggests questions and directions for future study. We discuss ten questions representing rich areas for problem-posing research: (a) Why is problem posing important in school mathematics? (b) Are teachers and students capable of posing important mathematical problems? (c) Can students and teachers be effectively trained to pose high-quality problems? (d) What do we know about the cognitive processes of problem posing? (e) How are problem-posing skills related to problem-solving skills? (f) Is it feasible to use problem posing as a measure of creativity and mathematical learning outcomes? (g) How are problem-posing activities included in mathematics curricula? (h) What does a classroom look like when students engage in problem-posing activities? (i) How can technology be used in problem-posing activities? (j) What do we know about the impact of engaging in problem-posing activities on student outcomes?

Contents

J. Cai (✉) • S. Hwang • S. Silber
Department of Mathematical Sciences, University of Delaware, Newark, DE, USA
e-mail: jcai@udel.edu; hwangste@udel.edu; spsilber@udel.edu

C. Jiang
Faculty of Education, University of Macau, Macau, China
e-mail: cljiang@umac.mo

F.M. Singer et al. (eds.), *Mathematical Problem Posing*,
Research in Mathematics Education, DOI 10.1007/978-1-4614-6258-3_1

Introduction

There is a long history of integrating mathematical problem solving into school curricula (Stanic & Kilpatrick, 1988). In the past several decades, there have been significant advances in the understanding of the affective, cognitive, and metacognitive aspects of problem solving in mathematics and other disciplines (e.g., Cai, 2003; Frensch & Funke, 1995; Lester, 1994; McLeod & Adams, 1989; Schoenfeld, 1985, 1992; Silver, 1985). In contrast, problem-posing research is a relatively new endeavor (Brown & Walter, 1993; Kilpatrick, 1987; Silver, 1994). Nevertheless, there have been efforts to incorporate problem posing into school mathematics at different educational levels around the world (e.g., Chinese National Ministry of Education, Office of School Teaching Materials and Institute of Curriculum and Teaching Materials, 1986; Hashimoto, 1987; Healy, 1993; Keil, 1964/1967; Ministry of Education of China, 2011; National Council of Teachers of Mathematics (NCTM), 1989; van den Brink, 1987). These efforts indicate interest among many practitioners in making problem posing a more prominent feature of classroom instruction.

Despite the interest in integrating mathematical problem posing into classroom practice, our knowledge remains relatively limited about the cognitive processes involved when solvers generate their own problems, the instructional strategies that can effectively promote productive problem posing, and the effectiveness of engaging students in problem-posing activities. In the discussion below, we synthesize the current state of knowledge in problem-posing research and suggest some directions for future study. In particular, we discuss the following questions:

1. Why is problem posing important in school mathematics?
2. Are teachers and students capable of posing important mathematical problems?
3. Can students and teachers be effectively trained to pose high-quality problems?
4. What do we know about the cognitive processes of problem posing?
5. How are problem-posing skills related to problem-solving skills?
6. Is it feasible to use problem posing as a measure of creativity and mathematical learning outcomes?
7. How are problem-posing activities included in mathematics curricula?
8. What does a classroom look like when students engage in problem-posing activities?
9. How can technology be used in problem-posing activities?
10. What do we know about the impact of engaging students in problem-posing activities on student outcomes?

Each of these questions represents a rich area for problem-posing research. As we explore each question, we begin by examining the work that has been done and by summarizing what we know as a field. We then consider, for each overarching question, some related questions that remain unanswered and which we feel merit further attention from the research community.

Why is Problem Posing Important in School Mathematics?

Problem posing has long been recognized as a critically important intellectual activity in scientific investigation. According to Einstein, the formulation of an interesting problem is often more important than its solution (Einstein & Infeld, 1938). However, whereas the case for problem solving in school mathematics has seemed relatively clear, the importance of problem posing in school mathematics has required slightly more explanation. As we noted above, problem solving has long been a fundamental part of mathematics education (Stanic & Kilpatrick, 1988). Although 30 years ago Getzels (1979) lamented that, compared to problem solving, problem posing was a neglected area of research, in recent years both educators and researchers have begun to give problem posing concerted attention.

Kilpatrick (1987) observed that in real life, problems must often be created or discovered by the solver. Thus, the onus of noticing a problem and subsequently framing it in a productive way is squarely on the solver. Indeed, in his analysis of invention in mathematics, the mathematician Jacques Hadamard (1945) considered the identification and posing of good problems to be an important part of doing high-quality mathematics. Thus, if a goal of education is to prepare students for the kinds of thinking they will need, it seems reasonable that problem posing should be an important part of the curriculum. Moreover, approaches to mathematics instruction that attempt to engage students in experiences that are more authentic to inquiry within the discipline of mathematics should provide students with opportunities to explore, make conjectures, and pose meaningful problems (Bonotto, 2013).

Problem posing is also a critical aspect of the work of teachers, both in posing problems for students and in helping students develop into better problem posers (Crespo, 2003; Olson & Knott, 2013). Teachers regularly must formulate and pose worthwhile problems for their students, even when they are working with problems given in curriculum materials (NCTM, 1991). The problems that a teacher poses can shape the mathematical learning in their classes and "further their mathematical goals for the class" (NCTM, 2000, p. 53). In addition, teachers can use problem-posing tasks to gain greater insight into their students' understandings of mathematics (Cai et al., in press; Kotsopoulos & Cordy, 2009; Leung, 2013; Silver, 1994).

As we will discuss in greater depth below, the theoretical arguments supporting the importance of problem posing in school mathematics are bolstered by a growing body of empirical evidence. Researchers are actively exploring links between problem posing and other aspects of mathematical ability including conceptual understanding, problem solving, and creativity (e.g., Cai et al., in press; Cai & Hwang, 2002; Ellerton, 1986; Silver & Cai, 1996; Singer & Moscovici, 2008; Van Harpen & Sriraman, 2013). Given its potential to enhance the teaching and learning of mathematics, it is clear that problem posing is an important part of research and practice in school mathematics.

Are Teachers and Students Capable of Posing Important Mathematical Problems?

If we recognize problem posing as an important intellectual activity in school mathematics, then we must determine if teachers and students are capable of posing important and worthwhile mathematical problems. In fact, a fundamental line of research in problem posing has been exploring the kinds of problems that teachers and students can pose. In this line of research, researchers typically design a problem situation and ask subjects to pose problems which can be solved using the information given in the situation. Different types of problem situations have been used, some of which are knowledge-free and others of which are knowledge-rich (see Figure 1.1 for four examples of such problem situations). Some situations are quite structured (Situation 1), whereas others are relatively open (Situation 3). Stoyanova and Ellerton (1996) classified the degree of structure in problem situations as free, semi-structured, and structured.

Mathematical problem-posing research has explored the performance of school students, prospective teachers, and in-service teachers (e.g., Cai, 1998; Cai et al., in press; Cai & Hwang, 2002; Crespo, 2003; English, 1998; L. Ma, 1999; Silver & Cai, 2005; Stickles, 2011). In general, the findings have supported the claim that both

Situation 1. Children were to pose problems based on the following statements about Rufus the dog: Rufus managed to get into the Bradley house one afternoon. He chewed up four of Amy's shoes, three of her toys, and six of her socks. He also chewed up five of Brad's shoes, seven of his toys, and two of his socks. Mrs. Smith baked two dozen biscuits. Rufus made off with twelve biscuits. He buried eight of them before Mrs. Smith discovered him. (This situation was used for elementary school students in English, 1998.)

Situation 2. Ann has 34 marbles, Billy has 27 marbles, and Chris has 23 marbles. Write and solve as many problems as you can that use this information. (This situation was used for middle school students in Silver & Cai, 2005.)

Situation 3.

1

3 5

7 9 11

13 15 17 19

21 23 25 27 29

The pattern continues. I wanted to make up some problems that used this pattern for a group of high students/college freshmen. Help me by writing as many problems as you can in the space below. (This situation was used for prospective secondary mathematics teachers in Cai, 2012.)

Situation 4. Imagine that you are teaching division with fractions. To make this meaningful for kids, something that many teachers try to do is relate mathematics to other things. Sometimes they try to come up with real-world situations or story problems to show the application of some particular piece of content. What would you say would be a good story or model for $1\frac{3}{4} \div \frac{1}{2}$? (This situation was used for in-service elementary teachers in L. Ma, 1999.)

Figure 1.1. Four sample problem situations used in research on problem posing.

students and teachers are capable of posing interesting and important mathematical problems. For example, for Situation 2 in Figure 1.1, middle school students were able to pose problems such as the following (Silver & Cai, 2005):

How many marbles do they have altogether?

How many more marbles does Billy have than Chris?

How many more marbles would they need to have together to have as many marbles as Sammy, who has 103?

Can Ann give marbles to Billy and Chris so that they all have the same number? If so, how can this be done?

Suppose Billy gives some marbles to Chris. How many marbles should he give Chris in order for them to have the same number of marbles?
Suppose Ann gives some marbles to Chris. How many marbles should she give Chris in order for them to have the same number of marbles?

For Situation 3 in Figure 1.1, prospective secondary teachers were able to pose problems such as these (Cai, 2012):

What is the first number on the *n*th row?
What is the number on the *i*th row and *j*th column?
What is the last number on the *n*th row?
What is the sum of the numbers in the *n*th row?
How many numbers are there in the *n*th row?
What is the sum of the numbers in the first *n* rows?
What is the pattern of each of the numbers in each diagonal line?
What is the sum of $1^3+2^3+3^3+4^3+\cdots+(n-1)^3+n^3$?
What is the middle number in an odd row?

And, L. Ma (1999) found that in-service elementary teachers could pose problems in response to Situation 4 in Figure 1.1, such as:

Cut an apple into four pieces evenly. Get three pieces and put them together with a whole apple. Given that ½ apple will be a serving, how many servings can we get from the 1¾ apples?
A train goes back and forth between two stations. From Station A to Station B is uphill and from Station B back to Station A is downhill. The train takes 1¾ hours going from Station B to Station A. It is only ½ time of that from Station A to Station B. How long does the train take going from Station A to Station B?
Given that we paid 1¾ Yuan to buy ½ of a cake, how much would a whole cake cost?
We know that the area of a rectangle is the product of length and width. Let's say that the area of a rectangle board is 1¾ square meters, its width is ½ meters, what is its length?

However, the ability to pose valid problems appears to be connected to other factors. For example, in her comparison of US and Chinese elementary teachers' understanding of elementary mathematics, L. Ma (1999) found that the teachers' abilities to pose problems like the ones cited above for the given fraction division was associated with their understanding of the meaning of fraction division. The US teachers in her study were unable to produce appropriate problems, and their difficulties were rooted in their inadequate conceptions of fraction division. In contrast, the Chinese teachers were generally able to pose at least one problem for the given fraction division based on one of three understandings of the concept (measurement model, partitive model, factors, and product). Stickles (2011) also found that secondary and middle school teachers were capable of posing problems, but that their success was partial and was related to experience and background. Specifically, the teachers in her study were prolific problem posers when presented with a given set of information, but struggled with crafting valid problems. The teachers were more successful when reformulating problems that were given to them.

Unanswered Question 1

Even though research has shown that students and teachers are capable of posing interesting and important mathematical problems, researchers have also found that some students and teachers pose nonmathematical problems, unsolvable problems, and irrelevant problems (e.g., Cai & Hwang, 2002; Silver & Cai, 1996; Silver, Mamona-Downs, Leung, & Kenney, 1996). For example, Silver and Cai (1996) found that nearly 30% of problems posed by middle school students were either nonmathematical problems or simply nonproblem statements (even though the directions clearly asked for problems). This suggests the following question: Why do students pose nonmathematical, trivial, or otherwise suboptimal problems or statements? Crespo and Sinclair (2008) hypothesized that these difficulties might be related to a lack of opportunity for students to explore a problem situation adequately before and during the posing process. There is clearly a need to investigate how students and teachers interpret and parse problem situations when engaging in problem posing.

Unanswered Question 2

Researchers have used many different types of problem situations to investigate problem posing, ranging from simply deleting a question from a textbook problem to very open-ended problem situations. With respect to mathematical problem-solving research in the past several decades, researchers have explored the effects of various task variables on students' problem solving. For example, several classifications of task variables related to problem solving are considered in Goldin and McClintock (1984): syntax variables, content and context variables, structure variables, and heuristic behavior variables. Syntax variables are factors dealing with how problem statements are written. These are factors that may contribute to ease or difficulty in reading comprehension, such as problem length and numerical and symbolic forms within the problem. Content variables refer to the semantic elements of the problem, such as the mathematical topic or the field of application, whereas context variables refer to the problem representation and the format of information in the problem. Structure variables refer to factors involved in the solution process, such as problem complexity and factors relating to specific algorithms or solution strategies. Finally, heuristic process variables refer to the interactions between the mental operations of the problem solver and the task. Considering heuristic variables separately from subject variables (factors that differ between the individuals solving the problem) is difficult, as heuristic processes involve the problem solver interacting with the task. However, the interaction between heuristic processes and the other task variables can have a significant impact on problem-solving ability.

Less is known about how problem situations influence students' problem-posing responses. How do different characteristics of problem situations affect subjects' problem posing? Leung and Silver (1997) developed and analyzed a Test of Arithmetic Problem Posing (TAPP), which they then used to examine how the presence of numerical information impacted preservice teachers' problem-posing abilities. Results from the TAPP indicated that the preservice teachers performed better on problem-posing tasks that included specific numerical information than on tasks without specific numerical information. This result provides some insight into how task variables can impact problem posing, yet more research must be done on the impact of various other variables. Adapting the TAPP to examine how different characteristics of problem situations affects subjects' problem posing could offer a way to study the effect of other task variables.

Can Students and Teachers Be Effectively Trained to Pose High-Quality Problems?

Although students and teachers are able to pose problems, even when those problems are mathematically sound they are not always of high quality. Thus, some studies have addressed the question of how to improve the abilities of teachers and students to pose better problems. Researchers have noted the importance of opportunities for exploration of mathematical situations in developing students' problem-posing abilities. Crespo and Sinclair (2008) suggested that without the opportunity to explore the limits of the mathematical situation in which students are working, the students are limited in the types of problems they can pose. Similarly, Koichu and Kontorovich (2013) found that the successful prospective teachers in their study posed the most interesting problems when blending exploration and problem solving with their problem posing. It would appear that students are able to improve the breadth and level of challenge of the problems they pose when they have experience solving such problems, and are prompted by informal contexts such as pictures, which may leave more room for exploration, instead of formal symbolic contexts (Crespo, 2003; English, 1998).

Indeed, with respect to formal symbolic contexts, Isik and Kar (2012) identified several types of difficulties experienced by prospective elementary teachers when posing problems related to daily life situations that could be solved using given linear equations or systems of two linear equations. These included conceptual difficulties, such as incorrectly translating the meaning of mathematical operations in the equations into corresponding verbal problem statements or posing separate problems for each equation in a system, contextual difficulties, such as assigning unrealistic values to the unknowns, and violations of the conventions of word problems, such as using symbolic representations in the problems posed. These difficulties suggest that, in order to pose high-quality problems that are based in formal symbolic contexts, teachers will need to build their conceptual understanding of the underlying mathematics (L. Ma, 1999) and their pedagogical understandings.

Some researchers have explored the characteristics of practice in the discipline of mathematics in order to identify and propose various collections of strategies to facilitate high-quality problem posing. Contreras (2007) discussed how to use the "fundamental mathematical processes" (p. 16) of proving, reversing, specializing, generalizing, and extending to pose new problems from a given problem. Moore-Russo and Weiss (2011) similarly described how to apply five "generative moves" that mathematicians use in determining what could be done next to spawn new, related geometry problems from an existing problem under consideration. The five generative moves (strengthening/weakening hypothesis, strengthening/weakening conclusion, generalize, specialize, consider converse) are consonant with the processes described by Contreras.

Unanswered Question 3

It would appear to be feasible to improve the quality of problems that students and teachers pose. Existing research suggests that strategies matter in how we train students and teachers to pose problems. However, it is not clear which strategies are most effective for teaching problem posing, nor is it clear which strategies are best for problem posers to use in particular problem situations. Further exploration of these strategies and their productiveness for problem posing in different mathematical situations is warranted. What strategies and ways of thinking are most productive for posing problems, and under what types of mathematical situations are different strategies effective?

What Do We Know About the Cognitive Processes of Problem Posing?

There are many potential processes involved in posing problems, and they may vary depending on the type of problem posing under consideration. These can involve techniques for reformulating existing problems, heuristics, or strategies for generating problems from given situations, and processes for exploring a mathematical context and testing its boundaries to develop a "feel" for the kinds of questions that can be asked. Researchers have worked to gain better understandings of these processes and to document the kinds of strategies that are used in problem posing.

In their study of middle school students' problem posing, Silver and Cai (1996) found that many students produced responses that consisted of a series of related problems, often generated by varying a single element, and that the complexity of the problems tended to increase within a series. Their results suggested that there were distinct processes that guided (and perhaps constrained) the students' problem posing. English (1998) observed that third graders' ability to pose multiple problems appeared limited to tinkering with the contexts of an original problem. Cai and Hwang (2002) suggested a potential parallel between students' thinking when posing and solving problems. Specifically, they observed that the sequence of pattern-based

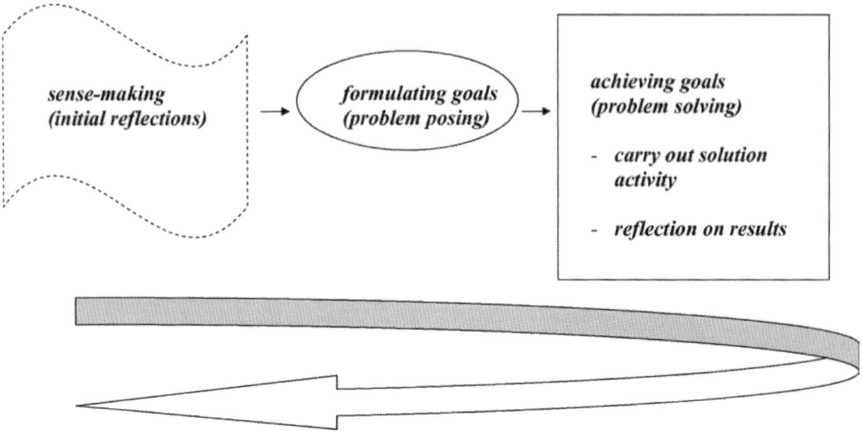

Figure 1.2. The recursive process of problem posing and solving proposed by Cifarelli and Cai (2005).

problems posed by students appeared to reflect a common sequence of thought when solving pattern problems (gathering data, analyzing the data for trends, making predictions). Thus, students might have a solution process in mind when thinking about posing problems.

Cai and Cifarelli (2005; Cifarelli & Cai, 2005) further refined this link between problem solving and problem posing, describing a recursive process of chains of solving and posing (Figure 1.2). Cai and Cifarelli (2005) examined how two college students posed and solved their own problems in an open-ended computer simulation task that involved the path of a billiard ball. They identified two different levels of reasoning strategies—hypothesis-driven and data-driven—that students appeared to incorporate in their posing and solving processes. They observed that problem solvers' self-generated questions reframed the problems they were working on and significantly changed the strategies they were using. Therefore, Cai and Cifarelli considered the posing and solving process to be mathematical exploration. Indeed, in a follow-up study, Cifarelli and Cai (2005) described mathematical exploration as structured by this recursive process. This cycling and entwining of posing and solving corresponds with the observations of Christou, Mousoulides, Pittalis, and Pitta-Pantazi (2005) about prospective teachers' use of dynamic geometry software to solve problems. Christou et al. found that the dynamic geometry software acted as a mediation tool that supported the processes of modelling, conjecturing, experimenting, and generalizing. In using the software to explore problem situations and extract meaning from them, the prospective teachers generated new problems as part of their problem-solving processes. For example, in their explorations of the figure formed by the bisectors of the interior angles of a parallelogram, the prospective teachers engaged in problem posing through experimenting with special cases (e.g., a rectangle) and making and checking conjectures based on the evidence they were gathering.

Pittalis, Christou, Mousoulides, and Pitta-Pantazi (2004) proposed a model of cognitive processes involved in problem posing. The model encompasses four processes: filtering quantitative information, translating quantitative information from one form to another, comprehending, and organizing quantitative information by giving it meaning or creating relations between provided information, and editing quantitative information from the given stimuli. Based on empirical testing, Pittalis et al. asserted that these processes correspond to different types of problem-posing tasks, and that the filtering and editing processes were most important in posing problems.

Christou, Mousoulides, Pittalis, Pitta-Pantazi, and Sriraman (2005) built on this model to develop a taxonomy of problem-posing processes related to different types of tasks. Tested with 143 sixth graders from Cyprus, their taxonomy also includes four processes. Tasks that involve posing problems from situations without restrictions involve the process of editing quantitative information. Tasks that involve posing problems that have specified answers involve the process of selecting quantitative information. Tasks that require students to pose problems corresponding to given equations or computations involve the process of comprehending and organizing quantitative information. And, tasks that involve posing problems from given graphs, diagrams, or tables involve the process of translating quantitative information from one form to another. Based on this model, the researchers found that students were more successful when first posing problems involving comprehending, then translation, and finally editing and selecting.

Although theories of the cognitive processes of problem posing are relatively new, there is a longer history of attention to strategies that may be useful in posing problems. Building on Polya's "looking back" stage in problem solving, Brown and Walter (1990) proposed the well-known "What if not" strategy. Along the same lines, Abu-Elwan (2002) and Cai and Brook (2006) suggested posing problems through a process of extending or generalizing an already-solved problem. Indeed, Gonzales (1998) even referred to this process as a fifth step to Polya's four-step method. Lavy and Bershadsky (2003) described the use of the "What if not" strategy for mathematical problem posing, dividing the activity into two stages. In the first stage, all the attributes included in the statement of the original problem are listed. In the second stage, each of the listed attributes is negated by asking "what if not attribute k?" and alternatives are proposed. Each of the offered alternatives creates a new problem situation.

Unanswered Question 4

Although we know that students and teachers are capable of posing mathematical problems, we have a considerably less fine-grained understanding of how they go about posing those mathematical problems in any given situation. Some researchers have identified general strategies students may use to pose problems. Others have explored some of the variables that may have an impact on students' problem posing. However, there is not yet a general problem-posing analogue to well-established general frameworks for problem solving such as Polya's (1957) four steps. Much more research is needed to develop a broadly applicable understanding of the fundamental processes and strategies of problem posing.

Unanswered Question 5

A better understanding of the cognitive processes of problem posing can also inform teaching. Ideally, the more that teachers know about their students' thinking, the better equipped they are to help their students develop (Cai, 2005). However, there is much work needed to connect research-based understandings of student cognition to teachers' practice. Much as Cognitively Guided Instruction (CGI) has provided a theoretical and empirical framework that has helped teachers understand their students' mathematical thinking and problem solving (Carpenter, Fennema, & Franke, 1996; Fennema et al., 1996), research that illuminates cognitive models of students' problem posing has the potential to improve teaching. In that vein, we ask the following question: How can an understanding of students' problem-posing cognition help teachers to improve student learning?

How Are Problem-Posing Skills Related to Problem-Solving Skills?

One important direction for research on problem posing is probing the links between problem posing and problem solving (see, e.g., Cai, 1998; Cai & Hwang, 2002; Ellerton, 1986; Kilpatrick, 1987; Silver & Cai, 1996). Kilpatrick (1987) provided a theoretical argument that the quality of the problems subjects pose might serve as an index of how well they can solve problems. In addition to this theoretical argument, several researchers have conducted empirical studies examining potential connections between problem posing and problem solving. Ellerton (1986) compared the mathematical problems generated by eight high-ability young children with those generated by eight low-ability young children, asking each to pose a mathematical problem that would be quite difficult for his or her friends to solve. Ellerton reported that the more able students posed problems that were more complex than those posed by less able students.

Silver and Cai (1996) analyzed the responses of more than 500 middle school students to a task that asked them to pose three questions based on a driving situation. The student-posed problems were analyzed according to their type, solvability, and complexity. In addition, Silver and Cai used eight open-ended tasks to measure the students' mathematical problem-solving performance. They found that problem-solving performance was highly correlated with problem-posing performance. Compared to less successful problem solvers, good problem solvers generated more, and more complex, mathematical problems.

Silver and Cai (1996) measured problem-solving performance using tasks that were rarely related to the problem-posing tasks. In other studies, Cai and his associates (Cai, 1998; Cai & Hwang, 2002) examined Chinese and US students' problem-solving and problem-posing performances using closely related problem-posing and problem-solving tasks. Cai and Hwang (2002) found differential relationships between posing and solving for US and Chinese sixth-grade students. There was a stronger link between problem solving and problem posing for the Chinese sample, whereas the link was much weaker for the US sample. Posing a variety of problem types appeared to be strongly associated with abstract strategy use in the Chinese sample. Cai and Hwang indicated that the differential nature of the relationships for US and Chinese students should not be interpreted as implying a lack of generality in the link between problem solving and problem posing. Rather, the stronger link between the variety of posed problems and problem-solving success for the Chinese sample could be attributable to the fact that the US students almost never used abstract strategies. Indeed, in a follow-up analysis that included data from seventh-grade US students, Cai and Hwang (2003) identified a corresponding link between the students' use of abstract problem-solving strategies and their ability to pose problems that extended beyond the given information.

Unanswered Question 6

Cross-national and cross-regional comparative studies provide unique opportunities to understand students' mathematical thinking and reasoning. Although there is a large body of cross-national studies of mathematical problem solving, there have been few attempts to use problem posing in such cross-national studies (e.g., Cai, 1998; Cai & Hwang, 2002; Yuan & Sriraman, 2011). How do students in different countries and regions pose mathematical problems? Observations of differences in problem posing across regions, such as in the study of Cai and Hwang, may provide fertile ground for further research. Analyzing, for example, differences in the magnitude of the relationship between problem solving and problem posing for students from different regions, may offer insights into the nature of the relationship. Indeed, in their analysis of problem posing among students from the United States and from two distinct regions of China, Van Harpen and Sriraman (2013) have found differences that suggest a strong link between mathematical knowledge and problem-posing success. In the future, we hope that more researchers around the world will engage in mathematical problem-posing research in cross-cultural contexts.

Is it Feasible to Use Problem Posing as a Measure of Creativity and Mathematical Learning Outcomes?

Student outcomes in mathematics classes are typically assessed by having the students solve problems. However, as noted above, researchers have found that students' success in problem solving is associated with their problem-posing abilities

(Cai et al., in press; Cai & Hwang, 2002; Silver & Cai, 1996). Moreover, there is some evidence that asking students to pose problems may provide additional useful insights into what mathematics students have learned and what students have learned about doing mathematics. For example, in a study of prospective elementary teachers' conceptual understanding of fractions, Tichá and Hošpesová (2013) used problem posing as a diagnostic tool to gauge the prospective teachers' understanding. By analyzing the problems that the prospective teachers posed, Tichá and Hošpesová were able to identify conceptual flaws and confusion that needed to be addressed. Similarly, Kotsopoulos and Cordy (2009) made use of their seventh-grade students' journal records of problem posing as a type of formative assessment to gauge the progress the students were making. This allowed these teacher-researchers to determine whether they "were on-track with our learning objectives for the four experiments" (p. 272).

As part of a large-scale study, Cai et al. (in press) investigated the feasibility of using problem posing to measure curricular effects on student learning. In particular, they compared the effects of a *Standards*-based middle school mathematics curriculum with those of more traditional curricula on students' algebra learning. Using parallel problem-solving and problem-posing tasks, they confirmed the association between students' abilities to solve and pose problems, and found that this relationship held for students using both types of curriculum. In addition, by using qualitative rubrics to assess different characteristics of students' responses, Cai and his colleagues found that students whose posed problems exhibited positive characteristics (such as reflecting the linearity of a given graph in their posed problem or embedding their posed problems in real-life contexts) were also strong problem solvers. However, student performance in general was poor on the problem-posing tasks in this study, suggesting that the students might need more experience with problem posing in order to have broader success on posing-oriented measures.

Given the generative qualities of problem posing, one might expect that problem-posing activities might be valid measures of students' creativity. Indeed, Silver (1997) has proposed a relationship between engaging students in problem posing and their development of creative fluency, flexibility, and novelty. Studying elementary children in Taiwan, Leung (1997) developed an 18-task instrument that was useful in measuring the students' general problem-posing competence as well as in highlighting their creative problem posing. Similarly, Van Harpen and Sriraman (2013) used a problem-posing test to examine US and Chinese high school students' problem-posing creativity along the three dimensions of fluency, flexibility, and novelty. Generally, performance on such tests has revealed weaknesses in problem posing. However, Voica and Singer (2012) have suggested that there are important nuances in the relationship between problem posing and creativity. Specifically, in their study of fourth to sixth graders' modifications to problems, they found that students who stayed close to the given problem's context displayed deeper understanding of the mathematics than those who posed modified problems that were ostensibly more creative because they strayed further from the original. Nevertheless, Voica and Singer (2013) have found that, with sufficiently careful analysis of students' cognition during problem modification, problem posing can provide useful evidence of students' cognitive flexibility.

Despite the theoretical feasibility of using problem-posing tasks as measures of student outcomes, it seems that students will need further experiences and preparation in order for problem-posing measures to provide the most useful information. The low levels of success students display may be due to a general lack of experience with problem-posing tasks. In addition, Crespo and Sinclair (2008) emphasize the need for students to develop aesthetic criteria for judging the mathematical quality of posed problems. The development of such criteria and the disposition to apply them may also be part of the experiences prerequisite for problem posing to be practically feasible as an outcome measure.

Unanswered Question 7

Given the potential for problem-posing tasks to be used as measures of creativity and other mathematical learning outcomes, it is incumbent on the mathematics education research community to develop and validate suitable problem-posing instruments. What kinds of problem-posing tasks best reveal students' creativity and their mathematical understandings and misunderstandings? Given the results of the LieCal problem-posing assessment (Cai et al., in press), in order for problem-posing measures to provide useful information, it will also be important for researchers to investigate the kinds of preparation students will need to perform adequately on them.

How Are Problem-Posing Activities Included in Mathematics Curricula?

If problem-posing activities are to play a more central role in classrooms, they must be more prominently represented in curricula. As noted above, researchers have adapted several kinds of materials in order to generate problem-posing situations for research purposes. Similarly, if teachers are to engage students in problem posing in the classroom, they must have sources for problem-posing activities. Such sources may be supplements to curricula, as in the case of the materials developed by Lu and Wang (2006). Lu and Wang and their associates (Lu & Wang, 2006; Wang & Lu, 2000) launched a project focused on developing and implementing a set of teaching materials about mathematical situations and problem-posing tasks. The teaching materials, including mathematical situations and problem-posing tasks, were not intended to replace textbooks; instead, they were used to supplement regular textbook problems. By 2006, more than 300 schools in ten provinces in China had participated in the project. Teachers received training to use mathematical situations and problem-posing tasks along with their regular curriculum.

However, education reform movements have also recommended that problem-posing activities be included in mathematics curricula themselves. Internationally, school mathematics reforms have recommended that students be able to "formulate interesting problems based on a wide variety of situations, both within and outside of mathematics" (NCTM, 2000) and that instructional activities should emphasize learning problem-posing skills. In the United States, the NCTM Principles and Standards for School Mathematics (2000) emphasized the use of problem generation activities, where problems are "posed out of a situation or experience" (Stickles, 2011).

Similarly, reforms to curriculum standards in China have increased the prominence of problem posing. The 9-year compulsory education mathematics curriculum standards call for providing students opportunities to pose problems, understand problems, and apply the knowledge and skills learned to solve real-life problems (Basic Education Curriculum Material Development Center, Chinese Ministry of Education, 2003). Similarly, the curriculum standards for senior high school mathematics also call for developing students' abilities to pose, analyze, and solve problems from mathematics and real life (Basic Education Curriculum Material Development Center, Chinese Ministry of Education, 2003). Indeed, in the reform standards, students are encouraged to discover and pose problems in order to prepare them to think independently and be inquirers.

However, the implications for the inclusion of problem posing in the curriculum are not necessarily clear. Ellerton (2013) has pointed out that although the Common Core State Standards—currently the most widely accepted US standards—call for problem-posing activities to be included in mathematics curricula, primarily the emphasis has been on problem-solving activities. In the Common Core State Standards, problem-posing activities are explicitly mentioned once (National Governors Association Center for Best Practices, 2010, p. 7), whereas problem solving is explicitly stated throughout the standards. The Common Core State Standards do recommend emphasizing the ability to "recognize and describe situations" for third-, fifth-, sixth-, and seventh-grade mathematics, which can be interpreted as problem posing (National Governors Association Center for Best Practices, 2010), but do not provide any recommendations on how to incorporate such activities into teaching plans (Ellerton, 2013).

This ambivalence is reflected in the available research on problem posing and curricula. Although reform movements have called for problem-posing activities to be included in mathematics curricula, there has not yet been a substantial body of research examining whether and how curricula incorporate problem posing. There is some evidence that more recent versions of textbooks emphasize problem posing more than previous versions. For example, an analysis of all problem-posing tasks in two editions of the Chinese elementary mathematics textbook series published by the People's Education Press found that between the 1994 edition and the 2004 edition, there was an increase in the percentage of problem-posing tasks (Cai, Jiang, Hwang, Nie, & Hu, in press). Notably, this problem-posing increase appears to have been related to an accompanying increase in curricular focus on data and statistics.

Unanswered Question 8

The lack of a robust body of research in this area leads us to call for greater attention to the textbooks that students and teachers actually use, not merely to the curriculum frameworks on which those textbooks are based (Cai et al., in press). How do the actual textbooks include problem posing? There are many ways to include problem posing, and it is not clear what choices textbook writers and curriculum developers have made in creating the existing materials. Given the emphasis on mathematical modelling in current curriculum frameworks, it would be helpful in particular to know what role problem posing might play in mathematical modelling tasks in textbooks.

Unanswered Question 9

If curriculum designers intend to integrate problem posing into textbooks and teaching materials, what are the best ways to do so? In the analysis of Chinese elementary mathematics textbooks mentioned above, Cai and his colleagues gave special attention to three types of problem-posing tasks: tasks which included a sample problem, tasks that required students to pose problems corresponding to given operations, and tasks that required students to pose problems based on data charts (Cai et al., in press). They found significant differences with respect to these types of tasks between the 1994 and 2004 editions of textbooks. However, it is not clear whether these shifts are reflective of an attempt to utilize problem posing more effectively in the curriculum, and if so, what criteria were used to make those judgments. Further work is needed to understand the effectiveness of different ways of building problem posing into curricula.

What Does a Classroom Look Like When Students Engage in Problem-Posing Activities?

Even when problem posing is included in textbooks and curriculum materials, there remains the significant work of implementation in actual classrooms. Classrooms are complex by their very nature, with students and teachers establishing patterns of practice and norms that can influence student learning (Boaler, 2003; Yackel & Cobb, 1996). Indeed, Crespo and Sinclair (2008) have pointed out that classroom activity around problem posing will involve the negotiation of sociomathematical norms, such as in determining criteria for what counts as a mathematically interesting problem. Researchers must therefore consider how the intended curriculum is realized by teachers and students and what factors influence implementation (Ball & Cohen, 1996; National Research Council, 2004). With respect to understanding how problem posing can be enacted in classrooms, there is a need for both theoretical frameworks and careful analyses of practice.

To that end, Ellerton (2013) has proposed an Active Learning Framework that situates the processes of problem posing in the broader processes of mathematics classrooms. Arranged along a spectrum from passive student processes to active student processes, Ellerton's framework suggests that classrooms that do not include problem posing, stopping instead at problem solving, cut short students' mathematical experiences. In particular, students are deprived of opportunities to reflect, critique, and question. Thus, this framework portrays problem posing in classrooms as a capstone activity that allows students to consolidate and think critically about the knowledge they have gained.

Although not specifically an analysis of problem posing in classrooms, Singer and Moscovici (2008) have described a learning cycle in constructivist instruction that includes problem posing as an extension and application of problem solving. In an example of instruction with ninth graders, Singer and Moscovici describe three phases of inquiry: immersion, structuring, and applying. In the third, applying phase, students use the patterns they have developed in earlier phases in related and unrelated situations and create new situations that need solving. Parallel to the role of problem posing in Ellerton's (2013) framework, Singer and Moscovici characterize the role of problem posing in a constructivist approach to instruction as that of consolidating and extending what they have learned.

Looking more specifically at the collective activities of students in classrooms, Kontorovich, Koichu, Leikin, and Berman (2012) have proposed a theoretical framework to help researchers handle the complexity of students' mathematical problem posing in small groups. This framework integrates five facets: task organization, students' knowledge base, problem-posing heuristics and schemes, group dynamics and interactions, and individual considerations of aptness. The last facet refers to the posers' comprehensions of implicit requirements of a problem-posing task and reflects their assumptions about the relative importance of these requirements. Kontorovich et al. applied their framework to analyze the problem-posing processes and decision making of two groups of high school students with similar backgrounds who were given the same problem-posing task.

In implementing the supplementary problem-posing curriculum materials designed in their project, Lu and Wang and their associates aimed to help teachers learn how to develop mathematical situations and to pose problems (Lu & Wang, 2006). As supplementary material for the regular mathematics curriculum, a series of teaching cases was developed by mathematics educators across grade levels and across content areas. Figure 1.3 presents a sample teaching case for *Making a Billboard* from Lu and Wang (2006, p. 359). The teaching materials given to teachers included different problem situations together with examples of problems which students might be expected to pose. Figure 1.3 shows a problem situation with six such sample problems. These sample problems were given to teachers as guidelines in much the same way as worked examples might be given in textbooks. When students were given the problem situations, they were encouraged to pose as many problems as they could.

After students had posed several problems, the teacher would show them how to solve some of the posed problems. Figure 1.4 shows a sample solution to problem 3.

Mathematics content: Linear equation with one unknown (for junior high school students).

Situation: A factory is planning to make a billboard. A master worker and his apprentice are employed to do the job. It will take 4 days by the master worker alone to complete the job, but it takes 6 days for the apprentice alone to complete the job.

Students' Task: Please create problems based on the situation. Students may add conditions for problems they create.

Problem 1. How many days will it take the two workers to complete the job together?

Problem 2. If the master joins the work after the apprentice has worked for 1 day, how many additional days will it take the master and the apprentice to complete the job together?

Problem 3. After the master has worked for 2 days, the apprentice joins the master to complete the job. How many days in total will the master have to work to complete the job?

Problem 4. If the master has to leave for other business after the two workers have worked together on the job for 1 day, how many additional days will it take the apprentice to complete the remaining part of the job?

Problem 5. If the apprentice has to leave for other business after the two workers have worked together for 1 day, how many additional days will it take the master to complete the remaining part of the job?

Problems 6. The master and the apprentice are paid 450 Yuan after they completed the job. How much should the master and the apprentice each receive if each worker's payment is determined by the proportion of the job the worker completed?

Figure 1.3. Sample teaching case and examples of problems posed by students in response to the task.

Suppose the two workers worked together for x days, the master worker did $(x+2)$ days.

$\frac{1}{4}(x+2)+\frac{1}{6}x=1$, and $x=\frac{6}{5}$;

So the master worked: $x+2=2+\frac{6}{5}=\frac{16}{5}$ days.

Figure 1.4. Solution presented by a teacher to posed problem 3 in Figure 1.3.

Once students had solved each of the posed problems, they were encouraged to pose new problems. Additional problems posed by students are shown in Figure 1.5. The teacher would then show students how to solve these problems.

Cai (2012) provided another example of problem posing in classroom instruction in a study of 14 preservice teachers engaging in the problem-posing activity shown in Situation 3 of Figure 1.1. The preservice teachers were divided into four groups and given 30 min to pose as many problems as they could. Then the class used another 70 min to solve the posed problems. During the process of solving the posed

Problem 7. The apprentice started the work by himself for 1 day, and then the master joined the effort, and they completed the remaining part of the job together. Finally, they received 490 Yuan in total for completing the job. How much should the master and the apprentice each receive if each worker's payment is determined by the proportion of the job the worker completed?

Problem 8. The master started the work by himself for 1 day, and then the apprentice joined the effort, and they completed the remaining part of the job together. Finally, they received 450 Yuan in total for completing the job. How much should the master and the apprentice each receive if each worker's payment is determined by the proportion of the job the worker completed?

Figure 1.5. Additional problems posed by students.

problems, each preservice teacher could pose additional problems. The preservice teachers posed a total of nine different mathematical problems after the first 30 min. Two groups posed the same question, "What is the sum of the numbers in the first n rows?", and the ensuing discussion produced an unanticipated result.

The first group of students answered the question based on the fact that the sum of the numbers in the first n rows is the "sum of the sum" of the numbers in each of the first n rows. Since the sum of the numbers in the nth row is n^3, the sum of the numbers in the first n rows should be $1^3+2^3+3^3+4^3+\cdots+(n-1)^3+n^3$. Then they posed the following question: What is the sum of $1^3+2^3+3^3+4^3+\cdots+(n-1)^3+n^3$?

The second group used a different approach to answer the original question. After some observations, students realized that the first row has one odd number which is 1, the second row has two odd numbers which are 3 and 5, the third row has three odd numbers which are 7, 9, 11, and so on. The nth row should have n odd numbers. Therefore, the sum of the numbers in the first n rows of the pattern should be the sum of the first $(1+2+3+4+\cdots+n)$ odd numbers. Since $1+3+5+\cdots+(2m-1)=m^2$, the sum of the numbers in the first n rows in the pattern should be $(1+2+3+4+\cdots+n)^2$.

After the two groups of students presented their answers to the class, they integrated their findings and realized that $1^3+2^3+3^3+4^3+\cdots+(n-1)^3+n^3=[n(n+1)/2]^2$ because $1+2+3+4+\cdots+n=n(n+1)/2$. This was not a result that the students had expected, nor was its development from this activity anticipated by the instructor beforehand. This example from empirical research showed that collective problem posing in the classroom context could lead to surprising results. Classrooms that include problem-posing activities may therefore allow students' voices to become relevant in the development of the mathematics they are learning and provide spaces to foster creativity and mathematical power.

Unanswered Question 10

Although we have discussed a few examples of classroom instruction involving problem posing, few researchers have tried to describe carefully the dynamics of classroom instruction where students are engaged in problem-posing activities.

Because classroom instruction is generally complex, with many salient features that can be investigated, researchers will need to identify those features that are most relevant for problem posing and which may be most influenced by the introduction of problem-posing activities. This leads to our tenth unanswered question: What are the key features of effective problem-posing and problem-posing instruction in classrooms?

Unanswered Question 11

In addition to identifying and describing the distinctive features of classrooms in which students engage in problem posing, it is also important to consider how teachers might change their practice and their classroom cultures to make problem posing an accepted practice (Leung, 2013). Indeed, the prevailing norms that shape school mathematics teaching are rooted in both teachers' and students' understandings of what is expected of them (Brousseau, 1984, 1997; Herbst, 2002) and in the practical rationality (Herbst & Chazan, 2003) that guides teachers' judgments about what actions are appropriate in the classroom. Moore-Russo and Weiss (2011) point out the potential difficulty in challenging and altering these norms and expectations, asking "Is it normative to encourage students to modify a problem or to introduce their own assumptions when solving problems?" and "Do teachers commonly encourage students to pose their own problems?" Thus, it is important to investigate the practical questions of whether and how problem posing can fit into the obligations teachers feel in their practice. What are the dynamics of negotiating a classroom culture in which posing is an expected behavior, and what supports do teachers need to be able to reposition themselves and their students for problem posing?

How Can Technology Be Used in Problem-Posing Activities?

The use of technology in the teaching and learning of mathematics has been a topic of interest for researchers in mathematics education. In particular, the flexibility of computer-based technologies for facilitating exploration and experimentation seems relevant to problem posing. Indeed, NCTM (1991) highlighted the promise of technology for problem posing (and solving) "in activities that permit students to design their own explorations and create their own mathematics" (p. 134). For example, Cai and Cifarelli (2005) made use of a computer microworld to allow students to explore a mathematical situation involving the motion of a billiard ball. The microworld provided the students with relative autonomy and freedom in exploring the relationships and boundaries of the mathematical situation. These explorations facilitated the students' generation of multiple questions and conjectures. Thus, by increasing opportunities for students to explore a problem situation and test its boundaries (Crespo & Sinclair, 2008), computer-based technologies may ultimately help students to extend given problems by posing related questions (Santos-Trigo & Diaz-Barriga, 2000) and to pose higher quality problems overall.

Computer-based systems have been particularly well suited to providing students with opportunities to explore dynamic visualizations of geometric situations. Christou, Mousoulides, Pittalis, and Pitta-Pantazi (2005) found that the use of dynamic geometry software facilitated the generation of new problems during the problem-solving process. Students were able to use the dynamic features of the software, and "dragging" in particular, to make and check conjectures, experiment, and generalize. Similarly, Chazan (1990) described how teachers could use the *Geometric Supposers* to increase student exploration and develop students' inquiry skills: verifying, conjecturing, generalizing, communicating, proving, and making connections. The *Supposers* are software programs that facilitate geometric constructions which can then be recorded and repeated with new initial conditions. Chazan found that the use of these programs could help students to pose very good problems by drawing auxiliary lines or systematically varying aspects of problems.

Although geometric situations appear to be particularly well suited to the dynamic visualization power of computer-based tools to aid in problem posing, some researchers have also investigated technological tools in other mathematical contexts. For example, Abramovich and Norton (2006) described the use of graphing software to explore the behavior of quadratic functions, in particular using the locus approach to investigate questions about quadratics with varying parameters. They posit that the use of graphing technology allows for the posing of problems that would be too difficult or abstract for prospective secondary teachers to formulate or solve purely algebraically. Abramovich and Cho (2006) further extended the range of technological tools for problem posing, investigating the use of spreadsheet-based environments to enable elementary preservice teachers and students to pose and solve money sharing and money changing problems. As with the geometric environments, the spreadsheet allows problem posers to explore the consequences of varying parameters of the problem situation. In addition, Abramovich and Cho noted that the spreadsheet tool helped the poser to generate data that ensured the solvability of the posed problems.

Taking advantage of the power of computers to engage students in games, Chang, Wu, Weng, and Sung (2012) implemented a problem-posing system that asked students to pose and refine problems which would then be presented in one of six computer game contexts. The mathematical focus of this project was on elementary word problems. By engaging students in this problem-posing game system, the researchers sought to improve the students' problem-posing and problem-solving skills as well as their flow experience. In particular, Chang et al. found that students using the technology-based activity were more engaged and challenged than students receiving traditional problem-posing instruction in the control group, who became tired of the tasks.

The recent rise of sophisticated web-based technologies has also had an impact on mathematical problem posing. Researchers have begun to investigate how web-based environments can facilitate the work of students and teachers to pose problems, discuss the solutions, and evaluate and improve the problems and solutions. For example, Beal and Cohen (2012) used a web-based content-authoring and sharing system in which middle school students posed mathematics and science problems and solved problems authored by their peers. The system included social media aspects, in which students could compliment or criticize their peers' problems. Beal and Cohen found that students were able to create problems successfully, generating four problems each on average. However, the students engaged in problem-solving activities much more often than authoring new problems, despite being given more points for posing than for solving problems. Nevertheless, both students and teachers responded positively to the activity.

Lan and Lin (2011) developed a web-based Question-Posing Indicators Service (QPIS) system which they used with first year college students in a programming course. Analogous to the social elements of the system used by Beal and Cohen (2012), the QPIS system has a question-posing module where students can pose questions on course content or for reflective thinking, a tool module where students can search problems posed by their peers and give comments to their peers, and an assessment module where students/teachers can evaluate the question-posing abilities of individual students. In particular, the quality of the posed questions was evaluated in a number of ways, such as:

- Content usefulness (whether a question helps students increase their understanding and/or learning)
- Content richness (multimedia content is taken as richer than text-based mode)
- Level of thinking skills reflected by question type (lower order such as true/false questions, intermediate order such as multiple choice questions, and higher order such as matching and short answer questions)
- Self and peer assessment modules
- Expert assessment modules

Lan and Lin found that the QPIS system could serve as both a learning and assessment tool in higher education by encouraging students to carry out active learning, constructive criticism, and knowledge sharing.

Unanswered Question 12

The rapid evolution of technology means that new tools are always becoming available. For purposes of improving educational outcomes, it can be difficult to keep pace with these developments. Of particular concern is the tendency in education to adopt technologies without having a clear picture of their impacts and effectiveness. This raises a key and persistent unanswered question. Are particular technological tools effective, and how do they affect students' problem posing? Some of the studies mentioned above (e.g., Chang et al., 2012; Lan & Lin, 2011) have measured changes in students' problem solving and problem posing. However, this question is not simply about looking for improved performance on existing tasks. As Abramovich and Norton (2006) point out, technological tools can not only enhance the curriculum, but also change it. Thus, studies of the effects of introducing technological tools for problem posing must consider how these tools may change the tasks and the learning goals of mathematics instruction.

What Do We Know About the Impact of Engaging in Problem-Posing Activities on Student Outcomes?

The ultimate goal of educational research is to improve students' learning. Research on problem posing is no exception. The NCTM *Principles and Standards for School Mathematics* (NCTM, 2000) suggest that problem-posing activities should be beneficial for both students and teachers, with students learning to pose problems in both school and out-of-school contexts (Bonotto, 2013), and teachers using problem posing to promote and challenge students' thinking (Stickles, 2011). Indeed, there are at least two reasons to expect that engaging students in problem-posing activities should have a positive impact on their learning. First, problem-posing activities are usually cognitively demanding tasks with the potential to provide intellectual contexts for students' rich mathematical development. Doyle (1983) argued that tasks with different cognitive demands are likely to induce different kinds of learning. Cognitively demanding problem-posing activities can promote students' conceptual understanding, foster their ability to reason and communicate mathematically, and capture their interest and curiosity (NCTM, 1991). Indeed, researchers (e.g., Silver, 1994) have suggested that student-posed problems are more likely to connect mathematics to students' own interests, something that is often not the case with traditional textbook problems. Second, problem-solving processes often involve the generation and solution of subsidiary problems (Polya, 1957). Previous studies (e.g., Cai & Hwang, 2002) have suggested that the ability to pose complex problems might be associated with more robust problem-solving abilities. Thus, encouraging students to generate problems is not only likely to foster student understanding of problem situations, but also to nurture the development of more advanced problem-solving strategies.

Even though theoretical arguments suggest that engaging students in problem-posing activities in classrooms should have a positive impact on students' learning and problem posing, there are relatively few empirical studies that systematically document this effect. English (1997) developed a problem-posing program and found in her post-interview that fifth graders in the problem-posing program did, in fact, pose quantitatively more, as well as more complex, problems. Similarly, Crespo (2003) examined the changes in the problem-posing strategies of a group of elementary preservice teachers as they posed problems to students. She found that, after teachers had engaged in problem-posing activities, they were able to pose more problems with multiple approaches and solutions, as well as pose problems that were more open-ended, exploratory, and cognitively complex.

Given the documented association between students' problem-solving and problem-posing abilities (e.g., Ellerton, 1986; Silver & Cai, 1996), some researchers have specifically investigated the effects of engaging in problem-posing activities on problem-solving performance. Traylor (2005) used a pretest–posttest design to compare the posing and solving performance of eighth-grade algebra students who engaged in both types of activities for the first 9 weeks of the school year to that of students in control classes who had not engaged in posing activities.

The results were mixed, with no clear benefit to engaging in problem posing. However, Traylor suggested that these results may have been influenced by the participants' comparative lack of effort on the posttest, which she attributed to the fact that the test did not "count" toward the students' grades and that, 9 weeks into the school year, students were no longer so eager to please their teachers.

Other researchers have found somewhat more positive effects of problem posing. Abu-Elwan (2002) conducted an experiment with 50 student–teachers, half of whom were given opportunities to pose problems as an extension of Polya's (1957) fourth problem-solving step. The experimental instruction was based on the suggestion of Gonzales (1994) to extend Polya's four steps to include a fifth stage in which students posed related problems. The control group received instruction based only on Polya's original four steps. Abu-Elwan found that the experimental group performed significantly better than the control group in both problem solving and problem posing.

In a study of the effects of problem-posing instruction on Turkish 10th graders' learning of probability, Demir (2005) found that students who had been taught using a problem-posing approach performed significantly better on a probability achievement test. Moreover, Demir documented significant positive effects on affect. Specifically, students who had experienced problem-posing instruction developed more positive attitudes toward probability and mathematics.

Similarly, researchers have investigated the effect of problem posing on various mathematics outcomes for prospective teachers. Positive impacts have been documented of problem posing on the prospective teachers' mathematical knowledge and understanding with respect to fraction concepts (Toluk-Uçar, 2009) and concepts from geometry (Lavy & Shriki, 2010). In addition, problem posing has been found to have positive impacts on other types of mathematics outcomes. For example, Toluk-Uçar found that problem posing had a positive effect on prospective teachers' views of understanding in mathematics, and Akay and Boz (2010) found that instruction integrated with problem posing had resulted in more positive attitudes toward mathematics and greater mathematics self-efficacy in prospective elementary mathematics teachers. Lavy and Shriki noted that, in addition to the prospective teachers' gains in geometric knowledge, problem posing was associated with gains in metamathematical knowledge about definitions, argumentation, and proof.

Unanswered Question 13

Over a decade ago, English (1997) observed that, as a field, we knew little about the relationship between students' problem-posing abilities and their competence in other areas of mathematics. It is clear that progress has been made on this front. Although some of the studies described above have focused specifically on students' problem-posing behavior after engaging in problem-posing activities, others have begun to explore connections between problem posing and broader student outcomes. However, no large-scale validation or efficacy studies have been carried out to examine the effect of engaging problem-posing activities more generally on students' learning of mathematics. Thus, the next unanswered question we raise is: What is the impact of engaging in problem-posing activities on students' mathematics achievement?

Research in reading has shown that engaging students in problem posing can lead to significant gains in reading comprehension. The results from one meta-analysis showed that the effect sizes were .36 using standardized tests and .86 using researcher-developed tests (Rosenshine, Meister, & Chapman, 1996). Although it is theoretically sound to engage students in problem-posing activities in an attempt to understand and improve their learning, more empirical studies are needed to demonstrate any actual effects on mathematics learning. The research in reading can serve as a model for systematically investigating the effect of mathematical exploration in general and problem-posing activities in particular on students' learning of mathematics.

Unanswered Question 14

Engaging in problem posing has the potential to influence more than just the mathematics that students learn, but also their dispositions toward mathematics. Silver (1994) argued that problem posing could influence students' attitudes, affect, and beliefs about mathematics. However, Silver carefully pointed out that, although studies did not typically report negative student reactions to problem posing, the influence of problem posing could be either positive or negative. The findings of Akay and Boz (2010) and Demir (2005) do provide some evidence that problem-posing activities may foster positive views of mathematics and greater self-efficacy. These affective gains may also be reinforced by the use of innovative technologies to stimulate student engagement, as in the work of Chang et al. (2012) and Beal and Cohen (2012). Given that many students suffer from anxiety that interferes with their achievement when solving mathematics problems (X. Ma, 1999; McLeod, 1992), problem posing may therefore offer a more approachable path to problem solving. Yet, the research basis for such a claim remains thin, and the question remains. How does problem posing influence affective aspects of students' mathematics learning? Systematic studies of the effects of problem posing on students' attitudes, affect, and beliefs about mathematics are needed.

Looking to the Future

Although research on mathematical problem posing is comparatively new in the field of mathematics education, researchers have gained some key footholds. Current curriculum frameworks and the curriculum materials that are built on those frameworks include problem posing, if somewhat peripherally to problem solving. We know that students and teachers are capable of posing problems. We have recognized that problem posing offers potential benefits for what mathematics students learn and what students learn about the practice of mathematics. Although students likely need more experiences and preparation with problem posing, it seems reasonable to assert that problem-posing tasks can provide useful measures of various student outcomes. Problem posing has found its way into some curriculum materials and some mathematics classrooms, though much work remains to understand how to encourage this process and produce the best results. And, there are encouraging signs that students who engage in mathematical problem posing seem to develop positive outcomes with respect to their mathematical understandings and dispositions.

Acknowledging both the work that has been done and the many questions that remain unanswered, we conclude our survey of the state of research on mathematical problem posing with a final, very broad unanswered question: How might we understand problem posing? This area of research, though comparatively new within mathematics education, has produced a number of empirical results. Yet, it remains ripe for theoretical work that will provide a cohesive framework for understanding these empirical results and the overall phenomenon of problem posing. This is not necessarily a call for a single, overarching theory of problem posing. Indeed, researchers have focused on many potentially distinct forms of problem posing, such as the kind of problem posing teachers do for their students and the kind of problem posing individuals do when reflecting, and we may need multiple frameworks to understand problem posing in all its guises. Nevertheless, there is a clear need for more robust theory-building so that we may better understand problem posing. A journey of a thousand miles begins with a single step. The journey of problem-posing research in mathematics has taken its first step, but many more steps need to follow.

References

Abramovich, S., & Cho, E. (2006). Technology as a medium for elementary preteachers' problem-posing experience in mathematics. *Journal of Computers in Mathematics and Science Teaching, 25*(4), 309–323.

Abramovich, S., & Norton, A. (2006). Equations with parameters: A locus approach. *Journal of Computers in Mathematics and Science Teaching, 25*(1), 5–28.

Abu-Elwan, R. (2002). Effectiveness of problem posing strategies on prospective mathematics teachers' problem solving performance. *Journal of Science and Mathematics Education in Southeast Asia, 25*(1), 56–69.

Akay, H., & Boz, N. (2010). The effect of problem posing oriented analyses-II course on the attitudes toward mathematics and mathematics self-efficacy of elementary prospective mathematics teachers. *Australian Journal of Teacher Education, 35*(1), 6. Retrieved from http://ro.ecu.edu.au/ajte/vol35/iss1/6

Ball, D. L., & Cohen, D. K. (1996). Reform by the book: What is—or might be—the role of curriculum materials in teacher learning and instructional reform? *Educational Researcher, 25*(9), 6–8, 14.

Basic Education Curriculum Material Development Center, Chinese Ministry of Education. (2003). *National mathematics curriculum standards at the compulsory education level* (draft for consultation) [in Chinese]. Beijing, China: Beijing Normal University.

Beal, C. R., & Cohen, P. R. (2012). Teach ourselves: Technology to support problem posing in the STEM classroom. *Creative Education, 3*, 513–519.

Boaler, J. (2003). *Studying and capturing the complexity of practice—The case of the "dance of agency."* Paper presented at the 27th International Group for the Psychology of Mathematics Education Conference held jointly with the 25th PME-NA Conference, Honolulu, HI.

Bonotto, C. (2013). Artifacts as sources for problem-posing activities. *Educational Studies in Mathematics, 83*, 37–55.

Brousseau, G. (1984). The crucial role of the didactical contract in the analysis and construction of situations in teaching and learning mathematics. In H. G. Steiner (Ed.), *Theory of mathematics education* (pp. 110–119). Occasional paper 54. Bielifeld, Germany: Institut fur Didaktik der Mathematik.

Brousseau, G. (1997). *Theory of didactical situations in mathematics* (N. Balacheff, M. Cooper, R. Sutherland, & V. Warfield, Eds. & Trans.). Dordrecht, The Netherlands: Kluwer Academic.

Brown, S. I., & Walter, M. I. (1990). *The art of problem posing* (2nd ed.). Hillsdale, NJ: Erlbaum.

Brown, S. I., & Walter, M. I. (1993). Problem posing in mathematics education. In S. I. Brown & M. I. Walter (Eds.), *Problem posing: Reflections and application* (pp. 16–27). Hillsdale, NJ: Erlbaum.

Cai, J. (1998). An investigation of U.S. and Chinese students' mathematical problem posing and problem solving. *Mathematics Education Research Journal, 10*, 37–50.

Cai, J. (2003). Singaporean students' mathematical thinking in problem solving and problem posing: An exploratory study. *International Journal of Mathematical Education in Science and Technology, 34*(5), 719–737.

Cai, J. (2005). U.S. and Chinese teachers' knowing, evaluating, and constructing representations in mathematics instruction. *Mathematical Thinking and Learning: An International Journal, 7*(2), 135–169.

Cai, J. (2012). *Problem posing as lenses of improving students' learning in classroom*. Newark, DE: University of Delaware.

Cai, J., & Brook, M. (2006). New perspectives of looking back in mathematical problem solving. *Mathematics Teaching, 196*, 42–45.

Cai, J., & Cifarelli, V. V. (2005). Exploring mathematical exploration: How two college students formulated and solved their own mathematical problems. *Focus on Learning Problems in Mathematics, 27*(3), 43–72.

Cai, J., & Hwang, S. (2002). Generalized and generative thinking in U.S. and Chinese students' mathematical problem solving and problem posing. *The Journal of Mathematical Behavior, 21*, 401–421.

Cai, J., & Hwang, S. (2003). A perspective for examining the link between problem posing and problem solving. In N. A. Pateman, B. J. Dougherty, & J. T. Zilliox (Eds.), *Proceedings of the 2003 Joint Meeting of PME and PMENA* (Vol. 3, pp. 103–110).

Cai, J., Moyer, J. C., Wang, N., Hwang, S., Nie, B., & Garber, T. (2013). Mathematical problem posing as a measure of curricular effect on students' learning. *Educational Studies in Mathematics, 83*, 57–69.

Cai, J., Jiang, C., Hwang, S., Nie, B., & Hu, D. (in press). Mathematical problem-posing tasks in U.S. and Chinese standards-based textbooks. In P. Felmer, J. Kilpatrick, & E. Pehkonen (Eds.), Problem solving in mathematics education: New advances and perspectives. Springer.

Carpenter, T. P., Fennema, E., & Franke, M. L. (1996). Cognitively guided instruction: A knowledge base for reform in primary mathematics instruction. *The Elementary School Journal, 97*, 3–20.

Chang, K. E., Wu, L. J., Weng, S. E., & Sung, Y. T. (2012). Embedding game-based problem-solving phase into problem-posing system for mathematics learning. *Computers & Education, 58*(2), 775–786. doi:10.1016/j.compedu.2011.10.002.

Chazan, D. (1990). Students' microcomputer-aided exploration in geometry. *Mathematics Teacher, 83*, 628–635.

Chinese National Ministry of Education, Office of School Teaching Materials and Institute of Curriculum and Teaching Materials. (1986). *A collection of mathematical syllabuses (1949–1985)*. Beijing, China: Author.

Christou, C., Mousoulides, N., Pittalis, M., & Pitta-Pantazi, D. (2005). Problem solving and problem posing in a dynamic geometry environment. *The Montana Mathematics Enthusiast, 2*, 125–143.

Christou, C., Mousoulides, N., Pittalis, M., Pitta-Pantazi, D., & Sriraman, B. (2005). An empirical taxonomy of problem posing processes. *Zentralbatt für Didaktik der Mathematik (ZDM), 37*, 149–158.

Cifarelli, V. V., & Cai, J. (2005). The evolution of mathematical explorations in open-ended problem-solving situations. *The Journal of Mathematical Behavior, 24*, 302–324.

Contreras, J. (2007). Unraveling the mystery of the origin of mathematical problems: Using a problem-posing framework with prospective mathematics teachers. *The Mathematics Educator, 17*(2), 15–23.

Crespo, S. (2003). Learning to pose mathematical problems: Exploring changes in preservice teachers' practices. *Educational Studies in Mathematics, 52*, 243–270.

Crespo, S., & Sinclair, N. (2008). What makes a problem mathematically interesting? Inviting prospective teachers to pose better problems. *Journal of Mathematics Teacher Education, 11*, 395–415.

Demir, B. B. (2005). *The effect of instruction with problem posing on tenth grade students' probability achievement and attitudes toward probability* (Unpublished master's thesis). Middle East Technical University, Ankara, Turkey.

Doyle, W. (1983). Academic work. *Review of Educational Research, 53*, 159–199.

Einstein, A., & Infeld, L. (1938). *The evolution of physics*. New York, NY: Simon & Schuster.

Ellerton, N. F. (1986). Children's made-up mathematics problems—A new perspective on talented mathematicians. *Educational Studies in Mathematics, 17*, 261–271.

Ellerton, N. F. (2013). Engaging pre-service middle-school teacher-education students in mathematical problem posing: Development of an active learning framework. *Educational Studies in Mathematics, 83*, 87–101.

English, L. D. (1997). The development of fifth-grade children's problem-posing abilities. *Educational Studies in Mathematics, 34*, 183–217.

English, L. D. (1998). Children's problem posing within formal and informal contexts. *Journal for Research in Mathematics Education, 29*(1), 83–106.

Fennema, E., Carpenter, T. P., Franke, M. L., Levi, L., Jacobs, V. R., & Empson, S. B. (1996). A longitudinal study of learning to use children's thinking in mathematics instruction. *Journal for Research in Mathematics Education, 27*, 403–434.

Frensch, P., & Funke, J. (1995). *Complex problem solving: The European perspective*. Hillsdale, NJ: Erlbaum.

Getzels, J. W. (1979). Problem finding: A theoretical note. *Cognitive Science, 3*, 167–172.

Goldin, G. A., & McClintock, C. E. (Eds.). (1984). *Task variables in mathematical problem solving*. Philadelphia, PA: Franklin Institute Press.

Gonzales, N. A. (1994). Problem posing: A neglected component in mathematics courses for prospective elementary and middle school teachers. *School Science and Mathematics, 94*(2), 78–84.

Gonzales, N. A. (1998). A blueprint for problem posing. *School Science and Mathematics, 98*(8), 448–456.

Hadamard, J. W. (1945). *Essay on the psychology of invention in the mathematical field*. Princeton, NJ: Princeton University Press.

Hashimoto, Y. (1987). Classroom practice of problem solving in Japanese elementary schools. In J. P. Becker, & T. Miwa (Eds.), *Proceedings of the U.S.-Japan Seminar on Mathematical Problem Solving* (pp. 94–119). Carbondale, IL: Southern Illinois University.

Healy, C. C. (1993). *Creating miracles: A story of student discovery*. Berkeley, CA: Key Curriculum Press.

Herbst, P. G. (2002). Engaging students in proving: A double bind on the teacher. *Journal for Research in Mathematics Education, 33*, 176–203.

Herbst, P., & Chazan, D. (2003). Exploring the practical rationality of mathematics teaching through conversations about videotaped episodes: The case of engaging students in proving. *For the Learning of Mathematics, 23*(1), 2–14.

Isik, C., & Kar, T. (2012). The analysis of the problems the pre-service teachers experience in posing problems about equations. *Australian Journal of Teacher Education, 37*(9). Retrieved from http://ro.ecu.edu.au/cgi/viewcontent.cgi?article=1568&context=ajte

Keil, G. E. (1964/1967). Writing and solving original problems as a means of improving verbal arithmetic problem solving ability (Doctoral dissertation, Indiana University, 1964). *Dissertation Abstracts International, 25*(12), 7109.

Kilpatrick, J. (1987). Problem formulating: Where do good problems come from? In A. H. Schoenfeld (Ed.), *Cognitive science and mathematics education* (pp. 123–147). Hillsdale, NJ: Erlbaum.

Koichu, B., & Kontorovich, I. (2013). Dissecting success stories on mathematical problem posing: A case of the Billiard Task. *Educational Studies in Mathematics, 83*, 71–86.

Kontorovich, I., Koichu, B., Leikin, R., & Berman, A. (2012). An exploratory framework for handling the complexity of mathematical problem posing in small groups. *The Journal of Mathematical Behavior, 31*(1), 149–161. doi:10.1016/j.jmathb.2011.11.002.

Kotsopoulos, D., & Cordy, M. (2009). Investigating imagination as a cognitive space for learning mathematics. *Educational Studies in Mathematics, 70*, 259–274.

Lan, Y. F., & Lin, P. C. (2011). Evaluation and improvement of student's question-posing ability in a web-based learning environment. *Australasian Journal of Educational Technology, 27*(4), 581–599.

Lavy, I., & Bershadsky, I. (2003). Problem posing via "what if not?" strategy in solid geometry—A case study. *The Journal of Mathematical Behavior, 22*(4), 369–387.

Lavy, I., & Shriki, A. (2010). Engaging in problem posing activities in a dynamic geometry setting and the development of prospective teachers' mathematical knowledge. *The Journal of Mathematical Behavior, 29*(1), 11–24.

Lester, F. K. (1994). Musing about mathematical problem-solving research: 1970–1994. *Journal for Research in Mathematics Education, 25*(6), 660–675.

Leung, S. S. (1997). On the role of creative thinking in problem posing. *Zentralbatt für Didaktik der Mathematik (ZDM), 29*, 81–85.

Leung, S. S. (2013). Teachers implementing mathematical problem posing in the classroom: Challenges and strategies. *Educational Studies in Mathematics, 83*(1), 103–116. doi:10.1007/s10649-012-9436-4.

Leung, S. S., & Silver, E. A. (1997). The role of task format, mathematics knowledge, and creative thinking on the arithmetic problem posing of prospective elementary school teachers. *Mathematics Education Research Journal, 9*(1), 5–24.

Lu, C., & Wang, B. (2006). *Research on mathematics teaching through using mathematical situations and posing problem in high school and primary school* [in Chinese]. Guiyang, China: Guizhou People's Publishing House.

Ma, L. (1999). *Knowing and teaching elementary mathematics*. Mahwah, NJ: Lawrence Erlbaum.

Ma, X. (1999). A meta-analysis of the relationship between anxiety toward mathematics and achievement in mathematics. *Journal for Research in Mathematics Education, 30*, 520–540.

McLeod, D. B. (1992). Research on affect in mathematics education: A reconceptualization. In D. A. Grouws (Ed.), *Handbook of research on mathematics teaching and learning* (pp. 575–596). New York, NY: Macmillan.

McLeod, D., & Adams, V. (Eds.). (1989). *Affect and mathematical problem solving: A new perspective*. New York, NY: Springer.

Ministry of Education of China. (2011). *Mathematics curriculum standard of compulsory education (2011 version)* [in Chinese]. Beijing, China: Beijing Normal University Press. Retrieved from http://www.moe.gov.cn/publicfiles/business/htmlfiles/moe/moe_711/201201/xxgk_129268.html

Moore-Russo, D., & Weiss, M. (2011). Practical rationality, the disciplinary obligation, and authentic mathematical work: A look at geometry. *The Mathematics Enthusiast, 8*(3), 463–482.

National Council of Teachers of Mathematics. (1989). *Curriculum and evaluation standards for school mathematics*. Reston, VA: Author.

National Council of Teachers of Mathematics. (1991). *Professional standards for teaching mathematics*. Reston, VA: Author.
National Council of Teachers of Mathematics. (2000). *Principles and standards for school mathematics*. Reston, VA: Author.
National Governors Association Center for Best Practices. (2010). *Common core state standards for mathematics*. Washington, DC: National Governors Association Center for Best Practices, Council of Chief State School Officers.
National Research Council. (2004). *On evaluating curricular effects: Judging the quality of K-12 mathematics education*. In J. Confrey & V. Stohl (Eds.). Washington, DC: National Academy Press.
Olson, J. C., & Knott, L. (2013). When a problem is more than a teacher's question. *Educational Studies in Mathematics, 83*, 27–36.
Pittalis, M., Christou, C., Mousoulides, N., & Pitta-Pantazi, D. (2004). A structural model for problem posing. In M. J. Hoines & A. B. Fuglestad (Eds.), *Proceedings of the 28th Annual Meeting of the International Group for the Psychology of Mathematics Education* (Vol. 4, pp. 49–56). Bergen, Norway: Bergen University College.
Polya, G. (1957). *How to solve it*. Princeton, NJ: Princeton University Press.
Rosenshine, B., Meister, C., & Chapman, S. (1996). Teaching students to generate questions: A review of the intervention studies. *Review of Educational Research, 66*(2), 181–221.
Santos-Trigo, M., & Diaz-Barriga, E. (2000). Posing questions from proposed problems: Using technology to enhance mathematical problem solving. *Mathematics Teacher, 93*(7), 578–580.
Schoenfeld, A. H. (1985). *Mathematical problem solving*. Orlando, FL: Academic Press.
Schoenfeld, A. H. (1992). Learning to think mathematically: Problem solving, metacognition, and sense making in mathematics. In D. A. Grouws (Ed.), *Handbook of research on mathematics teaching and learning* (pp. 334–370). New York, NY: Macmillan.
Silver, E. A. (1985). *Teaching and learning mathematical problem solving: Multiple research perspectives*. Mahwah, NJ: Erlbaum.
Silver, E. A. (1994). On mathematical problem posing. *For the Learning of Mathematics, 14*(1), 19–28.
Silver, E. A. (1997). Fostering creativity through instruction rich in mathematical problem solving and problem posing. *Zentralbatt für Didaktik der Mathematik (ZDM), 97*(3), 75–80.
Silver, E. A., & Cai, J. (1996). An analysis of arithmetic problem posing by middle school students. *Journal for Research in Mathematics Education, 27*(5), 521–539.
Silver, E. A., & Cai, J. (2005). Assessing students' mathematical problem posing. *Teaching Children Mathematics, 12*(3), 129–135.
Silver, E. A., Mamona-Downs, J., Leung, S. S., & Kenney, P. A. (1996). Posing mathematical problems: An exploratory study. *Journal for Research in Mathematics Education, 27*, 293–309.
Singer, F. M., & Moscovici, H. (2008). Teaching and learning cycles in a constructivist approach to instruction. *Teaching and Teacher Education, 24*(6), 1613–1634.
Stanic, G. M. A., & Kilpatrick, J. (1988). Historical perspectives on problem solving in mathematics curriculum. In R. I. Charles & E. A. Silver (Eds.), *Research agenda for mathematics education: The teaching and assessing of mathematical problem solving* (pp. 1–22). Reston, VA: National Council of Teachers of Mathematics.
Stickles, P. (2011). An analysis of secondary and middle school teachers' mathematical problem posing. *Investigations in Mathematics Learning, 3*(2), 1–34.
Stoyanova, E., & Ellerton, N. F. (1996). A framework for research into students' problem posing in school mathematics. In P. C. Clarkson (Ed.), *Technology in mathematics education* (pp. 518–525). Melbourne, Australia: Mathematics Education Research Group of Australasia.
Tichá, M., & Hošpesová, A. (2013). Developing teachers' subject didactic competence through problem posing. *Educational Studies in Mathematics, 83*(1), 133–143.
Toluk-Uçar, Z. (2009). Developing pre-service teachers understanding of fractions through problem posing. *Teaching and Teacher Education, 25*(1), 166–175.

Traylor, K. C. (2005). Effect of problem posing on eighth grade students' mathematical problem solving skills. *Instructional Technology Monographs, 2*(2). Retrieved March 11, 2013, from http://itm.coe.uga.edu/archives/fall2005/ktraylor.htm

van den Brink, J. F. (1987). Children as arithmetic book authors. *For the Learning of Mathematics, 7*, 44–48.

Van Harpen, X. Y., & Sriraman, B. (2013). Creativity and mathematical problem posing: An analysis of high school students' mathematical problem posing in China and USA. *Educational Studies in Mathematics, 82*(2), 201–221.

Voica, C., & Singer, F. M. (2012). *Problem modification as an indicator of deep understanding*. Paper presented at Topic Study Group 3, Activities and Programs for Gifted Students, ICME-12, Seoul, Korea.

Voica, C., & Singer, F. M. (2013). Problem modification as a tool for detecting cognitive flexibility in school children. *Zentralbatt für Didaktik der Mathematik (ZDM), 45*, 267–279.

Wang, B., & Lu, C. (2000). Innovation and mathematical education of the primary and middle school [in Chinese]. *Journal of Mathematics Education, 9*(4), 34–37.

Yackel, E., & Cobb, P. (1996). Sociomathematical norms, argumentation, and autonomy in mathematics. *Journal for Research in Mathematics Education, 27*, 458–477.

Yuan, X., & Sriraman, B. (2011). An exploratory study of relationships between students' creativity and mathematical problem posing abilities. In B. Sriraman & K. Lee (Eds.), *The elements of creativity and giftedness in mathematics* (pp. 5–28). Rotterdam, The Netherlands: Sense.

Chapter 2
Problem Posing from a Modelling Perspective

Ragnhild Hansen and Gert M. Hana

Abstract In this chapter, we consider how problem posing forms an integral part of mathematical modelling and consider its placement during modelling processes. The problem and its formulation is an essential part of modelling, and a modelling process is usually associated with a continual adjustment and reformulation of the main problem. In addition, one may formulate conjectures, ask monitoring and control questions, and have a critical stance toward the model and its results. We consider how the educational intention of the modelling activity and the placement in the modelling cycle relates to the problems and questions being posed. We briefly consider how problem posing may be implemented in mathematical modelling through the use of students' conjectures and by students acting as consultants and clients.

Contents

R. Hansen (✉) • G.M. Hana
Department of Mathematics, Bergen University College, Bergen, Norway
e-mail: Ragnhild.Hansen@hib.no; Gert.Monstad.Hana@hib.no

F.M. Singer et al. (eds.), *Mathematical Problem Posing*,
Research in Mathematics Education, DOI 10.1007/978-1-4614-6258-3_2

Introduction

We see mathematical modelling and problem posing as promoting essential skills necessary for involvement in a democratic society and as integral parts of a balanced mathematics curriculum. The topics of mathematical modelling and problem posing are closely related as modelling is concerned with using mathematics to solve or gain further insight into real-world problems. Our own experience as teacher educators is that posing problems that make good mathematical tasks is no trivial matter, and that student teachers often find it difficult to pose and implement appropriate modelling tasks in their teaching practice. A further layer of difficulty is added when one wishes pupils to take an inquiring stance, where they pose problems related to mathematical modelling. This chapter will look at problem posing from the perspective of the pupil, but much of it will be relevant to teachers' problem posing as well.

In this chapter, we discuss why we see modelling and problem posing as a potentially fruitful combination. This will be followed by short discussions on how problem posing relates to different perspectives on modelling and to the modelling process.[1] We end by sketching some ways to implement problem posing in mathematical modelling. The chapter tries to give some pointers to the many issues present in this under-researched intersection of mathematical modelling and problem posing.

Mathematical Modelling and Problem Posing: Possible Obstacles

An initial example will be presented to illustrate possible obstacles one can meet when trying to combine mathematical modelling and problem posing. In this example, a group of four student teachers were starting a lesson sequence with eighth grade pupils based on mathematical modelling. The student teachers decided to choose the general topic of mathematical modelling themselves. Since one of the student teachers had experience in biology, the topic chosen was plants. In the initial lesson, the student teachers encouraged the pupils to formulate as many questions as possible concerning plant growth. The student teachers planned this as a pure problem-posing lesson. A purpose of this lesson was to provide the student teachers with ideas of authentic mathematical modelling problems to use with the pupils within the context of plant growth, although the student teachers had not considered how to follow-up the problems posed by the pupils. In the continuation of the lesson sequence, the pupils' problems were not, in fact, used. Instead, the pupils were supposed to seed their own plants in small boxes in the classroom and work with modelling and growth prediction in accordance with the schedule made by the student

[1] Another relevant theme, which we do not pursue, is how different goals such as decision making, system analysis and design, and trouble shooting (OECD, 2004) affect the type of problem posing relevant for mathematical modelling.

teachers. The student teachers' main mathematical focus was on the measure and prediction of plant height, using scatter plots and linear functions. Commenting on the problem-posing stage, the student teachers noted that the problems posed by the pupils were to a large degree nonmathematical:

> Student teacher A: "I have been discussing … about ten questions like this, I think: Why a plant is able to grow up through the asphalt, why leaves are yellow in autumn, why some plants are poisonous, why some have thorns, why do flowers need water … it was a lot of that."

And that they had difficulties distinguishing appropriate problems:

> Student teacher B: "Are we likely to ask then what the largest plant in the world may be? Will that be a good enough question?"

Based on this example, we have identified five types of difficulties that the student teachers encountered when they attempted to combine mathematical modelling and problem posing. In particular, we will see the importance of teachers being able to assist in the refinement and reformulation of problems so that they become manageable for the students. For example, a question like "Why do flowers need water?" may be reformulated as a mathematical modelling problem, where one quantifies growth of flowers under the influence of different external conditions.

Five Types of Difficulties

The five types of difficulties faced by the student teachers were:

1. **Posing mathematically relevant problems**. In most cases, the pupils posed problems that were not mathematically relevant, i.e., the problems were stated in such a way that using mathematics to solve them would be contrived. This can in part be explained by the student teachers not being explicit in stating to their pupils that they were mainly interested in problems that could be handled using mathematics. However, we also see an underlying difficulty in posing problems that have mathematical relevance.

   Mathematical modelling is always interdisciplinary. This has several advantages, but also the disadvantage that pupils are not necessarily able to distinguish between problems that are mathematically relevant and problems that are not. Being able to distinguish problems that can be mathematized and being able to reformulate nonmathematical problems so that they can be handled using mathematical tools is part of the learning process. As these are competencies for which the pupils, and the beginning student teachers, were not proficient, this added a level of difficulty to the problem-posing activity.
2. **Posing mathematically suitable problems**. The student teachers had difficulties distinguishing which problems were mathematically suitable for the pupils, i.e., which problems would be neither too easy nor too difficult for the pupils and at the same time would enable the pupils to engage in significant mathematical modelling. Posing problems of an appropriate level of difficulty is, of course, a potential obstacle in any scenario, where one poses nontrivial

problems. However, this obstacle is enhanced in mathematical modelling since it is not always obvious from the initial problem formulation what mathematics will be needed. A needed skill here is to be able to reformulate and adjust problems in appropriate ways so that they attain a reasonable degree of mathematical sophistication.

In mathematical modelling, it is frequently the case that first attempts at problem posing give problems that are unmathematizable or too difficult as stated. It is the norm that repeated adjustments and reformulations of the problem are necessary before one arrives at a problem which is both mathematizable and mathematically manageable.

3. **Posing problems such that the pupils feel ownership of the problems**. In this example, the student teachers encountered pupils just "going through the motions"—that is to say, pupils just spurting out lots of similar looking problems without reflecting on them, or posing pseudo-problems for which they did not really anticipate an answer. Here, one needs to be aware that problem posing is an ongoing process, where reformulations and adjustments of the problem are frequently required. This is especially the case in mathematical modelling, where one continually refers back to the problem situation during the modelling process.
4. **Making problem posing a relevant part of the learning trajectory**. In this case, the pupils' problems were left hanging; they were not reflected upon at the end of the lesson nor were they used in the lessons that followed. If problem posing is to be seen as a mathematically significant activity for the pupils, it needs to be connected to other mathematical activities in the classroom. In particular, if problem posing is to be seen as an integral part of modelling, then the pupils should at times model problems they have posed.
5. **Incorporating the teaching of mathematical content with problem posing and mathematical modelling**. In this example, the student teachers wanted the mathematical content to be connected to scatter plots and linear regression. Two main difficulties can be associated with this: First, this intent had not been communicated to the pupils in the problem-posing lesson, and second, there is an inherent difficulty in posing problems to an unknown or little-known mathematical topic, especially in posing problems where the topic is to be connected to a specific real-world situation.

Why Mathematical Modelling and Problem Posing?

Although some research has been conducted on problem posing in mathematical modelling (e.g., Bonotto, 2010, 2011; English, 2010; English, Fox, & Watters, 2005), in general the topic of problem posing has tended to be peripheral in mathematical modelling research. For example, Goldstein and Pratt (2001) remarked that problem posing "falls outside the classical modelling cycle [of mathematization, transformation, interpretation and validation]" (p. 49). There are several remarks in the literature pointing to the importance of the problem in modelling, to reformulations of problems and to asking appropriate questions throughout the modelling process, although these are mostly incidental. In particular, Ottesen (2002) has drawn attention to the influence of working with mathematical modelling on the question of what makes a problem mathematical. Ottesen wrote:

> [Through working with mathematical modelling] students learn to ask certain types of questions that can only be answered by means of mathematics, as well as types of questions that can only be posed by means of mathematics. (p. 344)

This statement was also used by Swan, Turner, Yoon, and Muller (2007) who saw modelling as promoting "the asking and answering of mathematical questions" (p. 281). Mousoulides, Sriraman, and Christou (2007) drew attention to the potential of ongoing problem-posing activities throughout the modelling process:

> During modeling cycles involved in model eliciting activities students are engaged in problem posing, that is, they are repeatedly revising or refining their conception of the given problem. (p. 35)

Problem posing, in a wide sense, appears in multiple guises in modelling: posing and reformulation of the main problem, making of conjectures, and meta-questions (monitoring and control questions related to the mathematics and/or to the modelling process; or questions taking a critical stance to the model and/or its result). This ongoing problem posing throughout the modelling process makes modelling a natural arena for students' problem posing. According to English et al. (2005):

> Modeling activities promote problem posing as well as problem solving primarily because they evoke repeated asking of questions and posing of conjectures. ... Given a rich problem situation, such as mathematical modelling, in which generating problems and questions occurs naturally, numerous opportunities abound for learning by both child and teacher. (p. 156 and p. 158)

In professional modelling, we see problem posing as an essential component that initiates the modelling process as well as defining its parameters and goals. Formulation and reformulation of the problem are necessary throughout the whole modelling process. Being able to pose and adjust a problem appropriately for the data and mathematical tools available is a vital part of using mathematics in real-world situations. In particular, this implies that problem posing is important in the experience of authentic modelling processes.

Several reasons have been given for including problem posing (see other chapters in this book) and mathematical modelling (e.g., Kaiser & Sriraman, 2006; Maaß, 2010) in the mathematics classroom. Potentially, having a focus on problem posing while working on mathematical modelling will give the best from both worlds. Our motives for considering problem posing in conjunction with teaching and learning mathematical modelling is related to problem posing being a vital component of experiencing authentic modelling. Problem posing is helpful in understanding the decisions made during modelling (especially with respect to limitations and possibilities offered by mathematical modelling). Problem posing is also seen as a useful experience to help equip pupils for later engagement in modelling outside the school environment. Finally, problem posing can give students increased ownership of their learning environment, since it is a natural component of inquiry-oriented instruction and is grounded in the belief of giving priority to the question over the answer.

The Priority of the Question

One of the reasons for our interest in problem posing as a topic in mathematics education is the priority of the question over the answer (Hana, 2012). It is questions that drive our search for knowledge, not answers. The problems we engage in determine what knowledge and understanding it is possible to reach. To investigate or explore, there needs to be something to investigate, some kind of problem which lays the groundwork for the investigative and explorative activity. It may be a vague problem of a general nature; maybe one is only somewhat curious about a phenomenon; it may be a specific closed problem. In any case, the problem is there and gives us a goal and a lens through which we make and interpret our inquiries. Popper (1963) expressed this as "It is the problem which challenges us to learn; to advance our knowledge; to experiment; and to observe" (p. 301).

Likewise, several authors have stressed the connection between understanding and the underlying problem which forms our quest for understanding. As Gadamer (2004) wrote: "To understand meaning is to understand it as the answer to a question" (p. 368). This is not to say that answering a question always leads to understanding:

> Understanding starts with a question; not any question, but a real question. … [A] real question expresses a desire to understand. This desire is what moves the questioner to pursue the question until an answer has been made. (Bettencourt, cited in Wells, 2000, p. 64)

The importance of the question for the type of understanding one achieves is well illustrated by Collingwood (1939):

> Experience soon taught me that under these laboratory conditions one found out nothing at all except in answer to a question; and not a vague question either, but a definite one. That when one dug saying merely, 'Let us see what there is here,' one learnt nothing, except casually in so far as casual questions arose in one's mind while digging: 'Is that black stuff peat or occupation soil? Is that a potsherd under your foot? Are those loose stones a ruined wall?' That what one learnt depended not merely on what turned up in one's trenches but also on what questions one was asking: so that a man who was asking questions of one kind learnt one kind of thing from a piece of digging which to another man revealed something different, to a third something illusory, and to a fourth nothing at all. (pp. 24–25)

To take into account the priority of the question over the answer has significant pedagogical consequences. It implies that the goal of education should shift from pupils being able to answer question to pupils also being able to pose questions. To pose real problems is in general at least as difficult as answering them, for to pose one needs to know what one wants to know and, in particular, one needs knowledge of what one does not know (cf. Gadamer, 2004). It is an educational goal to educate citizens who can use and develop mathematics through posing problems which enables them to act and to further their understanding of the world we live in.

Problem Posing and Different Perspectives on Modelling

Within the mathematics education research community, the topic of mathematical modelling has been considered from different perspectives (Barbosa, 2006; Kaiser & Sriraman, 2006). Barbosa (2006), extending Julie (2002), considered three different perspectives: "modelling as content" (modelling competencies and modelling processes are themselves seen to be part of school mathematics); "modelling as vehicle" (modelling is seen as a vehicle for learning and teaching mathematical concepts and procedures); and "modelling as critic" (modelling is seen as essential for critical reflection of mathematics in society). Though rather coarse, we have previously found the classification of Barbosa (2006) to be a useful tool in discussing with student teachers how one's perspective on modelling affects one's implementation of modelling in the classroom and the type of learning one intends to achieve (Hansen & Hana, 2012). In relation to problem posing, we noted that the difficulties observed when student teachers posed modelling tasks were in part due to their perspective on modelling.

Modelling as Content

From the perspective of *modelling as content*, insight into models and the modelling process is seen in itself as a legitimate goal for mathematics teaching (see, for example, the overview of modelling competencies given in Maaß, 2006). This may include the study of mathematical models without the requirement that the models necessarily have to include specific mathematical concepts or techniques. Problem posing within this perspective includes posing problems from a real-world situation (see section "Problem Posing and the Modelling Process" for more details).

Modelling as Vehicle

Another perspective is to consider modelling as a vehicle for learning mathematical content. The aim is not to construct a mathematical model, but rather to use models as a tool to learn about mathematical themes, techniques, procedures, and concepts. Within this perspective modelling is used for both the development and application of mathematical content. If one wants to apply already known, or at least partially known, mathematical content, then it seems possible to ask students to pose problems within a real-world situation, where the specific mathematics content is applicable. If the aim is to develop new mathematical content, there is a serious obstacle in posing problems related to unknown mathematics. This comment is mainly related to posing and reformulation of the main problem. Making conjectures and posing meta-questions should still be manageable in this situation.

Modelling as Critic

Here, one wishes to "create situations in which students are able to identify, interpret, evaluate and critique the mathematics embedded in social and political systems and claims" (Mousoulides et al., 2007, p. 25). A necessary skill then is to be able to pose the questions needed to identify, interpret, evaluate, and critique. In many cases, the mathematical models used are mathematically sophisticated and involve mathematics that would not be accessible to the ordinary citizen. By posing relevant questions, such as questions pertaining to the assumptions and simplifications made in the model, or to the uncertainty of the model, one may be able to engage in meaningful discussion about the models on a meta level.

Problem Posing and the Modelling Process

There have been many descriptions of the modelling process. Here, we follow Galbraith and Stillman (2006). They considered the following transitions as key in the modelling process:

1. From messy real-world situation to real-world problem statement
2. From real-world problem statement to mathematical model
3. From mathematical model to mathematical solution
4. From mathematical solution to real-world meaning of solution
5. From real-world meaning of solution to revising model or accepting solution (p. 144)

It is clear that the first transition is one involving problem posing. Galbraith and Stillman (2006) identified this stage as consisting of clarifying the context of the problem, making simplifying assumptions, identifying strategic entities and specifying the correct elements of strategic entities. In educational modelling contexts, one often starts with the modelling problem, giving little or no attention to its creation. This removes valuable experiences related to modelling assumptions and specifications from students.

The second transition also involves problem solving. This transition hinges on being able to reformulate a real-world problem as a mathematically manageable problem. Problem posing in this transition is then about mathematizing problems through refinement and reformulation. This requires an understanding of the real-world problem and the possible ways it can be mathematized. It also involves control questions about whether the mathematization makes sense from a real-world perspective. Crouch and Haines (2004) pointed out that students often have problems in making transitions between the real world and the mathematical model, indicating that this transition requires more attention in educational research. Sometimes, students and teachers jump directly to the mathematical model, not paying attention to translation processes.

The third transition is within a purely mathematical content area, although it also includes asking control questions about whether the mathematical operations and techniques used are applicable in the real-world situation.

The fourth transition is one of demathematization. If the problem has been revised during the mathematical stages of the modelling process, this includes demathematizing the problem and comparing it with the original problem.

In the fifth transition, a necessary skill is being able to pose critical questions to analyze the model and solution.

Implementation of Problem Posing in Mathematical Modelling

Modelling Through Conjecturing

A proposed method of problem posing is for students to state conjectures pertaining to a real-world situation that are to be critically examined and refined in attempts to validate or falsify them. To make a conjecture is to move outside the obvious and to test the limits of one's knowledge. As such, conjecturing is a natural way to increase understanding and knowledge of a problem area. To conjecture is to ask "What if…?" and to sharpen one's inquiry toward a concrete statement. In particular, the concreteness of conjectures lessens the chance of one posing vague questions. Furthermore, in trying to validate or falsify a conjecture one necessarily has to pose the types of critical questions which are essential in the verification and examination of mathematical models. When working with conjectures, the focus is automatically moved to reasoning for and against the conjecture, in contrast to the type of problems where students are only interested in finding a numerical answer before moving on to the next question.

As an illustration of using conjectures in modelling, we have provided an example of three student teachers in their practice teaching. The example is related to the "modelling as critic" perspective and is concerned with making dubious conjectures. In general, we see it as beneficial for pupils to engage in making authentic conjectures for which they really wish to determine the validity, but this exercise of making dubious conjectures also seemed to engage the pupils in mathematics in positive ways.

The student teachers wanted to let the eighth grade class experience being critical of the mathematics to which pupils are exposed in society. They decided to implement this by encouraging pupils to make conjectures indicating unusual views or arguments pertaining to real-world situations of the pupils' own choosing. These conjectures were to be presented to the rest of the class, together with some sort of mathematical data and statistical model that supported the claimed conjecture. It was expected that in the ensuing discussion of the conjectures that their fellow pupils would make many critical comments, especially since the pupils were invited to make conjectures that could rather easily be attacked. Through critically evaluating the statistical models it was hoped that the pupils would gain insight into some types of critical questions pertaining to mathematical models and to engage in mathematical reasoning.

One group of students chose the conjecture "The local football team Brann Bergen will beat Barcelona." To support this conjecture, the students used an argument based on the number of goals the two teams scored. This conjecture resulted in a lively discussion in the classroom, where critical comments played an important role. The pupils were invited by the student teachers to dwell on questions such as "What is it that makes this diagram/argument so convincing/misleading?"

The student teachers' decision to use obviously dubious conjectures about different real-world scenarios seemed to activate the pupils' critical engagement in a positive manner. The phases where pupils were "inventing" mathematical models supporting the conjectures and when they presented and compared their conjectures seemed to inspire the pupils to pose critical questions relating to the validity of the conjectures and the mathematical models used.

Pupils as Consultants and Clients

The type of task used in mathematical modelling is one that calls for a mathematical model to be used by an identified client (Mousoulides, 2009). This type of task is intended to give pupils a justification for describing their thinking and considering different possible solutions. In a school–industry partnership where a class collaborated with an oil-valve company, and the pupils were given an authentic consultancy task from the company, we have witnessed firsthand some of the potential inherent in pupils taking on such a role while working on mathematical modelling (Hana, Hansen, Johnsen-Høines, Lilland, & Rangnes, 2011; Lilland, 2012). An important aspect of the activity was that, effectively, the pupils had to define the problems themselves, and that they needed to communicate with the company in order to define the task and gather additional data.

In a similar fashion, Crespo and Sinclair (2008) wrote "there is evidence to suggest that school students are able to generate less narrow and familiar types of problems … when they are invited to pose problems to an audience outside the classroom" (p. 396). In the example above, they were writing about an audience outside the classroom, and it seemed that the essential component was that the students experienced a genuine sense of purpose with the problem-posing task.

We propose combining these two strands of research—mathematical modelling and problem posing. A way to implement this would be to divide a class into groups such that every group has a dual role as a fictional company employing another group as consultants and as consultants to another group's company. As a fictional company, the groups would set the scene and pose a problem for the group of consultants to work on, within some parameters defined by the teacher. As a consultant, the group would work on and refine the problem given by the company group. During this stage, we would envisage that communication between the groups would be essential so that the problem could be refined and data obtained to help refine conclusions. To conclude the activity, the company groups would critically evaluate the solutions found by the consultancy groups.

Conclusion

Problem posing as a pedagogical tool and as an integral part of mathematical modelling has not yet been systematically investigated. In this chapter, we have sketched some of the different ways problem posing offers opportunities and challenges to mathematical modelling. Further work is needed in this area, especially with respect to implementation in the classroom. All in all, we see some golden opportunities in combining problem posing and mathematical modelling in school contexts.

References

Barbosa, J. C. (2006). Mathematical modelling in classroom: A critical and discursive perspective. *Zentralblatt fur Didaktik der Mathematik, 38*(3), 293–301.

Bonotto, C. (2010). Realistic mathematical modeling and problem posing. In R. Lesh, P. Galbraith, C. Haines, & A. Hurford (Eds.), *Modeling students' mathematical modeling competencies (ICTMA 13)* (pp. 399–408). New York, NY: Springer.

Bonotto, C. (2011). Engaging students in mathematical modeling and problem posing activities. *Journal of Mathematical Modeling and Application, 1*(3), 18–32.

Collingwood, R. G. (1939). *An autobiography*. Oxford, United Kingdom: Oxford University Press.

Crespo, S., & Sinclair, N. (2008). What makes a problem mathematically interesting? Inviting prospective teachers to pose better problems. *Journal of Mathematics Teacher Education, 11*, 395–415.

Crouch, R. M., & Haines, C. R. (2004). Mathematical modelling: Transitions between the real world and the mathematical model. *International Journal of Mathematics Education in Science and Technology, 35*(2), 197–206.

English, L. (2010). Young children's early modeling with data. *Mathematics Education Research Journal, 22*, 24–47.

English, L. D., Fox, J. L., & Watters, J. J. (2005). Problem posing and solving with mathematical modeling. *Teaching Children Mathematics, 12*(3), 156–163.

Gadamer, H.-G. (2004). *Truth and method* (2nd ed.) (J. Weinsheimer & D. G. Marshall, Trans.). London, United Kingdom: Continuum Press. (Originally published 1960)

Galbraith, P., & Stillman, G. (2006). A framework for identifying student blockages during transitions in the modelling process. *Zentralblatt fur Didaktik der Mathematik, 38*(2), 143–162.

Goldstein, R., & Pratt, D. (2001). Michael's computer game: A case of open modelling. In M. van den Heuvel-Panhuizen (Ed.), *Proceedings of the 25th Conference of the International Group for the Psychology of Mathematics Education (PME)* (Vol. 3, pp. 49–56). Utrecht, The Netherlands: Utrecht University.

Hana, G. M. (2012). Undersøkende virksomhet, koordinering og spørsmålets forrang [Inquiry, negotiation and the priority of the question]. In M. Johnsen-Høines & H. Alrø (Eds.), *Læringssamtalen i matematikkfagets praksis—Bok I* (pp. 65–88). Bergen, Norway: Caspar Forlag.

Hana, G. M., Hansen, R., Johnsen-Høines, M., Lilland, I. E., & Rangnes, T. E. (2011). Learning conversation in mathematics practice—school-industry partnerships as an arena for teacher education. In *Conference Proceedings Educational Interfaces Between Mathematics and Industry* (pp. 281–290). Lisbon, Portugal: ICMI/ICIAM Study, 2010.

Hansen, R., & Hana, G. M. (2012). "But it is not possible to do this until ..."—The sequencing of teaching mathematical modeling. In G. H. Gunnarsdottir et al. (Eds.), *Proceedings of NORMA 1—The 6th Nordic Conference on Mathematics Education* (pp. 299–308). Reykjavik, Iceland: University of Iceland Press.

Julie, C. (2002). Making relevance in mathematics teacher education. In I. Vakalis, D. Hughes Hallett, D. Quinney, & C. Kourouniotis (Eds.), *Proceeding of the 2nd International Conference on the Teaching of Mathematics*. New York, NY: Wiley.

Kaiser, G., & Sriraman, B. (2006). A global survey of international perspectives on modelling in mathematics education. *Zentralblatt fur Didaktik der Mathematik, 38*(3), 302–310.

Lilland, I. E. (2012). Oppdragets betydning for læringssamtalen i matematikkfagets praksis [The importance of the task for the learning conversation in mathematics practice]. In M. Johnsen-Høines & H. Alrø (Eds.), *Læringssamtalen i matematikkfagets praksis—Bok I* (pp. 125–142). Bergen, Norway: Caspar Forlag.

Maaß, K. (2010). What are modelling competencies? *ZDM, 38*(2), 113–142. doi:10.1007/BF02655885.

Maaß, K. (2010). Classification scheme for modelling tasks. *Journal fur Mathematik-Didaktik, 31*(2), 285–311. doi:10.1007/s13138-010-0010-2.

Mousoulides, N. G. (2009). Mathematical modeling for elementary and secondary school teachers. In A. Kontakos (Ed.), *Research and theories in teacher education*. Rhodes, Greece: University of the Aegean.

Mousoulides, N., Sriraman, B., & Christou, C. (2007). From problem solving to modelling: The emergence of models and modelling perspectives. *Nordic Studies in Mathematics Education, 12*(1), 23–47.

OECD (Organization for Economic Co-Operation and Development). (2004). *Problem solving for tomorrow's world—First measures of cross curricular competencies from PISA 2003*. Paris, France: Program for International Student Assessment.

Ottesen, J. T. (2002). Do not ask what mathematics can do for modelling. In D. Holton (Ed.), *The teaching and learning of mathematics at university level: An ICMI study* (pp. 335–346). Dordrecht, The Netherlands: Kluwer Academic.

Popper, K. (1963). *Conjectures and refutations: The growth of scientific knowledge*. New York, NY: Routledge.

Swan, M., Turner, R., Yoon, C., & Muller, E. (2007). The roles of modelling in learning mathematics. In W. Blum, P. L. Galbraith, H.-W. Henn, & M. Niss (Eds.), *Modelling and applications in mathematics education* (pp. 275–284). New York, NY: Springer.

Wells, G. (2000). Dialogic inquiry in education: Building on the legacy of Vygotsky. In C. D. Lee & P. Smagorinsky (Eds.), *Vygotskyan perspectives on literacy research* (pp. 51–85). Cambridge, United Kingdom: Cambridge University Press.

Chapter 3
Conceptualizing Problem Posing via Transformation

Jasmina Milinković

Abstract The goal of this chapter is to outline an approach for developing teachers' proficiency in posing problems. Reasons why it is important for a mathematics teacher to be good problem poser are investigated. Links between knowing mathematics and knowing how to pose problems are also discussed. Training students in problem-solving techniques does not necessarily end in their learning mathematics. In this chapter, problem-posing activities based on the idea of transformation are described—two kinds of transformations are proposed and analyzed successively. The first is transforming problems from routine to advanced ones by changing elements in the problem space. The second is posing problems by transformation of representation. Developing problem-posing skills, from posing routine tasks to posing more complex mathematics problems, encourages student–teachers to think about problem posing as a creative professional activity. Lastly, the possibility to developing pupils' capacity to pose problems via transformation is presented.

Contents

J. Milinković (✉)
Teacher Education Faculty, University of Belgrade, Belgrade, Serbia
e-mail: milinkovic.jasmina@yahoo.com

F.M. Singer et al. (eds.), *Mathematical Problem Posing*,
Research in Mathematics Education, DOI 10.1007/978-1-4614-6258-3_3

Introduction

An old saying states that there are no unintelligent questions but there are indeed unintelligent answers. As our contribution, I offer a view on how to make intelligent mathematics problems. The word question comes from the Latin word *quaestio* which has the following meanings: examination, seeking, searching, research, scrutiny, and problem. In the words of our time, when we pose questions we are calling for an examination of what we already know, for seeking an answer while putting under scrutiny what we have already found, and finally, for solving the problem.

There is extensive research and professional support literature on problem solving (e.g., Bonotto, 2007; Lesh, 1981; Michalewicz & Fogel, 2004; Polya, 1973; Schoenfeld, 1992; Wilson, Fernandez, & Hadaway, 1993). Problem posing, on the other hand, is a comparatively new issue for the educational community. Ask yourself, "Why is it important for mathematics educators to study problem posing? Is it important for teachers? Is it a worthy activity for students as well?"

First, let us try to answer why it is important for mathematics teachers to be good problem posers. An opponent to that idea may argue that we should instead teach them where to find the best resources for problems (books, Internet sites, etc.). Indeed, the task of finding good resources appears to be less of a challenge than ever before. Imagine that we decide to do exactly that, so we instruct teachers how to search through problem resources. But then a new issue comes to light. Can a teacher decide autonomously which set of problems to give children, when to use a particular problem, or in which order to present the problems to pupils? So, we are back to the beginning, and in fact it starts to look easier to teach teachers how to create problems of their own. We expect that as an additional achievement, teachers will not only learn to pose problems but also how and when to use problems, in which order, and how to present them. Moreover, in the course of studying how to pose diverse problems from routine tasks to mathematics problems, we may expect, as Schoenfeld (1992) remarked, that preservice teachers will gain deeper insight into the structure of elementary school mathematics.

Let us try to answer the second question: Why should pupils learn how to pose problems? There are at least two good reasons. One is that in real life we are not dealing with textbook tasks, but rather with more or less complex situations. Then, formulating a mathematics problem which reflects a (non)mathematical situation becomes an important part of the modeling process which, in turn, could lead us to a solution of a real life problem. The other reason is the well-known fact that formulating a problem implies an understanding of content matter. Currently, the activity of creating (more or less simple) problems is a part of regular activities even in traditional mathematics classes. For example, a teacher may ask young pupils to create a textual problem which can be represented by the equation $x + 3 = 5$. A pupil might create a problem like this: "Mark had few toy cars. When his mum gave him three more, he had five. How many toy cars did he have before getting the present?" Here, we have developed a course schema for successful problem posing primarily for teachers. To a lesser extent the same may apply for young pupils.

The arguments supporting the significance of problem solving are relevant for problem posing as well. Stanic and Kilpatrick (1988) identified three themes in problem solving: (a) problem solving as context; (b) problem solving as skill; and (c) problem solving as art. They also identified five roles that mathematics problems play: (a) as a justification for teaching mathematics; (b) as specific motivation for subject topics; (c) as recreation; (d) as a means of developing new skills; and (e) as practice. Similar themes may be recognized in problem posing. Indeed, problem posing may contribute to students' skills or provide context for learning, but may also be considered as an artful activity.

Proficiency in problem posing is, by some in the educational research community, considered to be part of pedagogical knowledge, whereas for others it is closer to subject matter knowledge. In literature which discusses possible relationships between subject matter knowledge, pedagogical (didactical) knowledge, and curricular knowledge (see, for example, Hiebert & Carpenter, 1992; Kennedy, 1998; Leinhardt, 1989; Peterson, 1988; Schulman, 1986), we find that problem posing is an underestimated issue. "Subject matter knowledge consists of an understanding of the key facts, concepts, principles and explanatory frameworks in a discipline … as well as the rules of evidence and proof within that discipline" (Schulman & Grossman, 1988, cited in Brown & Borko, 1992, p. 211). Pedagogical content knowledge, on the other hand, implies understanding of how to represent subject matter in ways suitable to the needs and abilities of learners. Shulman pointed out that pedagogical knowledge includes "the most useful forms of representation of concepts, the most powerful analogies, illustrations and examples, explanations, and demonstrations. It also includes an understanding of what makes the learning of a specific topic easy or difficult" (Schulman, 1986, p. 9). The model of pedagogical thinking developed by Wilson and colleagues depicts common components of teaching: (a) comprehension; (b) transformation; (c) instruction; (d) evaluation; (e) reflection; and (f) new comprehension (Schulman, 1986; Wilson et al., 1993; Wilson, Shulman, & Richert, 1987). Somewhere in between is Kilpatrick's (1987) statement that problem formulating should be seen not only as a means of instruction but as a goal of instruction.

Educators have long recognized the importance of problem-posing activities for children (Freudenthal, 1972; Polya, 1973). On the most global level, the OECD framework provides a modeling schema which highlights the need for students to work in the world of mathematics. But, the PISA framework leans toward mathematical applications and the search for solutions in real-world contexts (OECD, 2003). In both frameworks we can identify components: concepts/processes, competences, and contexts. In problem-posing activities, teaching competences such as fluency and flexibility of subject matter knowledge as well as inventiveness become visible.

In addition, problem posing has important role in applied mathematics as well as in various applications of mathematics concepts, methods, and achievements. Therefore, it should not be neglected by either mathematics teachers or by students. Indeed, if we glance into areas where mathematics is used as a tool (the sciences, engineering, etc.), we will recognize the importance of problem-posing skills. There, the process of applying mathematics begins with the recognition of a (mathematics) problem that one needs to answer. Pollak (1988, p. 31) identified two types of mathematical needs in engineering practice: (a) elementary needs "to set up the right problem, to have a good idea how big the answer should be, and to get the right answer by any available means whatsoever-mentally, calculator, paper-and-pencil, computer whatever"; and (b) advanced needs: "we need employees who know that there is a large variety of forms of mathematical thinking, and what these various forms can do."

Before I proceed to a description of my approach to the development of problem-posing skills, I would like to point to a somewhat neglected link between knowing mathematics and knowing how to pose problems (Nodding, 1992). Others have also recognized the significance of content domain knowledge. The process of problem posing is intertwined with each teacher's range of skills and competences and is influenced by the context (Figure 3.1). According to Wake (2010), the process of developing tasks "is inevitably fuzzy as the different factors are brought to bear on classroom experiences" (p. 7).

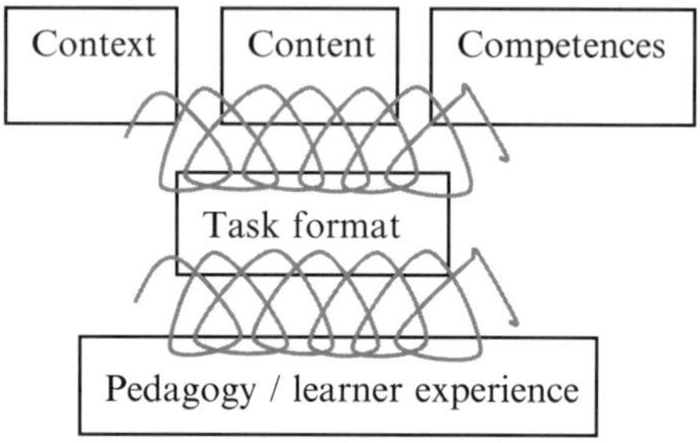

Figure 3.1. School mathematics domain (Adapted from Wake, 2010).

Is "knowing mathematics" a prerequisite for problem-posing proficiency? In other words, connections between a teacher's pedagogical and content knowledge need to be explored. In a chapter on teachers' professional development, Nodding (1992) argued that regardless of plausibility of the importance of "knowing mathematics," evidence of the significance of subject matter knowledge for teaching was lacking. Hiebert and Carpenter (1992) discussed important elements in helping student–teachers implement programs designed to develop students' understanding of mathematics. They acknowledged intuitive and formal psychological and pedagogical content knowledge about pupils and mathematics teaching and pointed to research findings about the lack of connections among those elements. Research on the development of children's addition and subtraction strategies provided a classification scheme for distinguishing among problems in terms of basic principles and children's strategies (Carpenter, Fennema, Franke, Levi, & Empson, 2000). Here, the need for connecting knowledge of mathematics and knowledge of children's thinking is obvious. Some researchers deduced from interviews with teachers that they fail to recognize the full complexity of concept of probability (Liu, 2005; Liu & Thompson, 2004). Others reported that teachers often experience great difficulty dealing with concepts of division and fraction (Ball, 1990). Similarly, researchers noticed that even textbook authors demonstrate a lack of proficiency in problem posing. Nesher (1980) and Reusser (1988) have pointed out that problems found in textbooks rarely address misconceptions related to the mathematical concept of multiplication, and that authors do not attend to varying numbers in tasks, tending to use only "easy" numbers which produce "clean" results. Likewise, we believe that when someone teaches prospective teachers to be good problem posers, he or she actually needs to ask them to reflect on their insights into mathematics. Our claim about the importance of having a conceptual understanding in mathematics as a prerequisite to problem-posing proficiency needs to be studied in the future.

Transformations in Problem Posing

To begin, I define the term *mathematical problem*. Mathematical tasks are anything that requires mathematical tools to be used. What kinds of mathematical tasks can be called problems? I distinguish between "mathematical problems" and other "mathematical tasks" by the level of cognitive demand. Problems are mathematical tasks whose solution is not immediately achievable for problem solver. Thus, even the so-called routine calculation exercises may be called problems if the solver needs to perform multiple steps in order to reach a solution. Besides, a problem at one level of schooling may become a routine task in the next level. Thus, the identification of a "problem" is linked to the problem solver's knowledge and abilities in the moment of solving the problem. From now on, we will assume that we are able to distinguish routine tasks from advanced problems, the ones for which a pathway to a solution is not obvious or known to the problem solver.

In what follows, I will begin by defining transformations in problem posing. Then, I will discuss examples of ranges of problems created by transformations within the same context. Three strategies for problem posing will be discussed. Next, I will present problem-posing techniques that are based on the transformation of problem representations. Again I will propose an additional strategy for problem posing based on the analysis of several examples. Each of the strategies discussed are tools for successful problem posing.

However, before considering different strategies for problem posing based on transformations, the concept of transformation itself, and its significance, will be described. In their research on the ways that student–teachers develop mathematical content knowledge within practical training, Thwaites, Huckstep, and Rowland (2005) distinguished the "knowledge quartet"—foundation, transformation, connection, and contingencies. The category "transformation" relates directly to the issue of representations because it refers to the ability of a teacher to transform the content knowledge into different forms. In the following pages I will explain my ideas on how to train teachers to pose problems. I will focus on two simple ideas, both of which are based on this process of *transformation*.

One of the key ideas in learning is schematization. Schematization is a process of gradual building up of mental schemes toward formal schemes of mathematics.

Schematization in mathematics is a result of mathematization. Treffers (1987) distinguished two types of mathematization in the educational context: horizontal and vertical. Horizontal mathematization involves going from the real world into the world of mathematical symbols. In other words, students go through a process of solving a problem from a real-life situation with mathematical tools. Vertical mathematization is the process of building up and reorganization within the mathematical structure by discovering connections and relationships among concepts and finding shortcuts. Note that a process of transformation is involved in both types of mathematization.

I propose the need for attending to different levels of schematization in problem posing. While students are learning mathematics, they pass through various levels of understanding. At the beginning of the learning process they start off with simple problem solving and the development of the ability to find informal context-dependent solutions. Students gradually build schemes of underlying principles and even broader relationships. The ability to reflect on previous activities signifies the next level in the process of learning. Progressive schematization is a product of horizontal and vertical mathematization. Thus, formal schemas are reached in several consecutive stages from horizontal mathematization to vertical mathematization.

Closely related to my conception of transformation is Kilpatrick's description (1987) of cognitive mechanisms which might help in the production of problems such as reasoning by analogy and Silver's idea of reformulation of problems. Silver, Mamona-Downs, Leung, and Kenney (1996) conducted a study on preservice and in-service teachers' ability to generate problems in a complex context (Billiard Ball Mathematics). They identified groups of associated generated problems. Some of the clusters were "chaining," with a sequentially linked character; others had "symmetry" in that the goals and conditions of one problem were symmetrically exchanged in another problem. The third cluster of problems was a group of problems in open-ended form, based on the tendency to challenge constraints. Silver et al. suggested that there was a distinction between teachers who generated problems while keeping the given constraints (and changing goals), and teachers who generated problems by challenging the givens. Research should reveal whether these differences were due to difference in subject matter knowledge or due to (un) familiarity with context.

Transformations by Changing Problem Space

Any problem can be described in terms of its context, of givens and unknown elements, and of the relationships between the elements. Any problem is defined within a problem space. The McGraw-Hill Science and Technology Dictionary (2003) defines *problem space* in psychology as "a mental representation of a problem that contains knowledge of the initial state and the goal state of the problem as well as possible intermediate states that must be searched in order to link up the beginning and the end of the task."

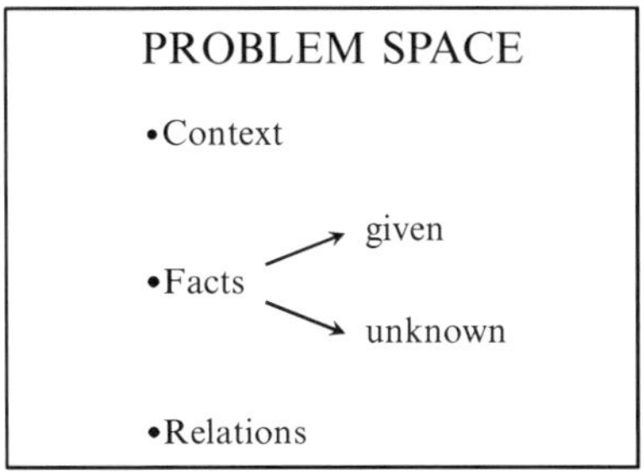

The givens in a problem space correspond to the answers to Polya's questions "What are the data?," "What is the given?," "What are the conditions?." The elements named "unknown" are those that we are asked to determine. Relationships between the elements (given and unknown) are also parts of Polya's "conditions." Finally, the context in which these elements are placed may be abstract as well as realistic. In an investigation on problem-posing skills in children, Stoyanova and Ellerton (1996) discussed problem-posing situations in terms of the source of ideas such as classroom activities or textbook problems. For us, the problem-solving situation or "context" creates the boundaries of the space under scrutiny. Facts (either given or unknown) and relationships define the structure of the problem space. Often a problem may be read as "Find a fact." Or in other problems, it may refer to relationships: "Determine whether *A* and *B* are related or not" (Figure 3.2).

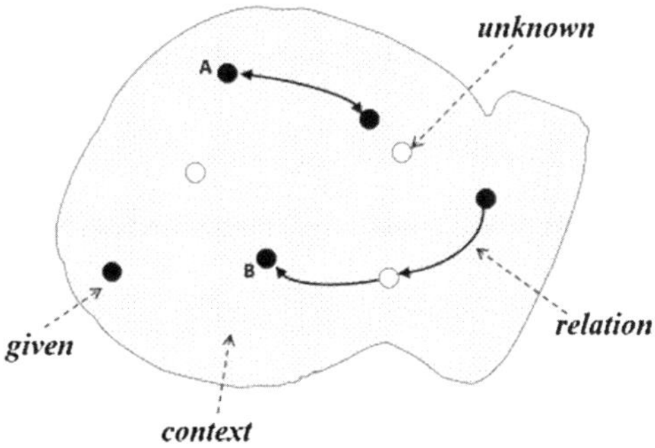

Figure 3.2. Problem space.

Transforming a problem into a new one means that some (one or more) of the elements of the problem space are changed while the others remain the same. One of our ideas is that problems can be transformed so that they reflect a changing mathematical structure. A new problem may be posed by changing: (a) what is given, (b) what is searched for (unknown), or (c) the context. The transformed problem may be more or less difficult than the initial problem. Note that sometimes the transformed problem may turn out to be unsolvable. But this is a result of the teacher's subject matter knowledge or lack of it.

Transforming Problems by Varying Unknown and Known Elements Within the Same Problem Space

Polya (1973) argued that "simple" problems are suitable for "simple" concepts, whereas "complex" problems serve well to examine "complex" concepts or connections between concepts, procedures, or strategies of reasoning. I believe that we can and we should learn how to transform simple problems into complex problems in a sequence of steps. A good example of a sequence of problems starting from the simplest to a complex case within the same context might be found in combinatorics.

Problem 1. How many 2-digit numbers may be written using the digits 2 and 4? Write them down.

Problem 2. Using digits 2, 4, and 8 write all 2 digit numbers, so that no digit in the number is repeated.

Problem 3. Write all 2 digit numbers using digits 2, 4, and 8. How many are there?

Problem 4. Write all 3 digit numbers using digits 1 and 2. How many are there?

Problem 5. How many different four digit numbers can you get by putting digits in place of the stars?

(a) 1**7 (b) **43 (c) ***5

Problem 6. How many 4 digit odd numbers can you get using digits 1, 2, 3, 4, 5 so that no digit in the number is repeated.

The strategy of transformation works well with problems in other fields as well. Here are three problems with tessellations.

Problem 7. The rectangle has sides of lengths 1 and 2. Divide it into two parts which can be compiled to form a right triangle.

Problem 8. The rectangle has sides of lengths 1 and 2. Divide it into three parts which can be compiled to form a square.

Problem 9. Divide the given figure into six equal triangles without removing the pencil from the sheet of paper.

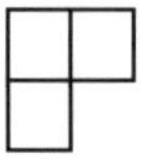

Erich Wittman (2005) in his work often followed the idea of transformation (though he did not name it as such). He proposed the same context for a set of problems that can be sequenced from the easiest to the most difficult one. In an example, the context was a game-like activity of finding missing element(s) in a special triangle called an Arithmogon. Wittmann got the inspiration for the game from McIntosh and Quadling (described in Wittman, 2005). An Arithmogon is divided into three distinct areas A, B, and C. The number of objects in A and in B add up to the number X, the number of objects in B and in C add up to Y, and finally the number of objects in C and in A add up to Z (Figure 3.3).

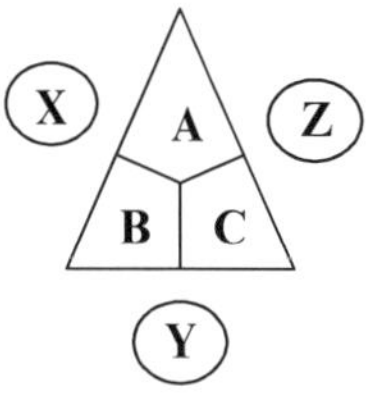

$$A + B = X$$

$$B + C = Y$$

$$C + A = Z$$

Figure 3.3. Context for a set of problems from Wittman (2005).

The missing numbers in the Arithmogon can be found by calculation (addition and subtraction). The easiest problem would be to find one missing inside number when two inside numbers and one outside number are given. The problem may be transformed into a new one: to find a missing inside number if one inside and one outside number is given. The new problem is as difficult as the first one. In later years this problem may be transformed into a new, more difficult one: finding a pattern for the Arithmogon instead of solving case by case. In the next level of transformation, students could be asked to find a solution by solving linear equations. An example of such a problem is in Figure 3.4.

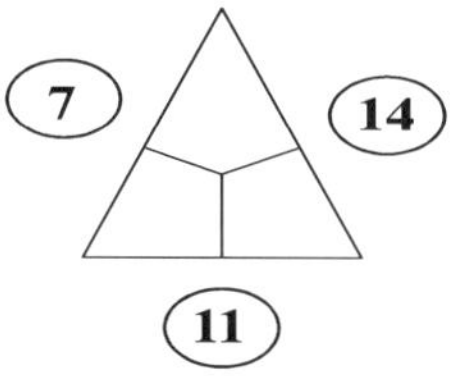

Figure 3.4. Arithmogon, nontrivial problem.

In the next transformation, a problem might be posed in a different context. Instead of an Arithmogon, the student could be asked to fill in the missing numbers in a quadrilateral (Figure 3.5). A problem in this context may have more than one solution or no solution. Here, the teacher's subject matter knowledge may contribute to his ability to pose a solvable problem.

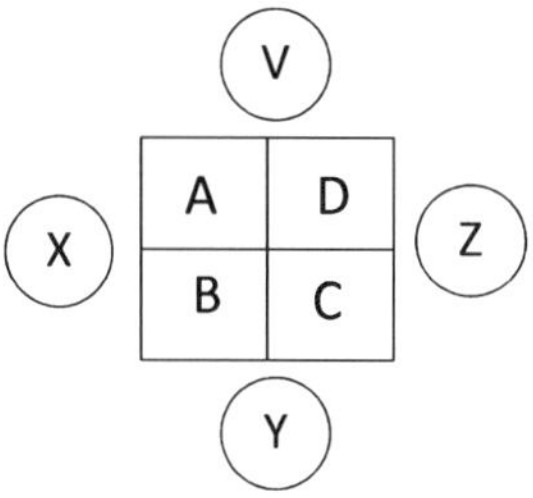

$$A + B = X$$

$$B + C = Y$$

$$C + D = Z$$

$$D + A = V$$

Figure 3.5. Complex problem.

Arithmogon problems can be transformed into a general case for n-sided polygons. Finally, it can be generalized to "arithmohedra" problems for college students.

What can be concluded from the examples seen so far? First, the idea of teaching teachers to transform simple tasks into more complex ones while keeping the same context or form of problem starts to become plausible and hopefully possible. Increased sensitivity of teachers to problem difficulty would also be expected to be an outcome of working with transformations.

Some of the simple strategies in transforming problems used in the previous examples were:

Strategy 1. Transform from problems with a smaller number of simpler parameters (e.g., "easy" numbers) to problems with more (often more complicated) parameters (e.g., "difficult numbers").

Strategy 2. Transform problems by adding new parameters.

Let us analyze the next example and find the strategy used for transforming the initial problem. The problem to be discussed is a geometrical one, and it might be judged that even the simplest problem in the cluster is intended for an advanced class of children. The problem belongs to a group of problems dealing within inscribing (placing figure *A* into figure *B*, where the vertices of the figure *A* are located on the outer boundary of figure *B*).

Problem 10. Inscribe a triangle in a circle.

Problem 11. Inscribe a triangle in a square.

Problem 12. Inscribe a rhombus (equilateral parallelogram) which is not a square in a circle.

Problem 13. Inscribe the triangle with the largest area in a given square.

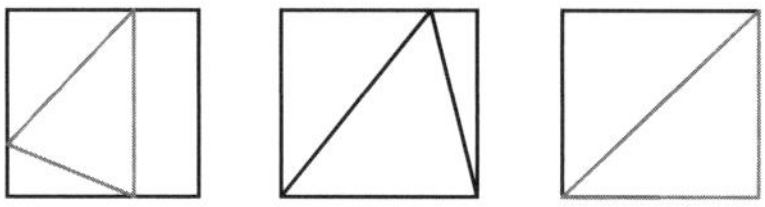

Problem 14. Inscribe the triangle with the maximum area in a polygon.

The class of problems involving placing in and inscribing polygons is a good source of the so-called extremal problems. Examples of such problems are drawing a figure with the greatest length, area, and so on. Additionally, the class of problems within this context is inexhaustible. Yet, teachers need to be cautious when posing problems of this type to be sure they are giving problems with a solution. It is easy to create an unsolvable problem in this context. For example, it is not possible to inscribe a rhombus in a circle if it is not a square. But it is possible to inscribe a quadrilateral in a circle. Further, a circle cannot be inscribed in a quadrilateral which is not a rhombus. But it is possible to inscribe a circle in any rhombus. Again, the importance of a teacher's subject matter knowledge is self-evident.

What was the origin of this sequence of five geometric problems? We transformed one problem into another with the following strategy:

Strategy 3. Transform from a case problem to a generalized one by removing some conditions.

(Case problems are most often simpler than generalized ones.)

Going back to the initial idea of making changes within a given problem space, we need to consider the possibility of transforming a problem by changing its context—mathematical modeling. However, a discussion about mathematical modeling lies beyond the scope of this chapter. However, the following provides an example of a transformation from a contextual to a realistic modeling task.

Problem 15. Pizzeria "Grand Roma" has a fixed price for a slice of pizza with cheese and tomato of 1.2 euro. A side order of vegetable (tomato, mushrooms, leek, and pepper) costs 30 cents, and of ham and sausage 40 cents. Ernie bought three slices of pizza with mushrooms and ham. How much did he pay?

Problem 16. Create a price structure for a pizzeria.

Whereas the first problem is a traditional word problem (although not simple), the second problem is an open task which requires a full modeling cycle during problem solving.

At the end of this section I will suggest how the training of teachers in problem posing might begin. A good starting point can be the transformation of games. For example, the game of Mankala is an old African game using seeds or stones. Players can reflect on the strategies involved in this game, and they can try to make it either simpler or more complex.

Players have seeds and a Mankala board (Figure 3.6). First the seeds are "planted" into alternate holes. The objective of the game is to capture more seeds than one's opponent, to leave the opponent with no legal move, or to finish with an empty side. Players need to discover a winning strategy. Teachers need to play the game before thinking about how to transform it.

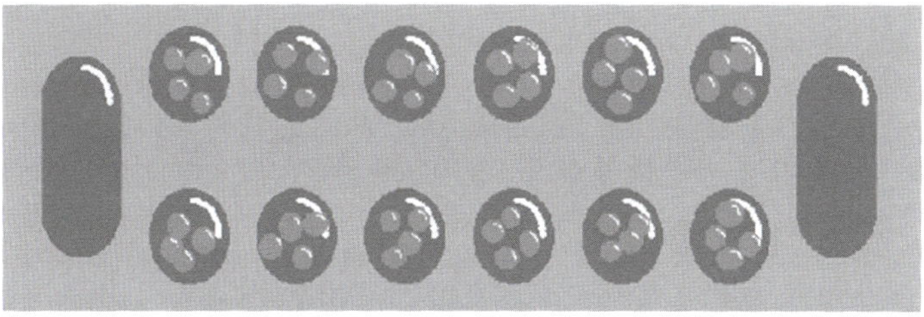

Figure 3.6. Mankala board.

By the same token, in an undergraduate course for preservice teachers called Math Games, our students have a similar activity. They have to find a game of their choice, and then offer a transformed game. But, they need to explain what kinds of changes they made to the game, and why they changed the game in the ways they did.

Transforming Problems by Changing Representation

The other training route in problem posing via transformations is based on a representational approach. The idea that representations are "tools in thinking" is well documented in the literature (Cobb, Yackel, & Wood, 1992; Couco & Curcio

2001; Dufour-Janvier, Bednarz, & Belanger, 1987; Janvier, 1987; Kaput, 1987; Lesh, 1981; Michalewicz & Fogel, 2000). But, this view can be extended into the field of problem posing. Goldin and Shteingold (2001) explained that:

> Effective mathematical thinking involves understanding the relationships among different representations of "the same" concept as well as the structural similarities (and differences) among representational systems. (p. 9)

When mathematicians speak about different representations, they usually think about different ways we can represent a problem. A concept somewhat close in meaning to representations is schematizations (which have already been introduced). "Representation" is taken to refer to the presentation of a certain problem at different levels of abstraction. In other words, it means thinking about the problem in different paradigms. Freudenthal (1983) stated that historical phases in the development of mathematics signified that knowledge achieved by discovery at one moment had been transformed by schematization (or coding) into new skills and/or understandings on a higher level of abstraction. For example, a big step forward in the development of mathematics was the introduction of symbolic (numerical) representations of numbers. Earlier phases of development were characterized by iconic representations. Nevertheless, Freudenthal insisted that schematization should not be viewed as a historical necessity but as humans' psychological development in understanding their surroundings. It would not be wrong to say that Bruner's theory was a theory of representations. He understood a representation as a final product of processing and coding information. Bruner's theory of three types of representations (active, iconic, and symbolic) provided a flow chart of progressive schematization and formalization that occurs in the learning process (Bruner, 1960). It should be noted that, contrary to his initial writing, in his later work Bruner claimed that three types of representations were not hierarchical but culturally bounded.

Knowledge of representations, as was noted earlier, is particularly important in problem solving (Goldin & Shteingold, 2001; Polya, 1957). We believe that one way to help students to become confident in using different representations in problem solving is to confront them with different forms of problems. Friedlander and Tabach (2001) maintained that a teacher's presentation of a problem situation in different representations could encourage flexibility in students' choice of representations, stating that "the presentation of a problem in several representations gives legitimatization to their use in the solution process" (p. 176). Similarly, Singer, Pelczer, and Voica (2011) emphasized that the task format underlined a sequence of transfers from external to internal representation. Wittman (2005) introduced the term "informal" representations to describe the presentation of abstract mathematical concepts in a "quasi-reality." He supported the idea of using them as a mediating tool which is more appropriate than symbolic representations. Some representations such as counters, number line, or place value table prove to be good contexts for posing various problems.

There is no agreement among researchers, however, regarding the benefits of diverse representations in teaching. While some researchers oppose it, others make strong argument for their use. On the one hand, Hiebert and Carpenter (1992) referred to the research of Cobb (1988), Erlwanger (1973), and Lawler (1981) which showed how some students experienced difficulties in recognizing the relationship between different solutions of identical problems which were presented in different contexts.

On the other hand, there are several researchers who call for the use of various representations in teaching. For example, extending the idea of representations as a medium for cognition, Arcavi (2003) identified visual representations as a "cognitive technology aid" (p. 216) for thinking, learning, and problem-solving activities in technology-driven communication. He and others have highlighted the sociocultural value of visual representations (e.g., Arcavi, 2003). In addition, Arcavi identified three functions of visual representations: (a) as a support and illustration of symbolic representations; (b) as a tool for resolving conflict between intuition and symbolic solution; and (c) as a tool to reorganize and recuperate conceptual understanding. Arcavi went on to suggest that "seeing things" sharpens our understanding and serves as a springboard for questions which we would not pose otherwise. Other researchers have made similar recommendations, calling for a range of situations to be modeled (e.g., Greer, 1992). In a similar way, Nunes (1992) advocated that "understanding several different situations involving the same invariant could lead to the abstraction and generalization of the core concept (the invariant), and to the enrichment of the concept by extending the set of situations to which it applies" (pp. 571–572).

In recent studies I have explored aspects of using representations in problem posing. First, preservice teachers' proficiency in using representations will be discussed. Then, to conclude this chapter, details of two studies will be presented, in which we have investigated the effects of using different representations for problem posing in school classrooms.

We studied preservice teachers' preferences in using representations of multiplication (Milinković, 2012a). The survey questions examined: (a) students' knowledge of representations of multiplication and of the commutative law; and (b) students' competence in using different representations in problem posing.

The request to use visual representations in problem posing proved to be a challenge for the preservice teachers. The analysis of preservice teachers' questionnaires revealed that they preferred concrete models which supported the idea of multiplication as repeated addition (sets and equal group representation). Students opted for grouping representations R1, R2, and R3 (see the top row in Figure 3.7) and simplified contexts. For example, they posed tasks of distributing flowers in vases or describing in numbers pictures with a clown having two dark and three white balloons in two hands, etc. Somewhat incomplete findings from a question on problem posing pointed to possible weaknesses in preservice teachers' readiness to pose problems. Consequently, they were limited to using simpler contexts and unsophisticated choices of representations in problem posing.

Without doubt, the most common representation of a problem situation to be modeled is through word problems. Fuson (1992) defined posing a mathematics problem as a task of "translating from the natural language representation of a problem to the mathematical–language representation of the model" (p. 285). A competent student is able to construct an appropriate mathematical formulation as an

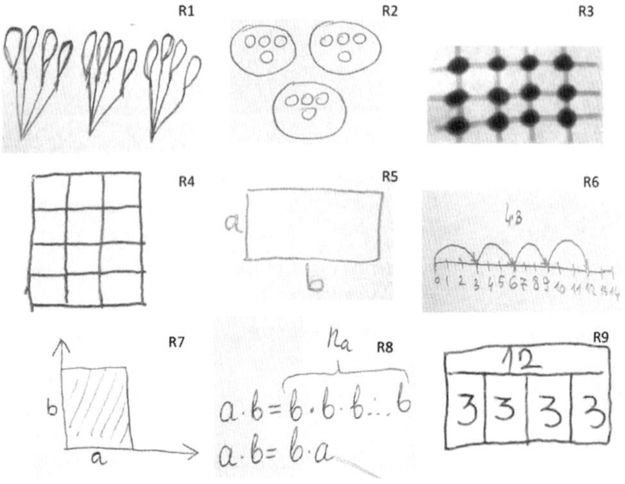

Figure 3.7. Preservice teachers' representations of multiplicative situations (Milinković, 2012a).

intermediate representation of the situation and then to move directly to a mathematical expression on the basis of syntactical surface tools.

Another way of representing problems is in pictures. Diagrams, graphs, or tables are often used as ways to present data. In this case, problems are related to analyzing pictures and understanding what the information given in pictures is telling them and how it can be used to find a solution.

In the problem shown in Figure 3.8 students needed only to read the graph and fill in the table. In the next problem (Figure 3.9) students needed to use the picture to deduce whether the train should stop or proceed.

Finally, we come to the least used representation in problems: action. One exception is the relatively common practice of posing problems with counters which are often used for representing numbers in early grades. In this context, problems can be posed in the form of a game. Problems might be as simple as these:

Problem 17. Put two counters together. Which number did you get?
Problem 18. Create number X with three counters.
Problem 19. Create an even number with two counters.

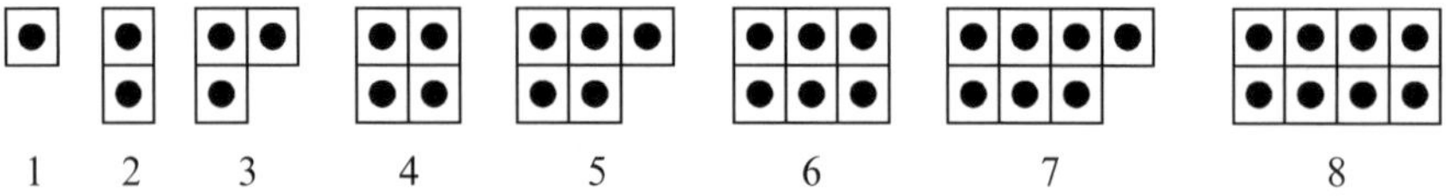

Problem 20. Discover what is common for all even numbers.

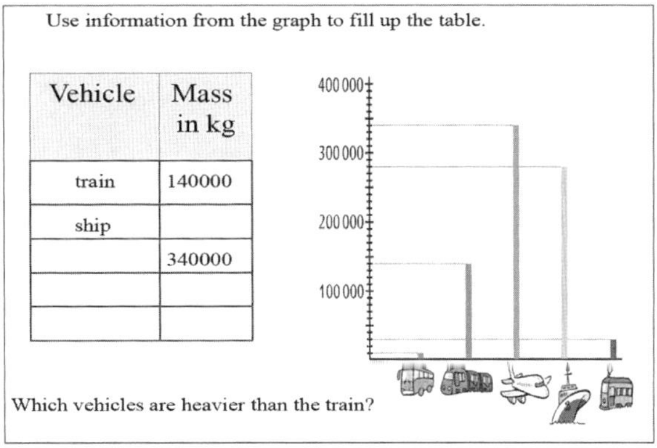

Figure 3.8. Reading a graph, adapted from a Fourth Grade Mathematics Textbook (Dejić, Milinković, & Djokić, 2005).

Figure 3.9. Problem in a picture, adapted from a Fourth Grade Mathematics Textbook (Dejić et al., 2005).

Another problem with an action representation comes from geometry. The "flexible springs" (Problem 21 shown in Figure 3.10) can be used to discuss the important idea of rigidness in geometry (Milinković & Micic, 2008). Such problems illustrate how (rarely used) action representations may contribute to students' development of mathematical understanding and skills.

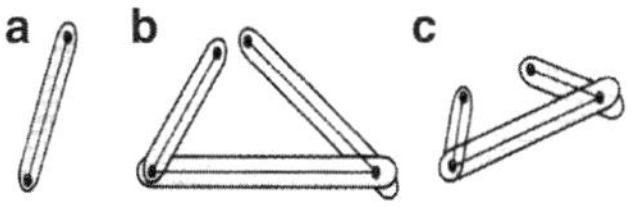

Figure 3.10. "Flexible springs."

Problem 21. (a) Can you make a triangle from any three segments?
(b) Can you make a quadrilateral from any four segments?

Manipulating strips as representations of fractions (Figure 3.11) could be a suitable context for posing problems at different levels of difficulty. Problems 22 and 23 may be solved by using strips representation. They are meant to be used for practicing addition and subtraction of fractions.

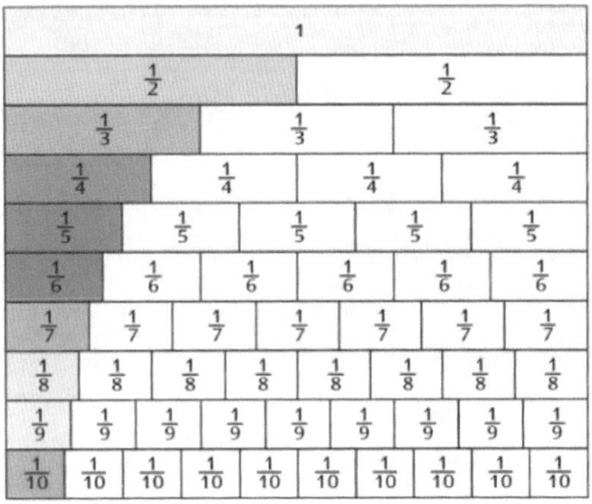

Figure 3.11. Fraction strips.

Problem 22. Compare 2/7 and 3/8.
Problem 23. Combine strips in 2 colors to make 5/6.

Now, when we are aware that different types of representations may provide good contexts for defining a problem space, we can consider transformations from one to another. Wittman (2005) with reference to Jean Piaget noted that "when it is intuitively clear that the operations applied to a special object can be transferred to all objects of a certain class to which the special object belongs then the relationships based on these operations are recognized as generally valid" (p. 20). Examples of representational transformation of problems will now be summarized; again we begin with combinatorics. These examples are of obviously different problems, with different representations, but which have the same underlying mathematical ideas in their solutions.

Problem 24. Friends are shaking hands when they meet each other. How many handshakes happened if there were:
(a) 2 friends (b) 3 friends (c) 4 friends (d) 5 friends?

Problem 25. How many segments can be drawn through the given points?

Problem 26. How many two letter combinations without paying attention to order can you make out of:
(a) 2 letters (b) 3 letters (c) 4 letters (d) 5 letters?

Problem 27. Determine the number of roads connecting cities if each two cities are connected.
(a) cities A and B (b) cities A, B, and C
(c) cities A, B, C, and D (d) cities A, B, C, D, E

A second set of example problems were designed for the *Mathematics in Context* series for middle-grade students. The *Patterns and Symbols* unit (Romberg, 1997) was designed to help students develop understanding of the idea of patterns. In particular, they learned how to express a pattern with mathematical symbols. In one activity, they studied the growth pattern of a snake with red (R) and black (B) rings. They were asked to explain her pattern of growth and to predict how she would look like at the *n*th iteration.

The drawing could be studied, or students could develop a model with rings and pose the same question. Alternatively, the problem could be posed in a symbolic representation: "What is the *n*th number in a sequence of numbers 1, 2, 4, 8, ... (or any other listed in the table for Figure 3.12)?"

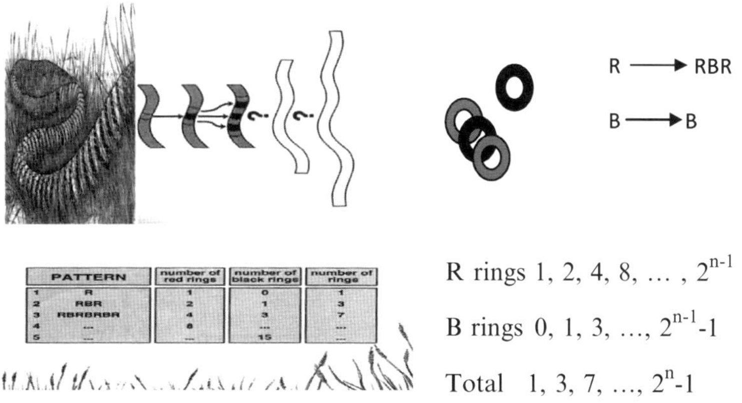

Figure 3.12. Adapted from *Mathematics in Context Patterns and Symbols* (Romberg, 1997).

Another example of problems transformed by changing representations can be taken from our research on learning probability and statistics (Milinković, 2007). In this study, I examined how different representations of tasks might affect students' learning of the concept of chance. Three groups of students were given different sets of problems in an attempt to develop their initial intuitive understanding of concept of chance. The sets of problems differed only in the dominant representation chosen for the problems. One group of students was involved in designing, conducting, and analyzing results of an experiment with dice, the other group studied different pictorial representations. Finally, the third group focused on calculating chance by using a given formula.

The first of the selected problems was a variation of the Piaget–Inhelder experiment of choosing chips out of a box. The scenario was as follows: In front of pupils were three boxes. The first box contains one white die and one red die; the second contains nine white dice and one red die; the third contains two white and two red dice (see Figure 3.13). The objectives were (a) to determine the chance of pulling a red die out of the box without looking; and (b) to compare the chances of drawing red dice from the first, the second, and the third boxes.

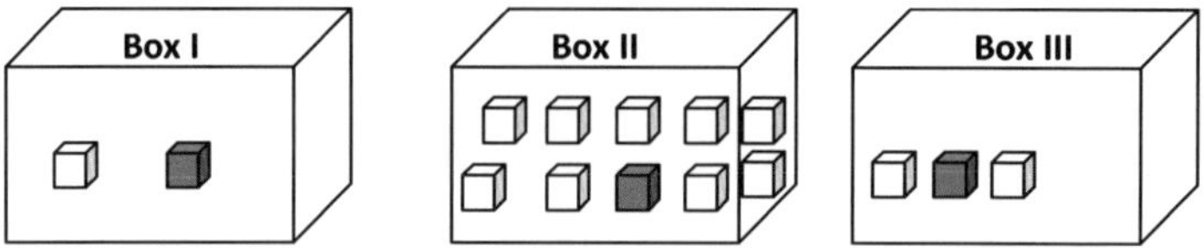

Figure 3.13. Three boxes with dice.

In the approach which we called *the action approach*, students were trying experimentally to determine the probability of pulling out dice of a particular color from the boxes. Each student picked out a die and wrote down the results. Then, they summarized results collectively and represented them on the board. The students then discussed the probability of drawing red dice, intuitively using a statistical definition of probability. The same procedure was repeated for all three boxes. Finally, the groups compared the results. As a result of the activity, the class came up with the concept of posterior (statistical) probability.

In the second approach, which we called *the iconic approach*, students were given the results of the game played by three imaginary children in the form of a graph analogous to the one obtained in the action approach. The class analyzed the graph. Students first expressed their expectations and then analyzed the results achieved. They discussed the child's chance of drawing a red die from the box. In this case, students also came to establish the idea of statistical probability.

In the third approach, which we called *the symbolic approach*, students wondered about the following game: In the box there is a certain, known number of dice in two colors. Pupils brainstormed about the chance of choosing a red die from each of the boxes without any additional information. Discussion with the class led them to the conclusion that the ratio of red dice to the total number of dice gives the answer. In this case pupils came to the idea of a priori probability, unlike the other two groups of students who were developing an initial understanding of posterior probability. Thus, different representations of the problem led students to successful, although diverse, understandings of the idea of chance.

Pupils' Problem Posing

Finally, I will briefly illustrate how problem posing may be more than ordinary routine practice in the classroom for pupils. Again this involves the idea of transformation. In one recent study on the integration of mathematics with a technical education class I observed children dealing with an ill-defined problem which emerged to be mathematical as much as technical (Milinković, 2012b). Children were asked to determine which one of their handmade paper planes was "the best." Students had to define what they meant by "the best plane" and in accordance with the meaning they needed to transform the initial question into a new one. Learning to define important aspects of a problem (and creating new problems as well), delineating a path to solve them, and representing and understanding the results were all significant stages in the activity. While attempting to resolve the initial problem, pupils found themselves posing a new set of problems either by reformulating/transforming the old one, or in some phases decomposing it. In the context of constructing different models of their paper planes, the children needed to decide what characteristics of paper planes were significant and to find appropriate procedures to measure those characteristics rationally.

This was an example of what I described at the beginning of this chapter as a genuine realistic problem. Some characteristics of genuine "problem solving activity" are that children should not think within the boundaries of a particular school subject and that there should be no learned procedure (algorithm) for solving the problem that could be applied in the given context. In the problem space of constructing different models of paper planes, the children needed to decide what characteristics of a paper plane were noteworthy and to find appropriate procedures to measure those characteristics rationally. Negotiation between students brought a collectively accepted procedure with a problem-solving strategy consisting of an array of problems which appeared to be easier. Students gradually came to the understanding that they needed to look for technical features accessible for testing rather than appearance. The students decided to organize a competition so they could check the performance of the paper planes. Then their problem turned into new one: how to organize a competition. Whereas the choice of important characteristics of paper models which were going to be examined fell within the domain of technical education, the process of organizing a paper plane competition belonged to both mathematics and technical education. Much of the time was spent on planning the competition as a part of the problem-solving algorithm (Polya's planning phase). The process of planning required determining different aspects: (a) structural characteristics of plane (length of flight, height of flight, speed of plane, time spent in air, or something else); (b) performance characteristics; (c) method of recording and presenting data; (d) criteria for winning (number of trials, etc.); and (e) organization (including control).

But the activity would not have succeeded without a teacher who possessed not only good pedagogical skills but also content knowledge in mathematics and technical education. When pupils did not have an idea about how to proceed, the teacher intervened. For example, the idea of graphically representing data did not emerge until the teacher suggested it. Then pupils posed themselves a new problem regarding how to make pictures of data. Further details about these activities may be found elsewhere (Milinković, 2010, 2012b). But what we have reviewed here illustrates how genuine problem posing involves rephrasing or transforming an initial problem into problem(s) which can be grasped.

Conclusion

In this chapter I have outlined an approach to develop teachers' proficiency in posing problems via transformations. I have described series of problems linked together by the idea of transformation. First, I defined transformations in problem posing. Then I proposed and analyzed examples of two kinds of transformations.

The first is transforming problems by changing elements in a given problem space. A new problem may be posed by changing what is given, what is searched for (unknown), or the context. I discussed examples of series of problems created by transformations within the same context and proposed three strategies for problem posing: (a) transform problems with a smaller number of (simpler) parameters (e.g., "easy" numbers) to problems with more (often more complicated) parameters (e.g., "difficult numbers"); (b) transform problems by adding new parameters; and (c) transform problems by removing some conditions.

The second kind of transformation is posing problems by transformation of representation. I pointed out that an important element of mathematics competence involves understanding the relationships among different representations. I extended the representational approach into the field of problem posing. I argued that diverse types of representations may provide good contexts for defining a problem space, and that brainstorming about transformation from one to another. We provided examples of apparently different problems (because of different representations) with the same underlying mathematics ideas in their solutions. Putting aside word problems, I discussed three ways of representing problems: action representations, iconic representations, and symbolic representations. Through examples, I showed how a choice of different representations of tasks affected students' learning of concepts of chance. There, I briefly described a problem-based teaching experiment involving elements of probability and statistics based on different representations.

Finally, I illustrated how problem posing may be a fruitful, out-of-the-ordinary activity for students. My argument is that multiple representations can help the development of flexibility of reasoning and can deepen understanding of mathematical concepts and procedures.

One of the objectives of this chapter was to develop an appreciation of problem-posing activities. The other objective was practical—to present a framework for training teachers in problem posing via transformations. My perception is that the vast majority of students as novices have great difficulty in creating problems, and my aim in developing this framework is to support the development of preservice teachers' problem-posing skills. The students should come to an understanding that they need to attend to numerous components in designing a problem, from choosing the content to finding an appropriate context, and then to formulating the problem. Finding solutions(s) or predicting possible solutions is a necessary element of problem posing. With this approach, prospective teachers can gain powerful means for helping their pupils learn mathematics. For the teacher, it is just as important to become flexible and to be able to adapt problems into new ones as it is to recognize the value of each particular example. The potential benefits of conceptualizing problem posing via transformation need to be investigated.

Throughout the chapter I have pointed to a link between knowing mathematics and knowing how to pose problems. I have repeatedly stressed how important teachers' content knowledge was in problem-posing activities, and my examples support this conclusion. Indeed, I am convinced that there is no way that anyone can become a good problem poser *without* sufficient domain knowledge.

My approach indirectly highlights the idea that mathematics is a science of patterns. This program is based on the belief that we can and we should recognize that there are patterns in the world of mathematics problems. We, as educators, need to elicit recognition of those patterns.

References

Arcavi, A. (2003). The role of visual representations in the learning of mathematics. *Educational Studies in Mathematics, 52*, 215–241.

Ball, D. L. (1990). The mathematical understandings that prospective teachers bring to teacher education. *The Elementary School Journal, 90*, 449–466.

Bonotto, C. (2007). How to replace the word problems with activities of realistic mathematical modelling. In W. Blum, P. Galbraith, M. Niss, & H. W. Henn (Eds.), *Modelling and applications in mathematics education* (New ICMI studies series no. 10, pp. 185–192). New York, NY: Springer.

Brown, C. A., & Borko, H. (1992). Becoming a mathematics teacher. In D. A. Grouws (Ed.), *Handbook of research on mathematics teaching and learning* (pp. 107–121). Reston, VA: NCTM.

Bruner, J. (1960). *The process of education*. Cambridge, MA: Harvard University Press.

Carpenter, T. P., Fennema, E., Franke, M. L., Levi, L., & Empson, S. B. (2000). *Cognitively guided instruction: A research-based teacher professional development program for elementary school mathematics*. Madison, WI: University of Wisconsin–Madison. Retrieved from http://ncisla.wceruw.org/publications/reports/RR00-3.PDF.

Cobb, P. (1988). The tension between theories of learning and instruction in mathematics education. *Educational Psychologist, 23*, 87–103.

Cobb, P., Yackel, E., & Wood, T. (1992). A constructivist alternative to the representational view of mind in mathematics education. *Journal for Research in Mathematics Education, 23*(1), 2–33.

Couco, A. A., & Curcio, F. R. (Eds.). (2001). *The roles of representation in school mathematics*. Reston, VA: National Council of Teachers of Mathematics.

Dejić, M., Milinković, J., & Djokić, O. (2005). *Matematika za 4. razred osnovne skole* (1st ed.). Beograd, Serbia: Kreativni centar.

Dufour-Janvier, B., Bednarz, N., & Belanger, M. (1987). Pedagogical considerations concerning the problem of representation. In C. Janvier (Ed.), *Problems of representation in the teaching and learning of mathematics* (pp. 109–122). Hillsdale, NJ: Erlbaum.
Erlwanger, S. H. (1973). Benny's conception of rules and answers in IPI mathematics. *Journal of Children's Mathematical Behavior, 1*(2), 7–26.
Freudenthal, H. (1972). *Mathematics as an educational task.* Heidelberg, Germany: Springer.
Freudenthal, H. (1983). *Didactical phenomenology of mathematical structures.* Dordrecht, The Netherlands: Reidel.
Friedlander, A., & Tabach, M. (2001). Promoting multiple representations of algebra. In A. A. Cuoco & F. R. Curcio (Eds.), *The roles of representations in school mathematics* (pp. 173–185). Reston, VA: NCTM.
Fuson, K. C. (1992). Research on whole number addition and subtraction. In D. A. Grouws (Ed.), *Handbook of research on mathematics teaching and learning* (pp. 243–275). New York, NY: Macmillan.
Goldin, G., & Shteingold, N. (2001). System of representations and the development of mathematical concepts. In A. A. Cuoco & F. R. Curcio (Eds.), *The roles of representation in school mathematics* (pp. 1–23). Reston, VA: NCTM.
Greer, B. (1992). Multiplication and division as models of situations. In D. A. Grouws (Ed.), *Handbook of research on mathematics teaching and learning* (pp. 276–295). New York, NY: Macmillan.
Hiebert, J., & Carpenter, T. (1992). Learning and teaching with understanding. In D. A. Grouws (Ed.), *Handbook of research in mathematics teaching and learning* (pp. 65–97). New York, NY: Macmillan.
Janvier, C. (Ed.). (1987). *Problems of representation in the teaching and learning of mathematics.* Hillsdale, NJ: Erlbaum.
Kaput, J. (1987). Representation systems and mathematics. In C. Janvier (Ed.), *Problems of representation in the teaching and learning of mathematics* (pp. 19–27). Hillsdale, NJ: Erlbaum.
Kennedy, A. (1998). Education reform and subject matter knowledge. *Journal for Research in Science Teaching, 35*(3), 249–263.
Kilpatrick, J. (1987). Problem formulating: Where do good problems come from? In A. H. Schoenfeld (Ed.), *Cognitive science and mathematics education* (pp. 123–147). Hillsdale, NJ: Erlbaum.
Lawler, R. W. (1981). The progressive construction of mind. *Cognitive Science, 5*, 1–30.
Leinhardt, G. (1989). Math lessons: A contrast of novice and expert competence. *Journal for Research in Mathematics Education, 20*(1), 52–75.
Lesh, R. (1981). Applied mathematical problem solving. *Educational Studies in Mathematics, 12*(2), 235–265.
Liu, Y. (2005). *Teacher' understanding of probability and statistical inference and their implications for professional development* (PhD dissertation). Retrieved from http://etd.library.vanderbilt.edu/available/etd-07152005114321/unrestricted/Dissertation.pdf
Liu, Y., & Thompson, P. W. (2004). An investigation of teachers' personal and pedagogical understanding of probability, sampling, and statistical inference. In D. E. McDougall & J. A. Ross (Eds.), *Proceedings of the 26th Annual Meeting of the International Group for the Psychology of Mathematics Education* (pp. 408–414). Toronto, ON: PME-NA. Retrieved from http://www.allacademic.com/meta/p117721_index.html
Mc Graw-Hill Science & Technology Dictionary. (2003). McGraw-Hill dictionary of scientific and technical terms. New York, NY: McGraw Companies. Retrieved from http://www.answers.com/topic/problem-space#ixzz2AnoDqZrw
Michalewicz, Z., & Fogel, D. B. (2000). *How to solve it: Modern heuristics.* Heidelberg, Germany: Springer.
Michalewicz, Z., & Fogel, D. B. (2004). *How to solve it: modern heuristics.* Berlin, Germany: Springer.
Milinković, J. (2007). *Metodicki aspekti uvoda u verovatnocu i statistiku* (in Serbian). Beograd, Bulgaria: Uciteljski fakultet.

Milinković, J. (2010). Pupils' active learning in integrated curriculum: A case study. *Student in contemporary learning and teaching* (pp. 97–109). Beograd, Bulgaria: Uciteljski fakultet.

Milinković, J. (2012a). Pre-service teachers' representational preferences. In T. Y. Tso (Ed.), *Proceedings of the 36th Conference of the International Group for the Psychology of Mathematics Education: Opportunities to Learn in Mathematics Education* (Vol. 3, pp. 209–217). Taipei, Taiwan: PME.

Milinković, J. (2012b). Problem solving in integrated curriculum: The results of an action research project. In F. Doyran (Ed.), *Research on teacher education and training* (pp. 165–176). Athens, Greece: Athens Institute for Education and Research.

Milinković, J., & Micic, V. (2008, June). Uloga didaktickih sredstava u osnovnoskolskoj geometriji. *Metodicki aspekti nastave matematike, Zbornik radova sa medjunarodnog naucnog skupa odrzanog* (pp. 38–44) (in Serbian). Jagodina, Serbia: Pedagoski fakultet, Univerzitet u Kragujevcu.

Nesher, P. (1980). The stereotyped nature of school word problems. *For the Learning of Mathematics, 1*(1), 41–48.

Nodding, N. (1992). Professionalization and mathematics teaching. In D. A. Grouws (Ed.), *Handbook of research in mathematics teaching and learning* (pp. 197–208). New York, NY: Macmillan.

Nunes, T. (1992). Ethnomathematics and everyday cognition. In D. A. Grouws (Ed.), *Handbook of research in mathematics teaching and learning* (pp. 557–574). New York, NY: Macmillan.

OECD. (2003). *Learning mathematics for life: A perspective from PISA*. Retrieved from http://www.oecd.org/pisa/pisaproducts/pisa2003/44203966.pdf

Peterson, P. L. (1988). Teachers' and students cognitional knowledge for classroom teaching and learning. *Educational Researcher, 17*(5), 5–14.

Pollak, H. O. (1988). Mathematics as a service subject—Why? In A. G. Howson, J. P. Kahane, P. Lauginie, & E. de Turkheim (Eds.), *Mathematics as a service subject* (ICMI study series, pp. 28–34). Cambridge, United Kingdom: Cambridge University Press.

Polya, G. (1957). How to solve it: A new aspect of mathematical method. Princeton, NJ: Princeton University Press.

Polya, G. (1973). *How to solve it*. Princeton, NJ: Princeton University Press.

Reusser, K. (1988). Problem solving beyond the logic of things: Contextual effects on understanding and solving word problems. *Instructional Science, 17*(4), 309–338.

Romberg, T. A. (1997). *Mathematics in context, patterns and symbols.* Chicago, IL: Encyclopedia Britannica Educational Corporation. Snake Patterns-s-s-s retrieved from http://www.pbs.org/teachers/mathline/lessonplans/pdf/atmp/SnakePatterns.pdf

Schoenfeld, A. E. (1992). Learning to think mathematically: Problem solving, metacognition and sense making in mathematics. In D. A. Grouws (Ed.), *Handbook of research in mathematics teaching and learning* (pp. 334–370). New York, NY: Macmillan.

Schulman, L. (1986). Those who understand knowledge growth in teaching. *Educational Researcher, 15*, 4–14.

Schulman, L. S., & Grossman, P. L. (1988). *Knowledge growth in teaching: A final report to the Spencer Foundation*. Stanford, CA: Stanford University.

Silver, E. A., Mamona-Downs, J., Leung, S., & Kenney, P. A. (1996). Posing mathematical problems in a complex task environment: An exploratory study. *Journal for Research in Mathematics Education, 27*(3), 233–309.

Singer, F. M., Pelczer, I., & Voica, C. (2011). In T. Rowland, & E. Swoboda (Eds.), *Proceedings of the 7th Conference of the European Society for Research in Mathematics Education (CERME 7)* (pp. 9–13). Rzeszów, Poland: University of Rzeszów.

Stanic, G., & Kilpatrick, J. (1988). Historical perspective on problem solving in the mathematics curriculum. In R. Charles & E. Silver (Eds.), *The teaching and assessing of mathematical problem solving* (pp. 1–22). Reston, VA: NCTM.

Stoyanova, E., & Ellerton, N. F. (1996). A framework for research into student problem posing in school mathematics. In P. C. Clarkson (Ed.), *Technology in mathematics education* (pp. 518–525). Melbourne, Australia: Mathematics Education Research Group of Australasia.

Thwaites, A., Huckstep, P., & Rowland, T. (2005). The knowledge quartet: Sonia's reflections. In D. Hewitt, & A. Noyes (Eds.), *Proceedings of the 6th British Congress of Mathematics Education, University of Warwick* (pp. 168–175). Retrieved from www.bsrlm.org.uk

Treffers, A. (1987). *Three dimensions. A model of goal and theory description in mathematics instruction—The Wiskobas Project*. Dordrecht, The Netherlands: Reidel Publishing.

Wake, G. (2010). *Connecting mathematics with reality—Connecting reality with mathematics*. London, United Kingdom: QCDA. Retrieved from http://www.qcda.gov.uk/resources/4466.aspx

Wilson, J. W., Fernandez, M. L., & Hadaway, N. (1993). Mathematical problem solving. In P. S. Wilson (Ed.), *Research ideas for the classroom: High school mathematics* (pp. 57–78). New York, NY: Macmillan.

Wilson, S. M., Shulman, L. S., & Richert, A. E. (1987). 150 ways different ways of knowing: Representations of knowledge in teaching. In J. Calderhead (Ed.), *Exploring teachers' thinking* (pp. 104–124). London, United Kingdom: Cassell.

Wittman, E. (2005, July). *Mathematics as the science of patterns*: A *guideline for developing mathematics education from early childhood to adulthood*. Plenary lecture presented at the International Colloquium "Mathematical Learning from Early Childhood to Adulthood" organized by the Centre de Recherche sur l'Enseignement des Mathematiques in collaboration with the Institut de mathématique de l'Université de Mons-Hainaut, Mons/Belgium.

Chapter 4
Using Digital Technology for Mathematical Problem Posing

Sergei Abramovich and Eun Kyeong Cho

Abstract This chapter demonstrates how the appropriate use of commonly available digital technology tools can motivate and support problem-posing activities. Informed by the authors' work with teacher candidates, the chapter underscores the importance of theoretical considerations associated with the use of computers in problem posing. The theory is illustrated by cases in elementary and secondary teacher education contexts. These cases vary in complexity from the formulation of problems for an elementary classroom context to discovering new knowledge within a familiar secondary education context. The use of graphing software as a medium for reciprocal problem posing is shown to be conducive for developing rather sophisticated questions about algebraic equations with parameters. As many traditional problems can be solved effectively by modern technology, modifying such problems to be not directly solvable by technology would open the whole new avenue for problem posing in the technological paradigm.

Contents

S. Abramovich (✉)
Department of Curriculum and Instruction, State University of New York at Potsdam, Potsdam, NY, USA
e-mail: abramovs@potsdam.edu

E.K. Cho
Department of Education, University of New Hampshire, Durham, NH, USA
e-mail: EunKyeong.Cho@unh.edu

F.M. Singer et al. (eds.), *Mathematical Problem Posing*,
Research in Mathematics Education, DOI 10.1007/978-1-4614-6258-3_4

In the real world, most of the time, an answer is easier than defining the question

(Dyson, 2012, p. 163)

Introduction

The condition of education has changed over time. Technology shapes "our material, intellectual, and cultural environments" (Organisation for Economic Co-operation and Development, 2011, p. 129). Modern students live in a technology-rich environment and are apt users of technology; their ways of communicating with teachers and peers, interacting with learning materials, and demonstrating knowledge of mathematical concepts and skills in using them are different from those of decades ago. Such changes in the social context of teaching and learning call for the innovative use of technology in mathematical education. Mathematics educators work with students in a global and digital society, which requires educators to "use their knowledge of subject matter, teaching and learning, and technology to facilitate experiences that advance student learning, creativity, and innovation in both face-to-face and virtual environments" (International Society for Technology in Education, 2008, p. 1). In its most recent position statement regarding the role of technology in the teaching and learning of mathematics, the National Council of Teachers of Mathematics [NCTM] (2011) states the following:

> It is essential that teachers and students have regular access to technologies that support and advance mathematical sense making, reasoning, problem solving, and communication. Effective teachers optimize the potential of technology to develop students' understanding, stimulate their interest, and increase their proficiency in mathematics. When teachers use technology strategically, they can provide greater access to mathematics for all students. (p. 1)

Seeing technology as a vehicle for success is not a new position; over two decades ago, NCTM (1991) already argued for the use of technology in the classroom "to enhance and extend mathematics learning and teaching" and suggested that "the most promising [ways] are in the areas of problem posing and problem solving in activities that permit students to design their own explorations and create their own mathematics" (p. 134). Earlier, problem posing was referred to by the Council as "an activity that is at the heart of doing mathematics" (NCTM, 1989, p. 138) and over the years it has been considered by researchers to be an important tool of mathematical education didactics (e.g., Brown & Walter, 1983; Crespo, 2003; Hoyles & Sutherland, 1986; Kilpatrick, 1987; Krutetskii, 1976; Noss, 1986; Silver, 1994; Silver & Cai, 1996; Singer & Voica, 2013).

The 1991 statement of NCTM regarding the use of technology in problem posing can be traced back to the pioneering ideas about the role of educational technologies in the development of new problems for mathematics education (Hoyles & Sutherland, 1986; Kilpatrick, 1987; Noss, 1986). In particular, in the context of teacher preparation, investigating the sources of problem formulation, Kilpatrick (1987) argued that electronic computers can be used effectively in fostering problem-posing skills among preservice teachers of mathematics. It is because these digital technological tools enable one to generate numerical and pictorial patterns that new problems can be created, and changes to the conceptual and syntactic structures of an existing problem statement can be facilitated. In other words, by using appropriately designed computer activities (such as the spreadsheet-based environments that, among other things, the authors will present in the sections below), many problematic situations manifesting different levels of mathematical complexity can emerge.

Although the mathematics education field's interest in and research on problem posing has been active (e.g., Akay & Boz, 2010; Cai & Hwang, 2002; Ellerton, 1986; English, 1997; Kar, Özdemir, İpek, & Albayrak, 2010; Kontorovich, Koichu, Leikin, & Berman, 2012; Lavy & Bershadsky, 2003; Leung & Silver, 1997; Voica & Singer, 2011), less focus has been on the study of the role of technology in facilitating and advancing skills in formulating problems. Further, published studies on problem posing with technology have been not only limited in number and scope but also in grade level. Most of the studies have been conducted at the secondary level with the main emphasis on developing conjectures in dynamic (or partially dynamic) geometry environments (Hoyles & Sutherland, 1986; Laborde, 1995; Lavy & Shriki, 2010; Noss, 1986; Yerushalmy, Chazan, & Gordon, 1993). The advent of the Internet as a pedagogical tool motivated studies on mathematical problem posing in web-based learning environments (Abu-Elwan, 2007; Hirashima, Nakano, & Takeuchi, 2000). However, the didactical potential of mathematical problem posing with electronic spreadsheets and computer algebra systems was not studied in detail until recently (Abramovich, 2012; Abramovich & Cho, 2006, 2008, 2009, 2012; Abramovich & Norton, 2006).

The aim of this chapter is to demonstrate how the *appropriate* use of *technology* can be integrated with problem-posing activities in a broad context of mathematics education towards the development of higher-order thinking skills in teacher education candidates as the learners of mathematics. The term *technology* refers to various commonly available software tools including an electronic spreadsheet, *Graphing Calculator 4.0* (Avitzur, 2011), *Maple* (Char et al., 1991) and online computational engine *Wolfram Alpha* (Dimiceli, Lang, & Locke, 2010). The term *appropriate* means that the tools of technology cannot be directly utilized for mathematical problem posing but rather, such utilization requires one's appreciation of their hidden educational potential and expertise in their use. It will be shown how the above-mentioned tools of technology facilitate the development of problems ranging from tasks for primary grades to an unsolved conjecture.

Before getting into the main body of this chapter, it will be helpful to describe the context and processes that have led the authors to pay attention to the conceptual aspects of problem posing. The context of problem-posing activities shared in this chapter consists of preservice K-12 teacher education courses. These include a mathematics content course for preservice teachers in elementary education, a capstone course for those preparing to teach secondary mathematics, and a course on the use of spreadsheets in teaching school mathematics for both groups of preservice teachers. While working with the preservice teachers in these courses and analyzing the problems posed by them, the authors were given unique opportunities to identify practical and theoretical issues in relation to problem posing with technology. Data sources that led to such conceptual understandings, which are shared in the sections below, include problems created by preservice teachers, observation notes of classroom interactions among students, and follow-up discussions with the preservice teachers. Specifically, problems posed by elementary pre-teachers were collected through portfolios filled with completed course assignments, one of which dealt with using a readymade spreadsheet in posing problems through the lens of didactical coherence. An initial analysis of the portfolios helped the authors develop ideas about didactical coherence of problems posed in the technological paradigm. The "bananas" problem posed by a preservice teacher in the context of the agent–consumer–amplifier (ACA) framework was presented as part of the portfolio for the course on using an electronic spreadsheet in teaching K-12 mathematics that emphasized the potential of the software for problem posing. Problems, shown in the section on reciprocal problem posing, were recorded by the course instructor during a classroom interaction between two preservice teachers as part of their joint final project for a technology-rich capstone course in secondary mathematics. Preliminary findings of the analysis of the data (i.e., problems created by preservice teachers and their thinking process reflected in their discussions and group interactions) emphasize that success of the teachers with technology-enabled problem posing requires practical experience with mathematical modeling and problem solving as well as theoretical preparation in pedagogical issues directly related to the development of skills in formulating new problems or modifying the existing ones. Conceptual themes identified by the authors from working with preservice teachers will be presented in the following four sections: (a) problem posing through the lens of didactical coherence, (b) problem posing through the lens of ACA framework, (c) reciprocal problem posing, and (d) problem posing as a discovery experience.

Problem Posing Through the Lens of Didactical Coherence

In posing a problem for their own students, that is, preparing to teach "with the learner in mind" (Thompson, Carlson, & Silverman, 2007, p. 416), teachers need to consider various aspects of mathematics pedagogy such as individual and group understandings of mathematical concepts and acquisition of process skills. In their work with preservice teachers on problem posing using technology, Abramovich and Cho (2006, 2008, 2012) emphasize the concept of didactical coherence. Didactical coherence of a problem refers to the problem's formal solvability, grade-level appropriateness, and other pedagogical features as well as sociocultural relevance. This section presents the concept of didactical coherence and its three interrelated, yet distinct, subconcepts: numerical, pedagogical, and contextual coherence. It shows that problem-posing activities need to consider didactical coherence which is established at the intersection of the three subconcepts. Below, each of the three subconcepts is explained using examples of problems either posed by or for (as illustrations) preservice elementary teachers using specifically designed spreadsheet environments within a mathematics content course (Abramovich, 2012).

Numerical Coherence

Numerical coherence of a problem refers to its "formal solvability within a given number system." In other words, "if the problem has a solution expressed by a number (or a set of numbers), it is numerically coherent" (Abramovich & Cho, 2008, p. 3). Alternatively, Hirashima et al. (2000), concerned with the fact that some of the problems posed by learners "may be wrong" (p. 745), called "adequate" (p. 746) arithmetical word problems that have a solution. As will be shown below, the word "adequate" as a characteristic of a problem includes more than just its formal solvability. Simply altering a number in a given problem may not result in creating a numerically coherent problem that has an answer (or a set of possible answers). Consider the following arithmetical word problem (presented to preservice teachers in a mathematics content course as an example of a possible use of spreadsheets in problem posing).

Problem 1: Using 2-cent, 4-cent, and 6-cent stamps only, find all ways to make a 25-cent postage.

In order to pose a problem such as Problem 1, one would need to select a set of four numbers—in this case (Figure 4.1), 2 (cell G1: denomination 2 cents), 4 (cell F1: denomination 4 cents), 6 (cell E1: denomination 6 cents), and 25 (cell A1: total postage)—and make sure that the data selected for posing the problem would yield an answer or a set of answers that make sense. In other words, the problem must have numerical coherence. In thinking about the problem's solvability, a problem poser (i.e., a preservice teacher) would realize that problem posing is another face of problem solving (Davis, 1985; Dunker, 1945; Kilpatrick, 1987; Silver, 1994). In the case of using a specifically designed spreadsheet environment to pose this kind of a problem, the preservice teacher has to know how to interpret the results of spreadsheet modeling. The emptiness of the range D4:J10 in the spreadsheet of Figure 4.1 (here the numbers in the ranges D3:H3 and C4:C10 are, respectively, the possible quantities of 6-cent, 4-cent, and 2-cent stamps to make up a total of 25 cents postage) indicate that Problem 1 does not have solutions. The software counts the number of nonempty cells in the range D4:J10 and, because all the cells in this range are empty, displays zero in cell A7. In other words, Problem 1 does not have the feature of numerical coherence.

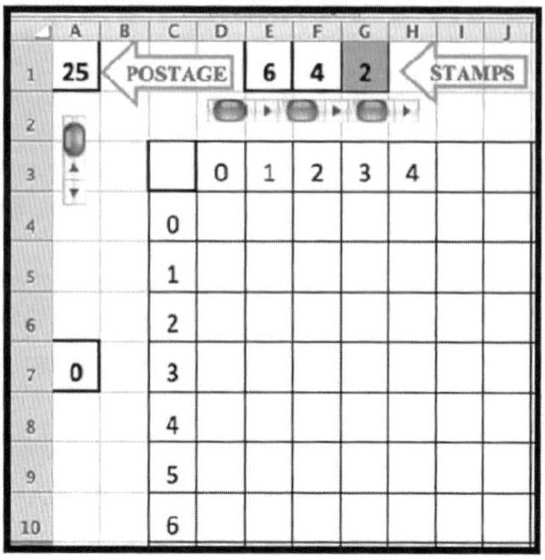

Figure 4.1. A spreadsheet environment showing a problem's numerical incoherence.

Nonetheless, the spreadsheet environment allows one to change the numbers involved by a single click, showing the results (solvability) instantly. Through exploring technology-supported problem-posing environments, preservice teachers can easily see that replacing 25 by 24 in Problem 1 (consequently, in cell A1) would make it numerically coherent (that is, solvable) as shown in Figure 4.2. More detailed discussion regarding this revised problem will be included later in this chapter when the concept of pedagogical coherence is discussed. Preservice teachers can also interpret the findings in mathematical terms; for example, the left-hand side of the equation $x_n = \frac{1}{2}(x_{n-1} - 1)$, $n = 1, 2, 3, \ldots$, is a multiple of two, its right-hand side is not.

Whereas the term Diophantine equation was not included in the discourse associated with equations of the above type, having quantities of the stamps as situational referents for the variables involved facilitated preservice teachers' understanding of algebraic formalism as part of their mathematics content coursework. Preservice teachers with such levels of mathematical understanding were able to see the numbers involved in Problem 1 as parameters that could be changed and tested for numerical coherence by using the spreadsheet. In this way, the spreadsheet was designed to enable preservice teachers, in the spirit of Kilpatrick (1987), to vary numeric data in a conceptually informed way. Moreover, this generalized perspective helped preservice teachers develop a better understanding of how a mathematical experiment—an activity that "involves calculating instances of some general hypothesis" (Baker, 2008, p. 331), in our case, the solvability in integers of a three-variable linear equation with integer coefficients—works in posing numerically coherent problems by using a spreadsheet.

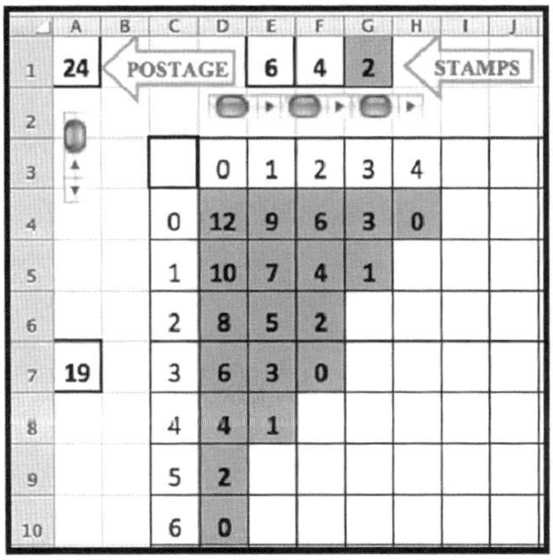

Figure 4.2. A modified problem is numerically coherent; however, too many (19) solutions point to its pedagogical incoherence.

Contextual Coherence

Contextual coherence of a problem means its consistency with the sociocultural background of a heterogeneous group of pupils. Teachers who pose problems for their students need to be aware that, just like the learning of arithmetic involves the mastery of the numeration system as a cultural tool (Cobb, 1995), arithmetical word problems often reflect other cultural systems such as measuring units (e.g., inches versus centimeters, gallons versus liters) and money systems (e.g., dollars, pounds, pesos, and yens) and, therefore, should learn to pose problems that facilitate rather than complicate students' learning (Singer & Voica, 2013). Abramovich and Cho (2008) explain the concept of contextual coherence (which to a larger extent is a cultural notion) in the following way:

> Generally speaking, contextual coherence of a problem is a variable attribute. Just as without the mastery of base ten system—a cultural tool designed to support one's counting abilities—one cannot understand the numerical meaning of a multi-digit number, without the mastery of another cultural tool—a currency system of a particular country—one cannot solve a problem which context does not relate well to one's cultural background. (p. 6)

For example, a word problem which contains culturally specific information such as the names of the US coins (e.g., NCTM, 2000, p. 52) may not be well understood by newly immigrated students from countries which use different systems. As an illustration, consider the following problem suggested by a preservice teacher for a (fictitious) second-grade classroom using the spreadsheet in a problem-posing assignment.

Problem 2: How many ways can one make a 35-cent postage using 10-cent, 8-cent, and 3-cent stamps?

Although this problem is numerically coherent, a second grader might argue that the current domestic postage in the United States is not 35 cents and that, as a collector of stamps, he has never seen an 8-cent stamp. In other words, for this student Problem 2 might not be contextually coherent. So, when posing a problem, teachers should also pay attention to the context within which the problem is posed and to the cultural relevance of the problem to individual students or to the whole class. However, any change of context that alters the conceptual and syntactic structure of a problem is a delicate proposition in the technological paradigm. Indeed, in order to address the issue of contextual coherence (revealed through the above analysis by the course instructor), the preservice teacher, without changing numerical data, modified Problem 2 as follows:

Problem 3: Find all ways of arranging 35 marbles into boxes of 10, 8, and 3 marbles each so that all three types of boxes are used and each box is full.

As the spreadsheet of Figure 4.3 shows, it appears that there exist four integer solutions to both problems given by the triples (0, 1, 9), (0, 4, 1), (1, 2, 3), and (2, 0, 5). That is what the preservice teacher, in fact, has claimed. However, what the teacher did not realize is the fact that these problems (i.e., Problems 2 and 3) not only deal with context but also have different conceptual and syntactic structures that the spreadsheet does not recognize. Requiring all three types of boxes to be used implies that the triples with zero elements are extraneous solutions. Thus, only the triple (1, 2, 3) satisfies the conditions of Problem 3 (requiring that the boxes may not be empty). At the same time, being extraneous for Problem 3, the three triples containing zero(s), that is, (0, 1, 9), (0, 4, 1), and (2, 0, 5), satisfy the conditions of Problem 2 (allowing for not all types of the stamps to be used). Indeed, one can make the 35-cent postage out of one 8-cent stamp and nine 3-cent stamps $(35 = 1 \times 8 + 9 \times 3)$; four 8-cent stamps and one 3-cent stamp $(35 = 4 \times 8 + 1 \times 3)$; one 10-cent stamp, two 8-cent stamps, and three 3-cent stamps $(35 = 1 \times 10 + 2 \times 8 + 3 \times 3)$; or two 10-cent stamps and five 3-cent stamps $(35 = 2 \times 10 + 5 \times 3)$. This example of

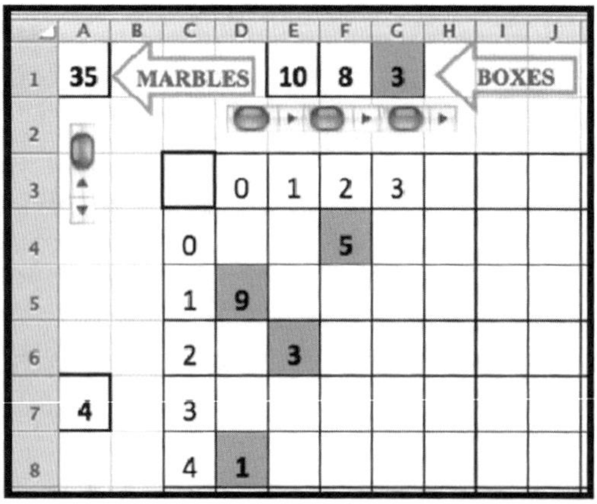

Figure 4.3. A spreadsheet environment sensitive to conceptual structure of Problem 2 and 3.

how changing context may inadvertently yield extraneous solutions, shows not only the complexity of posing problems when using technology but also the importance of accurate interpretation of modeling data by a problem poser, a preservice teacher in our case. Whereas the teacher might expect students to find four solutions to Problem 3 depending on his or her particular interpretation of modeling data, the students, in turn, could rightly insist on the existence of one solution only as they will *not* be using the spreadsheet in *solving* Problem 3. Familiarizing preservice teachers with these hidden pitfalls of posing problems in a technological paradigm elevates their mathematical and pedagogical competences to a higher level.

Pedagogical Coherence

Pedagogical coherence of a problem refers to its appropriateness for a specific grade, developmental level, or interests of pupils. A problem with more than three or four correct answers (recall, these problems are to be solved without technology) may not be appropriate for young children (see the example of a problem with 19 answers shown in Figure 4.2). Therefore, pedagogical considerations dealing with on-task behavior, interest, motivation, and discovery become important factors in problem posing. To illustrate the notion of pedagogical coherence, consider a similar problem posed by another preservice teacher.

Problem 4: How many ways can one make 20 dollars by using 1-dollar bills, 5-dollar bills, and 10-dollar bills only?

Figure 4.4 shows the existence of nine solutions[1] to the problem (note: the problem *is* numerically and contextually coherent). However, one may wonder: Is Problem 4 pedagogically coherent to be offered to young children? It appears that the children would hardly become motivated to stay on-task that requires adding the numbers 1, 5, and 10 over and over to reach 20 without "seeing light at the end of the tunnel."[2]

[1] There are mathematical methods, not studied at the pre-college level, allowing one to answer the question "how many?" without actually finding all solutions. One such method is to calculate the value of $D(n;\ a_1a_2,\ldots,a_k)$ referred to in Comtet (1974) as the denumerant of n with respect to the sequence $a_1,a_2,\ldots,a_k$. In the case of Problem 4, one has to calculate $D(20;1,5,10)$. Another method is to find the coefficient of x^{20} in the expansion of the product $\sum_{i=0}^{\infty}x^i \times \sum_{i=0}^{\infty}x^{5i} \times \sum_{i=0}^{\infty}x^{10i}$. For more information on the use of technology in calculating denumerants or coefficients in the expansion of the products of geometric series, see Abramovich and Brouwer (2003).

[2] A solution strategy that can be introduced to preservice teachers in the context of Problem 4 is to reduce it to three simpler problems each of which depends on the quantity of $10 bills used. As shown in Figure 4.4, the range for $10 bills is [0, 2] (row 3); the range for $5 bills is [0, 4] (column C); the numbers below and to the right of these ranges represent the corresponding quantities of $1 bills. For example, using two $10 bills yields no possibilities for other bills; using one $10 bill yields three possibilities for other bills: ten $1 bills, one $5 bill, and five $1 bills, or two $5 dollar bills. The number of possibilities to use these $1 and $5 bills increases as the number of $10 bills used to pay $20 decreases. The spreadsheet of Figure 4.4 shows how the number of $1 bills decreases by five vertically (counting by five) and by ten horizontally (counting by ten).

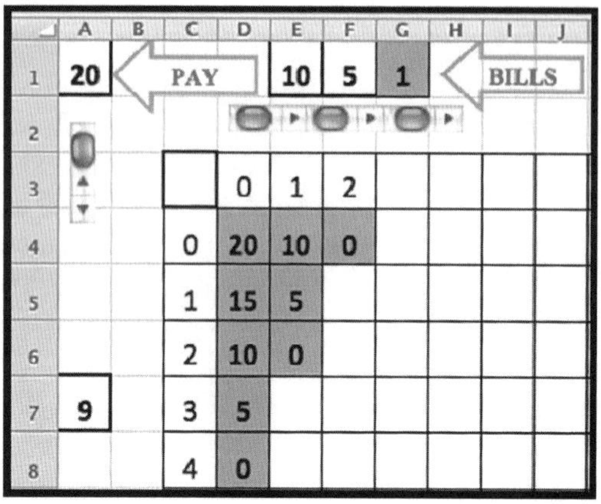

Figure 4.4. An example of a pedagogically incoherent problem.

A pedagogically coherent problem is one that considers individual and groups of students' developmental level, interests, capabilities, and strengths, which then can kindle students' interest, facilitate on-task behavior, promote systematic reasoning, and stimulate their cognitive development. A pedagogically coherent problem can be easily created through teachers' experimentation with numbers (parameters) enabled by spreadsheet's interactive computations.

Pedagogical Coherence as a Relative Concept

Another issue associated with problem posing deals with the relative nature of pedagogical coherence. Consider a case when students ask questions about problems they pose and solve. An approach to learning mathematics through problem solving, being a signature pedagogy of the modern mathematics classroom (Ernie, LeDocq, Serros, & Tong, 2009), is a "pedagogy of uncertainty … [which] render classroom settings unpredictable and surprising" (Shulman, 2005, p. 57). This uncertainty may lead to the emergence of the phenomenon of classroom instability in the sense that a slight modification of a simple problem may lead to a qualitatively new level of mathematical complexity, something that a preservice teacher may not be able to handle appropriately. Typically, students' interest toward mathematics is supposed to be in the state of a stable equilibrium, controlled by the teacher's ability to provide qualified assistance when answering questions that arise in the classroom. Thus, pedagogical coherence of a problem includes the teacher's awareness of the possibility that students' interest toward mathematics could bifurcate into a state of unstable equilibrium in the sense that once interest and motivation are lost, these traits may not come back (Abramovich, Easton, & Hayes, 2012).

Teachers have to be trained in recognizing the difference between questions that do have and do not have easy answers. A classic example of that kind is a legend (e.g., Dunham, 1991) about Carl Friedrich Gauss who, at an early age, was able to avoid the straightforward summation of the first 100 natural numbers (a pedagogically incoherent problem for a 10-year-old student) by recognizing a pattern that the numbers equidistant from the beginning and the end of this sequence follow. While such a problem without insight of genius does not have an easy solution at that grade level, solving such problems at a higher-grade level in the age of technology becomes so easy that now it might be difficult to motivate students to try to solve the problem. Indeed, typing in *Wolfram Alpha*—an open source software tool available on any computer with an Internet connection—the quest, "What is the sum of the first 100 natural numbers?" yields the answer 5,050. Therefore, it is necessary to think about reformulation of such computationally solvable problems. For example, one may be asked to find the smallest square number, which is equal to the sum of consecutive natural numbers starting from one as well as to the sum of consecutive odd numbers starting from one. In that way, even if a student uses technology, the nature of such a reformulated task has a definite mathematical flavor. Indeed, while the answer, 36, is not difficult to obtain, the possibility of representing 36 in three qualitatively different forms—$36 = 6 \times 6 = \sum_{n=1}^{8} n = \sum_{n=1}^{6} (2n-1)$—points to the very nature of numbers, something that, through posing new mathematical inquiries and resolving them, has stimulated the development of mathematics over the centuries.

It should also be noted that pedagogical coherence of a problem depends on the expected method of solution. Often, as students learn to use more and more sophisticated mathematical tools, a pedagogically incoherent problem for a lower-grade level becomes pedagogically coherent for a higher-grade level. By the same token, a pedagogically coherent problem for a lower-grade level may become pedagogically incoherent for a higher-grade level. For example, for a 6-year-old (who uses concrete materials—a noncomputational technology—as a means of problem solving) the tasks of arranging 24 students and 25 students into four groups to do team work are at the same level of complexity; however, for a 10-year-old the latter case is conceptually more difficult as it requires the interpretation of the meaning of remainder in the equality $25 = 4 \times 6 + 1$.

As shown in this section, in order for teachers to have robust problem-posing skills when using digital technology, they need to appreciate and pay attention to the notions of didactical coherence. The significance of paying attention to these concepts for problem posing is being recognized in recent literature (e.g., Bonotto, 2010, p. 21). Didactical coherence in problem posing is achieved when all the three types of coherences—numerical, contextual, and pedagogical—intersect (Figure 4.5) and the issue of extraneous solutions is addressed.

Finally, given a problem, preservice teachers may be asked to decide its place in the diagram (Figure 4.5) or to pose problems to fill in each of the seven areas formed by the three overlapping circles. Assignments of that kind would also allow them to

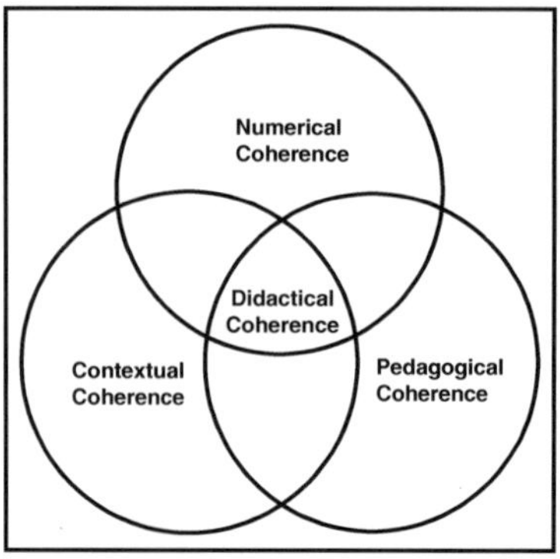

Figure 4.5. Venn diagram showing the concept of didactical coherence and its relationship with three subconcepts.

integrate the aforementioned pedagogical ideas with basic mathematical concepts of set theory (union, intersection, complement) and logic (Venn diagram). It is important to note that the concepts of coherence are bound by time, place, and learners involved; in other words, a contextually coherent problem today may not be contextually coherent in 10 years' time and a pedagogically coherent problem for high school students is unlikely to be pedagogically coherent for elementary students. In this case, technology allows users to reformulate a problem to make it contextually and pedagogically as well as numerically coherent. However, as seen from the example of Figure 4.3, by modifying the conceptual structure of a problem, preservice teachers can achieve its contextual coherence yet overlook the existence of extraneous solutions generated by a spreadsheet that served a different context.

Problem Posing Through the Lens of Agent-Consumer-Amplifier Framework

Many technology-enhanced mathematical activities, including problem posing, can be conceptualized in terms of the ACA framework (Abramovich, 2006). In what follows, there are brief descriptions of the ACA framework and cases of applying this framework in the context of a spreadsheet.

ACA Framework

During the first stage of the framework, technology serves as an agent of a mathematical activity in the sense that one's engagement in doing mathematics is motivated by the need to construct a computational environment for solving a specific problem. This motivation can stem from the applied nature of a mathematical activity.[3] During the second stage, technology serves as a consumer of a mathematical activity when a variety of similar problems can be posed and solved; that is, the constructed computational environment is utilized as a problem-solving tool applicable to more than one problem. It is in that sense that technology, being originally an agent of a mathematical activity (method), turns into a consumer of the method. During the third state, technology functions as an amplifier of a mathematical activity, which extends to a new dimension of problem posing, solving, and reformulating activities in a way that is hard to realize without the support of technology. Each of the three functions of the ACA framework is described below, starting with the role of technology as an agent.

One can recognize a dualism that exists between using technology as agency for mathematics and doing mathematics to enable this agency. Indeed, whereas a problem in question can determine one's choice of a tool in support of the required problem-solving method, the very structure of the tool determines a method through which the problem can be solved. Such dualism implies that any technology-enhanced mathematical activity is underpinned by one's expertise in the use of technology and knowledge of specific features of a particular technological tool. In the context of a computer-supported mathematics teacher education course, there are several directions through which the duality of mathematics and technology can be revealed already at the first stage of the triad: the creation of a new computational environment to enable experimentation with a particular concept, the appropriate modification of an old environment, or the development of efficient modeling techniques aimed at the sophisticated use of instructional computing.

At the second stage of the triad, once a computational environment is constructed, the computer can start functioning as a consumer of the mathematical method that emerges from one's engagement in doing mathematics and enabled him or her to achieve the original goal, that is, to solve a problem. At this stage, new problems can be formulated within the computational environment and solved immediately by that environment. The consumption of the mathematical activity (or method) that underscored the first stage of the triad comprises the creation of a variety of new problems, which, nonetheless, could (and, often, should) be solved without using technology. It is only for posing a new problem with a didactically coherent structure that one uses technology as a consumer of mathematical activity.

---

[3] For example, in programming the spreadsheets shown in Figures 4.1, 4.2, 4.3, and 4.4, the following problem can be posed: How can one make the ranges in row 3 and column C dependent on the number in cell A1? To answer this application-oriented question, one has to use algebraic inequalities for which, thereby, the need to construct a computational environment serves as an agency. For example, when making the 24-cent postage (Figure 4.2, cell A1) the largest quantity of 6-cent and 4-cent stamps that one can use is four and six stamps, respectively. Therefore, the spreadsheet is designed not to generate numbers greater than four in row 3 and greater than six in column C. For more information on spreadsheet modeling as an agency for posing problems leading to the use of algebraic inequalities see Abramovich (2006).

Indeed, whereas all problems are expected to be numerically coherent once a mathematical activity has been set in a workable computational environment, the contextual and pedagogical aspects of new problems depend on the problem poser's knowledge of cognitive abilities and problem-solving skills of those for whom the problems are designed.

Finally, at the third stage of the triad, technology plays the role of the amplifier of the mathematical activity in the sense that it enables posing (and, typically, solving) of appropriately extended problems, something that would not be possible (or might be too cumbersome) otherwise. The major characteristic of this stage is that one has to reorganize the original activity by modifying the method that underscores it. It should be noted that the tool itself does not enhance one's cognitive efficiency; rather, it is the combination of a computational tool and formal method that enables such amplification. Also, the third stage contributes to the enhancement of numerical coherence of a posed problem through the design of a new tool. Such amplification in the context of spreadsheets used to illustrate the notion of didactical coherence may be an inquiry into using four types of stamps for a given postage, something that leads to reorganizing a spreadsheet to work as a four-dimensional modeling tool (Abramovich & Cho, 2008).

Application of the ACA Framework in the Context of a Spreadsheet

To illustrate how the ACA framework can be utilized within a specific context, consider the following well-known problem (e.g., Gardner, 1961; Pask, 1998) formulated here in the simplest form. (This problem was discussed with preservice teachers in a course on the use of spreadsheets in teaching school mathematics).

Basic "coconuts" problem

In the rainforest, two men and a monkey gather coconuts all day and then fall asleep. During the night, each man wakes up and, after giving one coconut to the monkey, removes and hides half of the remaining coconuts for himself. Assuming that each man wakes up only once during the night, find the smallest number of coconuts originally gathered to allow for the described situation to take place; in other words, to allow for the problem to be numerically coherent.

To solve this problem, one can again use a spreadsheet, this time for its remarkable facility of recurrent counting. With this in mind, a spreadsheet can be connected to the problem by recognizing the recursive structure of the men's behavior during the night—what the first man does with the original pile, the second man does exactly the same with the remaining pile. In that way, the recursive nature of the problem calls for a tool capable of recurrent counting. Thus, a spreadsheet becomes the tool of choice.

Turning to mathematics, let x_0 be the number of coconuts sought (i.e., the number of coconuts gathered during the day). Then the number of coconuts that the first man leaves for the second man is $x_1 = (x_0 - 1) - \frac{1}{2}(x_0 - 1) = \frac{1}{2}(x_0 - 1)$. Likewise, the number of coconuts left by the second man is $x_2 = (x_1 - 1) - \frac{1}{2}(x_1 - 1) = \frac{1}{2}(x_1 - 1)$. Now, using a spreadsheet and trial-and-error in selecting x_0, one can easily discover that, along with x_0, both x_1 and x_2 should be whole numbers and, therefore, it is necessary (but not sufficient) that x_0 and x_1 are odd numbers. The spreadsheet pictured in Figure 4.6 shows that when $x_0 = 7$ it follows that $x_1 = \frac{1}{2}(7-1) = 3$, $x_2 = \frac{1}{2}(3-1) = 1$. At the same time, choosing $x_0 = 9$ yields $x_1 = \frac{1}{2}(9-1) = 4$; yet, $x_2 = \frac{1}{2}(4-1) = \frac{3}{2}$ is not an integer and, thereby, indeed, it is not sufficient for numerical coherence of the problem for x_0 to be just an odd number.

B2 fx =(1/2)*(A2-1)

	A	B	C	D	E
1	x_0	x_1	x_2		
2	7	3	1		

Figure 4.6. Solving basic "coconuts" problem.

One can define the sequence

$$x_n = \frac{1}{2}(x_{n-1} - 1),\ n = 1, 2, 3, \ldots, \tag{4.1}$$

to allow for more than two men (or more than one wake-up for each man) to be considered. However, the trial-and-error approach may not be as effective as it was in the case of two men. As a remedy, one can rewrite recurrence (4.1) in the form

$$x_{n-1} = 2x_n + 1, n = 1, 2, 3, \ldots, \tag{4.2}$$

so that, by moving backwards, one can reach x_0 starting from x_n. Indeed, let $n=7$ and $x_7 = 1$. Then $x_6 = 3$, $x_5 = 7$, $x_4 = 15$, $x_3 = 31$, $x_2 = 63$, $x_1 = 127$, and, finally, $x_0 = 255$ (Figure 4.7). This completes the first stage of the triad when the spreadsheet serves as an agency for utilizing the notions of recursive reasoning, trial-and-error, necessary and sufficient conditions, and ascending and descending sequences. These notions have been discussed with secondary teacher candidates to address the Conference Board of the Mathematical Sciences (2001, 2012) recommendations for teacher preparation in the context of discrete mathematics.

	A	B	C	D	E	F	G	H	I	J	K
1	x_n	1	3	7	15	31	63	127	255	511	1023

Figure 4.7. Using relation (4.2) in spreadsheet programming.

Now, the spreadsheet can become a consumer of the mathematical activity based on these notions by formulating problems for a number of men (or wake-ups) greater than two. For example, using the spreadsheet based on relation (4.2) and changing the number of men from two to five, if the number of coconuts left by the fifth man is five (i.e., assuming $x_5 = 5$), one can pose a problem of finding the number of coconuts originally gathered. The answer is in the spreadsheet of Figure 4.8 (cell F2). Likewise, many other similar problems can be formulated. This problem-posing activity completes the second stage of the triad.

B2 fx =2*A2+1

	A	B	C	D	E	F
1	x_5	x_4	x_3	x_2	x_1	x_0
2	5	11	23	47	95	191

Figure 4.8. Sharing 191 coconuts among five men and a monkey.

The third stage of the triad arises when one attempts to pose a modification of the basic problem. For example, a possible modification would be to assume that each man hides one-third of the available coconuts (this assumes that a generalization of one of the inputs is possible). Thus, relations (4.1) and (4.2) should be replaced, respectively, by $x_n = \frac{2}{3}(x_{n-1} - 1)$ and

$$x_{n-1} = \frac{3}{2}x_n + 1. \tag{4.3}$$

One can see (Figure 4.9) that the spreadsheet constructed for the first stage ceases to work as the whole-number property of the right-hand side of relation (4.3) depends on the factors of x_n. So, already a slight modification of the basic problem requires a reorganization of the mathematical activity used in the construction of the spreadsheet.

	A	B	C	D	E
1	x_n	4	7	11.5	18.25

Figure 4.9. Numerically incoherent data.

Noting that in the case of hiding one-fourth (rather than one-third) of the available coconuts, relation (4.3) has to be replaced by $x_{n-1} = \frac{4}{3}x_n + 1$, thereby requiring new multiplicative requirements for x_n, one can pose a general problem and solve it in order for a spreadsheet to amplify problem-posing (and, consequently, problem-solving) opportunities. With this in mind, the following problem can be formulated. Note that formulation of this problem arose because the course instructor had the goal of creating a generalized problem-posing environment which could be utilized in a course on using spreadsheets in teaching K-12 mathematics.

Generalized "coconuts" problem

In the rainforest, ***n*** men and ***p*** monkeys gather coconuts all day and then fall asleep. During the night, each man wakes up and, after giving one coconut to each monkey, removes and hides **1/*n*** of the remaining coconuts for himself. Assuming that each man wakes up only once during the night, find the smallest number of coconuts originally gathered.

Let x_0 be the number of coconuts sought. Under the general conditions, one can compute recursively in succession

$$x_1 = (x_0 - p) - \frac{x_0 - p}{n} = \frac{n-1}{n}(x_0 - p),$$

$$x_2 = (x_1 - p) - \frac{x_1 - p}{n} = \frac{n-1}{n}(x_1 - p) = \frac{n-1}{n}\left[\frac{n-1}{n}(x_0 - p) - p\right]$$
$$= \left(\frac{n-1}{n}\right)^2 x_0 - \left[\left(\frac{n-1}{n}\right)^2 + \frac{n-1}{n}\right] p,$$

$$x_3 = \left(\frac{n-1}{n}\right)^3 x_0 - \left[\left(\frac{n-1}{n}\right)^3 + \left(\frac{n-1}{n}\right)^2 + \frac{n-1}{n}\right] p,$$

and then generalize inductively that

$$x_n = \left(\frac{n-1}{n}\right)^n x_0 - \left[\left(\frac{n-1}{n}\right)^n + \left(\frac{n-1}{n}\right)^{n-1} + \cdots + \left(\frac{n-1}{n}\right)\right] p.$$

Simplifying the sum in the brackets by summing the first n terms of the geometric series $g_i = \left(\frac{n-1}{n}\right)^i$, $i \geq 1$, as follows,

$$\left(\frac{n-1}{n}\right)^n + \left(\frac{n-1}{n}\right)^{n-1} + \cdots + \left(\frac{n-1}{n}\right) = \frac{n-1}{n}\left[\frac{1-\left(\frac{n-1}{n}\right)^n}{1-\frac{n-1}{n}}\right] = (n-1)\left[1-\left(\frac{n-1}{n}\right)^n\right],$$

leads to the formula

$$x_n = \left(\frac{n-1}{n}\right)^n x_0 - (n-1)\left[1-\left(\frac{n-1}{n}\right)^n\right] p \tag{4.4}$$

which expresses x_n in terms of x_0, n, and p.

Formula (4.4) can be easily proved by mathematical induction using the recursive relation $x_{n+1} = \frac{n-1}{n}(x_n - p)$ for which the action of each man can serve as a situational referent. In turn, it follows from formula (4.4) that

$$x_0 = \left(\frac{n}{n-1}\right)^n \left\{ x_n + (n-1)\left[1 - \left(\frac{n-1}{n}\right)^n\right] p \right\} = \left(\frac{n}{n-1}\right)^n x_n + (n-1)\left(\frac{n}{n-1}\right)^n p$$

$$-(n-1)p = \frac{n^n\left[x_n + (n-1)p\right]}{(n-1)^n} - (n-1)p.$$

That is,

$$x_0 = \frac{n^n\left[x_n + (n-1)p\right]}{(n-1)^n} - (n-1)p, \tag{4.5}$$

from where it follows that $x_n + (n-1)p$ should be divisible by $(n-1)^n$. In particular, when $n=2$ and $p=1$ we have $x_0 = 4x_2 + 3$, confirming the solution to the original "coconuts" problem (Figure 4.6). The case $n=3, p=1$ is computed as follows:

$$x_0\big|_{n=3,p=1} = \frac{27}{8}\left[x_3 + 2\left(1 - \frac{8}{24}\right)\right] = \frac{27(x_3 + 21)}{8}.$$

Note that the smallest value of x_3 for which the sum $x_3 + 2$ is divisible by 8 is equal to 6. In that case, we have $x_0 = 25$. One can program a spreadsheet using formula (4.5) and then pose a variety of problems for different values of n and p, something that would be too cumbersome or too computationally involved to use pencil-and-paper alone. In that way, a spreadsheet amplifies problem posing as a result of integration of mathematical machinery and the unique capability of the software. This spreadsheet is shown in Figure 4.10.

B4 fx =(n^n)*(A4+p*(n-1))/(n-1)^n-(n-1)*p

	A	B	C	D	E	F	G	H
1	n	p						
2	4	2						
3	x_n	x_0						
4	1	16.1235						
78	75	250						
79	76	253.16						
159	156	506						
160	157	509.16						

Figure 4.10. Solving a special case of generalized "coconuts" problem.

Obviously, such a (numerically coherent) problem would not be possible to formulate without using the spreadsheet of Figure 4.10 based on formula (4.5). It is in that sense that a spreadsheet played the role of an amplifier of mathematical activity which originated in the context of solving the basic coconuts problem.

> **Problem.** *Four men and a monkey gather bananas all day and then fall asleep. During the night each man wakes up in turn and, after giving two bananas to the monkey, he removes and hides one fourth of the pile of bananas for himself. Assuming that each man wakes up only once during the night, find a number of bananas originally gathered so that all divisions came out in integers and give a minimum number of remaining bananas.*
>
> *The answer to this problem would be 75 bananas. The only other numbers that fit the conditions of the problem are those in the set that begins with 75 and have a difference 81 between the next closest integer.*

Figure 4.11. A teacher-posed problem: modifying the original "coconuts" problem.

Analysis of a Preservice Teacher's Problem

Figure 4.11 shows a problem posed by a preservice teacher using the spreadsheet of Figure 4.10. The teacher provided the answer: 75 bananas. However, not only would this problem be difficult to pose without technology, but also the problem would be difficult to solve without using formula (4.5). In the case of $n=4$ and $p=2$ formula (4.5) yields

$$x_0 = \left(\frac{4}{3}\right)^4 (x_4 + 6) - 6 \tag{4.6}$$

from where the values for x_4 and x_0 follow without difficulty if one notes that $x_4 + 6$ should be divisible by 81 and the smallest $x_4 = 75$ whence $x_0 = 250$. However, the ease of this solution is due to formula (4.5), the derivation of which requires rather advanced algebraic skills motivated by the need to amplify problem posing. This example, in particular, illustrates Pólya's (1957) argument "more general problem may be easier to solve" (p. 109).

Thus, the problem shown in Figure 4.11 may not be pedagogically coherent for a regular secondary mathematics classroom unless it is designed with the development of formula (4.5) in mind. One way to modify the teacher's problem in order to make it less challenging and solvable with paper-and-pencil would be to include the answer 75 as a given and ask only for the number of bananas originally gathered. Then an expected method of solution could be to write down the following four relations

$$\frac{3}{4}(x_0 - 2) = x_1,\quad \frac{3}{4}(x_1 - 2) = x_2,\quad \frac{3}{4}(x_2 - 2) = x_3,\quad \frac{3}{4}(x_3 - 2) = x_4 \tag{4.7}$$

and, setting $x_4 = 75$ in (2.7), find x_0 in four steps:

$$x_3 = \frac{4}{3} \times 75 + 2 = 102, \quad x_2 = \frac{4}{3} \times 102 + 2 = 138,$$
$$x_1 = \frac{4}{3} \times 138 + 2 = 186, \quad x_0 = \frac{4}{3} \times 186 + 2 = 250.$$

Note that until we reach 250, each of the numbers 75, 102, 138, and 186 is divisible by three.

Likewise, the number 250 may be given (instead of 75) so that, setting $x_0 = 250$ in (4.7), would allow one to reach 75 in four steps noting along the way that each of the numbers 248, 184, 136, and 100 is divisible by four.

The preservice teacher who posed the problem in Figure 4.11 also provided a rule for finding the quantity of all remaining bananas had it be different from 75. The rule is that these quantities are in arithmetic progression with the first term 75 and difference 81. Although the teacher established this rule by analyzing data generated by the spreadsheet of Figure 4.10, formula (4.6)—a special case of formula (4.5)—yields this rule immediately. The rule, in turn, makes it possible to pose many new problems with $x_4 \in \{81k - 6,\ k = 2, 3, 4, \ldots\}$ to be solved by using the chain of relations (4.7). In that case, x_4 does not represent the smallest number of the remaining bananas.

Reciprocal Problem Posing: A Case in Secondary School Algebra

One can extend the concept of reciprocal teaching, originally introduced by Palincsar and Brown (1984, 1988) in the context of reading instruction, to entertain the idea of reciprocal problem posing towards the goal of developing and assessing preservice teachers' skills in creating didactically coherent curriculum materials. This idea can be integrated into a technology-supported learning environment by arranging teachers in pairs, and asking each pair, by using appropriate computational tools introduced by the instructor, to pose a problem for an associate pair. Each pair of preservice teachers needs to solve a problem given to them using both pencil-and-paper setting and digital technology. Note that in the elementary classroom the idea of using reciprocity in mathematical problem posing has been used by a number of researchers (e.g., Ellerton, 1986; Richardson & Williamson, 1982; Van den Brink, 1987).

The instructional goal of reciprocal problem posing, besides developing and assessing problem-posing skills and encouraging cooperative learning in a technology-rich setting is to highlight problem posing and problem solving not as dichotomized but as closely related mathematical activities. Through active engagement in reciprocal problem posing, teachers can indeed begin viewing problem posing as "a platform from which further development proceeds" (Davis, 1985, p. 23).

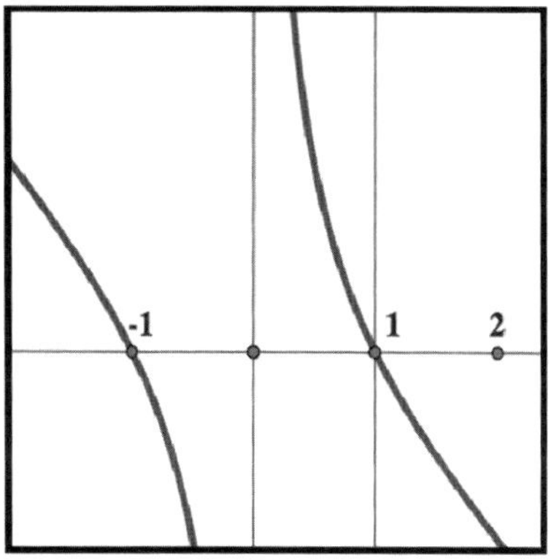

Figure 4.12. Locus of the equation $x^2 + bx - 1 = 0$.

This development includes preservice teachers' use of the notion of didactical coherence with a focus on avoiding the emergence of pedagogical instability when a posed problem turns out to be didactically incoherent due to the lack of necessary problem-solving skills on the part of those to whom the problem was offered. Consequently, a solution found by a problem solver might be different from the one expected by a problem poser and this could (and should) lead to the exchange of ideas. By experiencing reciprocity in posing problems, one can better appreciate the notion that, whereas a problem is supposed to challenge learners to a certain extent, it should not develop their sense of frustration with mathematics. Furthermore, designing a problem *to be solved* should eliminate any perception that there exists a kind of dichotomy between problem posing and solving.

Finally, each pair of preservice teachers can then be asked to extend the problem given to them, that is, to pose a closely related problem, and offer such an extension to another pair. This process, guided by the instructor, has the potential to increase the complexity of extended problems. It also allows for preservice teachers (working either in pairs or individually) to continue to improve their problem-posing/problem-solving performance, and encourages the development of mathematical ideas that are far beyond the originally posed problem. While, in general, reciprocal problem posing does not require the use of technology, the latter allows one to elevate the art of mathematical problem posing to a higher level.

As an illustration, consider the case of interaction between two secondary preservice teachers, Alice and Bob, working with *Graphing Calculator 4.0*, computer graphing software capable of plotting loci of two-variable equations and inequalities.[4] Using a computer-generated graph of the equation $x^2 + bx - 1 = 0$ in the plane (x, b) where b is a real parameter (Figure 4.12), Alice formulated

Alice's Problem #1: *For which values of parameter b are both roots of the equation* $x^2 + bx - 1 = 0$ *smaller than one*?

[4] Note that preservice teachers had experience in exploring equations with parameters as described by Abramovich and Norton (2006).

Using the graph (Figure 4.12), Bob found the answer in the form of the inequality $b > 0$.

In turn, Bob modified Alice's Problem #1 as follows:

Bob's Problem #1: *For which values of parameter b are both roots of the equation* $x^2 + bx - 1 = 0$ *greater than negative one*?

Alice solved Bob's Problem #1 in the form $b < 0$ (Figure 4.12) and reciprocated with

Alice's Problem #2: *For which values of parameter b are both roots of the equation* $x^2 + bx - 1 = 0$ *smaller than two*?

Bob solved Alice's Problem #2 in the form $b > -1.5$ (Figure 4.13) and then posed

Bob's Problem #2: *For which values of parameter a are both roots of the equation* $ax^2 + x + 2 = 0$ *greater than negative one*?

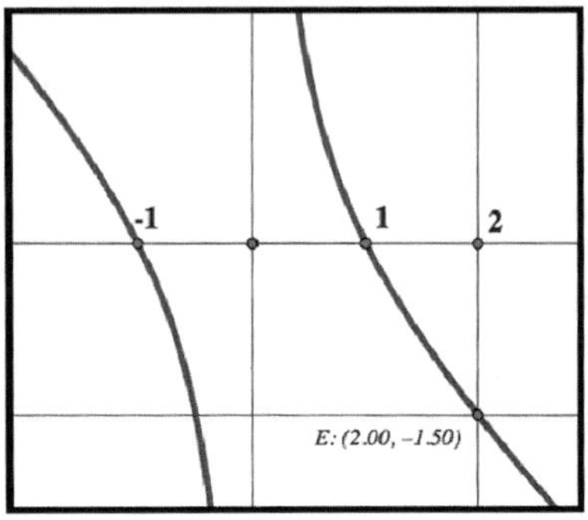

Figure 4.13. Posing (and solving) Alice's Problem #2.

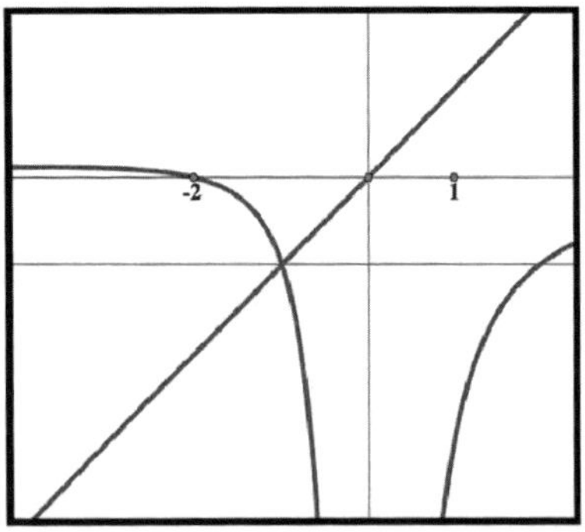

Figure 4.14. Locus of the equation $ax^2 + x + 2 = 0$ *crossed* by the angle bisector $a = x$.

This required Alice to construct a qualitatively different graph (Figure 4.14) and then to offer the answer in the form $a < -1$ along with the explanation of how the answer was found. Consequently, Bob used the new graph (Figure 4.14) and formulated for Alice

Bob's Problem #3: *For which values of parameter a are both roots of the equation* $ax^2 + x + 2 = 0$ *greater than a*?

One can see that such process of reciprocal problem posing can continue like a never ending experimentation with computer-generated mathematical objects through which problem posing motivates problem solving and vice versa, provided that each problem is didactically coherent and that the teacher's interest towards mathematics remains in a state of stable equilibrium.

The following is a reflection by a preservice teacher produced during a secondary mathematics education capstone course (taught by the first author) that emphasized the value of reciprocal problem posing with technology in promoting experimental mathematics approach (Baker, 2008):

> Throughout participation and engagement in this course, I learned a significant amount regarding experimental and theoretical knowledge regarding mathematics. The primary relationship between these categories was the natural relationship between them in terms of one driving the other in a cyclic fashion. An individual explores a problem or situation by experimental analyzing, until an experimental result is reached. This result drives theoretical mathematics to provide rigorous reasoning and backing to validate the conclusion. This then opens the door to more questions and problem situations, which are subjective to experimental exploration, and so the cycle continues until the practicality or individuals' sanity is lost. Such a relationship embodies the explorative and discovery method of research and educational viability in the classroom in both learning and teaching itself (p. 334).

The preservice teacher's recognition of the value of technology as a medium within which a new problem can be born through solving an already existing problem suggests that the approach is conducive to the development of higher-order thinking skills and to providing teachers with research-like experiences in experimental mathematics methodology and pedagogy.

Problem Posing with Technology as a Discovery Experience

The use of technology in the context of education can lead to the formulation of new problems that can be given the status of conjecture in the real mathematical sense of this word. Put another way, by experimenting with technology one can come across an opportunity to pose a problem for which no solution can be found even when the problem is offered to a professional mathematician. This aspect of problem posing with technology is very important for it has the potential, by using an experimental mathematics approach with secondary mathematics teacher candidates, to open a window to new mathematical knowledge. One such problem (Abramovich & Leonov, 2009) can be explained in very simple terms although at the time of writing this chapter it does not have a formal solution and remains a technology-motivated conjecture. Therefore, this section, drawing on the ideas included in a capstone course for secondary pre-teachers, will demonstrate how problem posing and discovery experiences could be connected.

A well-known mathematical structure is Pascal's triangle (Figure 4.15), each line of which represents coefficients in the expansion of $(x+y)^n, n = 0,1,2,3,\ldots$. The elements of Pascal's triangle can be rearranged in a form resembling a diagonally shaped table (the spreadsheet of Figure 4.16) in such a way that the rows of the triangle are turned into the diagonals of this table as shown in Figure 4.17. Consider now, for example, the polynomial of degree seven

$$P_7(x) = x^7 + 13x^6 + 66x^5 + 165x^4 + 210x^3 + 126x^2 + 28x + 1, \qquad (4.8)$$

the coefficients of which are the entries of row 13 of the spreadsheet shown in Figure 4.16. Using *Wolfram Alpha*, one can discover (Figure 4.18) that this polynomial has exactly seven real roots. The absence of complex roots in this polynomial is not an isolated fact. Indeed, none of the polynomials so constructed with the

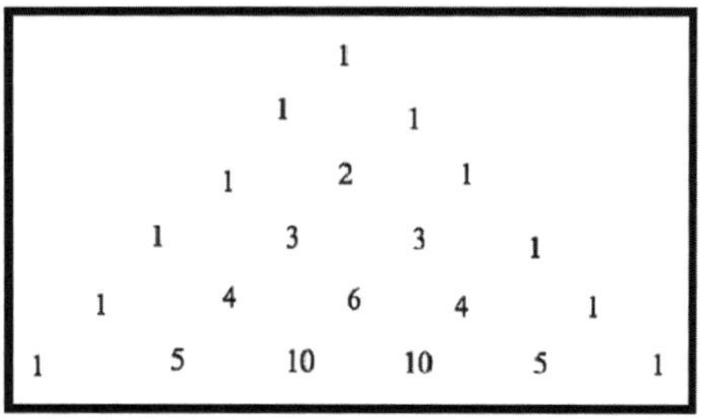

Figure 4.15. Pascal's triangle.

	A	B	C	D	E	F	G	H	I	J	K	L	M
1	1	1											2
2	1	2											3
3	1	3	1										5
4	1	4	3										8
5	1	5	6	1									13
6	1	6	10	4									21
7	1	7	15	10	1								34
8	1	8	21	20	5								55
9	1	9	28	35	15	1							89
10	1	10	36	56	35	6							144
11	1	11	45	84	70	21	1						233
12	1	12	55	120	126	56	7						377
13	1	13	66	165	210	126	28	1					610

Figure 4.16. Rearranged entries of Pascal's triangle.

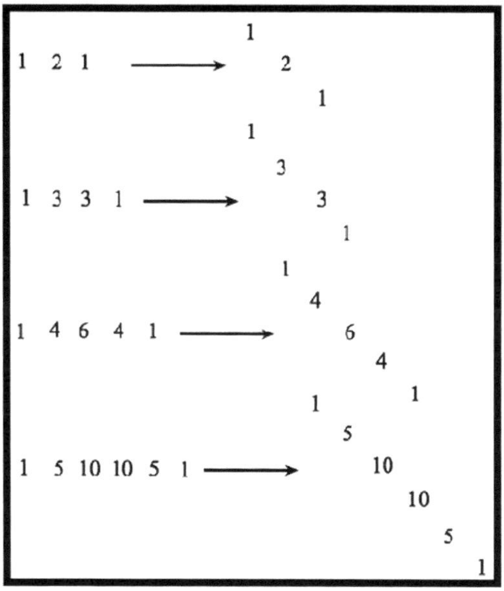

Figure 4.17. Turning rows into diagonals.

coefficients taken from the rows of the rearranged Pascal's triangle has complex roots (polynomials of degree 6—rows 11, 12; polynomials of degree 5—rows 9, 10; polynomials of degree 4—rows 7, 8; polynomials of degree 3—rows 5, 6; polynomials of degree 2—rows 3, 4). These polynomials are called Fibonacci-like polynomials (Abramovich & Leonov, 2009) because the sums of their coefficients are consecutive Fibonacci numbers (Figure 4.16, column M). The polynomials can be defined recursively as

$$P_n(x) = x^{\mathrm{mod}(n,2)} P_{n-1}(x) + P_{n-2}(x),\ P_0(x) = 1,\ P_1(x) = x+1 \tag{4.9}$$

where mod$(n, 2)$ is the remainder of n divided by 2. The following problem can be posed:

Problem. Prove that Fibonacci-like polynomials defined by recursive relation (4.9) do not have complex roots.

In order to explain how this problem originated from the use of technology, note that the roots of Fibonacci-like polynomials were found to be responsible for a cyclic behavior of the so-called generalized Golden Ratios generated by the orbits of a two-parametric difference equation

$$f_{k+1} = af_k + bf_{k-1},\ f_0 = f_1 = 1,\ k = 1, 2, 3, \ldots. \tag{4.10}$$

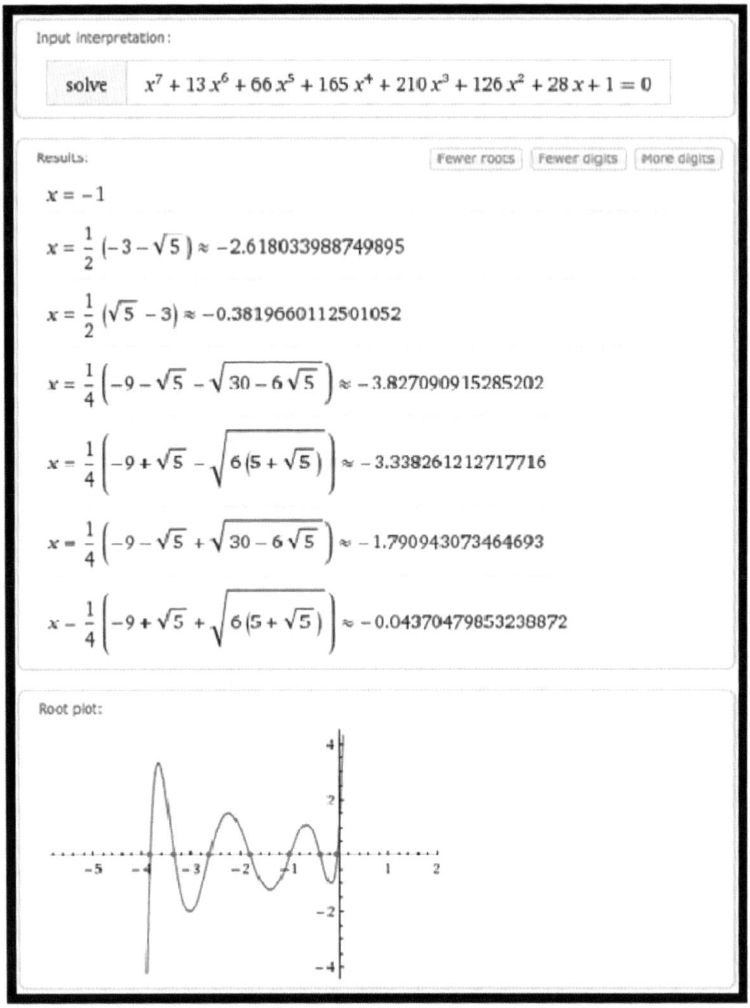

Figure 4.18. All the roots of (4.8) are real.

Equation (4.10)—another example of a mathematical model mentioned in connection with "the emerging importance of topics and methods in discrete mathematics" (Conference Board of the Mathematical Sciences, 2001, p. 140)—can be explored by using a spreadsheet; in the case $a=b=1$ it generates the celebrated sequence 1, 1, 2, 3, 5, 8, 13, …, the so-called Fibonacci numbers.

Using a spreadsheet, one can also see that when $\dfrac{a^2}{b} = -\dfrac{3+\sqrt{5}}{2}$, this number being the smallest root of polynomial (4.8), the ratios f_{k+1} / f_k form cycles of period

	A	B	C	D
1	f_n	f_{n+1}/f_n	a	b
2	1		3	-3.4376941
3	1	1		
4	-0.4376941	-0.4376941		
5	-4.75077641	10.85410197		
6	-12.7476708	2.683281573		
7	-21.9112963	1.718847051		
8	-21.9112963	1		
9	9.590445155	-0.4376941		
10	104.0956696	10.85410197		
11	279.3179921	2.683281573		
12	480.1049069	1.718847051		
13	480.1049069	1		
14	-210.139086	-0.4376941		
15	-2280.87106	10.85410197		
16	-6120.2193	2.683281573		
17	-10519.7209	1.718847051		
18	-10519.7209	1		

Figure 4.19. Ratios f_{k+1}/f_k form a 5-cycle in (4.10).

five as shown in Figure 4.19. Using *Maple*, the family of period five cycles corresponding to this root can be expressed as follows:

$$\left\{ 1, \frac{a\left(3+\sqrt{5}-2a\right)}{3+\sqrt{5}}, \frac{a\left(1+\sqrt{5}-2a\right)}{3+\sqrt{5}-2a}, \frac{-2a\left(-1-\sqrt{5}+a+\sqrt{5}a\right)}{\left(3+\sqrt{5}\right)\left(1+\sqrt{5}-2a\right)}, \frac{\left(-1+\sqrt{5}\right)a^2}{\left(-1-\sqrt{5}+a+\sqrt{5}a\right)} \right\}.$$

One can check to see that when $a=3$, the 5-cycle generated by the spreadsheet of Figure 4.19 results. This is a quite notable example when an unsolved problem stems from the educational use of commonly available technology. For more information on this topic and its use in a capstone course for preservice teachers of secondary mathematics, see Abramovich and Leonov (2009, 2011).

Conclusion

This chapter has described the role of modern tools of digital technology such as an electronic spreadsheet, computer-based *Graphing Calculator*, *Maple* and *Wolfram Alpha* in facilitating and advancing skills of preservice teachers in mathematical problem posing. Using the cases in elementary and secondary teacher education contexts, the chapter demonstrated the importance of theoretical considerations associated with the use of computers in problem posing. Problems posed in these technological contexts can vary from the "single question-multiple answers" word problems appropriate for primary grades to rather advanced exploratory tasks depending on parameters that extend mundane problems typically found in the traditional secondary school curriculum.

Theoretical constructs such as the concept of didactical coherence of a problem and the ACA framework for computer use in mathematics education were introduced and illustrated through the analysis of a number of genuine problems posed by preservice teachers. By paying attention to the numerical, contextual, and pedagogical coherence constructs when formulating problems from spreadsheet-generated data, the teachers were able to critically interpret and appropriately modify the computational results.

The use of graphing software as a medium for reciprocal problem posing is shown to be conducive for developing rather sophisticated questions about algebraic equations with parameters. The benefits of reciprocal problem-posing pedagogy made possible by the appropriate use of *Graphing Calculator 4.0* along with the potential of turning problem posing into discovery experiences were presented. In this respect, technology can play a dual role in problem posing: its presence has great potential to facilitate the development of new problems, and modern computational engines such as *Wolfram Alpha* can make already existing mathematical problems somewhat outdated for they can be easily computable by transforming a free-form question into an achievable solution. This, in turn, can open a whole new avenue for problem posing in the technological paradigm by replacing outdated problems by technology-immune, yet technology-enabled, mathematical explorations.

The proposed conceptual framework and its use in the practice of teaching mathematics have several important implications for K-12 teaching and teacher education. First, the use of technology in problem posing encourages open-ended classroom pedagogy, fosters mathematical reasoning and thinking skills of preservice teachers, and, consequently, has great potential to make K-12 students better problem solvers. Second, the concept of didactical coherence of a problem has important implication for teaching mathematics without technology. Preservice teachers' familiarity with the didactical coherence constructs and such notions as an extraneous solutions and contextual/pedagogic incoherence, learned in the context of using digital technology for problem posing, would help them during their inductive years and beyond to offer problems free from what can be seen as didactic flaws. The ability to minimize the presence of such flaws in curriculum materials would, in turn, help K-12 students to concentrate on mathematical aspects of the problems involved and not to be distracted by sometimes ill-designed and unimportant details of the problems. With the growth of mathematical reasoning and thinking skills, the ability of preservice teacher education students to recognize and eliminate didactic flaws as nonessential elements of problem structures will develop. For example, the ability of elementary preservice teachers to pose a mathematically appropriate question about a situation when two whole numbers have to be compared in context requires experience with the concept of didactical coherence. Depending on the operation to be used in comparison (subtraction when comparing through difference and division when comparing through ratio), the very context of comparison of two numbers can be selected in more than one way. So, experience in posing problems through the use of technology can be helpful in nontechnological contexts also, something that has important implications for the teaching of mathematics, especially at the elementary level.

This chapter set out to demonstrate that the creative use of commonly available digital technology tools can motivate and support problem-posing activities. In the context of preservice teacher education, the chapter illustrated the importance of preparing teacher candidates to be equipped with conceptual understanding of didactic issues related to problem posing with technology, and allowing them to participate actively in their own learning process and pose their own problems using technology. In this way, preservice teachers will be able to have an ownership of their learning experiences and a renewed understanding of what it means to be a student in a mathematics classroom, as a producer not just a consumer of knowledge. As noted in the beginning of this chapter, literature on problem posing with technology and especially the scope of research in this area are still limited and deserve more research attention both in theoretical and practical aspects. The authors hope that their experience shared above can motivate further research on using digital technology for mathematical problem posing.

References

Abramovich, S. (2006). Spreadsheet modeling as a didactical framework for inequality-based reduction. *International Journal of Mathematical Education in Science and Technology, 37*(5), 527–541.

Abramovich, S. (2012). Computers and mathematics teacher education. In S. Abramovich (Ed.), *Computers in education* (Vol. 2, pp. 213–236). Hauppauge, NY: Nova.

Abramovich, S., & Brouwer, P. (2003). Revealing hidden mathematics curriculum to pre-teachers using technology: The case of partitions. *International Journal of Mathematical Education in Science and Technology, 34*(1), 81–94.

Abramovich, S., & Cho, E. K. (2006). Technology as a medium for elementary pre-teachers' problem posing experience in mathematics. *Journal of Computers in Mathematics and Science Teaching, 26*(4), 309–323.

Abramovich, S., & Cho, E. K. (2008). On mathematical problem posing by elementary pre-teachers: The case of spreadsheets. *Spreadsheets in Education, 3*(1), 1–19.

Abramovich, S., & Cho, E. K. (2009). Mathematics, computers, and young children as a research-oriented learning environment for a teacher candidate. *Asia Pacific Education Review, 10*(2), 247–259.

Abramovich, S., & Cho, E. K. (2012). Technology-enabled mathematical problem posing as modeling. *Journal of Mathematical Modelling and Application, 1*(6), 22–32.

Abramovich, S., Easton, J., & Hayes, V. O. (2012). Parallel structures of computer-assisted signature pedagogy: The case of integrated spreadsheets. *Computers in the Schools, 29*(1–2), 174–190.

Abramovich, S., & Leonov, G. A. (2009). Spreadsheets and the discovery of new knowledge. *Spreadsheets in Education, 3*(2), 1–42.

Abramovich, S., & Leonov, G. A. (2011). A journey to a mathematical frontier with multiple computer tools. *Technology, Knowledge and Learning, 16*(1), 87–96.

Abramovich, S., & Norton, A. (2006). Equations with parameters: A locus approach. *Journal of Computers in Mathematics and Science Teaching, 25*(1), 5–28.
Abu-Elwan, R. (2007). The use of Webquest to enhance the mathematical problem-posing skills of pre-service teachers. *International Journal for Technology in Mathematics Education, 14*(1), 31–39.
Akay, H., & Boz, N. (2010). The effect of problem posing oriented analyses-II course on the attitudes toward mathematics and mathematics self-efficacy of elementary prospective mathematics teachers. *Australian Journal of Teacher Education, 35*(1), 59–75.
Avitzur, R. (2011). *Graphing calculator (Version 4.0)*. Berkeley, CA: Pacific Tech.
Baker, A. (2008). Experimental mathematics. *Erkenntnis, 68*(3), 331–344.
Bonotto, C. (2010). Engaging students in mathematical modelling and problem posing activities. *Journal of Mathematical Modelling and Application, 1*(3), 18–32.
Brown, S. I., & Walter, M. I. (1983). *The art of problem posing*. Hillsdale, NJ: Lawrence Erlbaum.
Cai, J., & Hwang, S. (2002). Generalized and generative thinking in US and Chinese students' mathematical problem solving and problem posing. *Journal of Mathematical Behavior, 21*(4), 401–421.
Char, B. W., Geddes, K. O., Gonnet, G. H., Leong, B. L., Monagan, M. B., & Watt, S. M. (1991). *Maple V language reference manual*. New York, NY: Springer.
Cobb, P. (1995). Mathematical learning and small-group interaction: Four case studies. In P. Cobb & H. Bauersfeld (Eds.), *The emergence of mathematical meaning: Interaction in classroom cultures* (pp. 25–129). Hillsdale, NJ: Lawrence Erlbaum.
Comtet, L. (1974). *Advanced combinatorics*. Dordrecht, Holland: D. Reidel Publishing.
Conference Board of the Mathematical Sciences. (2001). *The mathematical education of teachers*. Washington, DC: The Mathematical Association of America.
Conference Board of the Mathematical Sciences. (2012). *The mathematical education of teachers II*. Washington, DC: The Mathematical Association of America.
Crespo, S. (2003). Learning to pose mathematical problems: Exploring changes in preservice teachers' practices. *Educational Studies in Mathematics, 52*(3), 243–270.
Davis, P. J. (1985). What do I know? A study of mathematical self-awareness. *College Mathematics Journal, 16*(1), 22–41.
Dimiceli, V. A., Lang, A. S. I. D., & Locke, L. A. (2010). Teaching calculus with Wolfram|Alpha. *International Journal of Mathematical Education in Science and Technology, 41*(8), 1061–1071.
Dunham, W. (1991). *Journey through genius: The great theorems of mathematics*. New York, NY: Penguin.
Dunker, K. (1945). *On problem solving* (Psychological Monographs, Vol. 58(5)). Washington, DC: American Psychological Association.
Dyson, G. (2012). *Turing's cathedral: The origins of the digital universe*. New York, NY: Pantheon Books.
Ellerton, N. F. (1986). Children's made up mathematics problems: A new perspective on talented mathematicians. *Educational Studies in Mathematics, 17*(3), 261–271.
English, L. D. (1997). The development of fifth-grade children's problem-posing abilities. *Educational Studies in Mathematics, 34*(3), 183–217.
Ernie, K., LeDocq, R., Serros, S., & Tong, S. (2009). Challenging students' beliefs about mathematics. In R. A. R. Guring, N. L. Chick, & A. A. Ciccone (Eds.), *Exploring signature pedagogies* (pp. 260–279). Sterling, VA: Stylus.
Gardner, M. (1961). *More mathematical puzzles and diversions*. New York, NY: Penguin.
Hirashima, T., Nakano, A., & Takeuchi, A. (2000). A diagnosis function of arithmetical word problems for learning by problem posing. In R. Mizoguchi & J. Slaney (Eds.), *Topics in artificial intelligence, PRICAI 2000* (pp. 745–755). Berlin, Germany: Springer.
Hoyles, C., & Sutherland, R. (1986). Peer interaction in a programming environment. In L. Burton & C. Hoyles (Eds.), *Proceedings of the Tenth International Conference for the Psychology of Mathematics Education* (pp. 354–359). London, United Kingdom: University of London Institute of Education.

International Society for Technology in Education. (2008). *National educational technology standards (NETS) for teachers*. Retrieved from http://www.iste.org/standards/nets-for-teachers.aspx
Kar, T., Özdemir, E., İpek, A. S., & Albayrak, M. (2010). The relation between the problem posing and problem solving skills of prospective elementary mathematics teachers. *Procedia—Social and Behavioral Sciences, 2*, 1577–1583.
Kilpatrick, J. (1987). Problem formulating: Where do good problems come from? In A. H. Schoenfeld (Ed.), *Cognitive science and mathematics education* (pp. 123–147). Hillsdale, NJ: Lawrence Erlbaum.
Kontorovich, I., Koichu, B., Leikin, R., & Berman, A. (2012). An exploratory framework for handling the complexity of mathematical problem posing in small groups. *Journal of Mathematical Behavior, 31*(1), 149–161.
Krutetskii, V. A. (1976). *The psychology of mathematical abilities in school children*. Chicago, IL: The University of Chicago Press.
Laborde, C. (1995). Designing tasks for learning geometry in a computer-based environment. In L. Burton & B. Jaworski (Eds.), *Technology in mathematics teaching* (pp. 35–67). Bromley, United Kingdom: Chartwell-Bratt.
Lavy, I., & Bershadsky, I. (2003). Problem posing via "what if not" strategy in solid geometry—a case study. *Journal of Mathematical Behavior, 22*(4), 369–387.
Lavy, I., & Shriki, A. (2010). Engaging in problem posing activities in a dynamic geometry setting and the development of prospective teachers' mathematical knowledge. *Journal of Mathematical Behavior, 29*(1), 11–24.
Leung, S. S., & Silver, E. A. (1997). The role of task format, mathematics knowledge, and creative thinking on the arithmetic problem posing of prospective elementary school teachers. *Mathematics Education Research Journal, 9*(1), 5–24.
National Council of Teachers of Mathematics. (1989). *Curriculum and evaluation standards for school mathematics*. Reston, VA: Author.
National Council of Teachers of Mathematics. (1991). *Professional standards for teaching mathematics*. Reston, VA: Author.
National Council of Teachers of Mathematics. (2000). *Principles and standards for school mathematics*. Reston, VA: Author.
National Council of Teachers of Mathematics. (2011). *Technology in teaching and learning of mathematics: A position of the National Council of Teachers of Mathematics*. Reston, VA: Author.
Noss, R. (1986). What mathematics do children do with Logo? *Journal of Computer Assisted Learning, 3*(1), 2–12.
Organisation for Economic Co-operation and Development. (2011). *PISA 2009 assessment framework: Key competencies in reading, mathematics, and science*. Paris, France: Author.
Palincsar, A. S., & Brown, A. L. (1984). Reciprocal teaching of comprehension-fostering and comprehension-monitoring activities. *Cognition and Instruction, 1*(2), 117–175.
Palincsar, A. S., & Brown, A. L. (1988). Teaching and practicing thinking skills to promote comprehension in the context of group problem solving. *Remedial and Special Education, 9*(1), 53–59.
Pask, C. (1998). The monkey and the coconuts puzzle: Exploring mathematical approaches. *Teaching Mathematics and its Applications, 17*(3), 123–131.
Pólya, G. (1957). *How to solve it*. New York, NY: Anchor Books.
Richardson, J., & Williamson, P. (1982). Toward autonomy in infant mathematics. In *Research in mathematics education in Australia* (pp. 109–136). Melbourne, Australia: Mathematics Education Research Group of Australia.
Shulman, L. S. (2005). Signature pedagogies in the professions. *Daedalus, 134*(3), 52–59.
Silver, E. A. (1994). On mathematical problem posing. *For the Learning of Mathematics, 14*(1), 19–28.
Silver, E. A., & Cai, J. (1996). An analysis of arithmetic problem posing by middle school students. *Journal for Research in Mathematics Education, 27*(5), 521–539.

Singer, F. M., & Voica, C. (2013). A problem-solving conceptual framework and its implications in designing problem-posing tasks. *Educational Studies in Mathematics, 83*(1), 9–26. doi:10.1007/s10649-012-9422-x.

Thompson, P. W., Carlson, M. P., & Silverman, J. (2007). The design of tasks in support of teachers' development of coherent mathematical meanings. *Journal of Mathematics Teacher Education, 10*(4–6), 415–432.

Van den Brink, F. J. (1987). Children as arithmetic book authors. *For the Learning of Mathematics, 7*(2), 44–48.

Voica, C., & Singer, F. M. (2011). Creative contexts as ways to strengthen mathematics learning. *Procedia—Social and Behavioral Sciences, 33*, 538–542.

Yerushalmy, M., Chazan, D., & Gordon, M. (1993). Posing problems: One aspect of bringing inquiry into classrooms. In J. L. Schwartz, M. Yerushalmy, & B. Wilson (Eds.), *The geometric supposer: What is it a case of?* (pp. 117–142). Hillsdale, NJ: Lawrence Erlbaum.

Chapter 5
On the Relationship Between Problem Posing, Problem Solving, and Creativity in the Primary School

Cinzia Bonotto and Lisa Dal Santo

Abstract Problem posing is a form of creative activity that can operate within tasks involving semi-structured rich situations, using real-life artefacts and human interactions. Several researchers have linked problem-posing skills with creativity, citing flexibility, fluency, and originality as creativity categories. However, the nature of this relationship still remains unclear. For this reason, the exploratory study presented here sought to begin to investigate the relationship between problem-posing activities (supported by problem-solving activities) and creativity. The study is part of an ongoing research project based on teaching experiments consisting of a series of classroom activities in upper elementary school, using suitable artefacts and interactive teaching methods, in order to create a substantially modified teaching/learning environment. In addition, the study provides a method for analyzing the products of problem posing that teachers could use in the classroom to identify and assess both the activity of problem posing itself and students' creativity in mathematics.

Contents

C. Bonotto (✉)
Department of Mathematics, University of Padova, Padova, Italy
e-mail: bonotto@math.unipd.it

L. Dal Santo
Department of Mathematics, Istituto Comprensivo Valdagno 2, Valdagno, Italy
e-mail: lisads7@gmail.com

F.M. Singer et al. (eds.), *Mathematical Problem Posing*,
Research in Mathematics Education, DOI 10.1007/978-1-4614-6258-3_5

Introduction

Problems have occupied a central place in the school mathematics curriculum since antiquity. In fact, examples of mathematical and geometrical problems go back to the ancient Egyptians, Chinese, and Greeks. A common belief was that studying mathematics would improve one's ability to think, to reason, and to solve problems that one was likely to confront in the real world. Mathematics problems were a given element of the mathematics curriculum that contributed, like all other elements, to the development of reasoning power (Stanic & Kilpatrick, 1988).

However, traditional school word problems typically focus on the application of operational rules that involve a mapping between the structure of the problem situation and the structure of a symbolic mathematical expression. Often, solving these word problems is not a problem-solving activity for students; rather, it is an exercise that relies on syntactic cues for solution, such as key words or phrases in the problem (for example, "times," "less," "fewer"). While not denying the importance of these types of problems in the curriculum, they do not adequately address the mathematical knowledge, processes, representational fluency, and communication skills that our students need for the twenty-first century (English, 2009).

Furthermore, many researchers have documented that the practice of solving word problems in school mathematics actually promotes in students a suspension of sense-making (Schoenfeld, 1991), and the exclusion of realistic considerations. Primary and secondary school students tend to exclude relevant and plausible familiar aspects of reality from their observation and reasoning.

As a kind of minimal instructional response to this bridging problem, some scholars have made a plea for improving the quality of the word problems by making them resemble somewhat more the real-life problems encountered out-of-school, for example, by making the data, the question, and the contextual constraints more authentic or realistic (see, e.g. Palm, 2006; Verschaffel, Greer, & De Corte, 2000). As an even more radical response, other researchers have argued for the replacement of these word problems by *real* real-life problems that depart from existing (problematic) descriptions of the world (Chen, Van Dooren, Chen, & Verschaffel, 2011).

Our approach falls under the second type of response. If we want to help students to prepare to cope with natural situations they will have to face out of school, we need to rethink the type of problem-solving experiences we present to our students.

Almost all of the mathematical problems a student encounters have been proposed and formulated by another person—the teacher or the textbook author. In real life outside of school, however, many problems, if not most, must be created or discovered by the solver, who gives the problem an initial formulation (Kilpatrick, 1987).

In our opinion, the activities used to create an interplay between mathematics classroom activities and everyday-life experiences must be replaced with more realistic and less stereotyped problem situations, founded on the use of materials, real or reproduced, which children typically meet in real-life situations (Bonotto, 2005). In particular, we deem that classroom activities using suitable artefacts and interactive teaching methods could foster a mindful approach towards realistic mathematical modelling and problem solving, as well as a positive attitude toward problem-posing (Bonotto, 2009). In fact, we maintain that the problem-posing process represents one of the forms of authentic mathematical inquiry which, if suitably implemented in classroom activities, could move well beyond the limitations of word problems, at least as they are typically utilized.

Kilpatrick (1987) maintained that "problem formulation is an important companion to problem solving. It has received little explicit attention, however, in the mathematics curriculum until a few years ago" (p. 123). In the United States, for example, formally and for the first time "the inclusion of activities in which students generate their own problems, in addition to solving pre-formulated examples, has been strongly recommended by the National Council of Teachers of Mathematics" (English, 1998, p. 83). More recently, the Chinese *National Curriculum Standards on Mathematics* (Ministry of Education of Peoples' Republic of China (NCSM), 2001) has emphasized that students must be able to "pose and understand problems mathematically, apply basic knowledge and skills to solve problems and develop application awareness" (p. 7). Also, a document of the Italian Mathematics Union (UMI-CIIM, 2001) and of the Italian Ministry of Education (2007) recognized the importance of problem posing in the mathematics curriculum.

Given the importance of problem-posing activities in school mathematics, some researchers have started to investigate various aspects of the problem-posing process. Several have examined thinking processes related to problem posing (e.g. Brown & Walter, 1990; Christou, Mousoulides, Pittalis, Pitta-Pantazi, & Sriraman, 2005). In particular, Kontorovich, Koichu, Leikin, and Berman (2012) posited that the problem-posing process is constituted by a knowledge base, heuristics and schemes, group dynamics and interactions, individual considerations of aptness, and task organization. Others have underlined the need to incorporate problem-posing activities into mathematics classrooms and have reported approaches that included it in instruction. They have provided evidence that problem posing has a positive influence on students' ability to solve word problems (e.g. Leung, 1996; Silver, 1994). English (1998) asserted that problem posing improves students' thinking, problem-solving skills, attitudes and confidence in mathematics and mathematical problem solving, and contributes to a broader understanding of mathematical concepts.

Furthermore, problem posing is a form of creative activity that can operate within tasks involving structured "rich situations" in the sense of Freudenthal (1991), using real-life artefacts and human interactions (English, 2009). Creativity, understood as the cognitive ability to create and invent, is linked to the activity of mathematical problem posing. In fact, problem posing is a form of mathematical creation: the creation of mathematical problems in a specific context. In particular, Silver and other authors (Cai & Hwang, 2002; Kontorovich, Koichu, Leikin, & Berman, 2011; Silver, 1994; Silver & Cai, 2005; Yuan & Sriraman, 2010) have linked problem-posing skills with creativity, citing flexibility, fluency, and originality as creativity categories. Moreover, some authors have suggested that students' considerations of whether or not the created problems are appropriate could serve as another useful indicator of their creativity (Kontorovich et al., 2011, 2012; Mednick, 1962).

However, the nature of this relationship still remains unclear. For this reason, the exploratory study presented here begins to investigate the relationship between *problem-posing* activities (supported by problem-solving activities) and *creativity*. Also, the study provides a method for analyzing the products of problem posing that teachers could use in the classroom to identify and assess both the activity of problem posing itself and the students' creativity in mathematics.

Problem Posing

Students are usually asked to solve mathematical problems at school that have been presented by teachers or textbooks (Silver, 1994). Therefore, students only have the task of solving problems, while the teachers have to create them.

But, what is a problem? In discussing the nature of problems, Starko stated that "problems come in various shapes, sizes, and forms, some with more potential than others. A 'problem' is not necessarily difficult; it may be a shift in perspective or a perceived opportunity" (Starko, 2010, pp. 30–31). In his studies about problems and creative thinking, Getzels (1979) distinguished between three illustrative types of problems or problem situations: presented problem situations, discovered problem situations, and created problem situations. In the first type of problems, there are three components—a formulation, a method of solution, and a solution known to others if not yet to the problem solver. Most classroom problems are of this type. Problems of the second type "may or may not have a known formulation, known method of solution, or known solution" (Getzels, 1979, p. 169). In the last type of problems, there is no presented problem and someone must invent or create it. As explained by Starko (2010), "Type 1 problems primarily involve memory and retrieval processes. Type 2 problems demand analysis and reasoning. Only Type 3 problems, in which the problem itself becomes a goal, necessitate problem finding" (p. 31). And problem finding is the first step of the problem-posing process.

In mathematics education, after over a decade of studies which have focused on problem solving, researchers have slowly begun to realize that "developing the ability to *pose* mathematics problems is at least as important, educationally, as developing the ability to *solve* them" (Stoyanova & Ellerton, 1996). Problem posing, in fact, is of central importance in the discipline of mathematics and in the nature of mathematical thinking, and it is an important companion to problem solving (Kilpatrick, 1987). Kilpatrick believed that

> Problem formulating should be viewed not only as a *goal* of instruction but also as a *means* of instruction. The experience of discovering and creating one's own mathematics problems ought to be part of every student's education. Instead, it is an experience few students have today—perhaps only if they are candidates for advanced degrees in mathematics. (p. 123)

In recent years, in recommendations for the reform of school mathematics around the world, the results of many studies have supported the central role of problem posing. For example, *The Principles and Standards for School Mathematics* in the United States of America (National Council of Teachers of Mathematics, 2000) called for students to "formulate interesting problems based on a wide variety of situations, both within and outside mathematics" (p. 258) and recommended that students should make and investigate mathematical conjectures and learn how to generalize and extend problems by posing follow-up questions. In *The Interpretation of Mathematics Curriculum* (Mathematics Curriculum Development Group of Basic Education of Education Department, 2002) "it is pointed out that students' abilities in problem solving and problem posing should be emphasized and students should learn to find problems and pose problems in and out of the context of mathematics" (Yuan & Sriraman, 2010, p. 6).

Problem posing has been defined by researchers from different perspectives (Silver & Cai, 1996). The term problem posing has been used to refer both to the generation of new problems and to the reformulation of given problems (e.g. Dunker, 1945; Silver, 1994). Silver (1994) linked problem solving and problem posing and argued that problem posing could occur:

- *Prior* to problem solving when problems were being generated from a particular stimulus (such as a story, a picture, a diagram, a representation, etc.);
- *During* problem solving when an individual intentionally changes the problem's goals and conditions (such as in the cases of using the strategy of "making it simpler"); and
- *After* solving a problem when experiences from the problem-solving context are modified or applied to new situations.

Stoyanova and Ellerton (1996) identified three categories of problem-posing situations: free, semi-structured, or structured. In free situations, students pose problems without restrictions: students are simply asked to make up mathematics problems from a given situation. Semi-structured problem-posing situations refer to ones in which students are "given an open situation and are invited to explore the structure of that situation and to complete it by using knowledge, skills, concepts and relationships from their previous mathematical experiences" (p. 520). Finally, structured problem-posing situations refer to situations where students pose problems by reformulating already solved problems or by varying the conditions or the questions of given problems.

In this chapter, we shall consider mathematical problem posing as suggested by Stoyanova and Ellerton (1996): "the process by which, on the basis of mathematical experience, students construct personal interpretations of concrete situations and formulate them as meaningful mathematical problems" (p. 519). In the study presented here, this process is supported by the use of suitable social or cultural artefacts that, according to this framework, can become a meaningful source for problem-posing activities of the semi-structured type (Bonotto, 2013). A cultural artefact can support a semi-structured problem-posing situation, because it can become a concrete source for types of tasks and activities where the students are invited to explore the mathematical structure, find a problem, and by using knowledge, skills, concepts, and relationships from their previous mathematical experiences, create one or more new mathematical problems.

Problem posing, therefore, becomes an opportunity for interpretation and critical analysis of reality since: (a) the students have to discern significant data from immaterial data; (b) they must discover the relations between the data; (c) they must decide whether the information in their possession is sufficient to solve the problem; and (d) they have to investigate if the numerical data involved is numerically and/or contextually coherent. These activities, quite absent from today's Italian school context, are typical also of mathematical modelling processes and can help students to prepare to cope with natural situations they will have to face out of school (Bonotto, 2009).

A semi-structured situation, as well as an unstructured situation, invites the use of creative thinking inasmuch as it stimulates student sensitivity to a problem—to ideation (the creation of new ideas), originality, the ability to synthesize, and to reorganize the information in a new way, analytical skills, and evaluating ability.

The advancement of mathematics requires creative imagination, which is the result of raising new questions, new possibilities, and viewing old questions from new angles (Ellerton & Clarkson, 1996). Silver (1997) argued that inquiry-oriented mathematics instruction, which includes problem-solving and problem-posing tasks and activities, could assist students to develop more creative approaches to mathematics. It is claimed that through the use of such tasks teachers can increase their students' capacity with respect to the core dimensions of creativity, namely, fluency, flexibility, and originality. We believe in the didactic potential of using suitable artefacts, combined with particular teaching methods, as a source for types of tasks and activities that encourage problem posing and creativity processes—see Bonotto (2005, 2009) for a discussion on the use of artefacts in classroom activities.

Creativity

In the nineteenth and early twentieth centuries, creativity was identified with the genius of a few people of remarkable intelligence who revolutionized their fields. Therefore, early studies on creativity examined the characteristics of these outstanding personalities, such as Mozart and Einstein. These studies were based on three ideas: first, that creativity belonged to exceptional personalities; second, that a creative person was a break with the spirit of the time in which that person lived; and third, that sudden insight was involved. However, it is interesting to note the position of Poincaré (1908), later recaptured by Hadamard (1945), that inventing, at least in mathematics, meant to discern and to choose.

Afterwards, the psychological study of thought addressed aspects of intelligence, and in particular logical mathematical skills. As a result, creativity began to be identified with high intelligence. Beginning in 1950, Guilford dealt with creativity and noted that IQ and creativity could not be overlapped. He, therefore, hypothesized that a person could be creative without exceptional intelligence and vice versa. Then, Guilford hypothesized that there was a different way of thinking, subsequently called divergent thinking, characterized by the ability to imagine a range of solutions to a given problem. Guilford's ideas inspired subsequent research on creativity and the development of tests to measure people's creativity such as the Torrance Tests of Creative Thinking. Thus, creativity began to be recognized as an asset, even if present in different degrees and shapes, for each person. Today, there are many definitions and theories of creativity, each of which considers some aspect of creative thinking.

One of the main lines of research on creativity concerns the distinction between two types of thought proposed by Guilford (1950): productive (divergent thinking) and reproductive (convergent) thinking. Included in the divergent thinking category were the factors of fluency, flexibility, originality, and elaboration. Guilford saw creative thinking as clearly involving what he categorized as divergent production (Yuan & Sriraman, 2010) which he broke down into nine skills: sensitivity to problems, ideational fluency, flexibility of set, originality, the ability to synthesize, analytical skills, the ability to reorganize, span of ideational structure, and evaluation ability. All these skills influence each other and represent the related aspects of a dynamic and unified cognitive system. In particular, sensitivity to problems, flexibility of approach, ability to synthesize, application of analytical skills, and the ability to reorganize are all aspects that characterize mathematical thinking.

It is hardly surprising, therefore, that the main models used to describe the creative process emphasize the importance of sensitivity to the problems (problem finding) and their resolution (problem solving). Problem finding, in particular, may be associated with mathematical problem posing. Problem-posing and problem-solving activities are therefore used by several authors to promote and evaluate creativity (Leung, 1997; Silver, 1997; Silver & Cai, 2005; Siswono, 2010; Sriraman, 2009; Torrance, 1966). For example, in a recent study, Kontorovich et al. (2011) used fluency, flexibility, and originality as indicators of creativity in students' problem posing.

We must not forget that there is a distinction between mathematical creativity at the professional level and at the school level: it is certainly feasible to expect students to offer new insights into a mathematical problem rather than expecting works of extraordinary creativity and innovation (Nadjafikhah, Yaftian, & Bakhashalizadeh, 2012; Sriraman, 2005).

We believe that the creative process in school mathematics may be encouraged by the presence of semi-structured situations (defined by Stoyanova & Ellerton, 1996). These situations are similar to those encountered by professional mathematicians who are frequently engaged in problems which are full of vagueness and uncertainty; the use of appropriate cultural artefacts can help realize these situations.

Through the use of artefacts, children can be encouraged to recognize a great variety of situations as mathematical situations, or more precisely "mathematizable" situations, by asking them: (a) to select other artefacts from their everyday life; (b) to identify the mathematical facts associated with them; (c) to look for analogies and differences (e.g., different number representations); or (d) to generate problems (e.g., discover relationships between quantities) (Bonotto, 2009). These aspects are related to another line of research on creativity that highlights the importance of the process of association of ideas (e.g. Mednick, 1962; Starko, 2010).

In the study presented here, we focused on the analysis of the problem posing and creativity processes. These two processes were studied using a semi-structured situation. We also began to reflect on the relationship between mathematical knowledge and these two processes. Figure 5.1 presents possible relationships between the variables involved in the problem posing and creativity processes at the school level.

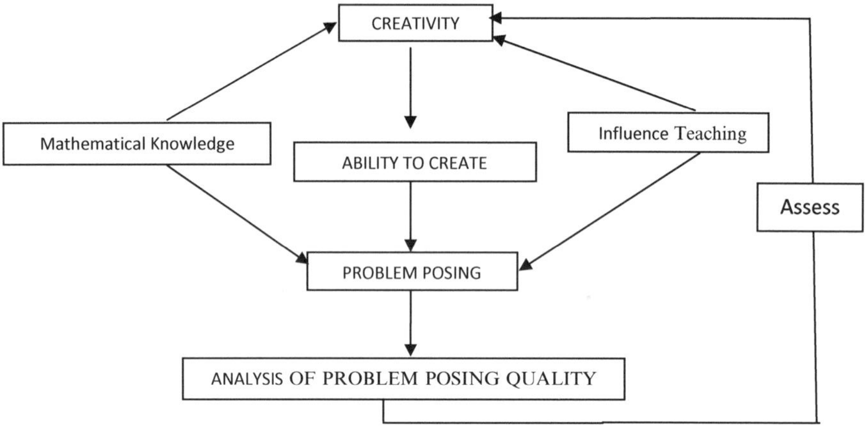

Figure 5.1. Possible relationships between problem posing and creativity.

The Study

The overall aim of this exploratory study, briefly described also in Bonotto (2013), was to examine the relationship between *problem-posing* activities (supported by problem-solving activity) and *creativity*, when these processes are implemented in situations involving the use of real-life artefacts. In particular, the study sought to continue to investigate:

- The role of suitable artefacts as sources of stimulation for the problem-posing process in semi-structured situations; and
- Primary school students' capacity to create and deal with mathematical problems (including open-ended problems).

Furthermore, the study sought to begin to investigate:

- The potential that these problem-posing activities have for identifying creative thinking in mathematics; and
- A method for analyzing the products of problem posing that the teacher could use in the classroom to identify and assess both the activity of problem posing itself and the creativity of the students.

Participants

This exploratory study involved four fifth-grade classes (10–11 years old) from two primary schools in northern Italy. The study was carried out by the second author of this paper in the presence of the official logic-mathematics teacher.

The first primary school was located in an urban area situated within a few miles of the centre of a city. This school participated in the study with two classes, one consisting of 14 students and the other of 16. The children were already familiar with activities using cultural artefacts, group work, and discussions.

The second primary school was located in a mountainous area. This school also participated in the study with two classes, one consisting of 20 students and the other of 21. The children were not already familiar with these types of activities, even though the teacher had once proposed a problem-posing activity where the situation was a drawing of the prices of different products in a shop.

The average marks in mathematics of the students from the two schools were classified into three categories: high, medium, and low. On the basis of this classification the two schools were not uniform; in particular, the second school had more students with averages in the medium-high range in mathematics. These data were obtained in order to make observations concerning the influence that mathematical knowledge can have on the creativity process.

Materials

To perform the problem-posing activity a real-life artefact was used as the initial situation. We wanted to create a semi-structured situation that was as rich and contextualized as possible for the students in order to permit them to use their extra-scholastic experience in the creation and resolution of problems. Thus, the artefact was a page of a brochure containing the special rates for groups visiting the Italian amusement park “Mirabilandia” (Figure 5.2) and shows the menu and applicable discounts, the cost for access to the beach, etc. This artefact was chosen with the belief that all students were already familiar with an amusement park because they had been to one. The page was full of information, including prices (some expressed as decimals), percentages, and constraints on eligibility for the various offers (Figure 5.2 shows part of the artefact). Finally, we gave pupils the individual rates.

Procedure

Assuming, for the reasons discussed previously, that problem posing can be an activity that highlights creativity, we structured a problem-posing activity supported by a problem-solving activity that could be evaluated with regard to creative thinking (in terms of fluency, flexibility, and originality). The experiment was structured

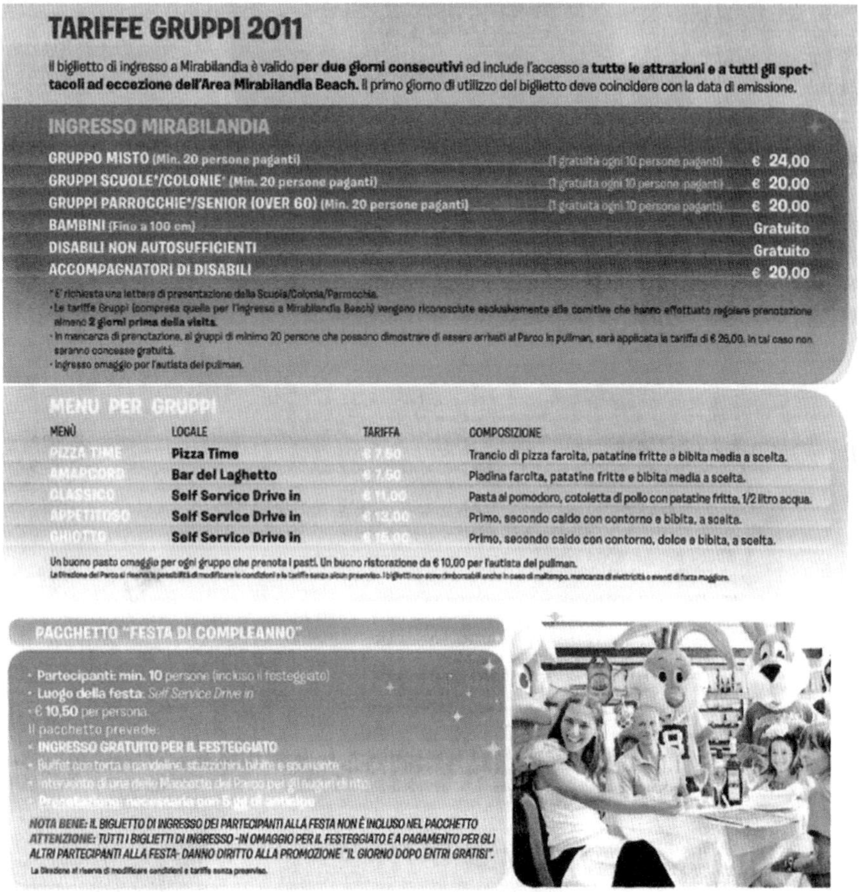

Figure 5.2. Artefact for semi-structured situation.

in three phases: (a) the presentation of the artefact used; (b) a problem-posing activity; and (c) a problem-solving activity. The activities took place on three different days, a few days apart. The students worked individually for part 2. For part 3, they were, at first, divided into groups of two or three students and then participated in a collective discussion. Students could use the artefact and its summary during all three activities.

The first phase, lasting about 2 hours, consisted of the analysis and synthesis of the artefact. This phase was preparatory to the problem-posing activity. After presenting the whole brochure, a copy of one of the pages was given to each student and then he/she was invited to write down everything they could see on that page. Following that, there was a discussion on the observations: the aim was to verify their understanding of the artefact and to create a summary of the mathematical concepts involved.

The second phase, lasting about an hour, consisted of an individual problem-posing activity in which the children had to create the greatest number of solvable mathematics problems (in a maximum time of 45–50 minutes), preferably of various degrees of difficulty, to bring to their partners in the other classroom. The children were not informed of the time limit in order to avoid generating anxiety. Rather, they were told that they would have plenty of time to do this activity and that problems would be collected when the majority of the students had finished. To allow for the pupils' self-assessment, they were given a sheet of paper for their calculations and solutions to the problems they had created.

Then, four problems for the next problem-solving activity were selected from among all the problems that had been created. To facilitate problem solving, problems that would have favoured a discussion among the students were chosen:

- One Multi-step problem, for example "Francesca decided to go to Mirabilandia. There are 15 people including 7 adults and 8 children. Each child spends 26 euro and each adult spend 31 euro. Then, they decide to go to the Mirabilandia beach and they pay an additional 7 euro. What is the total spent?"
- Two Open-ended problems (problems with insufficient information), for example "Luca and his 10 friends go to Mirabilandia to celebrate Luca's birthday. How much did they spend?"
- One Incorrect data problem, for example "A group of 20 people, children and adults, decide to go to Mirabilandia. In total, they spend 480 euro. How much will each person pay to enter?" (The total of 480 was included in the problem by the student. It is incorrect because all of the conditions of the artefact were not taken into consideration—in fact for every 10 entries, 1 entry was free).

For the classes at the second school, the selection criteria of the problems were the same for the first three problems; in the fourth problem, the topic of percentage was included because the students had not yet studied percentage problems. The modified criterion was used since we wanted to study the way in which "anticipatory learning" (Freudenthal, 1991) can be enhanced by the use of an artefact.

The third phase, lasting about 2 hours, consisted of a problem-solving activity by students and ended with a collective discussion. The students were asked to solve problems, to write the procedure that they had used, and to write considerations on the problem itself. Different solutions and ideas that emerged during the discussion were compared and, at the end of the activity, a collective text summarizing the students' conclusions was written.

Methodology and Data Analysis

Data from the teaching experiment included the students' written work, field notes from classroom observations, and audio recordings of the collective discussions.

All of the problems created by the students were analyzed with respect to their quantity and quality. To analyze the types of created problems, the methodology proposed by Leung and Silver (1997) was followed; for the analysis of the text of

Table 5.1
Examples of Each Category of Problem

Category	Example
Non-mathematical problem	Find the name of the following problem.
Implausible mathematical problem	A group of 20 children go to Mirabilandia with the school and each child pays 20 euro. The school children are 130. How much does the school pay to go to Mirabilandia?
Plausible mathematical problem with insufficient data	Giovanni goes to Mirabilandia with his dad. How much does Giovanni spend? How much does his dad spend?
Plausible mathematical problem with sufficient data	A group of 15 people enter in Pizza Time pub and every person pays 7.50 euro. What is total spent?

the problems we referred to the research of Silver and Cai (1996) and Yuan and Sriraman (2010).

Table 5.1 illustrates the first qualitative analysis of the created problems with an example from each category of problem.

In this work, non-mathematical problems are texts which cannot be considered problems or they are not solved through mathematical tools. The mathematical problems were analyzed and divided into implausible mathematical problems and plausible (can apparently be solved, with no discrepant information, and respects the conditions in the artefact) mathematical problems. The plausible mathematical problems were divided further into plausible mathematical problems with insufficient data and plausible mathematical problems with sufficient data.

Plausible mathematical problems with sufficient data were analyzed with respect to their complexity and were assessed by two aspects: the complexity of the text of the problem and the complexity of the solution. With regard to the complexity of the text of the problems, plausible mathematical problems with sufficient data were divided into problems with a question and problems with more than one question. The latter were divided into concatenated questions and non-concatenated questions. With regard to the complexity of the solution, these mathematical problems were divided into multi-step, one-step, and zero-step problems.

Furthermore, only the plausible mathematical problems with sufficient data were re-analyzed to evaluate their creativity. The problems developed by children were grouped taking into account the number and type of details extrapolated from the artefact, the type of questions posed, and the added data included by the students.

To evaluate their creativity in mathematics, three categories were taken into consideration—fluency, flexibility, and originality—as proposed by Guilford (1950) to define creativity, and as used in the tests by Torrance and in other studies such as that by Kontorovich et al. (2011).

When considering the fluency of a problem, the total number of problems created by the pupils of each school in a given time period, as well as the average number of problems created by each student, were taken into account. By contrast, flexibility refers to the number of different and pertinent *ideas* created in a given time period. In order to evaluate the flexibility of the students, the mathematical problems were categorized considering both the number of details present in the brochure (e.g., entrance fee, price of lunch, etc.) which were incorporated into the text of the problem posed, and the additional data introduced by the students (e.g., calculating the change due after a payment). Once the problems had been categorized in the above ways, the various types of problems that occurred in each class were counted.

The originality of the mathematical problems created by the students took into consideration the uniqueness of the problem compared to the others posed in each school. In order to evaluate the originality of a problem, it was considered original if it was posed by less than 10% of the pupils in each school (Yuan & Sriraman, 2010).

Therefore, two different analyses were conducted: one for problem posing and one for creativity. With regard to problem posing, a qualitative analysis was carried out to evaluate students' performance on problem posing and to analyze the structure of the texts of the problems and their solutions. Both quantitative and qualitative analyses were undertaken to evaluate student creativity. The number of problems created per student was counted, and then the texts of the problems associated with each of the problems created by students were analyzed.

Some Results and Comments

A total of 63 students in both schools participated in the problem-posing phase and they created a total of 189 problems. Students from the first school created 58 problems (57 were mathematical problems), while students from the second school created 131 (all mathematical problems).

More than half of the created problems—64% of the problems created by pupils at the first school and about 60% of those created by the pupils at the second school—were solvable mathematical problems (plausible mathematical problems with sufficient data). Table 5.2 shows the main quantitative results for both schools.

After analyzing these solvable mathematical problems we found that:

- 81% of the problems from the first school and 75% of the problems from the second school were multi-step problems; and
- 78% of the problems from the first school and 73% of the problems from the second school were problems with a question.

Table 5.2
Percentage of Problems Created in Each Category

Category	First school (%)	Second school (%)
Non-mathematical problem	1.7	0
Irrelevant mathematical problem[a]	6.9	0
Implausible mathematical problem	19.0	29.0
Plausible mathematical problem with insufficient data	8.6	10.7
Plausible mathematical problem with sufficient data	63.8	60.3

[a] *Note*: Irrelevant mathematical problems did not use any of the information provided in the artefact—the problems, therefore, did not relate to the artefact. The students who posed these problems, in fact, did not understand the presentation of the task.

For problems which involved more than one question, in the first school 62% had concatenated questions, and in the second school, about 43%.

We concluded, from the analysis of the above data, that the first school had better problem-posing performance because the children from first school created fewer implausible problems, more multistep problems, and more problems with concatenated questions than the children from the second school.

Most of the problems created by the pupils were similar to standard problems used in schools (for example: "A father and his son go to Mirabilandia. The adult pays 31 euro, and the child 26. How much do they pay?" And, "How much do they receive in change if they pay with two bills, one of 50 and one of 10 euro?"), although there were some cases (17%, corresponding to 32 out of 189 problems) of creative and open-ended problems.

An example of a creative problem is:

> A group of 50 students go to Mirabilandia. Everyone takes a Ghiotto meal. Then, 50% of this group decide to go to Mirabilandia beach while the other 50% remains in the park area and goes on the rides. The day after, 24 of these students return to the amusement park and 50% of them order a Ghiotto menu while the other 50% takes the Classico menu. 25% of this group wants to return to Mirabilandia. How much does the group pay to go to Mirabilandia beach? And for the food? And for the entrance? And in total?

The text of this problem did not include certain information (for example the entrance fee) because the students who created the problem knew that the other class had the artefact. This consideration also applied to many other problems created by the students.

As far as creativity is concerned, the second school was more successful in all three categories used to assess performance (fluency, flexibility, and originality). With regard to fluency, each student in the first school created two problems on average, while each pupil of the second school created three problems on average. With regard to flexibility, the problems created by the classes of the first school were divided into 11 categories, those of the second school into 16 categories. In evaluating originality, it was found that three original problems were created in the first school and ten original problems in the second school. Original problems included inverse problems and problems involving almost all the information from the artefact. Table 5.3 presents a summary of the creativity results.

In terms of the creativity indicators listed in Table 5.3 (fluency, flexibility, and originality), the students of the second school demonstrated better performance on the parameters used to evaluate fluency and flexibility. It should, however, be noted that the second school had more students with averages in the medium–high range in mathematics, as Table 5.4 shows. The results may suggest that there is a correlation between creativity and performance in mathematics; this aspect deserves to be investigated in a subsequent study.

The results obtained were consistent with those from another study we conducted (see e.g., Bonotto, 2005, 2009) and demonstrate that an extensive use of suitable cultural artefacts, with their associated mathematics, can play a fundamental role in

Table 5.3
Analysis of Problems for Creativity

Category	Method of analysis	Results
Fluency	The total number of problems created by the pupils of each school and the average of the problems created by each student is taken into account	57 mathematical problems were created (two problems per student, on average) in the first school, while 131 problems were created (three problems per student on average) in the second school
Flexibility	The plausible math problems with sufficient data were categorized according to the number and type of information of the artefact present in the text, the type of questions, and the addition of information from the student. Then, the number of categories produced by each school was counted	The problems created by the four classes were divided into 18 total categories, 11 for the first school, and 16 for the second school
Originality	The rarity of the answer was considered: an answer was considered original if it came from less than 10% of pupils in that school	There were three original problems in the first school and ten original problems in the second school. Original problems included inverse problems and involved almost all of the information in the artefact

Table 5.4
Academic Performance in Mathematics of Students in the Two Schools

	Pupils' academic performance in mathematics		
Category	Low (%)	Medium (%)	High (%)
Students of the first school	29	42	29
Students of the second school	12	51	37

bringing students' out-of-school reasoning and experiences into play by creating a new dialectic between school mathematics and the real world. As a paradigmatic example, we have included below some segments from the class discussion concerning the following problem:

> Giovanni decides to celebrate his birthday at Mirabilandia. There are 10 people in total, 6 adults and 4 children. Every 3 children pay 26 euro and each adult pays 31 euro. Giovanni is the birthday boy and he doesn't pay. Also, they decide to make use of the refreshments and they pay 10.50 euro. What is the total spent?

During the discussion, students justified their reasoning using everyday-life experiences and making estimates, as illustrated in this dialogue:

Student 1: This problem isn't written well. The 10.50 euro should be what every person pays for the refreshments, but I realize that the 10.50 euro is the total, because what is written is: Also, they decide to make use of the refreshments and they pay 10.50 euro.

Student 2: But no, because what should be written—and finally all pay 10.50 euro.

Student 3: In the brochure there is written—in the *pacchetto festa* (party package) every person pays 10.50 euro, and not "in total." If you read the brochure carefully, you can understand that the price is per person.

Student 1: But, if you don't have the brochure, how can you solve the problem?

Student 3: It's impossible that ten people pay only 10.50 euro for all the refreshments! It's more likely that the refreshments are more expensive.

[…]

Student 1: It's impossible that all the refreshments cost 105 euro ….

Student 4: If you do the count, 10 people: ten, twenty, thirty, forty [she shows the count with her fingers] fifty, sixty, eighty, ninety, one hundred!

Student 1: For me it's a bit too much.

Student 3: Too much … if you see the table with all the sandwiches, drinks … even the tablecloth has a cost! If there are all the towels, the dishes, the drinks, all these things, all the services cost!

Student 1: But, how can you understand all these things?

Student 5: I think that Martina's considerations about 105 euro are possible. In the brochure there are a lot of things that the children can eat!

Student 3: Then, I think that a drink costs about 3 euro. A drink is enough for two people, because you drink a lot. Then, we image that there are five bottles, therefore five bottles cost already 15 euro. Then there are other things, and each person takes different things; so, it's impossible that all costs 10.50 euro!

[…]

Student 6: Then, here the children are in Mirabilandia; it isn't just any place!

With regard to the problem-solving phase, this appears to be important and helpful in allowing a better understanding of the initial situation, fostering quality control of the problems created by the students themselves, and giving them a starting point for analyzing the structure of problems. We have included below some parts of the class discussion concerning the "incorrect-data problem," reported also in Bonotto (2013). The problem presented incorrect data (480 euro) and the students, during the problem-solving activity, found two different solutions discussed during the collective discussion:

Student 1: We didn't divide by 20. We divided by 18 because the Mirabilandia brochure stated that every 10 entries, 1 entry was free. Therefore, if there were 20 people together, there would be two free entries, and so we divided by 18.

Student 2: I believe that reasoning is wrong because the text of the problem says that they went to Mirabilandia and in total they spent 480 euro, but it doesn't specify if they paid only the entrance or if they went to other places, so the discount is only on the entrance fee and not on the other things.

[…]

Student 3: I think that both solutions are right

[…]

Student 4: One of them must be wrong, because one takes off two people, while the other does not!

[...]
Student 3: Probably, the writer of the problem didn't consider that every 10 entries, 1 entry was free.
Student 2: Practically, the student of the other classroom wrote this problem without realizing that the data was wrong, so we solved it incorrectly.

By solving problems created by their peers, the students became able to analyze them in a more detached and critical way. For example, students reflected on what information was really important and what was not and discovered that numerical information is not always the most important information contained in the text of a problem, as the following problem illustrates:

It's Giulia's birthday and she invited 9 people to her birthday party, but she didn't benefit from the *pacchetto festa* (party package), how much did she pay for the entrance?

During the discussion, almost all of the students did not read the words of the problem question carefully, because a lot of students calculated the total and not only Giulia's entrance cost. In fact, the total number of people (9) in the problem was superfluous.

Discussion

The specific artefact utilized in this study provided a particularly attractive context inasmuch as it referred to an amusement park known to the children and was desirable because it furnished conditions allowing the students to formulate hypotheses regarding the various possibilities offered. Students were therefore able to create diverse problems with various degrees of difficulty. This activity made it possible to assess problem posing itself and creative thinking in mathematics: children created both original and open-ended problems (in addition to the classic problems), demonstrating that the activity of problem posing can be an environment that fosters creative thinking.

The cultural artefact reflects the complexity of reality and so it offers a rich setting for raising issues, asking questions and formulating hypotheses. It is interesting to reflect on the fact that there were good results for students accustomed to using cultural artefacts (the classes from the first school) as well as those who were using them for the first time (the classes from the second school). In fact, pupils from both schools were able to use the artefact as a context to create problems. This indicates that an artefact provides a useful context for the creation of problems and the mathematization of reality as a result of its accessibility to all students (Bonotto, 2013).

In order to have better performance on the problem-posing task in terms of the greater number of plausible problems, with more complex texts and concatenated questions, it proved to be important to structure, organize, and summarize the information presented in the brochure. In fact, students who had previously performed this type of analysis outperformed the others in the problem-posing activity. With regard to this aspect, students from the first school, who were already familiar with this type of activity, produced fewer implausible problems and therefore appear to have constructed a better analysis and synthesis of the artefact. Instead, about one third of the problems produced by the second school students were implausible problems.

Overall the students involved in the study produced some original problems (13 problems) and open problems (19 open problems). This highlights the fact that pupils were able to deal with open-ended tasks. The problem-solving phase combined with group discussions allowed students to reflect on different types of problems and explore new possibilities (e.g., suggesting that mathematical problems do not always require a numerical answer or a unique solution, and that there are problems which are not solvable). Not only does this confirm the potential of students to create problems, but it also demonstrates the importance of educational action to support students in these kinds of processes.

In fact, almost all of the problems created by the pupils of both schools were classified as mathematically relevant (98% in the case of the first school, 100% in the case of the second school). Of these, more than half of the problems created by the pupils were solvable (about 64% of the problems created by the pupils of the first school and about 60% of those created by the pupils of the second school). This indicates that, at the end of primary school, pupils are not only aware of what mathematical problems are, but they are also able to create appropriate problems.

Furthermore, the results of the discussion in the classroom suggest that asking students to analyze the problems they created facilitated their critical thinking. In this context, students seemed to feel freer to discuss the validity of a given problem, to consider different assumptions, and to decide whether the problem had been solved or not (Bonotto, 2013).

Teachers can assess problem-posing activities and creative thinking several times during the year by applying the proposed method:

- Students are first engaged in problem-posing activities stimulated through a cultural artefact, and this is supported by a problem-solving activity and collective discussions.
- From all of the problems created by the students, plausible mathematical problems with sufficient data are selected for analysis. These can be initially analyzed, from the point of view of problem posing, with respect to complexity of the text and their solutions.
- Then, these same problems can be analyzed from the point of view of creativity with respect to fluency (counting the number of problems created by each student); flexibility (considering both the number of details present in the artefact which were incorporated into the text of the problem posed, and any additional data introduced by the students); and originality (uniqueness of the problem compared to problems created by other pupils).

If these activities are periodically offered to the class, the teacher can then assess changes and improvements over time.

Conclusion and Open Problems

The exploratory study presented here investigated the impact of *problem-posing* activities (supported by problem-solving activities) when these were implemented in meaningful situations involving the use of suitable artefacts. These situations fall under those defined by Stoyanova and Ellerton (1996) as *semi-structured situations*.

Furthermore, this study has allowed us to investigate the potential that problem-posing activities have for identifying critical and creative thinking in mathematics. A method for analyzing the products of problem posing and for assessing both the activity of problem posing itself and the creativity of the students was provided. Furthermore, the study investigated possible relationships between students' knowledge of mathematics, their problem-posing ability, and their creativity.

Two questions arose from the results obtained that require additional research in the future:

1. Does good academic performance in mathematics favour better performance in the three creativity categories (fluency, flexibility, and originality)?
2. How much do teaching practices and classroom experiences influence the creative processes?

Finally, we would like to look more deeply at how children respond over the long term to programs designed to develop their problem-posing skills in the form described here. In agreement with other researchers, we believe that the presence of problem-posing activities should not emanate from a specific part of the curriculum but should permeate the entire curriculum.

References

Bonotto, C. (2005). How informal out-of-school mathematics can help students make sense of formal in-school mathematics: The case of multiplying by decimal numbers. *Mathematical Thinking and Learning, 7*(4), 313–344.

Bonotto, C. (2009). Working towards teaching realistic mathematical modelling and problem posing in Italian classrooms. In L. Verschaffel, B. Greer, W. Van Dooren, & S. Mukhopadhyay (Eds.), *Words and worlds: Modelling verbal descriptions of situations* (pp. 297–313). Rotterdam, The Netherlands: Sense Publishers.

Bonotto, C. (2013). Artifacts as sources for problem-posing activities. *Educational Studies in Mathematics, 83*(1), 37–55. doi:10.1007/s10649-012-9441-7.

Brown, S. I., & Walter, M. I. (1990). *The art of problem posing*. Hillsdale, NJ: Erlbaum.

Cai, J., & Hwang, S. (2002). Generalized and generative thinking in U.S. and Chinese students' mathematical problem solving and problem posing. *The Journal of Mathematical Behavior, 21*(4), 401–421.

Chen, L., Van Dooren, W., Chen, Q., & Verschaffel, L. (2011). An investigation on Chinese teachers' realistic problem posing and problem solving ability and beliefs. *International Journal of Science and Mathematics Education, 9*, 919–948.

Christou, C., Mousoulides, N., Pittalis, M., Pitta-Pantazi, D., & Sriraman, B. (2005). An empirical taxonomy of problem posing processes. *Zentralblatt für Didaktik der Mathematik, 37*(3), 149–158.

Dunker, K. (1945). On problem-solving. *Psychological Monographs, 58*(5), 270.
Ellerton, N. F., & Clarkson, P. C. (1996). Language factors in mathematics teaching and learning. In A. J. Bishop, K. Clements, C. Keitel, J. Kilpatrick, & C. Laborde (Eds.), *International handbook of mathematics education* (pp. 987–1033). Dordrecht, The Netherlands: Kluwer.
English, L. D. (1998). Children's problem posing within formal and informal contexts. *Journal for Research in Mathematics Education, 29*(1), 83–106.
English, L. D. (2009). The changing realities of classroom mathematical problem solving. In L. Verschaffel, B. Greer, W. Van Dooren, & S. Mukhopadhyay (Eds.), *Words and worlds: Modelling verbal descriptions of situations* (pp. 351–362). Rotterdam, The Netherlands: Sense Publishers.
Freudenthal, H. (1991). *Revisiting mathematics education. China lectures*. Dordrecht, The Netherlands: Kluwer.
Getzels, J. W. (1979). Problem finding: A theoretical note. *Cognitive Science, 3*, 167–172.
Guilford, J. P. (1950). Creativity. *The American Psychologist, 5*(9), 444–454.
Hadamard, J. (1945). *The psychology of invention in the mathematical field*. Princeton, NJ: Princeton University Press.
Kilpatrick, J. (1987). Problem formulating: Where do good problems come from? In A. H. Schoenfeld (Ed.), *Cognitive science and mathematics education* (pp. 123–147). Hillsdale, NJ: Erlbaum.
Kontorovich, I., Koichu, B., Leikin, R., & Berman, A. (2011). Indicators of creativity in mathematical problem posing: How indicative are they? In M. Avotina, D. Bonka, H. Meissner, L. Sheffield, & E. Velikova (Eds.), *Proceedings of the 6th International Conference Creativity in Mathematics Education and the Education of Gifted Students* (pp. 120–125). Riga, Latvia: University of Latvia.
Kontorovich, I., Koichu, B., Leikin, R., & Berman, A. (2012). An exploratory framework for handling the complexity of mathematical problem posing in small groups. *The Journal of Mathematical Behavior, 31*(1), 149–161.
Leung, S. S. (1996). Problem posing as assessment: Reflections and reconstructions. *The Mathematics Educator, 1*(2), 159–171.
Leung, S. S. (1997). On the role of creative thinking in problem posing. *Zentralblatt für Didaktik der Mathematik, 97*(3), 81–85.
Leung, S. S., & Silver, E. A. (1997). The role of task format, mathematics knowledge, and creative thinking on the arithmetic problem posing of prospective elementary school teachers. *Mathematics Education Research Journal, 9*(1), 5–24.
Mathematics Curriculum Development Group of Basic Education of Education Department. (2002). *The interpretation of mathematics curriculum (trial version)*. Beijing, China: Beijing Normal University.
Mednick, S. A. (1962). The associative basis of the creative process. *Psychological Review, 69*, 220–232.
Ministry of Education of Italy. (2007). *Indicazioni per il curriculo*. Roma, Italy: Ministero della Pubblica Istruzione.
Ministry of Education of Peoples' Republic of China (NCSM). (2001). *Chinese national curriculum standards on mathematics*. Beijing, China: Beijing Normal University.
Nadjafikhah, M., Yaftian, N., & Bakhashalizadeh, S. (2012). Mathematical creativity: Some definitions and characteristics. *Social and Behavioural Sciences, 31*, 285–291.
National Council of Teachers of Mathematics. (2000). *Principles and standards for school mathematics*. Reston, VA: NCTM.
Palm, T. (2006). Word problems as simulations of real-world situations. A proposed framework. *For the Learning of Mathematics, 26*(1), 42–47.
Poincaré, H. (1908). L'invention mathématique. *Bulletin de l'Institut Général Psychologique, 3*(8), 175–187.
Schoenfeld, A. H. (1991). On mathematics as sense-making: An informal attack on the unfortunate divorce of formal and informal mathematics. In J. F. Voss, D. N. Perkins, & J. W. Segal (Eds.), *Informal reasoning and education* (pp. 311–343). Hillsdale, NJ: Erlbaum.

Silver, E. A. (1994). On mathematical problem posing. *For the Learning of Mathematics, 14*(1), 19–28.

Silver, E. A. (1997). Fostering creativity through instruction rich in mathematical problem solving and problem posing. *Zentralblatt für Didaktik der Mathematik, 29*(3), 75–80.

Silver, E. A., & Cai, J. (1996). An analysis of arithmetic problem posing by middle school students. *Journal for Research in Mathematics Education, 27*(5), 521–539.

Silver, E. A., & Cai, J. (2005). Assessing students' mathematical problem posing. *Teaching Children Mathematics, 12*(3), 129–135.

Siswono, T. Y. E. (2010). Leveling students' creative thinking in solving and posing mathematical problem. *Indonesian Mathematical Society Journal on Mathematics Education, 1*(1), 20–41.

Sriraman, B. (2005). Are giftedness and creativity synonyms in mathematics? An analysis of constructs within the professional and school realms. *The Journal of Secondary Gifted Education, 17*, 20–36.

Sriraman, B. (2009). The characteristics of mathematical creativity. *Zentralblatt für Didaktik der Mathematik, 41*(1–2), 13–27.

Stanic, G. M. A., & Kilpatrick, J. (1988). Historical perspectives on problem solving in the mathematics curriculum. In R. I. Charles & E. A. Silver (Eds.), *Research agenda for mathematics education: The teaching and assessing of mathematical problem solving* (Vol. 3, pp. 1–22). Hillsdale, NJ: Erlbaum.

Starko, A. J. (2010). *Creativity in the classroom. Schools of curious delight*. New York, NY: Routledge.

Stoyanova, E., & Ellerton, N. F. (1996). A framework for research into students' problem posing in school mathematics. In P. C. Clarkson (Ed.), *Technology in mathematics education* (pp. 518–525). Melbourne, Australia: Mathematics Education Research Group of Australasia.

Torrance, E. P. (1966). *Torrance tests of creative thinking*. Princeton, NJ: Personnel Press.

UMI-CIIM. (2001). *Matematica 2001. Materiali per il XXVII Convegno Nazionale sull'Insegnamento della matematica*. Ischia, Italy: UMI.

Verschaffel, L., Greer, B., & De Corte, E. (2000). *Making sense of word problems*. Lisse, The Netherlands: Swets & Zeitlinger.

Yuan, X., & Sriraman, B. (2010). An exploratory study of relationships between students' creativity and mathematical problem posing abilities. In B. Sriraman & K. Lee (Eds.), *The elements of creativity and giftedness in mathematics* (pp. 5–28). Rotterdam, The Netherlands: Sense Publishers.

Chapter 6
Beyond Routine: Fostering Creativity in Mathematics Classrooms

Vincent J. Matsko and Jerald Thomas

Abstract Mathematics is a creative endeavor. However, students typically think of mathematics as a body of knowledge used to solve well-defined problems in a unique way. Yet virtually all problems encountered in "real life" involve ambiguity and may not be solvable by a single approach. Expert problem solvers are original, creative thinkers who are able to devise novel approaches to solving ill-structured or ambiguously posed problems. Recently, research has been conducted on having students create and solve their own problems as assignments in mathematics classes in an attempt to give them an experience of interacting with mathematics problems beyond the routine and mechanical. Results suggest that such experiences could also be valuable in other disciplines at various levels, and that these experiences encourage students to be creative. In addition to a theoretical discussion of creativity, detailed examples of student work are presented, as well as a historical background of the assignment and practical implications for teachers interested in using the writing of original problems in their classrooms.

Contents

V.J. Matsko (✉)
University of San Francisco, San Francisco, CA, USA
e-mail: vince.matsko@gmail.com

J. Thomas
Doctor of Education (Ed.D.) Program, Graduate School of Education, Aurora University, Aurora, IL, USA
e-mail: jthomas@aurora.edu

F.M. Singer et al. (eds.), *Mathematical Problem Posing*,
Research in Mathematics Education, DOI 10.1007/978-1-4614-6258-3_6

Introduction

In a classic scene in the movie *Apollo 13*, the Apollo crew finds itself in danger because the lunar module cannot provide sufficient lithium hydroxide to remove carbon dioxide from the air of the module. To complicate matters further, the available lithium hydroxide designed for the command module is in canisters that are incompatible with the sockets on the lunar module, which houses the astronauts. In short, their survival depends on the ability to get a round canister into a square hole.

The problem (in the movie and, of course, in real life) was resolved by the ingenuity of the engineers on the ground. By making use of an array of available objects—duct tape, plastic bags, hoses from the astronauts' suits, etc.—the engineers were able to rig a functional "scrubber" that provided sufficient breathable atmosphere for the astronauts' return to earth.

While the drama of the scene leads us to see how ingenuity can help us address complex problems, it also poses several considerations for educators. What kinds of skills are at work in the resolution of an ill-structured, real-life problem? Originality? Creativity? Divergent thinking? Fluid intelligence? Each of these is evident, of course, but where, exactly, were these skills developed in the crew of engineers? In the current educational climate and practice, in what ways are these critical skills cultivated? Does educational practice prepare students for ill-structured problems with no unique "right answer?"

In this chapter, we present an approach to mathematics instruction that allows students the opportunity to create—rather than simply solve—their own conceptual mathematics problems. The article derives from an observation of one of Matsko's students (referenced in Matsko (2011)):

> Anyone can write tedious, difficult problems that review core math subjects, but to write problems in a novel, challenging, and refreshing manner, one must be imaginative. I feel that this creative side of math is an often overlooked aspect of the field as many believe math to be an extremely black-and-white, rigid, and boring subject.

Others have taken similar approaches (see for example, Blake, 1984; Brown & Walter, 1988; Goldenberg & Walter, 2003), but few reports include a detailed discussion of student work and classroom practice. One purpose of this chapter is to give a teacher potentially interested in using problem posing in the classroom a real sense of what shape such an assignment might take.

This exploratory study originated with the contention that if students were provided the opportunity to demonstrate their conceptual understanding of mathematics concepts (as opposed to responding to conceptually related problem sets), they would demonstrate deeper engagement, greater motivation, enhanced creativity, and improved transfer of mathematics concepts. Over the period of several academic years, this study developed from insights drawn from classroom practice to a solid evidence base of the efficacy of allowing students to create rather than simply respond.

Organizing Constructs: Creativity and Conceptual Understanding

Silver (1997) provided the context for this inquiry and suggested that mathematics in particular presents a disparity between student perception and the possibility for creativity and scholarly production. He wrote:

> Mathematics as an intellectual domain stands at or near the top of any hierarchical list of intellectual domains ordered according to the extent to which creativity is evident in disciplinary activity or production. Thus, it is ironic that for most students throughout the world, mathematics would almost certainly be among the set of school subjects least associated with creativity. (p. 75)

In a field that easily lends itself to creative and divergent thinking, the study of mathematics is popularly construed as a discipline in which we are expected to struggle to arrive at a single correct answer. But behind Silver's (1997) assertion are compelling questions for teachers: How might creativity be more deeply understood and more fully developed in a mathematics classroom? How can students be led to an understanding that mathematics is both relevant and rich with creative possibility?

A useful heuristic for understanding creativity is Csikszentmihalyi's (1990) distinction between little-c creativity and big-C Creativity. Little-c creativity is expressed as the types of creative tasks that are the product of specific contexts, such as the classroom or professional settings. Big-C creativity, however, refers to creative products that effectively alter a field or larger culture. For example, in a high school creative writing course, a talented student might compose a prize-winning poem that finds its way into a poetry collection or literary magazine (little-c creativity). But an eminent linguist such as E. E. Cummings is recognized for his exploration of new forms and sounds and redefined the possibilities of poetry in the twentieth century (big-C creativity).

Furthermore, according to Subotnik, Olszewwski-Kubilius, and Worrell (2011) there is lack of agreement about whether creativity comprises generalized abilities or whether it is domain-specific. These authors have suggested that the lack of consensus exists largely because of differences between childhood creativity, which is a person-centered trait, and adult creativity, which is often associated with products and contexts.

The literature on creativity is in broad agreement that creativity comprises the ability to generate new and novel products or insights, but there is not, however, consensus regarding other important conceptual relationships. For example, are creative children more likely to become big-C creative thinkers as they grow older? Is there a relationship between general intelligence and creativity? If children are creative in one domain (music, for example) are they likely to demonstrate creativity in another, such as language or mathematics? Does creativity necessarily first manifest itself in childhood as a precursor to adult creativity and eminence in a particular field?

Such questions have been explored in the research literature and provide the context for a second important consideration in this inquiry, namely conceptual understanding. In a widely referenced study conducted at the Harvard–Smithsonian Center for Astrophysics, graduates of Harvard and the Massachusetts Institute of Technology (MIT) were asked to perform several tasks that call on basic understandings of science. For example, one group of students was asked to light a small light bulb using a battery and a wire, and another group was asked to explain the earth's four seasons. When it became evident that the graduates were having difficulty answering such basic questions, a critical dimension of learning was exposed: if students lack conceptual understanding (say, of the idea of open vs. closed circuits in the battery and bulb problem), then all subsequent understanding was built on faulty knowledge.

So how can teachers assess for deep conceptual understanding? Research is virtually unanimous in the assertion that deep, conceptual understanding is related to achievement and further, that it is related to cognitive transfer. But are classroom assessments designed in such a way that students can demonstrate their conceptual understanding of mathematics, biology, physics, or economics?

In this study, we present a form of assessment that allows students to engage in meaningful, authentic, relevant mathematics problems and offers mathematics teachers insight into students' conceptual understanding as well as students' misconceptions. Further, as a departure point for future assessments in mathematics (and, we believe, other disciplines) this assessment may, in fact, lead to a longitudinal analysis of the development of creative thinking.

Finally, we note that the construct of conceptual understanding is subordinate to the construct of creativity. As we discuss in the next section, the prompt for students to write "conceptual" problems was intended to foster creativity. This prompt was specific to the student population in the study (gifted and talented high school students in a specialized mathematics and science school). Different assignments intended to foster creative thinking might very well emphasize different organizing constructs.

Problem Creation

What does it mean to "pose" or "create" a problem? Brown and Walter (2005) lay the groundwork in *The Art of Problem Posing*:

> Where do problems come from and what do we do with them once we have them? The impression we get from much of schooling is that they come from textbooks or from teachers, and that the obvious task is for students to solve them. (…) Problem posing can help students see a standard topic in a sharper light and enable them to acquire a deeper understanding of it as well. (p. 1)

With Silver's (1997) and Matsko's (2011) observations about mathematics instruction and assessment and Brown and Walter's (2005) justification for problem posing in mind, in this semester-long exploratory study, the instructor administered three assessments in which students were asked to generate original mathematics problems in an area of interest to them. These Original Problem assignments were administered in addition to an array of student assessments. Students included comprised sophomores enrolled in a three-year (sophomore through senior), residential high school for students identified as talented in mathematics and science.

For this chapter, we use the term "conceptual problem" to mean a problem whose creation, statement, and solution demonstrate conceptual understanding. Although it is difficult to give a precise definition of "conceptual understanding," it is not strictly necessary here. The significance of the prompt to write a conceptual problem was to emphasize that the problem should not be a routine problem which can be solved simply by applying a known solution method, or a problem which is simply a restatement of another existing problem with the numbers changed, for example. The prompt was intended to encourage students to think in novel ways. It is important to note, however, that abundant evidence suggests that conceptual understanding of mathematics is related to higher achievement, persistence, and motivation. (See Donovan & Bransford, 2005, for extensive supporting research).

For the first assignment, some examples were presented from a previous exam to give students an idea of what constitutes a "conceptual problem." Examples of previous student work were intentionally not presented so as to avoid the possibility that students would mimic a successful example rather than create something fresh. For the Original Problem assignment, we were interested in understanding the students' motivation in the development of the problem, so we did not instruct them to construct problems that they thought would be engaging or motivating to others. Rather, we asked them the source of their motivation to create their original problems.

In addition to the discussion of conceptual problems, the Original Problem assignment included the following prompts:

1. Motivation: How did you come up with the problem? Was it based on a problem on the worksheets? An exam? A Problem Set? Were you doodling? Did it come to you in a dream? In the shower? Just a sentence or two will suffice here. But, importantly: acknowledge your source! It's OK to look at other problems, just cite them if you use them.
2. Problem Statement: Fairly self-explanatory. But a caution: give it to someone else to proofread! One of the most common traps to fall into is to write a problem which can be interpreted in more than one way. Is your problem stated absolutely clearly, so that someone else can understand it perfectly without needing to ask you any questions about interpretation?

3. Problem Solution: Again, self-explanatory. But your solution should be in paragraph form, using complete sentences! And if you only have a partial solution, you should explain where you are stuck and those questions whose answers could enable you to make further progress.
4. Reflection: Only a few sentences are necessary here. What did you learn? What did you observe about yourself as a problem-writer? At the end of the semester, you will need to write an essay about your growth as a mathematician and problem-writer, so making notes along the way would be a good idea.

Although the students who completed this assessment had been identified for their interest and ability in mathematics, science, or both, it is important to note that the students were not "creative producers" in mathematics (Subotnik et al., 2011). Instead, they demonstrated a developing degree of expertise over their three years, and we suggest that this assessment is viable across levels and abilities.

Historical Background

The subjects of this study were first-semester sophomores (entering in Fall 2011) who placed out of the first semester of precalculus. However, the first use of the Original Problem assignment by the instructor (Matsko) was in an Advanced Problem Solving course in the Fall 2008 and Fall 2009 semesters. This elective course was designed for those students intensely interested in solving more advanced problems of all types, and so the students in this course were self-selected as having an interest in problem solving in general.

The success of the assignment in this course motivated the instructor to use the Original Problem assignment in other courses, beginning with an accelerated calculus sequence (Spring 2010 and Fall 2011), then the traditional calculus sequence (Spring 2011 and Fall 2011) as well as the precalculus sequence (Fall 2011 and the focus of the current chapter).

What surprised the instructor was that the Original Problem assignment was as accessible and successful when given to the atypical student (self-selected in the Advanced Problem-Solving course and the students in the accelerated calculus sequence) as when given to the more typical student (traditional calculus and precalculus students). The positive responses of students at all levels were the primary motivation for undertaking the current study.

Analysis of Students' Responses

We now take a detailed look at examples of student work. The purpose here is to illustrate the broad range of problems which teachers might expect to encounter in assigning Original Problems in their classroom, as well as give a few comments on assessment. Furthermore, the diversity of problems devised indicates that students can challenge themselves in ways that the teacher cannot always anticipate, and suggests that the writing of Original Problems is an effective way to implement differentiation in the classroom. Several students wrote problems at a level far too challenging to be given as assignments for the entire class.

Keep in mind that the prompt to write a conceptual problem was to encourage students to think creatively. When encountering a road block with a prospective problem, students were encouraged to think about altering important features of the problem, much like the "What if not?" approach described in (Brown & Walter, 2005). Students interacted with the instructor more frequently than with a typical assignment, and the ensuing discussions were richer.

In addition, at the end of the semester, students were asked to write a Reflection on their problem-writing experience. They were prompted to answer the following two questions: (a) How did you grow as a problem-writer this semester? and (b) Was this type of assignment valuable? Why or why not? The excerpts taken from these Reflections and the Original Problems are slightly edited to correct grammatical errors.

If students had difficulty coming up with an idea for an Original Problem, they were urged to take a concept discussed in class and extend it in a novel way, especially if there was a topic troubling them. For example, during the unit on polynomials, one student found a problem in a precalculus textbook which examined the influence of the coefficients of the polynomial $f(x)=x^3+bx^2+cx+d$ on its graph. This was too much like our in-class work; a suggestion that the absolute value function be incorporated resulted in the following problem: Compare the cubic polynomial $f(x)=x^3+bx^2+cx+d$ and the function $g(x)=|x^3|+bx^2+c|x|+d$ with respect to the number of points they intersect the y-axis, and the number of points they intersect the x-axis.

The first part of the problem is trivial, while the second involves some subtleties. What was remarkable was that the student's analysis of the second part was not only completely correct, but organized in a natural manner according to the number of zeros of $f(x)$. There was a level of sophistication in this student's thinking that would not have been evident in having them complete a more routine assignment.

During the unit on rational functions, one student looked at the graph of the function $f(x)=\dfrac{(x-1)^2(x-2)}{|x|(x-1)}$. This is not a rational function (as the denominator is not a polynomial), and we did not graph functions with $|x|$ terms in class. But the interesting feature of the graph of this function is that there are two distinct oblique asymptotes, which is *not* possible with rational functions. Motivated by the fact that $y=\arctan(x)$ has two horizontal asymptotes, the student commented, "I wanted to find a function that had two *oblique* asymptotes!" Another student addressed the issue of two oblique asymptotes by studying the function $f(x)=\left|\dfrac{1}{x}+x\right|$.

One student decided to see how the graph of $f(x)=\dfrac{(x+3)^2}{x^2+4x+A}$ changed with the parameter A. The graphing utility Winplot is used in class, which allows the creation of a "slider" so that a student can dynamically see how the graph changes as they move the slider. Some work was done with sliders during class, but in a different context.

In the trigonometry unit, one student used trigonometric ideas to evaluate $\sqrt{i}$. Although this problem was given in an earlier problem set, with students being prompted to use an algebraic approach, this student was interested in applying trigonometry to the same problem. He did some research, found De Moivre's Theorem, and successfully applied it to solve the problem. We do in fact use De Moivre's Theorem to solve such problems, but not until the following semester. This student was motivated to go beyond the classroom material, anticipating a topic which would not be covered until some months later.

A few other students looked at combinations of trigonometric functions not discussed in class. The main difference was in using function composition; two examples from student work are $y=\cos(\tan(x))$ and $y=\sin(3\sin(2x))$. Such combinations are too involved to be included as routine classroom exercises, and often specific features of the graphs can only be described approximately. One student commented:

> Seeing as I had done poorly on the trigonometry quiz, I was eager to work on graphing equations easily. I put a lot of thought and effort into the problem and spent even more time on the solution. Even though I did not realize it then, just two hours of concentrated effort on one problem had more of an impact on my graphing skills than two hours of effort on a few worksheets.

Done thoughtfully, Original Problems can deepen an understanding of important course concepts.

Some students used the Original Problem assignment as a means to explore entirely new mathematical ideas—at least new to them. One student wrote:

> I was able to explore topics that have never been covered in class and connect them to topics we were covering in class. These connections made looking at my original problems and my nightly homework was interesting because I had a bigger picture of what was happening.

Another student commented that the assignment "rekindled my curiosity in math."

One student explored the pentagonal numbers, illustrated in Figure 6.1.

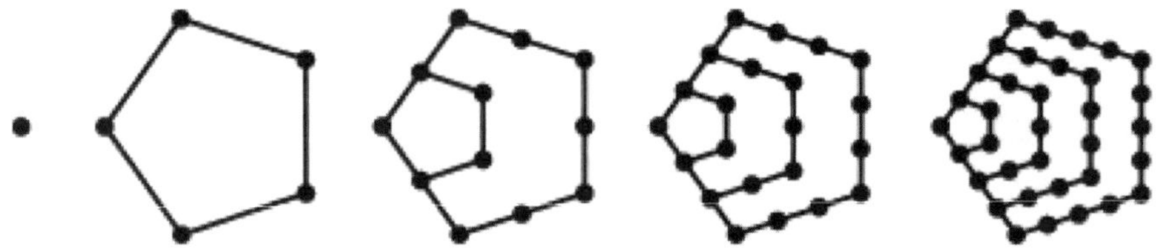

Figure 6.1. Pentagonal numbers.

The student researched the explicit formula for this sequence of numbers beginning 1, 5, 12, 22, 35, … namely $P_n = \dfrac{3n^2 - n}{2}$, but was unable to derive this formula. While such an exploration typically earns an *A* or *A*− even if the problem solution is not complete (owing to its difficulty), the student earned a *B*+ on the assignment since it was possible to derive the formula by fitting a parabola to the points (1, 1), (2, 5), and (3, 12). Since such an exercise was discussed in class earlier in the semester, the student would be expected to address this point in the problem solution.

One student was playing around with the Winplot utility and stumbled upon the "Tube" function, which produces objects like the one shown in Figure 6.2.

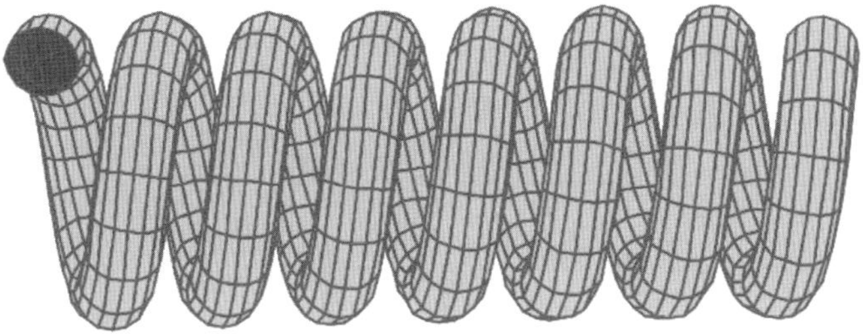

Figure 6.2. An object produced by the "Tube" function.

There are many different parameters one may give to this function, and the student explored these while creating several interesting three-dimensional graphics.

One creative problem involved exploring trigonometry at a much deeper lever than done in class. The problem as stated is: Find the exact value of

$$\sum_{k=1}^{90}\cos^2(K)+\sum_{k=1}^{22}\tan^2(2K)-\sum_{k=1}^{44}\frac{1}{\tan^2(2K)}+\sum_{k=1}^{22}\frac{1}{\tan^2(2K)}.$$

While apparently daunting, the solution involved a careful and methodical calculation of the sums involved. The sum is 44½.

Students explored various other topics, including number theory (Euler's totient function) and calculus (integration by substitution and Riemann sums). One of the students who wrote a calculus problem sat in on a friend's calculus class, while the other got help from his roommate. The first student earned an *A*− as there were some errors in the application of calculus principles, while the second earned a *B*+ since he solved his problem using Wolfram Alpha rather than actually computing the sums he had written. In any case, it is evident that given the opportunity, some students will go far afield and explore topics well beyond the classroom material. One student reflects, "Being able to explore whatever topics we choose, and making it into something more complex wows me since I did not know sophomores in high school are capable of something like this."

Other students approached the assignment by creating what, in their minds, are "real-world problems." Examples from physics were most common, as they involved applying formulas for phenomena like gravitational attraction, projectile motion, or refraction.

One student considered the problem of finding the optimum viewing angle, say, of a piece of artwork in a gallery. When looking up at a painting, the viewing angle is that angle made by the upper edge of the painting, the viewer's eye, and the bottom edge of the painting. As it happens, when walking backwards from a painting, there is precisely one point where this angle is at its maximum. This student found and researched this problem online.

This begs the frequently asked question of how it may be determined if a student's problem is truly "original," or if the problem was culled from somewhere else and modified. The Motivation section usually provides guidance. It is important that a student's problem is "original to them"—since, in general, it is very difficult to determine whether a problem is completely original. In the case just mentioned, the student wrote up a clear and complete solution, indicating that he thoroughly understood the problem he was investigating. This type of problem is within the spirit of exploration and application, and even though not a "new" problem, it was certainly novel to the student.

Another student created his own mathematical model, inventing the recurrence relation $A_k = \frac{A_{k-1}}{2} + \frac{A_{k-2}}{4}$ when $k \geq 2$ for the amount of toxic waste produced by an automobile company in the year k, where $k=0$ corresponds to the year 2012 and appropriate initial conditions are given. An environmental agency requires that the total amount of toxic waste produced after January 1, 2010 not exceed a certain limit, and the problem is to determine whether or not the company meets this guideline. (The actual problem statement is considerably more elaborate, but is abbreviated here).

The solution of such recurrence relations is not part of the current curriculum, and so finding a solution was challenging in itself. Moreover, the student's written work was truly exceptional for a sophomore in high school—it would not have appeared out of place in an expository mathematics journal. Such work underscores the value of writing Original Problems for challenging especially bright students; this level of work could never have been successfully assigned to the entire class, but the assignment allowed this motivated student to work at a very high level.

Some students created models derived from their personal experience, as well. One student who wrote a problem about runners on a track wrote in their Motivation section,

> The next day I was in the fitness center and I started running on the treadmill. Then about five minutes later someone hopped on the treadmill next to me and started running at a higher speed than I was. I kept looking over to see when he would catch up. Then I thought, this was the perfect idea for my original problem!

The student then posed a problem about two runners, each running at different speeds, with the faster runner behind and trying to catch up.

Another student wrote, "It was a cold, blustery day. In a city which knows how to keep its secrets, seven acquaintances huddled in a back room of a glorious institution, discussing what mattered in life: waffles." This scenario did actually occur, and the student then went on to create a problem determining how much maple syrup a waffle could hold if an accurate and detailed description of the waffle's geometry was given. Prompted to be creative, some students are able to devise interesting problems from simply looking at the world around them.

The open-ended nature of the prompt also allowed students to create problems unrelated to course material or real-world scenarios. For example, after looking at some mathematics problems online, a student came up with the following problem. "There are three different rational numbers that can be expressed as: $1, A+B$ and A. Similarly, these three numbers can also be expressed as 0, B/A and B. What is the value of $A^{1999}+B^{2000}$?" The solution is not unusually difficult ($A=-1$ and $B=1$), but the problem statement is highly creative. The student comments, "I wanted the numbers to be rather simple (−1, 1, 0), since that is often the case pertaining to many challenging math problems I have encountered before." Thus, the problem design involved making sure the solution involved small numbers.

Another student posed the following problem (after a brief story introducing the question): "So, given a bishop and a knight that are both placed on two different squares randomly on a chessboard, what is the probability that one of the pieces will threaten the other?" The solution is very involved, and the student made one slight calculation error (but still earned an *A* due to the substantive nature of the problem); such a problem would be challenging even in an undergraduate discrete mathematics course. This student would not be aware of this, naturally—but she created a problem which interested her (she was inspired while watching two students play chess in the library) and proceeded to tackle it with her knowledge at hand.

It should be remarked that, in this study, there was no attempt to classify problems into different categories. Such an additional level of analysis should be the focus of a further study whose aim is to correlate the type of problems a student writes with their performance in class and previous mathematical knowledge. Results of such a study could inform the use of Original Problems as a tool for incorporating differentiation into the classroom in a more formal, organized way.

Creativity

Writing Original Problems fosters creativity in the classroom. This is sorely needed in contemporary education, and even more so in mathematics education. Students commented, "Never before in my life had I ever written or made up anything original for math," and "The Original Problem assignments let us use our imagination which is not frequently used in mathematics." Such comments illustrate the fact that most students do not view mathematics as a creative endeavor, but rather a fixed body of facts and knowledge to master. One student went so far as to write, "The Original Problem experience revolutionized my view of mathematics."

Why is this perspective so important? One student, in responding to the prompt "Has the exercise of creating mathematics problems enhanced your ability to think creatively? If so, in what ways?" wrote, "No, because I have never been creative in the first place." It is unfortunate that this student's educational experience has not allowed him to perceive himself as creative—but if education has become a matter of memorizing facts and mastering procedures, this comes as no surprise. It is important to note that this response was not typical, but was nonetheless provocative.

Practical Implications for Classroom Learning

The preceding examples serve to illustrate the broad range of problems generated by an open-ended prompt. And although it is tempting to select exceptional examples for publication purposes, it would not have been difficult to include several more pages also filled with intriguing examples. But teachers new to this type of assignment might consider a more restrictive prompt, such as "Write a word problem whose solution involves solving a system of linear equations," or "Create a geometry problem involving triangles." It is important that the assignment is at a level comfortable to the instructor.

Three primary purposes of the Original Problem assignment are to have students create, write, and reflect. There are many possible ways to accomplish these purposes—and not all necessarily need to be accomplished within the same assignment.

What about grading? Useful guidelines are: *A*—exceptional, *B*—satisfactory, and *C*—lack of effort. A paper which includes all the necessary sections of the assignment with correct mathematics typically earns a grade in the *B* range. A particularly creative problem, or one which stretches a student significantly, typically earns a grade in the *A* range unless there are issues with correctness. An unusually exceptional paper will earn an *A*+ (such as the toxic waste problem discussed earlier). A paper which earns a *C* is often easily seen to be a last-minute effort, or a paper which is well below a student's potential. In other words, a problem which one student might earn an *A* for writing might result in a *C* for another student. This might seem problematic for some instructors, but must be considered for this type of assignment. A student who struggles with the course material may come up with a creative, amusing word problem and successfully solve it—even though the mathematics may not be at a high level. However the same problem, written by a brilliant student, might simply indicate laziness.

Although such a system of grading—it is fairly easy to earn a *B* or higher—is meant to encourage creativity, one student commented, "I feel that I would have been much more creative in this project if I was not restricted by the fact that my Original Problem affected my grade." This was not a typical written response; most students felt free to be creative. But not all students were entirely comfortable with the idea.

The attitude of the teacher is critical. Conveying an attitude that "This is a really hard assignment! It's tough to create math problems!" is very different from the attitude that "Of course you can create an interesting math problem! Everybody's creative!" It is well known that students will rise to a teacher's expectations—but it is important to make those expectations clear.

There are a few obstacles which may be encountered with this type of assignment. Foremost is the time involved—students usually ask for more help on this type of assignment, and grading takes more time. Another is the more subjective nature of the grading, which is not typical for mathematics assignments. Moreover, whatever the prompt for the assignment, the instructor should have a reasonable sense that students' work is their own.

Reflections of the Instructor

As a mathematics teacher, I have been surprised and inspired by the creativity of my students in writing Original Problems from the very beginning. As an avid problem-poser, I have always imagined that writing problems must certainly improve my ability to solve problems—and so I introduced the Original Problem assignment when I taught Advanced Problem Solving. Perhaps my satisfaction and surprise over the quality of students' problems derives from an instructor's traditional perspective of students as problem solvers. If we expect them to "create" in writing and art classes, why not in mathematics courses?

The end-of-semester reflections of the students were revealing. Students reported that they saw mathematics as a creative endeavor for the first time, they wrote problems they never imagined they could, and they found engaging in the creative process highly satisfying. As an instructor, I wish I could say that the Original Problem assignments were intended to produce these changes in attitudes—but the results were serendipitous.

Nonetheless, they inspired me to continue experimenting with the Original Problem assignment with students at all levels. I was continually surprised at the success of the assignment, and am now a firm believer that such (or similar) assessments can be instrumental in altering student attitudes toward mathematics.

From my perspective, the main barrier to working more with Original Problems is the time needed both to consult with students and read their papers. As every paper is different, there is no rhythm to the grading process (as when grading a large stack of identical assignments or exams). However, the results of the assignment are routinely satisfying, and there are always those few students whose work is really inspiring—and it is rewarding to realize that I played a role in motivating them to achieve at a level beyond my expectations.

The larger question—Does problem posing positively impact problem solving?—is difficult to answer. To study this question, the prompt for the Original Problem would likely have to be narrowed. The results of such a study would be interesting—yet for now, there is ample evidence of the benefits of having students write Original Problems regardless of its impact on problem solving.

Conclusion

Having students write Original Problems may successfully be used to foster creativity in the mathematics classroom. Students comment everywhere from "I really hate Original Problems," to "I love writing Original Problems!!"—although the majority of students think the assignment is valuable and enjoy it. Results are often surprising—not just the problems themselves, but comments such as "I am actually very surprised of myself of how far I have come in becoming a problem-writer. It makes me feel really good about myself," or "I believe that problem-writing is probably the best way to go if one is trying to actually make their students comprehend and improve their mathematical skills." But more importantly, students encounter mathematics in a different way, and as a result appreciate aspects of mathematics they may not have encountered before in their education. Leaving students with a deeper appreciation for and a more positive attitude toward mathematics is an important step in improving mathematical literacy.

Amidst a discussion of "twenty-first century skills," it is becoming more critical that students develop the skills necessary to compete in a workforce with a strong emphasis on innovation and invention. Real-world problems are now of a global nature, and their solutions require problem solvers with flexible, fluid minds. Having students write Original Problems, or undertake similar assignments, stimulates the development of skills necessary for solving complex problems. Of course, not all will become engineers for space missions, but we can certainly do more to insure that an increasing number of our students develop a skill set which would enable them to make such a career choice.

Prompting students to write "conceptual" problems was successful in stimulating creative thought. As remarked earlier, this prompt is by no means required, but was well suited to a cadre of students singled out as particularly talented in mathematics. What is important is that students are encouraged to go beyond the routine and engage in the creative process (in the sense of little-c creativity). Also important are the written and reflective components of the assignment. There is no "one size fits all" assignment here; rather, the type of assignment must be tailored to the topic under study, the students in the classroom, and the teacher's background.

There is much to be done—refinement of the assignment and its assessment, and deeper analysis of student responses, for example. Hopefully the results of our work will encourage other educators, in both mathematics and other disciplines, to work with assignments similar to the Original Problem assignment in their classrooms. It is also important to see whether skills developed by having students engage in such assignments are transferable, and to what degree. We hope that the examples of student work shown above demonstrate that given the opportunity, students can do creative work at a high level—a level often far beyond what might otherwise be expected of them. Finally, we suggest that we may, rather than be surprised by the occasional demonstration of creativity in our mathematics classrooms, expect students to be creative as a matter of routine.

References

Blake, R. N. (1984). 1089: An example of generating problems. *Mathematics Teacher, 77*(1), 14–19.

Brown, S. I., & Walter, M. I. (1988). Problem posing in mathematics education. *Questioning Exchange, 2*(2), 123–131.

Brown, S. I., & Walter, M. I. (2005). *The art of problem posing*. Manwah, NJ: Erlbaum.

Csikszentmihalyi, M. (1990). *Flow: The psychology of optimal experience*. New York, NY: Harper & Row.

Donovan, M. S., & Bransford, J. D. (2005). *How students learn: Mathematics in the classroom*. Washington, DC: National Academies Press.

Goldenberg, P., & Walter, M. (2003). Problem posing as a tool for teaching mathematics. In H. Schoen & R. Charles (Eds.), *Teaching mathematics through problem solving* (pp. 69–84). Reston, VA: National Council of Teachers of Mathematics.

Matsko, V. (2011). *On problem writing*. Retrieved December 12, 2011, from http://www.vincematsko.com/Talks/IPST2011/ProblemWriting.pdf.

Silver, E. (1997). Fostering creativity through instruction rich in mathematical problem solving and problem posing. *Zentralblatt für Didaktik der Mathematik, 29*(3), 75–80.

Subotnik, R. F., Olszewwski-Kubilius, P., & Worrell, F. C. (2011). Rethinking giftedness and gifted education: A proposed direction forward based on psychological science. *Psychological Science in the Public Interest, 12*(1), 3–54.

Chapter 7
Is Problem Posing a Tool for Identifying and Developing Mathematical Creativity?

Florence Mihaela Singer and Cristian Voica

Abstract The mathematical creativity of fourth to sixth graders, high achievers in mathematics, is studied in relation to their problem-posing abilities. The study reveals that in problem-posing situations, mathematically high achievers develop cognitive frames that make them cautious in changing the parameters of their posed problems, even when they make interesting generalizations. These students display a kind of cognitive flexibility that seems mathematically specialized, which emerges from gradual and controlled changes in cognitive framing. More precisely, in a problem-posing context, students' mathematical creativity manifests itself through a process of abstraction-generalization based on small, incremental changes of parameters, in order to achieve synthesis and simplification. This approach results from a tension between the students' tendency to maintain a built-in cognitive frame, and the possibility to overcome it, which is constrained by their need to devise mathematical problems that are coherent and consistent.

Contents

F.M. Singer (✉)
Department of Educational Sciences, University of Ploiesti, Bucharest, Romania
e-mail: mikisinger@gmail.com

C. Voica
Department of Mathematics, University of Bucharest, Bucharest, Romania
e-mail: voica.cristian@gmail.com

F.M. Singer et al. (eds.), *Mathematical Problem Posing*,
Research in Mathematics Education, DOI 10.1007/978-1-4614-6258-3_7

Introduction

Is problem posing a tool for identifying and developing mathematical creativity? This is an intriguing question. Apparently, problem posing refers to generating something new or to revealing something new from a set of data, therefore somehow involving creativity. However, for a more structured answer, we have to investigate the nature of (mathematics) creativity. Consequently, we start by addressing some related issues.

Researchers have used the term "creativity" to characterize quite different behaviors, and this diversity of definitions inevitably led to ambiguity and controversy. Creativity comprises many discrete abilities that often do not correlate very much with each other (Guilford, 1967). Although the general public commonly associates creativity with novelty and surprise, many researchers defined creativity by highlighting two characteristics: originality and appropriateness (e.g., Amabile, 1989; Baer, 1993; Sternberg & Lubart, 1999). Baer and Kaufman (2005) identified some prerequisites that condition the ability to express creative behavior; they are intelligence, motivation, and suitable environments. Creativity can be latent—and, it may not show in the absence of an environment in which it can be nurtured and valued (Csikszentmihalyi, 1996; Gardner, 1993, 2006).

The above-mentioned authors based their definitions of creativity on the conclusions drawn from experiments/observations carried out across a variety of domains, such as arts, genetics, physics, or journalism. The diversity of domains poses the following important question: Is creativity domain specific, or general? This issue is strongly related to the idea of transfer: if creativity is domain general, an adequate training in a domain might be transferable to other domains, helping the trainees to solve any problem more creatively. However, a large body of research suggests that this may not be the case (e.g., Baer, 1993, 1998; Lubart & Guignard, 2004; Nickerson, 1999). In fact, the situation is even worse; the transfer might not work even within the same domain. For example, Baer (1996) investigated the effect of a training course focused on divergent-thinking skills related to poetry writing for middle school students. When after training, these students were asked to write poems, their poems were significantly more creative than those written by their peers from a control group. However, when the same students were asked to write short stories, both groups were nearly as creative. So, as the above researchers and many others (e.g., Dow & Mayer, 2004) have shown, the transfer of creative abilities does not automatically occur even within related sub-domains of the same domain. Therefore, it does make sense to speak about mathematics-specific creativity and to try to understand its specificity as compared to general creativity.

Mathematicians and mathematics educators alike have formulated various solutions to this issue over time. For example, Hadamard (1945/1954) associated creativity with an intuitive mathematical mind and believed that creative expression requires ample time for reflection and incubation of ideas. Hadamard mainly envisaged the creative behavior of the expert mathematician. This option corresponds to the idea that the essence of mathematics is what mathematicians do (Poincaré, 1913). We are equally concerned, however, with what students do when they behave creatively within a mathematical context and what the limits of these behaviors are. This is relevant for learning as we assume that the essence of mathematics is creative thinking, rather than just the identification of the right answer (Dreyfus & Eisenberg, 1996; Ginsburg, 1996).

When comparing students to experts, another dilemma emerges: does mathematical creativity only occur through the discovery of a completely new result, or can it also occur when re-discovering a fact already known by the scientific community? In other words, we can ask what relevance "novelty" has in the students' case. Which aspects are specific to the mathematical creativity of students? How can this be studied? Can creativity be developed?

The answers we arrive at in this study are strictly related to our target population: young students 10–13 years old, proficient in mathematics. We studied their creativity by using problem posing (PP) activities. We further explain the choice of PP as a tool in our research.

Traditionally, the context used to study student's creativity is problem solving (PS). Might it be the case that problem posing is more relevant than problem solving to study creativity? According to Sternberg and Lubart (1991), creative individuals not only solve problems, but also pose the right problems; therefore, the capacity to pose problems might be a sign of creativity. In a series of articles on discovering problems in art contexts, Getzels and Csikszentmihalyi highlighted differences in thinking between the case where the starting point was an already formulated problem, compared to situations in which the problem must be discovered or created (Csikszentmihalyi & Getzels, 1971; Getzels, 1975, 1979; Getzels & Csikszentmihalyi, 1976). In these articles, the ability to "discover" a problem was used as a primary category for analyzing creative processes. Extrapolating these findings to mathematics, we can infer that PP might be more significant than the PS in the study of mathematical creativity, even if a common definition of mathematics refers to it as a problem-solving domain. Other studies confirm this conclusion. For example, Smilansky (1984) showed that there is very low correlation between the abilities of mathematical PS and PP in a group of high school students and undergraduate students. Smilansky's conclusion is that PP is a task more significant than PS for the study of creativity. Starting from Smilansky's findings, we offered our students a context in which they posed and/or modified problems, in order to study their creativity.

If we agree that PP is relevant for the study of creativity, a second question is what taxonomy would be more effective to reveal creative behaviors. Typically, mathematical creativity is studied and assessed through the lenses of: fluency, flexibility, and novelty, the parameters conceptualized by Torrance (1974). We consider that both the study and the development of school creativity should be aligned to new scope and purpose. If in the eighth decade of the twentieth century the creativity focus was on theoretical studies, today, the knowledge society—characterized by complex dynamics and over-information—needs individuals, and especially leaders, able to anticipate changes and to take knowledgeable decisions under varying conditions that are hardly predictable (e.g., European Commission, 2003/2004, 2005; Hargreaves, 2003; Singer, 2006; Singer & Sarivan, 2006).

In other words, more than ever before, today's schools should help students to develop creative approaches as part of leadership qualities, especially in those who are promising high achievers. Previous studies on mathematical creativity in a PP context (Singer, 2012; Singer, Pelczer, & Voica, 2011; Singer & Voica, 2011, 2013; Voica & Singer, 2012, 2013) have concluded that a framework highlighting social integration and leadership could provide better information about students' creativity in a PP context.

We have seen that the transfer of creative abilities from one domain to another is less likely to appear spontaneously. We assume that the study of creativity in a broader, socially-oriented framework that faces opportunities for transfer within the training could offer more relevant data for contemporary research on creativity development.

We support this claim based on the conclusions of a study of Yuan and Sriraman (2011), regarding the achievements in PP activities of groups of students from the USA and China. These students performed several types of tests, including: a mathematics content test; a mathematical problem-posing test; Verbal *Torrance Tests of Creative Thinking* (TTCT) (where students were asked to think with words); and Figural TTCT (where students were asked to express their ideas by drawing pictures).

Yuan and Sriraman (2011) maintained that US students performed much better than Chinese students from the point of view of fluency, flexibility, and originality on the Verbal TTCT. This result is not surprising, if we relate it to the features of the teaching practice in the two countries. US students often work in groups, are involved in projects, and are encouraged to ask questions, to experiment, and to provide explanations (see, for example, National Council of Teachers of Mathematics, 2000). Therefore, US students seem more capable of expressing their ideas in words. Conversely, in the education system in China, where typical lessons are characterized by "order and routine" (Lim, 2007, p. 80), and teachers often maintain control by directly teaching to the whole class (Huang & Leung, 2004), the communication and interaction between students are not important, and the focus is mainly on factual knowledge. In addition, the Chinese language—based on ideograms—offers support for the recourse to drawings and pictures in explaining ideas—a hypothesis taken into account by some psychologists (e.g., Demetriou et al., 2005).

On the other hand, Yuan and Sriraman's study found "no significant correlations" between general Torrance creativity and PP abilities for US students, while for the students from China, PP abilities are "significantly correlated" with Verbal TTCT scores (Yuan & Sriraman, 2011, p. 25). This lack of consistencies between the two groups led us to two major ideas, which we will try to convey in this chapter.

A first claim is that Torrance's criteria do not represent the most suitable framework for the study of creativity in the context of PP. It seems that parameters related to classroom management activities and to students' communication skills are not highlighted enough in such tests. Therefore, we assume that a social-oriented framework is more appropriate for analyzing students' mathematical creativity.

A second claim is that mathematical creativity is of a special nature compared to creativity in general. This is used to explain why there were no significant correlations between TTCT results and PP abilities of the US students in the above-quoted study.

Our research tries to identify this special nature of mathematical creativity in students. Our preliminary studies led us to formulate the following hypothesis: in a PP context, students' mathematical creativity manifest itself through a process of abstraction-generalization based on small, incremental changes of parameters, in order to achieve synthesis and simplification. As a result, students expressed their creativity by making small-scale changes of the mathematical model of a problem, which resulted in maintaining control over the proposed problem.

In this chapter, we try to see if the above hypothesis is confirmed for our sample. More precisely, we seek an answer to the question: How does mathematical creativity manifest in 10- to 12-year-old students? If the hypothesis can be confirmed, it will once again result in the need for a new tool suitable for analyzing mathematical creativity.

Theoretical Background

Given the interdisciplinary nature of this study, we will discuss the theoretical background from four perspectives: mathematical creativity, problem posing, connections between mathematical problem posing and creativity, and cognitive flexibility.

Mathematical Creativity

The topic of mathematical creativity received much attention from researchers who focused on defining it, or on establishing criteria for its evaluation (see, for example, Ervynck, 1991; Freiman & Sriraman, 2007; Silver, 1997; Sriraman, 2004, 2009). The literature contains a variety of definitions and characterizations (e.g., Balka, 1974; Evans, 1964; Getzels & Jackson, 1962; Haylock, 1987; Jensen, 1973; Poincaré, 1948; Prouse, 1967). Earlier references to mathematical creativity came from the work of expert mathematicians like Poincaré and Hadamard (Hadamard, 1945/1954; Poincaré, 1948). Subsequently, various studies have identified certain behaviors that provide evidence of mathematical creativity in students. Haylock (1987) and Singh (1988) assessed mathematical creativity based on the three characteristics defined by Torrance (1974): fluency, flexibility, and novelty. The common interpretation is that these features represent, respectively: the number of identifiable changes in approaching a problem; the number of generated solutions; and the level of their conventionality (e.g., Ervynck, 1991; Leikin & Lev, 2007; Silver, 1997).

Balka (1974) synthesized another line of analysis: he considered convergent thinking—characterized by determining patterns, and divergent thinking—seen as formulating mathematical hypotheses, evaluating unusual mathematical ideas, sensing what is missing from the problem, and splitting general problems into specific sub-problems, as the main components of mathematical creativity. In this context, Haylock (1997) insisted that one of the key elements of creativity is the ability to overcome fixations in mathematical problem-solving (leading, for example, to breaking away from stereotyped solutions).

Problem Posing

There are different terms that are used in reference to problem posing, such as problem finding, problem sensing, problem formulating, creative problem-discovering, problematizing, problem creating, and problem envisaging (Dillon, 1982; Jay & Perkins, 1997). Because of this variety of meanings, different authors use different frameworks for studying PP activities. For example, Brown and Walter (1983/1990) looked at PP within a strategy focused on the phrase "what-if-not." This strategy assumes that, by discussing the significance of the problem components and by trying to modify this, students can come up with a deeper understanding of the problem, rather than just focusing on finding the solution.

Stoyanova and Ellerton defined PP as "the process by which, on the basis of mathematical experience, students construct personal interpretations of concrete situations and formulate them as meaningful mathematical problems" (Stoyanova & Ellerton, 1996, p. 518). In their paper, problem-posing situations were classified into three categories: free, structured, or semistructured (Stoyanova & Ellerton, 1996). In the present study, we adopt Silver's (1994) less restrictive definition, in accordance with which, problem posing refers to both the generation of new problems and the re-formulation of given problems.

Mathematical Problem Posing and Creativity

The literature on PP shows that this activity is important from various perspectives and emphasizes connections between PP and creativity. Some researchers have reported a positive relationship between mathematics achievement and problem-posing abilities (English, 1998; Leung & Silver, 1997). Other researchers (e.g., Cai & Cifarelli, 2005; Singer, Ellerton, Cai, & Leung, 2011; Singer, Pelczer, & Voica, 2011) claimed that instruction that includes problem-posing tasks (problem modification tasks included) can assist students to develop more creative approaches to mathematics.

There are also researchers who have expressed doubts regarding the connection between creativity and PP. For example, Yuan and Sriraman (2011) concluded that "there might not be consistent correlations between creativity and mathematical problem-posing abilities or at least that the correlations between creativity and mathematical problem-posing abilities are complex" (Yuan & Sriraman, 2011, p. 25). However, other studies, for instance Haylock (1997) and Leung (1997), who did not agree that there is a correlation between creativity and problem posing in mathematics, did not consider instruction. From an empirical perspective, Silver (1997) suggested a position that supports our hypothesis: that any relationship between creativity and problem posing might be the product of previous instructional patterns.

Cognitive Flexibility

Cognitive flexibility of a person can be defined as the dynamic activation and modification of cognitive processes in response to changes in task demands, which results in representations and actions that are well adapted to the altered task and context (Deák, 2004). In other words, cognitive flexibility addresses the readiness with which a person's concept system changes selectively in response to appropriate environmental stimuli (Deák, 2004; Scott, 1962). Pragmatically, cognitive flexibility refers to a person's ability to adjust his or her working strategies as task demands are modified (Krems, 1995; Spiro, Feltovich, Jacobson, & Coulson, 1992).

In an organizational context, cognitive flexibility is conceptualized as consisting of three primary constructs: cognitive variety, cognitive novelty, and change in cognitive framing (Furr, 2009). Cognitive variety refers to the diversity of mental templates for problem solving that exists in an organization (Eisenhardt, Furr, & Bingham, 2010), or to the diversity of cognitive pathways or perspectives (Furr, 2009). Cognitive novelty refers to the concepts pertaining to the subject of study and the overall mastery of content (Orion & Hofstein, 1994), or to the addition of external perspectives (Furr, 2009). One's previous experiences, particularly successful experiences, may lead to the phenomenon called cognitive framing: it manifests itself through a person's persistence in trying to solve a new problem by using a certain strategy, previously practiced (Goncalo, Vincent, & Audia, 2010). In certain situations, it denotes an algorithmic fixation (in terms of Haylock, 1997); in these cases, the only possible way to overcome this is to change the thinking frame.

Methodology

Sample

The participants in this research are students in grades 4–6 (10–13 year-olds), winners of a two-round national mathematics competition (the Kangaroo contest). Within this competition, the participants were supposed to choose and solve 30 out of 40 multiple-choice problems (with five possible answers, only one being correct) in 75 minutes. In the Kangaroo contest the problems are graded 3, 4, or 5 points, incorrect answers are penalized with a quarter of the score, and non-responses are ignored. These regulations are publicly available and are reinforced before the test session.

After the first round (involving approximately 60,000 students in grades 4–6, which represents approximately 10% of the Romanian school population for these grades), the top 10% of students attending the first round qualified for the second round (where the competition regulations are the same as in the first round, but the problems are much more difficult). The winners of the second round attended a summer camp.

Due to the selection process, we consider that the participants in the camp (280 students from a total of 60,000) are high achievers or excelling in mathematics. During the camp, the authors of this chapter—as invited professors—launched a call for problems to the students. The 48 students who voluntarily responded to this call represent our sample.

Data Collection

The 53 problems posed by the students of our sample were initially assessed by two reviewers (other than the authors), who worked independently. They graded the problems from 1 to 10, based on the following criteria: the statement completeness, the correctness of the posed solution, and the novelty (expressed as the "distance" between the proposal and the types of "usual" problems of school textbooks and auxiliaries). Subsequently, the two experienced teachers who served as problem reviewers shared the scores they had given and, in each case where they found significant differences, they discussed and reached a consensus.

Following this preliminary assessment, we chose to interview 19 of the students who responded to the call for problems; for this selection, we took into account the scores given by the reviewers, but also some surprising or interesting aspects we had noted in the students' comments and solutions. In some cases, we decided to interview a certain student even if his/her proposal was not highly ranked because a particular aspect of that proposal (e.g., an unusual context for a problem, or unusual comments) suggested that the student showed creative potential. Briefly, we chose the students for interviews either based on the intrinsic qualities of the highly ranked posed problems, or based on the hints that we found in students' proposals that might illuminate the mental mechanisms activated in PP.

We have included the texts of the 20 problems posed by the students invited to the interviews in the Appendix (one of the students suggested two problems). In the following sections, we refer to these problems using the Appendix ordering numbers, but we also quote the problem text when this is needed to enable the reader to follow the line of argument more easily.

Each interview lasted between 10 and 40 minutes. The interviews were video-recorded and subsequently transcribed. Before the interviews, we asked students to re-read the problem they initially posed. We structured the interview protocol around questions such as: What inspired you to pose this problem? How might you change your posed problem? Can you pose a simpler/more complicated problem? What did you change compared to your initially posed problem? How would you proceed to pose new variants of the problem? Therefore, during the interviews, students were given the opportunity to pose new problems, or to modify their initially posed ones. Thus, the interviewed students generated other 26 new problems.

We used the protocol for guidance during the interviews, but we encouraged students to express their ideas as freely as possible. In some cases, the interview departed from the protocol because we sought to identify students' thinking patterns. Thus, we got information about the models students used as starting points in a PP activity (if any), their strategies for generating and correlating problem givens, their perceptions concerning the difficulty and complexity of their posed problems, and finally, the metacognitive processes they activated when posing and solving problems. Based on these, we tried to outline a cognitive profile in problem posing and solving situations for each selected student. We then compared the conclusions formed from the interviews with the participants' behaviors in the Kangaroo national contest. For this, we analyzed students' answers from the contest, obtaining information on: series of correct/incorrect answers, types of wrong answers, types of mistakes, types of preferred/avoided problems. We then compared the results obtained by the students of our sample with the statistical results of all participants in the competition. These comparisons helped us to identify possible correlations between a student's PP-PS cognitive profile and his/her options for posing a certain type of problem, thus validating the identified profile over time.

Therefore, the data analyzed in this chapter come from the following sources: students' posed problems (initially posed problems, problems posed during the interviews, problems obtained by modifying the initial ones), interviews, and statistical databases. Each of the sources was analyzed from several perspectives. Following this multiple-level analysis, we gathered as much information as possible in relation to students' creative behavior by examining students' preferences for particular mathematical domains, their mathematical abilities, students' strategies in PP and PS, and their intra- and inter-personal approaches.

Data Analysis Framework

From our analysis of primary data, we found that, when children posed problems, they involuntarily resorted to their teacher's model. Inevitably, large parts of the students' proposals were tasks that had a specific target audience (of colleagues, friends, competitors, even taking into account different levels of competency of those audiences). As a result, the posed problems did not just represent students' theoretical approaches, but rather encompassed an ensemble of relationships between the poser and potential solvers, expressing a poser's need to feel integrated within a structured social ensemble. This finding, observed in students of different ages and with varying mathematical abilities, emphasizes the fact that, unlike PS, PP activities have an important component of inter-personal interaction which can significantly influence the quality of students' posed problems. In addition, from the perspective of contemporary society, we are interested in those capabilities that enable students to manage their own learning and to be able to identify, pose, and solve problems arising in unpredictable contexts (e.g., Singer, 2006, 2007).

For these reasons, we investigated the relationship between problem posing and mathematical creativity in terms of cognitive flexibility in organizational contexts. In a problem-posing context, we consider that a student exhibits cognitive flexibility when the following three conditions are fulfilled (Pelczer, Singer, & Voica, 2013a): the student poses different new problems starting from a given input (i.e., cognitive variety), generates new proposals that are far from the starting item (i.e., cognitive novelty), and he/she is able to change his/her mental frame related to the proposal, if necessary, in generating and solving problems (i.e., change in cognitive framing).

Criteria for Data Classification

We used the following criteria for classifying students' posed problems: the involved mathematical domain, the coherence, and the consistency of the problem. Further details of these will be provided in the sections which follow.

A first classification concerns the mathematical content of these posed problems. Within this criterion, we used the following categories:

- **Numerical computing**. This includes problems containing instruction(s) that refer to numerical calculations explicitly stated in the text. It may include percentage calculation or computation with fractions.
- **Relations**. Here are problems that use specific properties of sets of numbers, for example: divisibility on **N** or **Z**, or order relation on **Q**.
- **Equations**. Problems where equation solving is essential (even if this is not formalized) are included here. We also included here problems where unknown data can be found using a scheme or a graphical representation.

- **Algebraic computing**. Here are problems involving general features of numbers or abstract schemas for solving, which lead to generalizations that can be expressed by algebraic formulas.
- **Change of patterns**. This includes problems that need to be understood and analyzed in their kinematic development, as they assume successive stages and understanding transitions from one stage to another.
- **Handling data**. This category contains problems in which the analysis of data sets or their distribution is relevant for the solution.
- **Geometry**. Here there are problems in which the students effectively used specific geometric properties (such as parallelism, perpendicularity, and congruency).

The next categories for clustering students' proposals refer to the intrinsic qualities of a posed problem. Since a problem text is expressed in a specific language, we use two criteria—syntax and semantics—that are characteristic of language in a broad sense and used in both natural language and in artificial languages such as computer programming.

To characterize these two attributes, we have adopted the *problem-analysis framework* used by Singer and Voica (2013). According to this framework, the text of a problem contains, in general: a background theme, parameters, (numerical) data, one or more operating schemes (or, simply, operators), constraints over the data and operating schemes, and constraints that involve at least one unknown value of the parameter(s).

Concerning the syntax, we define *the coherence* of a problem, which refers to the rules and principles that govern the structure of a mathematical problem. Essentially, these rules and principles are:

- The following text components—givens, operations, constraints—are present;
- The following text components—givens, operations, constraints—are recognizable or identifiable;
- The givens are not redundant, or missing.

The syntax offers a formal valid shape of a problem, but does not provide any information about the meaning of the problem or the results of its solving. The meaning associated with a combination of text elements belongs to semantics.

Concerning the semantics, we define *the consistency* of a problem. This supposes the existence of meaningful links among the elements of the problem. More specifically:

- The problem data are not contradictory;
- The following text components—givens, operations, constraints are correlated;
- The components of the problem text satisfy a certain assumed mathematical model;
- The information provided leads to at least one solution of the problem (or to the proof that there is no solution).

Within the problems obtained by modifying a given problem, consistency also requires that:

- At least one of the mathematical elements of the starting problem is identifiable in the new problem.

We specify that not all syntactically correct problems are semantically correct. Many syntactically correct problems are nonetheless ill formed and are merely a combination of parts obeying some rules. Such problems may result in error-prone processing. In addition, it may not be possible to assign meaning to a syntactically correct problem, or the wording may be false.

Results

We used the above-mentioned criteria to analyze students' behavior in posing and solving problems based on the students' submitted problems and their answers during the interview sessions.

The Students' Posed Problems

Mathematical content. Table 7.1 presents the distribution of the 20 problems posed by the interviewed students according to the mathematical content criterion. (We remind the reader that the texts of these problems can be found in the Appendix.) For space reasons, the entire classification of the sample consisting of the 53 problems initially posed by all of the students can be found only in Figure 7.1. We relate our classification to the *National Assessment of Educational Progress 2011 Framework* (NAEP, 2011). The NAEP framework describes five mathematics content areas: number properties and operations, measurement, geometry, data analysis

Table 7.1
Distribution of the Problems Posed by the Interviewed Students, According to the Mathematical Content

Math content	Problem number (from the Appendix)	NAEP correspondent
Numerical computing	6, 8, 10	Number properties and operations
Relations	11	
Equations	3, 4, 5, 7, 9,19	Algebra
Algebraic computing	2, 12,15	
Change of patterns	13, 17, 20	
Handling data	1, 16	Data analysis and probability
Geometry	14, 18	Geometry

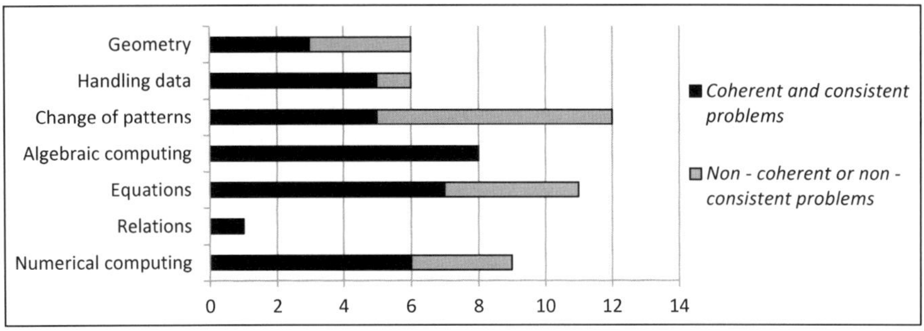

Figure 7.1. Classification of students' posed problems based on the criteria: coherence (mathematical syntax) and consistency (mathematical semantics).

and probability, and algebra. After analyzing students' posed problems, we concluded that the categories listed in Table 7.1, which subdivide some of the NAEP content areas, better describe students' proposals.

Some of the students' posed problems can be included in several content areas. To get a clearer picture of students' preferences for different content subareas, we sought to distribute the posed problems into disjoint classes. Therefore, for each problem, we tried to identify the depth of thought involved, considering both the wording and the problem solution. Subsequently, we checked (when necessary) if our framing matched the student's intention.

To illustrate the way we classified the problems according to their content, we provide below one significant example (problem #16, posed by Mihai, grade 6):

> Because the 6th grade students were the best, they received a prize consisting in one hour free on paintball field. The field has the dimensions 80 m × 120 m, and two people are able (and allowed) to shoot one another if they are at no more than 29 m distance. Prove that howsoever 26 students place themselves on the ground, at least 3 get shot.

Apparently, this problem is one of geometry. At a closer look, we may find that the essential element in its solution is to identify certain regularity in the arbitrary distribution of points inside a rectangle. Indeed, the interview revealed that Mihai has used a grid (he decomposed the rectangle into congruent squares) and made "order in disorder," applying the pigeonhole principle. Therefore, we classified this problem in the category *Handling data*.

Syntax and semantics. In line with recent perspectives on creativity, the outcomes of a creative process should have relevance within a community (scientific, cultural, organizational, etc.). More precisely, not every new product means creativity, unless it is socially valued (e.g., Gardner, 1993; Gardner, Csikszentmihalyi, & Damon, 2001). In particular, in a PP context, a creative mathematical product must be coherent and consistent, since these are minimal conditions for conventionally accepted "correctly formulated" problems.

Earlier in this chapter, we presented a brief description of *coherence* and *consistency* criteria; we now illustrate their application, taking as an example Problem 3 (posed by Malina, grade 4).

> In Princess Rose's jewelry box, there are sapphires, emeralds and rubies. 27 are not rubies, 31 are not emeralds and 32 are not sapphires. In total, there are 45 jewels. How many jewels of each kind does Princess Rose have?

We found that this problem was not coherent, since one of the givens (i.e., the total number of jewels) is redundant. On the other hand, we classified this problem as mathematically consistent because:

- The given data in the posed problem are not contradictory: for example, the numbers 27, 31, 32 are smaller than 45.
- The text elements satisfy a mathematical model: the sum of the numbers 27, 31, 32 is twice the number 45.
- The information given in the text lead to a solution of the problem: Princess Rose has 18 rubies, 14 emeralds, and 13 sapphires.

Figure 7.1 shows the classification of the 53 initially posed problems by coherent-consistent criteria, according to the mathematical content.

Most problems that are both coherent and consistent belong to the categories of Algebraic computing and Handling data, and the fewest are in the categories Change of Patterns and Geometry. (In this discussion, we did not take into account the category Relations, containing only one problem.)

What we hold from this classification is that some content areas seem "safer" in terms of the intrinsic qualities of the posed problems. In other words, coherent and consistent problems occur mainly in the areas of content that require certain formalism, precisely because this formalism provides some stability to a problem statement.

The Interviews

The analysis of the mathematical content and of the text characteristics (syntax and semantics) of the students' posed problems outlined a first overview of the PP *products*. To have a more nuanced understanding of the quality of these problems, we analyzed the interviews to get information on the PP *processes*.

Students' metacognitive strategies in problem posing. We carefully listened to the students' explanations about what they did to elicit a problem. In some cases, they were only able to explain how they had chosen the thematic context of the problem ("My roommates were talking about candies, so I came up with a problem about candies."). In most cases, however, we found that the students had adopted specific strategies for problem posing, which they managed to communicate. Two excerpts from the interviews that exemplify this fact are included below.

When asked how she came up with her problem (#3: "In Princess Rose's jewelry box there are sapphires, emeralds, and rubies. 27 are not rubies, 31 are not emeralds, and 32 are not sapphires. In total, there are 45 jewels. How many jewels of each kind does Princess Rose have?"), Malina explained:

> Malina (grade 4): "You must first establish the answer to the problem, and then you build the wording. You cannot go vice versa; no problem of this type can start otherwise."

Malina explained how she created her posed problem statement: she first decided on the numerical answer, and then she built the givens for the wording. Malina was very categorical in her claim probably because she was aware that the data cannot be random, they should verify certain constraints. In addition, although as a fourth grader she was not previously exposed to PP, she referred not only to her specific problem, but also to an entire class of problems that can be built in that way.

Radu started similarly in posing problem #18 (*Prove that any parallelogram can be divided in 16,384 congruent parallelograms*), but he went deeper into the eliciting process:

> Radu (grade 6): "A problem is made as follows: first we find a purpose: algebra, geometry … an idea … any problem must have a basic idea. After we find the idea, we develop: we add all sorts of tricks, we polish, we re-formulate, and we look for the right numbers. Unfortunately, we got tricks from experience: you cannot do your own problems if, in your turn, you didn't solve problems. We may borrow some ideas; we cannot do something 100% original."

Compared to Malina, Radu had a different approach: he said that the wording of a problem has to be built in successive steps, being modified by a kind of trial and error strategy ("we re-formulate, we look for suitable numbers"). The difference in approach might come from the age difference between the two students. Radu (grade 6) possessed mathematical knowledge certainly more developed than Malina's (grade 4). Radu was confident that, during the problem-solving process, he could anticipate constraints among the data, parameters, and operations, and that he could amend the wording to ensure problem consistency. Indeed, Radu's proposals showed that he had spent a lot of time in formulating and reformulating the problem (compared to other students in our sample). In addition, he engaged himself in qualitative analysis: in his problem, the choice of the number 16,384 was purpose-oriented—it is a perfect square big enough to remove any possibility of reasoning on a geometrical figure. The choice of this number shows that Radu actually developed a generalization (as confirmed during the interview). Radu's behavior in problem posing was of an expert type, as described by Silver and Marshall (1989). This was apparent also in his spontaneous development of explanations and "meta" comments.

As we have seen in the above examples, some students spontaneously shared their visions on what a problem should look like. These comments on the desirable qualities of a problem led us to the conclusion that these students developed metacognitive strategies, which they were able to make explicit. While expressing their opinions concerning the problem-posing process, students revealed a complex philosophy about PP. Some significant examples are included below.

> Radu (grade 6): "[The posed problem] must be original. But it's not worth to be original if it is too easy, so we have to give it difficulty."

Radu asserted that, for posed problems, originality was a necessary condition and, in addition, that a problem must have a degree of difficulty. Radu's claim, in correlation with his other assumptions, can be interpreted as an intuitive understanding of the need for consistency in a problem.

> Malina (grade 4): "Creation usually happens in an artistic composition. You can compose also math problems, but in this case, logic occurs, you have to be very logical and exact in explanations and calculations. It's complicated, because you must keep in mind certain rules of mathematics."

Malina compared PP with an artistic act. Actually, specific literature shows that metacognitive abilities can be related to creative thinking (e.g., Fasko, 2001). Starting from the art comparison, Malina emphasized the differences. She was aware that the wording of a problem must satisfy certain specific constraints that contribute to its mathematical consistency.

These examples serve to illustrate that, in the PP process, students in our sample frequently demonstrated metacognitive behaviors. This allowed them to *look from above* on how to pose a problem; as a result, students can develop strategies for selecting content and constraints to make the new posed problem consistent. Thus, they become able to evolve within the cognitive frames generated by their chosen problem models.

Students' need for social interaction in problem posing. Some students spontaneously included comments for possible collocutors in their posed problems. Others referred to such collocutors in their remarks during the interviews. This observation outlines a need for social interaction of these children through posing and solving problems.

For example, Cristiana (grade 6) seemed to be posing her problem (#13, the "look-and-say sequence") for a friend and displayed a protective role, revealed through her careful reflection on the problem difficulty. Cristiana added in Problem 13 an indication "you must empty your mind of all other mathematical information." When we asked her why she did this, she said:

> I thought that in this way a child would like to read so far.

Already at this step, she entered into the teacher's role and tried to bring both support and motivational elements to the potential solver. In that respect, she summed up how she came to understand the look-and-say sequence herself and tried to translate her own experiences into her proposal.

Cristiana formulated her proposals to provide some support to the solver. This case is not unique: other students also formulated questions keeping in mind the profile of potential solvers. The students who took into account the mathematical skills of their colleagues as potential solvers focused not only on the problem text, but also on how the other person was likely to decode the problem, a fact also noticed in other studies (e.g., Lowrie, 2002).

Unlike the cases described above, other students introduced some elements with the purpose of misleading the solver. Given some situations frequent in the teaching practice in Romania (see, for example, Pelczer, Singer, & Voica, 2013b), we concluded that, through the wording of their posed problems, these students were, in fact, mimicking their teachers' behavior and beliefs. One significant example of this type was Malina's comment about her problem:

> Malina (grade 4, referring to Problem 3): "Instead of saying directly that the yellow and the red ones are 39, I have complicated it, that the children think: we eliminate the blues, and we remain with the yellows and the reds."

In this example, we saw that Malina *intended* complicating this problem to mislead potential solvers, in contrast to those who assumed the role of "protecting" solvers by adding some points of support (as Cristiana did). We have thus highlighted two opposite behaviors exhibited by problem posers, with both emphasizing students' desire for social interaction.

Many educational researchers perceive social interaction as an important factor for stimulating mathematical creativity (e.g., Sfard, 1998; Sriraman, 2004). Most students in our sample spontaneously made connections to social interaction when discussing their posed problems. Their approach in this respect is an additional argument in favor of choosing an organizational framework to study creativity. In this way, we can capture specific aspects, especially related to the field of organizational learning, aspects that are irrelevant for other frameworks of mathematical creativity analysis.

Discussion

Our study focused on students who excel in problem solving, winners of a national contest. Usually, the students proficient in mathematics competitions are specifically trained for this purpose. We were interested to see if these students would be able to manage their own learning, and we provided them with a PP context. We consider that PP is a natural and simple situation where we can separate students' creative behaviors from behaviors learned through systematic practice.

Correlations Between PP and PS: Exploring Cognitive Frames

In posing problems, students showed preferences that influence the types of problems they pose. In this section, we will show that:

1. Students' preferences in PP correlate with their strengths in dealing with a certain mathematical content in problem-solving situations.
2. Students' focus on their strengths suggests that personal strengths are the main elements to build well-defined cognitive frames in PP.

We build the argumentation around three significant examples.

Example 1. Cosma (grade 5) initially posed the following problem (# 10):

> Two boys need £67 to buy a game. The price of the game decreases by 50%. If the first boy pays three times more than the second does, how much money should each pay?

When, during the interview, we asked Cosma where the idea of the problem came from, he said that he likes problems with fractions. To see if this comment is consistent with Cosma's problem-solving capacity, we analyzed his answer sheets from the two rounds of the Kangaroo contest. We noticed that he formulated responses to all 9 problems with fractions (of 80 problems). Cosma was wrong on 15 of the 60 problems he had chosen (so his proportion of mistakes is 25%), but none of these was related to fractions. At least two of the problems with fractions proposed in the second round and correctly solved by Cosma can be considered of high level of difficulty, being correctly solved by less than 20% of the participants. (We should also take into account that more than half of the co-participants were one year older than Cosma.) These observations confirm, on the one hand, the student's real preference for problems with fractions, and on the other hand, his high mathematical capacity in solving problems with this content. Therefore, his preference strongly correlates with the mastery of solving this category of problems.

Cosma proposed a coherent and consistent problem in which operations with fractions appeared as the main working tool and defined his cognitive frame for this problem. The fact that he explicitly claimed a preference for this area strengthens the persistence in this frame. Obviously, in Cosma's case, the cognitive frame correlates with his cognitive strengths.

Example 2. An interesting case is that of Victor (grade 4), who initially posed Problem 8. During the interview, Victor modified it, arriving at the following problem:

> At the "ABC" contest of numbers, each letter has received a number of points. Miss B exceeded Mr. A with 2 points, but D has exceeded Miss B. The Letter D scored so high that only the sum of scores obtained by A and B is equal to D's score. But D wasn't the best! E's score was double of that of D. However, F was the best. He got a score equal to the sum of the scores obtained by E and D. Knowing that if from the score of F we subtract 20 and then divide the result by 7 we get the half of the half of 16, how many points did each participant gain?

For both posed problems, Victor used a graphical method of solving; Figure 7.2, shows the solution he gave to Problem 8.

When we asked him how he came to pose these problems, he said:

> I didn't have a pattern; the idea with graphics that *come one after another* came to me randomly.

Even if he does not seem to be aware of this, Victor developed problems that can be modeled with systems of equations in *row echelon form*, in which the solving can be made "step by step" from the end to the beginning. For both problems, he actually used generating schemes similar to those shown in Figure 7.3. Therefore, Victor acted within a cognitive frame that he systematically used in his posed problems.

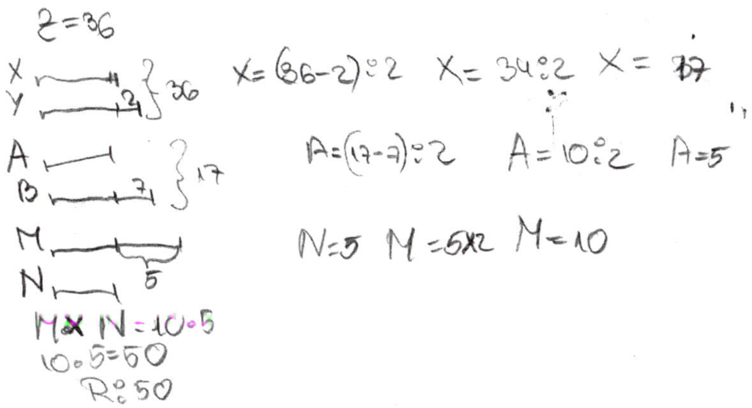

Figure 7.2. Victor's solution to his initially posed problem.

Problem 8, posed initially

Problem posed during the interview

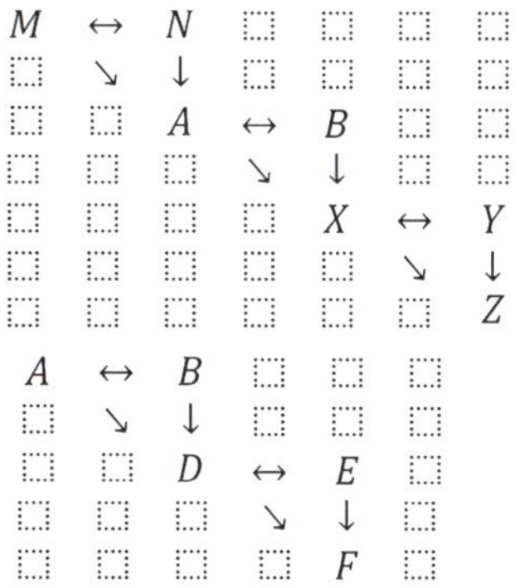

Figure 7.3. The problem-generating scheme used by Victor (grade 4).

Victor's model for creating problems is one for which a high degree of generalization is possible. However, although just a fourth grader, he was able to control the model for different cases, which demonstrates that this approach was a strength of his.

Example 3. Mihai (grade 6) posed the problems, classified in the category *Handling data*, which we analyzed at the beginning of the Results section of this chapter. We were interested to see if there is any connection between Mihai's preference for this category of content and his response pattern in the Kangaroo competitions. We noticed that the strategy used by Mihai (grade 6) in the competition allowed him to give wrong answers to only 6 problems (10%) of his 60 chosen problems (of the two rounds of the Kangaroo contest). Analysis of his pattern of choices in the competition showed that he jumped over high-complexity problems whose decoding at first sight appeared to be difficult. An example is provided by the following question, difficult to approach at a first glance (and in a relatively short time) by a sixth grader:

> On the number line, the segment between 1 and 100 is divided by points in 2011 equal parts. Find the sum of the coordinates of these points.

A problem of this type, in which many (seemingly unrelated) parameters occur, may generate chaos in a sixth grader's mind because without a culture of solving such problems, the given information could not be structured to minimize the number of independent variables. Mihai avoided this problem, and others of this type, probably because he failed to interpret the wording so as to diminish the number of parameters.

Mihai's preference for structured problems, where the relationship between the different variables of the problem were easily identifiable, is consistent with the model he chose for his posed problem (see Problem 16)—a model in which the identification of regularity in the random distribution of points was essential for solving.

Intuitively, Mihai knew that problems which do not display an immediate structure (or suppose a structure to which he had no access) were to be avoided in competitions. His target was to optimize his actions—a reasoning which is similar to that of a manager who analyzes his/her resources and makes the best knowledgeable decision.

As we have seen above, some students naturally displayed metacognitive abilities. They were able to describe their own approaches to problem posing—they were able to manipulate the constraints and the data, and they were successful in following, both consciously and systematically, a certain strategy in order to get an anticipated result. As in Mihai's case, we see that these students also applied metacognitive strategies in problem solving, in competitions where they had to solve a large number of problems in a short time. The analysis of such metacognitive behaviors of students confirmed our decision to use an organizational framework for analyzing creativity. The above three cases were not isolated examples in our study—for most of the students who posed coherent and consistent problems we found a correlation between their assumed strengths and the cognitive frames within which they built their problems.

The cases presented in this chapter demonstrate that, usually, students posed problems that were associated with their preferred mathematical content areas, and which were connected to their cognitive strengths. Thus, students intuitively felt the need to have a deep understanding of the chosen area in order to keep some control over the quality of their posed problems. This suggests that, when posing problems, students typically work within a well-defined cognitive frame.

Starting Points in Problem Posing: Exploring Cognitive Novelty

In this section, we claim the following:

1. Typically, in PP activities students start from known models, thus limiting cognitive novelty; and

2. When students do not start from a model or when they are not familiar with the model, their posed problems show cognitive novelty, but in most cases, at the expense of problem coherence and consistency.

Initial analyses of the students' posed problems suggest that, in most cases, students seem to use problems previously encountered as support for generating new ones. During the interviews, some students confirmed this assumption, as, for example, in the following excerpt from an interview with Malina:

> Malina (grade 4), with reference to Problem 3: "I had a similar problem in a contest in grade 2. I did not know how to handle it and I was very upset, because at that time I used to get very upset when I was finding something I do not know."

In other situations, we recognized "classical" problems, adapted and transformed. For example, Problem 13, posed by Cristiana (grade 6), is known in the literature as "the look-and-say sequence." Cristiana's contribution was to add some comments, designed to target possible approaches to solving the problem (and a possible collocutor).

The starting point is best visible in the problems generated during the interviews, where the model was clear—the problem originally posed by the student. For example, as shown in the cases presented earlier in this chapter, Mihai managed to generate new problems that did not depart significantly from his original problem, but were coherent and consistent. The same applies to Cristiana. Her Problem 13 was a classical one and Cristiana's contribution was minimal. When asked to generate a new problem, Cristiana proposed the following wording:

> Maria has to solve the following problem: "311311122112, 111312112, 132112, ... What is the next term of the string?"

Cristiana did not move away from the assumed model significantly: in the new problem, she just reversed the order of crossing "the look-and-say sequence." During the interview, Cristiana affirmed her belief that every natural number that has an even number of digits may be a term within the look-and-say sequence. Even if this claim is not true (her condition is necessary, but not sufficient), the new posed problem is coherent and consistent, but again, is not far from her model.

Among the posed problems, there were only a few for which we either did not identify a possible model, or uncover it during the interview. One of these exceptions was Problem 17, posed by Nandor (grade 6):

> Dan has a 24 hour-display digital clock that is broken: the first digit of the hours' counter and the last digit of the minutes' one get switched every 5 hours. Example: if a switch occurs at 17:42, the clock will show 24:71. The clock continues to run correctly after that, and stops at 99:99, when it gives an error (1 hour is transformed in 100 minutes). If the clock breaks when the correct time of the day is 10:10, what will be the time before giving the error?

This problem indicates cognitive novelty, because it was far from those in textbooks or school auxiliaries. Nandor's posed problem is, however, neither coherent nor mathematically consistent. The author himself failed to clarify the solution, saying only that "all the numbers should be written—there are about 100." Nandor did not start from a model, and his problem showed cognitive novelty but at the expense of consistency.

We systematically asked the interviewed students to formulate new problems starting from those they originally posed. Consequently, during the interviews, they generated 26 new problems in total (some students posed several new problems, and each offered at least one). For example, Mihai (grade 6) posed the following amendment to his problem (#16):

> Prove that if we have 801 points in a square 60×60, then there exists a triangle with an area smaller than 9, determined by three of the 801 points.

To solve his new problem, Mihai used a technique similar to one he used for his originally posed problem (#16, in the Appendix)—he determined the most dispersed distribution related to a square grid of 3×3 and applied the pigeonhole principle. Later, when he explained his solution, Mihai considerably improved his proposal by finding an optimal version; he replaced 9 with 4.5 without any suggestion or request from the interviewer.

These examples demonstrate that, in general, when a student modified a given problem, she/he changed only some of the elements of that problem. We analyzed those changes based on the problem-analysis framework of Singer and Voica (2013), looking at changes to the following: the background theme, the parameters, (numerical) data, the operating schemes, the constraints over the data and the operating schemes, and the constraints that involve at least one unknown value of the parameter(s). Compared to the problems initially posed, in the 26 new problems, the students in our sample most often changed the givens (in 14 cases), or the background theme (in 8 cases). In only one case was the operating-scheme changed.

Yet, most of the students in our sample posed problems starting from an already known model. The existence of a starting model seemed to prevent the students from showing cognitive novelty. Thus, the vast majority of students in our sample started from a model when they posed problems, and in most cases, the posed problems did not go far from the model. Therefore, in problem posing (and modification) situations, cognitive novelty is limited, probably because of the students' awareness of a predefined cognitive frame.

However, this limitation seems to be relevant beyond the creativity issue, because it seems to ensure coherence and consistency in the new posed problems. Conversely, students who are apparently more creative did not have or have not yet built a cognitive frame, a fact that prevents them, most likely, from offering mathematically consistent problems.

Limits and Challenges of Mathematical Creativity

Within the framework used in this chapter, cognitive flexibility is characterized by cognitive novelty, cognitive variety, and change in cognitive framing. As we saw above, in PP situations, cognitive novelty is limited, and the students feel the need to evolve within a well-defined frame, which corresponds to the outgoing model used for posing a problem. In this section, we focus on how the students made changes to their cognitive frame. More specifically, we present evidence to support the following claims:

1. In PP situations, students are cautious about making major changes to the assumed model;
2. Consequently, students adopt a strategy of "small steps" in changing the starting model; and
3. This strategy of "small steps" seems to characterize mathematical creativity in PP activities.

We will focus this discussion on three concluding examples.

Example 1. Malina (grade 4) originally posed Problem 3. (A discussion on this problem appears in the results section of this chapter.) During the interview, we wanted to see, on the one hand, if Malina could develop new problems starting from her original problem, and on the other hand, whether she understood the mathematical tools she used for solving her problem.

First, Malina modified the numerical data of her problem by proposing the numbers 39, 50, 61, and 75 (=total number of jewels). She later explained us how new wordings can be developed starting from this problem:

> Malina: "If I think about marbles of more colors … So in a box there are black, blue, red and yellow marbles. If I say: a defined number, for example, 13, are not blue, it means that they are yellow, red and black. Of total … it is the same thing, only that there are more numbers; of the total number, I subtract the sum of the three and I got exactly the needed number…"

Malina kept the background theme and the constraints of the original problem, but changed the givens and the number of parameters (she now considered four different objects—i.e., marbles of different colors, instead of the three types of jewelry in the original problem). Malina explained how she generated the new wording: she increased the number of parameters ("I think to marbles of more colors…") and applied the same strategy for solving. Malina actually got to a generalization process for the original problem ("it's the same thing, only that there are more numbers"). We were interested to see if Malina was aware of the constraints on the numerical values of the problem. The interview continued as follows:

Interviewer: So, how do we get the total number?

Malina: Oh, here comes a different kind of problem … if you know that some are not black, some are not blue, some are not yellow and some are not red, you have to add these amounts and you get three times the amount exactly.

I: How is that, 3 times when there are 4 colors?

M: Well, you collect the yellow, the red, the blue [she gestures], then collect the yellow and black and blue, then … and then what's left and every time you notice that each number comes out three times.

I: And if there were 100 colors?

M: Then we would get … uh … uh …

I: Let's not say 100, let's say 6 colors!

M: If there were …we would get 6 colors 6 times … no! 5 times the amount!

It seems that Malina has activated cognitive mechanisms to verify the correctness of this type of problem. These mechanisms allowed her to establish correlations between the data and parameters and to verify the mathematical consistency of the problem. In fact, the mathematical model of the problem described by Malina is a linear system of four equations with four unknowns. As a fourth grader, Malina had no formalized knowledge in solving mathematical systems of equations. Nevertheless, she controlled the system and determined conditions for compatibility. She not only showed the computational strategy to solve the problem, but she was able to generalize this method for other similar conditions, chosen at random. Problem 3, originally posed by Malina, was classified as non-coherent (because redundant data occurred in the wording). The explanations presented above, given by Malina during the interview, convinced us that this redundancy of data seemed to be rather a reassurance that the proposed data were compatible, than an expression of conceptual misunderstandings. Thus, in problem posing, Malina acted within a well-defined cognitive frame set up for her problem.

Further, we wanted to see how far Malina might make changes in her cognitive frame. Consequently, we asked her to pose problems as simple as possible, starting from her initial one. Malina's proposal was:

> In a box there are 75 balls, yellow, red and blue. Of these, 39 are red and yellow, 61 are blue and red and 50 are blue and yellow. How many balls are there in the box?

The interviewer expressed the opinion that this was, in fact, the same problem as one of her previous reformulated problems. Her answer was: "It's the same problem, but told differently, more clearly." The interviewer insisted and asked Malina to pose an even simpler problem. She needed a longer time to think, hesitated, and then posed the following wording:

> In a box there are 75 balls: red, yellow and blue. Of these, 39 are yellow and blue. Find the number of each color.

Malina posed a new problem by reducing the number of constraints and giving up two of the parameters. The change was now more extensive than in the previous cases, but this led to a problem that was neither coherent nor consistent. This was quite surprising, since we thought that Malina had showed deep understanding of her problem's pattern. Because she well understood the relationship between the components of her initial problem, she succeeded in making changes to her cognitive frame, and to keep control over the problems obtained by generalization or by changing the operating scheme, but only as long as the changes were not far. When these changes were wider, she ended up losing control, and posed problems that kept a general pattern, but did not prove consistency and/or coherence. Continuing the analysis, we find that Malina's intuitive attempt to keep control over the new posed problems limited cognitive novelty. Malina intuitively did not go too far from the assumed model, but when she was pushed to do so, her new problems, although simpler, became mathematically inconsistent.

Example 2. Maria (grade 6) initially posed the following problem (#15):

> A number is "special" if it can be written as both a sum of two consecutive integers and a sum of three consecutive integers. Prove that: (a) 2,001 is special, and 3,001 is not special; (b) the product of two special numbers is special; (c) if the product of two numbers is special, then at least one of them is special.

Maria managed to identify equivalent characterization for her so-called "special" numbers: a number is "special," if and only if it is an odd number, divisible by 3. Once she had this general characterization of algebraic nature, Maria could easily pose some new problems:

> Prove that the sum of three "special" numbers is "special."

> A number is "very special" if it is both special and perfect square. Give an example of a very special number.

In posing the first new problem, Maria largely kept the wording and varied only the constraints that involved at least one unknown value of the parameter (she changed the question). For the second problem, Maria included a new constraint (the number must also be a perfect square).

Maria worked in a well-defined cognitive frame: she transposed the problem algebraically and used a general characterization of the "special" numbers to identify new properties of these numbers. Maria did not change her cognitive frame associated with this problem; she always used the same initial properties and did not explore her problem in other directions. The changes she made for her new proposals were minimal, although her posed problems were highly abstract.

Example 3. Radu (grade 6) originally posed Problem 18 (*Prove that any parallelogram can be divided in 16,384 congruent parallelograms*). During the interview, Radu explained that, in posing this problem, he started from the observation that a given parallelogram can be divided into four or nine congruent parallelograms (by dividing each side in two or three equal parts and constructing parallel sides through the points of division). He chose the number 16,384 just to give difficulty to the problem ("There must be a big enough number, perfect square."). Thus, the relatively big distance between the initial model (i.e., for the particular cases 4 and 9) and his final proposal was given by his evolution within a well-internalized cognitive frame. For this proposal, Radu changed only one parameter (the number of congruent parallelograms) and thus obtained a new problem, which was coherent and consistent.

When we asked him to pose another problem of the same type, Radu made the following comment:

> I'd be a bit tempted to say that any triangle can be divided into 16,384 congruent triangles, but I am not sure of the solution. ... Yes, I would be tempted to do again with a parallelogram and to apply the same idea, just up here...

In the new posed problem, Radu kept the background theme and numerical data, but modified a parameter (he replaced the parallelogram with a triangle). In fact, Radu formulated a conjecture ("a triangle can be divided into 16,348 congruent triangles"). Radu tried to solve his new problem by completing the triangle to a parallelogram and applying the same idea for solving. In this phase of testing, he remained within the same cognitive frame. Analyzing some particular cases, Radu concluded that the problem required some additional assumptions (such as, for example, that the number of triangles must be even) and that, perhaps, the initial solution method could not be applied. Radu returned later (after 1 day) with new reformulations and attempted to solve this problem. Although his attempts were not entirely correct (probably because the solving of the new problem required knowledge about similarity, to which Radu had no access at that time as a sixth grader), he concluded that it was plausible that the number of triangles must be a perfect square. This showed that Radu was, in fact, able to reframe.

These examples, like others of a similar kind that we found in our sample, led us to conclude that, in problem-posing situations, a student acts within a definite cognitive frame that allows him/her to generate mathematically consistent problems. Further, some students succeed in making changes to those cognitive frames or even to reframe. These changes were not always spectacular because students intuitively tend to maintain coherence and consistency of the posed problems, and changes that are more extensive prevent them from keeping control over the shape of the problem. But when students make small-scale changes (usually by varying a single parameter), they can understand the impact of these changes on the constraints of the problem text and they can choose appropriate numerical data. Therefore, a student's capacity to generate coherent and consistent problems in the context of problem posing (and modifications) may indicate the existence of a strategy of IN-OUT functional type consisting of small changes followed by checking the outcomes, which seems specific to mathematical creativity. In more general terms, mathematical creativity seems to emerge from changes in cognitive framing, which express a tension between the maintenance within a frame and the possibility of overcoming it for generalizations, possibility constrained by the need for consistency.

Conclusions

This chapter presents an empirical study in which students in grades 4–6, with above-average mathematics abilities, posed problems. We tried to answer the question: How does mathematical creativity manifest in 10–13 year-old high achievers?

The results show that in the PP process, students develop a genuine philosophy, which refers both to practical actions—embodied in their problem-posing strategies—and to the qualitative form of the posed problems. Typically, students start from a model to which they apply certain constraints based on the philosophy they developed, and they then spontaneously try to get a problem that is mathematically consistent and coherent.

We noticed that both the problems posed by the students and the behaviors presented by the students highlighted a social dimension. We have several arguments to support this claim. On the one hand, in most cases, the background theme of the posed problems had a civic connotation: students go to paintball as a prize because they are in an advanced class, or some families do not pay their waste collection fee, and so on. Other posed problems simply involved friends, classmates, or neighbors. On the other hand, some students directly addressed some challenging areas, or provided some support for the solver. Thus, most students took into account possible collocutors within the PP activities. The PP context allowed students to seek ways to distort/alter the magnitude of the problem changes by adding text elements that referred to the author's interaction with a potential recipient of the problem. Students succeeded both in maintaining quality control over the new posed problems and in responding to a need for social interaction.

The social dimension of the PP process revealed by these children's options confirms the meaningfulness of the framework of analysis that we used in this study, in which we discuss the relationship between problem posing and mathematical creativity in terms of cognitive flexibility in an organizational framework.

The study provides evidence that of the three components of cognitive flexibility (i.e., cognitive variety, cognitive novelty, and change in cognitive framing), the last seems to be the most relevant for PP situations. More specifically, the majority of students in our sample started from a model for which they already had a well-defined cognitive frame and posed new problems within this frame.

Students were generally able to make changes to their cognitive frames as they succeeded in posing new problems starting from the initial ones, problems that displayed different approaches compared to the starting point. Yet, among these, it was significant to study the thinking patterns of those students whose proposals, issued either initially or during a modification process, were coherent and consistent.

In modifying a problem, students tried to vary a single parameter; the ones who succeeded to do this could control the consequences of the changes and managed to develop coherent and consistent mathematical problems. Their strategies revealed a kind of cognitive variety that was relatively limited by their desire to control the outcomes of the process. This also limited cognitive novelty.

Therefore, cognitive flexibility seems to be oriented towards finding generalizations and is constrained by the need to maintain the mathematical consistency of the problem. Consequently, students' approaches in the PP process seem to be of an "in-out" functional type, with a careful check of the variations induced over the outcomes.

The study brings evidence that this type of approach in posing/or modifying a problem, which allows for making generalizations, seems to be specific for mathematical creativity, at least for the sample analyzed in this chapter. More specifically, we show that in PP contexts, students tend to make incremental changes to some parameters in order to arrive at simpler and essential forms needed in generalizing sets of data. It follows that mathematical creativity is of a special type—one which requires abstraction and generalization. In addition, students showed awareness of the need for mathematical consistency, which made them persevere as they carefully controlled the changes.

Moreover, the interviews revealed that this awareness meant that most of the students were able to analyze their own proposals critically and their own thinking mechanisms in PP, thus reflecting their metacognitive skills. These metacognitive skills helped them to be aware of their strengths and to use these strengths to reinforce a well-defined cognitive frame for a problem.

On the basis of these conclusions, our study highlights some aspects that have consequences for effective teaching. We briefly present these below.

First, we have seen that students have preferences for some subareas of mathematics, or for some problem-solving strategies, which can be relatively easy identified through PP activities. Students' preferences reveal the strengths on which teachers can focus in order to develop students' mathematical competences.

Second, our data show that students need social interaction. Surprisingly, this need surfaced through problem posing—an individual type of activity. Our conclusion is that social interaction should be part of the teaching-learning process in the class in a consistent way, for example, by means of activities involving posing and solving problems organized in pairs or in teams.

Third, the study shows that PP stimulates metacognitive abilities in students. From this perspective, the use of PP in teaching is beneficial to students' personal development.

Finally, training for the development of mathematical creativity should include features that distinguish it from training for the development of creativity in general. Briefly said, while in the latter more general case, techniques are to be used for stimulating the free development of ideas, in mathematics the variation of parameters should be practiced within a variety of activities where the processes are mindfully controlled and oriented towards abstraction and generalization.

Appendix

Problem 1 (*posed by Diana, grade 4*): On the planet Zingo live several types of aliens: with two or three eyes, with two or three ears, and with five or six hands. They are green or red. How many aliens should shake hands with Mimo to be sure that he shook hands with at least two of the same type?

Problem 2 (*posed by Emilia, grade 4*): In the Infinite king's castle there are 43 corridors with 18 rooms each. Each room has 52 windows. At every window, there are three princesses. How many princesses are in the Infinite king's castle?

Problem 3 (*posed by Malina, grade 4*): In Princess Rose's jewelry boxes there are sapphires, emeralds, and rubies. 27 are not rubies, 31 are not emeralds, and 32 are not sapphires. In total, there are 45 jewels. How many jewels of each kind does Princess Rose have?

Problem 4 (*posed by Paul, grade 4*): If a group of students sits by two at their desks, seven students remain standing, and if they are placed by three at the same desks, seven desks remain free. How many students and how many desks are there in the classroom?

Problem 5 (*posed by Sabina, grade 4*): In the Fairies' Glade, live 60 unicorns and fairies. They have 160 legs in total. How many beings of each kind are there?

Problem 6 (*posed by Sergiu, grade 4*): One day, the plane leaves Cluj [*a city in Romania*] to go to Japan. It departs at sharp hour in the morning, when the hour and the minute hands of a clock form a right angle. The hour hand points to a number bigger than 4. This plane travels at 60 km per hour, and the distance Cluj-Japan is 540 km. In the plane climbed three times more men than women, who are 24,484 people. The cost for a man's tickets is the first odd number greater than 7. Women's tickets cost as double of 3 added with 4 and the result divided by 2. (A) When did the plane leave Cluj? (B) When did the plane arrive in Japan? (C) How many men boarded the plane? (D) How many women boarded the plane? How much money did the pilot receive, if he received all the money, without 3,000 of total?

Problem 7 (*posed by Tudor, grade 4*): On a farm, there are two cows, some geese and horses, a total of 86 heads and 328 feet. How many horses are there at the farm?

Problem 8 (*posed by Victor, grade 4*): In the world of letters, each letter represents a number. M is two times greater than N, and the difference between these two letters is equal to A. A is less than B by seven, and the sum of A and B is neither bigger nor smaller than X. If we add two to X, we get Y. The sum of X and Y equals Z. Knowing that $Z-(A{:}\ O+P{:}\ P+Q{:}\ Q+A{:}\ R)=30$, find $M\times N$.

Problem 9 (*posed by Alin, grade 5*): (A poem!) If one places three cakes in each box/There'll be three cakes left/If one places five cakes in each box/There'll be an empty box left. (…) How many boxes and how many cakes/Do I put on the shelves?

Problem 10 (*posed by Cosma, grade 5*): Two boys need £67 to buy a game. The price of the game decreases by 50%. If the first boy pays three times more than the second does, how much money should each pay?

Problem 11 (*posed by Andrei, grade 6*): On the planet Uranus in the T316B2 city, there are less than 101 and more than 49 aliens. 1/2 of them are red, 2/7 are green, 1/14 are yellow, and the rest are blue. How many aliens live in E943S4 city, the capital of the planet, if their number is 149 times greater than the count of blue aliens from T316B2?

Problem 12 (*posed by Cosmin, grade 6*): $P\cdot R\cdot I\cdot C\cdot E\cdot P\cdot I\cdot P\cdot R\cdot O\cdot B\cdot L\cdot E\cdot M\cdot A=x$. Knowing that different letters represent different digits, find the last digit of the number x. (*He multiplies the letters meaning YOU UNDERSTAND THE PROBLEM.*)

Problem 13 (*posed by Cristiana, grade 6*): Maria has to solve the following problem: "4, 14, 1114, 3114, 132114, 1113122114, … What is the next term of the sequence?" The mathematics teacher gave her some advice: "You must empty your mind of all other mathematical information." Can you help Maria to solve the problem?

Problem 14 (*posed by Cristiana, grade 6*): Maya the puppy has six bones. She wants to make four equilateral triangles out of these six bones, but she forgot one essential rule: one has to think out of the box. Can you help her?

Problem 15 (*posed by Maria, grade 6*): A number is "special" if it can be written as both a sum of two consecutive integers and a sum of three consecutive integers. Prove that: (a) 2,001 is special, and 3,001 is not special, (b) the product of two special numbers is special, (c) if the product of two numbers is special, then at least one of them is special.

Problem 16 (*posed by Mihai, grade 6*): Because the sixth-grade students were the best, they received a prize consisting in 1 h free on paintball field. The field has the dimensions 80 m $\times$ 120 m, and two people are able (and allowed) to shoot one another if they are at no more than 29 m distance. Prove that howsoever 26 students place themselves on the ground, at least 3 get shot.

Problem 17 (*posed by Nandor, grade 6*): Dan has a 24 hour display digital clock that is broken: the first digit of the hours' counter and the last digit of the minutes' one get switched every 5 hours. Example: if switch occurs at 17:42, the clock will show 24:71. The clock continues to run correctly after that and stops at 99:99, when it gives an error (1 hour is transformed in 100 minutes). If the clock breaks when the correct time of the day is 10:10, what will be the time before giving the error?

Problem 18 (*posed by Radu, grade 6*): Prove that any parallelogram can be divided into 16,384 congruent parallelograms.

Problem 19 (*posed by Teofil, grade 6*): In 2011, 300 students went on the field trip. Knowing that the percentage of girls was 45%, find the number of boys who participated.

Problem 20 (*posed by Vlad, grade 6*): A new quarter was built near a forest. The residents put their garbage in waste containers with a capacity of 750 kg each. At every 10 kg of garbage throw away by the residents, 4 kg disappear, being consumed by bears leaving in the forest. The residents produce 20 kg of garbage per hour. (a) Find out how long it take to fill a waste container; (b) Knowing that in the neighborhood live 500 families that fill 86 containers per month, that each family should pay 7.8 euros garbage fee, but only 400 families are fair and pay, calculate how much money is collected as garbage fees in a month.

References

Amabile, T. M. (1989). *Growing up creative: Nurturing a lifetime of creativity*. Buffalo, NY: Creative Education Foundation Press.

Baer, J. (1993). *Divergent thinking and creativity: A task-specific approach*. Hillsdale, NJ: Lawrence Erlbaum.

Baer, J. (1996). The effects of task-specific divergent-thinking training. *Journal of Creative Behavior, 30*, 183–187.

Baer, J. (1998). The case for domain specificity in creativity. *Creativity Research Journal, 11*, 173–177.

Baer, J., & Kaufman, J. C. (2005). Bridging generality and specificity: The amusement park theoretical (APT) model of creativity. *Roeper Review, 27*, 158–163.

Balka, D. S. (1974). The development of an instrument to measure creative ability in mathematics. *Dissertation Abstracts International, 36*(01), 98. (UMI No. AAT 7515965).

Brown, S. I., & Walter, M. I. (1983/1990). *The art of problem posing*. Hillsdale, NJ: Lawrence Erlbaum.

Cai, J., & Cifarelli, V. (2005). Exploring mathematical exploration: How two college students formulated and solved their own mathematical problems. *Focus on Learning Problems in Mathematics, 27*(3), 43–72.

Csikszentmihalyi, M. (1996). *Creativity, flow, and the psychology of discovery and invention*. New York, NY: Harper Collins.

Csikszentmihalyi, M., & Getzels, J. W. (1971). Discovery-oriented behavior and the originality of creative products: A study with artists. *Journal of Personality and Social Psychology, 19*(1), 47.

Deák, G. O. (2004). The development of cognitive flexibility and language abilities. *Advances in Child Development and Behavior, 31*, 271–327.

Demetriou, A., Kui, Z. X., Spanoudis, G., Christou, C., Kyriakides, L., & Platsidou, M. (2005). The architecture, dynamics, and development of mental processing: Greek, Chinese, or Universal? *Intelligence, 33*(2), 109–141.

Dillon, J. T. (1982). Problem finding and solving. *Journal of Creative Behavior, 16*, 97–111.

Dow, G. T., & Mayer, R. E. (2004). Teaching students to solve insight problems: Evidence for domain specificity in creativity training. *Creativity Research Journal, 16*(4), 389–398.

Dreyfus, T., & Eisenberg, T. (1996). On different facets of mathematical thinking. In R. J. Sternberg & T. Ben-Zeev (Eds.), *The nature of mathematical thinking* (pp. 253–284). Mahwah, NJ: Lawrence Erlbaum.

Eisenhardt, K. M., Furr, N. R., & Bingham, C. B. (2010). Microfoundations of performance: Balancing efficiency and flexibility in dynamic environments. *Organization Science, 21*(6), 1263–1273.

English, D. L. (1998). Children's problem posing within formal and informal contexts. *Journal for Research in Mathematics Education, 29*(1), 83–106.

Ervynck, G. (1991). Mathematical creativity. In D. Tall (Ed.), *Advanced mathematical thinking* (pp. 42–53). Dordrecht, The Netherlands: Kluwer Academic.

European Commission. (2003/2004). *Working group on basic skills, progress reports 2003, 2004*, European Commission interim working document for the objectives 1.2 (developing the skills for the knowledge society), 3.2 (developing the spirit of enterprise), and 3.3 (Improving foreign language learning) Annex 2 Domains of Key Competences—Definitions, Knowledge, Skills and Attitudes. Retrieved from http://europa.eu.int/comm/education/policies/2010/objectives_en.html#basic

European Commission. (2005). *Proposal for a recommendation of the European Parliament and of the Council on Key Competences for Lifelong Learning*, Council of The European Union, 11 November 2005.

Evans, E. W. (1964). Measuring the ability of students to respond in creative mathematical situations at the late elementary and early junior high school level. *Dissertation Abstracts, 25*(12), 7107. (UMI No. AAT 6505302).

Fasko, D. (2001). Education and creativity. *Creativity Research Journal, 13*(3–4), 317–327.

Freiman, V., & Sriraman, B. (2007). Does mathematics gifted education need a working philosophy of creativity? *Mediterranean Journal for Research in Mathematics Education, 6*(1–2), 23–46.

Furr, N. R. (2009). *Cognitive flexibility: The adaptive reality of concrete organization change* (Doctoral dissertation). Stanford University, Stanford, CA. Retrieved May 12, 2011, from http://gradworks.umi.com/3382938.pdf

Gardner, H. (1993). *Creating minds: An anatomy of creativity as seen through the lives of Freud, Einstein, Picasso, Stravinsky, Eliot, Graham, and Ghandi*. New York, NY: Basic Books.

Gardner, H. (2006). *Five minds for the future*. Boston, MA: Harvard Business School Press.
Gardner, H., Csikszentmihalyi, M., & Damon, M. (2001). *Good work: When excellence meets ethics*. New York, NY: Basic Books.
Getzels, J. W. (1975). Problem-finding and the inventiveness of solutions. *The Journal of Creative Behavior, 9*(1), 12–18.
Getzels, J. W. (1979). Problem finding: A theoretical note. *Cognitive Science, 3*(2), 167–172.
Getzels, J. W., & Csikszentmihalyi, M. (1976). *The creative vision: A longitudinal study of problem finding in art*. New York, NY: Wiley.
Getzels, J. W., & Jackson, P. W. (1962). *Creativity and intelligence: Explorations with gifted students*. New York, NY: Wiley.
Ginsburg, H. P. (1996). Toby's math. In R. J. Sternberg & T. Ben-Zeev (Eds.), *The nature of mathematical thinking* (pp. 175–282). Mahwah, NJ: Lawrence Erlbaum.
Goncalo, J. A., Vincent, L., & Audia, P. G. (2010). Early creativity as a constraint on future achievement. In D. Cropley, J. Kaufman, A. Cropley, & M. Runco (Eds.), *The dark side of creativity* (pp. 114–133). Cambridge, United Kingdom: Cambridge University Press.
Guilford, J. P. (1967). Creativity: Yesterday, today, and tomorrow. *Journal of Creative Behavior, 1*, 3–14.
Hadamard, J. (1945/1954). *Psychology of invention in the mathematical field*. Princeton, NJ: Princeton University Press.
Hargreaves, A. (2003). *Teaching in the knowledge society: Education in the age of insecurity*. New York, NY: Teachers College Press.
Haylock, D. W. (1987). A framework for assessing mathematical creativity in schoolchildren. *Educational Studies in Mathematics, 18*(1), 59–74.
Haylock, D. (1997). Recognizing mathematical creativity in school children. *International Reviews on Mathematical Education, 29*(3), 68–74. Retrieved March 10, 2003, from http://www.fiz-karlsruhe.de/fix/publications/zdm/adm97
Huang, R., & Leung, K. S. F. (2004). Cracking the paradox of Chinese learners: Looking into the mathematics classrooms in Hong Kong and Shanghai. In L. Fan, N. Y. Wong, J. Cai, & S. Li (Eds.), *How Chinese learn mathematics: Perspectives from insiders* (pp. 348–381). River Edge, NJ: World Scientific.
Jay, E. S., & Perkins, D. N. (1997). Problem finding: The search for mechanism. In M. Runco (Ed.), *The creativity research handbook* (pp. 257–293). Cresskill, NJ: Hampton Press.
Jensen, L. R. (1973). The relationships among mathematical creativity, numerical aptitude and mathematical achievement. *Dissertation Abstracts International, 34*(05), 2168. (UMI No. AAT 7326021).
Krems, J. F. (1995). Cognitive flexibility and complex problem solving. In P. A. Frensch & J. Funke (Eds.), *Complex problem solving: The European perspective* (pp. 201–218). Hillsdale, NJ: Lawrence Erlbaum.
Leikin, R., & Lev, M. (2007). Multiple solution tasks as a magnifying glass for observation of mathematical creativity. In J.-H. Woo, H.-C. Lew, K.-S. Park, & D.-Y. Seo (Eds.), *Proceedings of the 31st International Conference for the Psychology of Mathematics Education* (Vol. 3, pp. 161–168). Seoul, South Korea: PME.
Leung, S. S. (1997). On the role of creative thinking in problem posing. *Zentralbatt fur Didaktik der Mathematik (ZDM), 97*(2), 48–52.
Leung, S. S., & Silver, E. A. (1997). The role of task format, mathematics knowledge, and creative thinking on the arithmetic problem posing of prospective elementary school teachers. *Mathematics Education Research Journal, 9*(1), 5–24.
Lim, C. S. (2007). Characteristics of mathematics teaching in Shanghai, China: Through the lens of a Malaysian. *Mathematics Education Research Journal, 19*(1), 77–89.
Lowrie, T. (2002). Young children posing problems: The influence of teacher intervention on the type of problems children pose. *Mathematics Education Research Journal, 14*(2), 87–98.
Lubart, T., & Guignard, J. (2004). The generality-specificity of creativity: A multivariate approach. In R. J. Sternberg, E. L. Grigorenko, & J. L. Singer (Eds.), *Creativity: From potential to realization* (pp. 43–56). Washington, DC: American Psychological Association.

NAEP Mathematics Framework for the 2011 National Assessment of Educational Progress. (2012). Retrieved October 2012, from http://www.nagb.org/publications/frameworks.html

National Council of Teachers of Mathematics. (2000). *Principles and standards of school mathematics*. Reston, VA: Author.

Nickerson, R. S. (1999). Enhancing creativity. In R. J. Sternberg (Ed.), *Handbook of creativity* (pp. 392–430). Cambridge, United Kingdom: Cambridge University Press.

Orion, N., & Hofstein, A. (1994). Factors that influence learning during a scientific field trip in a natural environment. *Journal of Research in Science Teaching, 31*, 1097–1119.

Pelczer, I., Singer, F. M., & Voica, C. (2013a). Cognitive framing: A case in problem posing. *Procedia—Social and Behavioral Sciences, 78*, 195–199.

Pelczer, I., Singer, F. M., & Voica, C. (2013b). Teaching highly able students in a common class: Challenges and limits. In B. Ubuz, C. Haser, & M. A. Mariotti (Eds.), *Proceedings of CERME 8* (pp. 1235–1244). Antalya, Turkey: ERME.

Poincaré, H. (1913). *The foundations of science*. New York, NY: The Science Press.

Poincaré, H. (1948). *Science and method*. New York, NY: Dover.

Prouse, H. L. (1967). Creativity in school mathematics. *The Mathematics Teacher, 60*, 876–879.

Scott, W. A. (1962). Cognitive complexity and cognitive flexibility. *Sociometry, 25*, 405–414.

Sfard, A. (1998). On two metaphors for learning and the dangers of choosing just one. *Educational Researcher, 27*(2), 4–13.

Silver, E. A. (1994). On mathematical problem posing. *For the Learning of Mathematics, 14*(1), 19–28.

Silver, E. A. (1997). Fostering creativity though instruction rich in mathematical problem solving and problem posing. *International Reviews on Mathematical Education, 29*, 75–80.

Silver, E. A., & Marshall, S. P. (1989). Mathematical and scientific problem solving: Findings, issues and instructional implications. In B. F. Jones & L. Idol (Eds.), *Dimensions of thinking and cognitive instruction* (pp. 265–290). Hillsdale, NJ: Lawrence Erlbaum.

Singer, F. M. (2006). A cognitive model for developing a competence-based curriculum in secondary education. In A. Crisan (Ed.), *Current and future challenges in curriculum development: Policies, practices and networking for change* (pp. 121–141). Bucharest, Romania: Humanitas Educational.

Singer, F. M. (2007). Balancing globalisation and local identity in the reform of education in Romania. In B. Atweh, M. Borba, A. Barton, D. Clark, N. Gough, C. Keitel, C. Vistro-Yu, & R. Vithal (Eds.), *Internalisation and globalisation in mathematics and science education* (pp. 365–382). Dordrecht, The Netherlands: Springer.

Singer, F. M. (2012). Exploring mathematical thinking and mathematical creativity through problem posing. In R. Leikin, B. Koichu, & A. Berman (Eds.), *Exploring and advancing mathematical abilities in high achievers* (pp. 119–124). Haifa, Israel: University of Haifa.

Singer, F. M., Ellerton, N. F., Cai, J., & Leung, E. (2011). Problem posing in mathematics learning and teaching: A research agenda. In B. Ubuz (Ed.), *Proceedings of the 35th Conference of the International Group for the Psychology of Mathematics Education* (Vol. 1, pp. 137–166). Ankara, Turkey: PME.

Singer, F. M., Pelczer, I., & Voica, C. (2011). Problem posing and modification as a criterion of mathematical creativity. In M. Pytlak, T. Rowland, & E. Swoboda (Eds.), *Proceedings of the CERME 7* (pp. 1133–1142). Rzeszów, Poland: University of Rzeszów.

Singer, F. M., & Sarivan, L. (Eds.). (2006). *QUO VADIS ACADEMIA? Reference points for a comprehensive reform of higher education* (In Romanian, with a synthesis in English). Bucharest, Romania: Sigma.

Singer, F. M., & Voica, C. (2011) Creative contexts as ways to strengthen mathematics learning. In M. Anitei, M. Chraif, & C. Vasile (Eds.), *Proceedings PSIWORLD 2011. Procedia—Social and Behavioral Sciences* (Vol. 33, pp. 538–542). Retrieved from http://dx.doi.org/10.1016/j.sbspro.2012.01.179

Singer, F. M., & Voica, C. (2013). A problem-solving conceptual framework and its implications in designing problem-posing tasks. *ESM, 83*(1), 9–26. doi:10.1007/s10649-012-9422-x.

Singh, B. (1988). *Teaching-learning strategies and mathematical creativity*. Delhi, India: Mittal.

Smilansky, J. (1984). Problem solving and the quality of invention: An empirical investigation. *Journal of Educational Psychology, 76*(3), 377–386.

Spiro, R. J., Feltovich, P. J., Jacobson, M. J., & Coulson, R. L. (1992). Cognitive flexibility, constructivism, and hypertext: Random access instruction for advanced knowledge acquisition in ill-structured domains. In T. M. Duffy & D. H. Jonassen (Eds.), *Constructivism and the technology of instruction* (pp. 57–75). Hillsdale, NJ: Lawrence Erlbaum.

Sriraman, B. (2004). The characteristics of mathematical creativity. *The Mathematics Educator, 14*(1), 19–34.

Sriraman, B. (2009). The characteristics of mathematical creativity. *The International Journal on Mathematics Education [ZDM], 41*, 13–27.

Sternberg, R. J., & Lubart, T. I. (1991). Creating creative minds. *Phi Delta Kappan, 72*, 608–614.

Sternberg, R. J., & Lubart, T. I. (1999). The concept of creativity: Prospects and paradigms. In R. J. Steinberg (Ed.), *Handbook of creativity* (pp. 3–15). New York, NY: Cambridge University Press.

Stoyanova, E., & Ellerton, N. F. (1996). A framework for research into students' problem posing. In P. Clarkson (Ed.), *Technology in mathematics education* (pp. 518–525). Melbourne, Australia: Mathematics Education Research Group of Australasia.

Torrance, E. P. (1974). *Torrance tests of creative thinking: Norms, technical manual*. Lexington, MA: Ginn.

Voica, C., & Singer, F. M. (2012). *Problem modification as an indicator of deep understanding*. Paper presented to Topic Study Group 3 "Activities and Programs for Gifted Students," at the 12th International Congress for Mathematics Education (ICME-12). *ICME-12 Pre-proceedings* (pp. 1533–1542). Retrieved from http://icme12.org/upload/UpFile2/TSG/1259.pdf

Voica, C., & Singer, F. M. (2013). Problem modification as a tool for detecting cognitive flexibility in school children. *ZDM Mathematics Education, 45*(2), 267–279. doi:10.1007/s11858-013-0492-8.

Yuan, X., & Sriraman, B. (2011). An exploratory study of relationships between students' creativity and mathematical problem posing abilities: Comparing Chinese and U.S. students. In B. Sriraman & K. Lee (Eds.), *The elements of creativity and giftedness in mathematics* (pp. 5–28). Rotterdam, The Netherlands: Sense.

Part II
Mathematical Problem Posing in the School Mathematics Curriculum

Chapter 8
Problem Posing as Reformulation and Sense-Making Within Problem Solving

Victor V. Cifarelli and Volkan Sevim

Abstract This chapter examines a type of problem posing that has received little attention in the mathematics education literature. Silver (*For the Learning of Mathematics 14*:19–28, 1994) defined *within-solution* problem posing as "problem formulation or reformulation [that] occurs within the process of problem solving" (p. 19). Our analysis documents and explains the role that within-solution problem posing plays during problem solving, focusing on episodes of students from two grade levels: (a) Two fourth-grade students solving a multiplication task, and (b) A mathematics education graduate student solving a number array task. Our research examines: (a) How problem posing evolves from the students' ongoing interpretations of problematic situations, and (b) How these posed problems contribute to the students' problem solving. The results provide an explanation of how problem posing and problem solving coevolve in the course of solution activity and thus indicate the beneficial role that problem posing can play in the solution of mathematics problems.

Contents

V.V. Cifarelli (✉)
Department of Mathematics and Statistics, University of North Carolina at Charlotte, Charlotte, NC, USA
e-mail: vvcifare@uncc.edu

V. Sevim
School of Education, Virginia Commonwealth University, Richmond, VA, USA
e-mail: vsevim@vcu.edu

F.M. Singer et al. (eds.), *Mathematical Problem Posing*,
Research in Mathematics Education, DOI 10.1007/978-1-4614-6258-3_8

Introduction

The study of mathematical problem posing has been an important area of investigation by researchers in mathematics education (Cai et al., 2012; English, 1997a, 1997b; Kilpatrick, 1987; Silver, 1994; Silver & Cai, 1996). Underlying these studies is the view that having students generate and develop their own mathematical problems from particular situations may help them to become stronger problem solvers as they advance their inquiry-based activity. Proponents of problem posing advocate for their inclusion in the mathematics curriculum for several reasons. Among the different reasons, the most important is the view that having students make up their own problems encourages self-reflection on problem situations (Goldin, 1987; NCTM, 2000; Schoenfeld, 1994; Thompson, 1994). Reflection on problem situations that includes the planning of potential solution strategies has been associated with effective problem solving in several studies of problem solving (Carlson & Bloom, 2005; Cifarelli & Cai, 2005; Goldin, 1987; Schoenfeld, 1992). Hence, posing problems is viewed by many as a useful classroom activity that may help nurture the mathematical thinking, and particularly, the problem-solving actions of students.

Our view is that problem posing needs to be considered as occurring throughout problem solving. As students act to solve problems, we believe that they continually monitor the usefulness of current goals and revise or reorganize their goals and purposes as needed to solve the problem. Problem posing is then a series of transformations of the original problem, with each successive posed problem indicating progress towards a solution as well as providing possibilities for action to expand further the scope of the original problem.

Exemplary research on problem posing has been reported in a series of studies by English (1997a, 1997b) and Silver and his colleagues (Cai, 1998; Silver, 1994; Silver & Cai, 1996; Silver & Mamona, 1989; Silver & Stein, 1996). For example, English designed a comprehensive framework for developing young children's mathematical problem posing (English, 1997a) and assessed the effectiveness of using problem posing in middle-grades classrooms (English, 1997b). In addition, Silver and his colleagues conducted studies that encompass a range of important issues related to problem posing, including studies of the problem-posing activities of middle-grades students (Silver & Cai, 1996) and in-service teachers (Silver & Mamona, 1989), and the effectiveness of the use of problem posing in the middle-grades mathematics curriculum (Silver & Stein, 1996).

While these studies have undoubtedly added to our knowledge of problem posing as a productive mathematical activity, the research is less certain concerning the specific roles that problem posing plays in problem-solving situations. A particular issue concerns the ways in which problem posing and problem solving interact while a student is in the process of solving a problem. In what ways do the solver's initial problem formulations have an impact on his or her solution activity? This is an important question to address since the student may view a problem in a way that is different from what the teacher sees; and that view is likely to influence the goals he or she sees fit to develop and pursue.

Conversely, how do students' reflections on the results of carried-out solution activity help them to reformulate the current problem if needed, or pose additional problems to solve? According to Brown and Walter (1993), the process of solving a problem presents opportunities to the solver for new questions to emerge, that "we need not wait until after we have solved a problem to generate new questions; rather, we are logically obligated to generate a new question or pose a new problem in order to solve a problem in the first place" (Brown & Walter, 1993, p. 114). In this way, problem posing and problem solving may be viewed as naturally related in the sense that, in order to solve the original problem, the solver generates additional questions or problems that must be addressed. Silver (1994) referred to this kind of problem posing as "problem formulation or re-formulation [that] occurs within the process of problem solving" (Silver, 1994, p. 19). For example, students engaged in the solution of a problem may generate a result that, upon reflection, challenges or calls into question their prior goals and actions. In these situations, the ways that students act to resolve the new question often lead to a reformulation of the original problem, which may in turn lead to progress in finding a solution. While studies of the problem-solving actions of college students have documented this recursive property that involves successive and ongoing reformulations of problems (Carlson & Bloom, 2005; Cifarelli & Cai, 2005), studies are needed that document and explain how problem posing and solving interact at education levels throughout K-12 and beyond.

Theoretical Framework: Connections Between Problem Posing and Problem Solving

Although the research literature contains few studies of the specific connections between problem posing and problem solving as hypothesized by Brown and Walter (1993), we found several studies that provided further rationale for our study. In particular, we found studies that documented the structural character of problem solving as occurring within chunks or clusters of activity (Schoenfeld, 1985) that are both situational and episodic in structure (Hall, Kibler, Wenger, & Truxaw, 1989) and that unfold in the course of ongoing activity (Pirie & Kieran, 1994). These studies suggest how the solver's problem solving may involve a series of self-generated problematic situations within which particular goals and purposes are pursued. For example, Schoenfeld (1985) found that solvers developed localized goals and purposes within these episodes of activity and that a solver's solution of a task may be built upon several such episodes. Hall et al. (1989) focused on these episodes as situational expressions that unfold as the solver becomes engaged within the situation and begins to develop goals and purposes. Pirie and Kieran (1994) found that solvers develop their understanding in problem situations by unfolding their actions in the course of problem formulating and then reconstructing their actions at increased levels of understanding as they carry out the solution. These studies were helpful in identifying particular goals and purposes in our analysis and conclusions.

Goals and Purposes

The purpose of this chapter is to examine the role that within-solution problem posing plays as problems are solved. As educators, we know that new problems can come up or are posed by solvers in the least expected situations, often appearing as a surprise for the student to address and make sense of. We seldom think of problem posing as related to the solver's ongoing sense-making activity, either in research or classroom settings; rather, the problems we as teachers typically ask students to pose often correspond to particular questions that we have formulated and ask them to consider and answer. For example, in the primary grades, children are often asked to "make up" and solve problems about particular holidays such as Halloween and Christmas. While these kinds of questions can serve as useful prompts to stimulate the students' mathematizing of a situation, studies are needed which focus on problem posing as a sustained process that occurs and may reoccur throughout the solution of a problem. Of particular interest here is to examine how the solver proceeds from his or her initial interpretations to develop goals and purposes, and how this process may reoccur as the solver progresses towards a solution. In this way, the current study considered problem posing as problem formulation and reformulation that aids the solver's ongoing development of goals and purposes throughout problem solving. We address the following questions:

1. How does problem posing evolve from the solver's ongoing interpretations of a problematic situation?
2. How do these posed problems contribute to the solver's problem-solving activity?

Methodology

In addressing these questions, we examined episodes of students coming from a fourth-grade mathematics classroom and a student from a graduate course in mathematics education. Our rationale for choosing subjects at these levels was as follows. We thought it was important to observe and explain how students at different levels of mathematical sophistication and competence formulate and if necessary, reformulate problems based on their interpretations of the tasks we gave them.

Students at the younger age, to the extent that they posed problems about the situation they faced, would be expected to pose problems based on relatively simple questions about the situation. By including in our analysis an examination of the mathematical actions of a graduate student, we looked to illustrate and explain the useful role that problem posing can play in more advanced problem-solving activities. Specifically, we would expect that graduate students would demonstrate more sophisticated problem-solving activity than would younger students, particularly with regard to planning potential solutions, self-monitoring their actions, and reflecting on the results of their actions. So, in terms of answering the first research question, we would expect the graduate student to pose more mathematically sophisticated problems than the younger students in developing their solution activity. We believe that the consideration of problem posing in these different contexts will yield a broad-based explanation of the role of problem posing in problem solving.

We first present episodes from fourth-grade students who came from a classroom in which the first author served as a tutor. Students were assigned a worksheet of various multiplication problems. The tutor's role was to circulate among the students of the class and provide assistance as needed. The episodes we present involve a pair of students solving a series of multiplication problems. In analyzing the actions of the fourth graders, we focused mainly on data taken from the written verbal transcripts generated from the videotape of the students, and the observations and written records of the researcher who provided tutorial assistance to the students.

We then present episodes from an interview with a student who was enrolled in a graduate mathematics education program. During the interview, the graduate student solved a set of open-ended mathematics problems while thinking aloud. The data consisted of the videotaped records of the interview, the experimenters' field notes, and the student's written work. Written transcripts of the student's verbal responses were generated from the videotapes.

In conducting the interview with the graduate student, we followed the principles of teaching experiments established by Cobb and Steffe (1983). The interviewer's questions ranged from questions that asked the solver to clarify or explain an action performed to more elaborate questions that might induce the solver to consider a new problem. For example, the interviewer asked the solver, "what are you thinking?" whenever an extended period of silence was accompanied by an absence of paper-and-pencil activity. Research on the use of verbal self-reports as data suggests that such questions cause only minor interruption of the solver's ongoing actions and do not threaten the overall validity of the data (Schwarz, 1999). Moreover, these periods of self-reflection may indicate instances where the solver is monitoring and assessing his or her ongoing actions and thus can be seen as important indicators of knowledge development (Cobb & Steffe, 1983).

We utilized protocol analytic techniques in the analysis (Cai, 1994; Ericsson & Simon, 1993). Specifically, we examined the written and video protocols in order to (a) identify examples of problem posing in the students' actions, and (b) determine the significance of the problem posing in making progress towards solution. The use of videotaped records was crucial in making these determinations for the following reasons. The use of videotape proves more effective than sole reliance on written protocols when analyzing such diverse examples of solution activity. Second, an interview is a social interaction in which the interviewer and the solver participate in a dialogue. Hence, viewing videotape gives the researchers an opportunity to "step back," and analyze the dialogue from an observer's perspective, and allows for ongoing interpretation and revision of the subject's activity in the course of the analysis (Cobb & Steffe, 1983; Roth, 2005). Thus allows for continual communication between the theory and the data.

Results

We observed the students self-generating questions based on their initial interpretations of the task with which they were presented; these observations formed the basis for answers to the first research question. From these analyses as well as analyses of the students' subsequent solution activities, we answered the second research question by tracing how their problem posing evolved into sophisticated posing and solving. We elaborate on these results in the following sections. The first section reports on the problem-solving episodes of two fourth graders solving a multi-digit multiplication task; the second section reports on the episodes of a graduate student solving a number array task.

Fourth-Grade Students Solving the Multiplication Task

We first present episodes of two fourth-grade students solving a traditional computation task with two-digit multiplication. In preparation for an upcoming test, the students were working in pairs, completing a set of multiplication tasks that were presented in a standard vertical-algorithm format. Working as tutor, the first author moved around the classroom and provided tutorial assistance where it was necessary. As the students worked through this long list of exercises, it appeared that most were not working together as was intended by the classroom teacher. Rather, their activity took on the appearance of a sports competition, with each student racing to get an answer before the other.

As the tutor moved around the room, he came upon two students, Corrine and Austin, who were working to compute the product 15×15. Like the other students in the room, these students competed with each other, each trying to get the correct answer before the other. Knowing that the students were familiar with, and had been practicing, the standard multiplication algorithm, the tutor observed that they appeared to use different algorithms to complete the tasks: Austin correctly solved the problem using the standard algorithm while Corrine used a seemingly strange nontraditional algorithm, which Austin stated as "magical." Yet, she amazingly got the same correct answer and actually finished working on the problem well ahead of Austin (Figure 8.1).

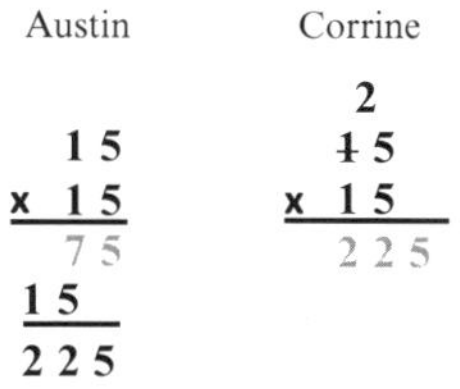

Figure 8.1. Austin and Corrine's multiplication algorithms.

When Austin had finished working, he peered over at Corrine's work and noticed both that she had noted the same answer as he had, and also that she had written down her answer before he had. "How did you do that?" he asked with a genuine sense of excitement in his voice. "Oh, my dad showed me how" responded Corrine with an air of bravado, appearing pleased by the fact that she was able to solve a problem faster than Austin. Austin repeated his question to Corrine with a sense of urgency, "But how did you do that?" to which Corrine replied "It's easy, it goes like this." Corrine proceeded to explain her method to Austin: "You need to draw a straight line so that these (*she points to the pair of fives*[1]) line up. Then you add one over here (*crossing out the one and entering a two just above it*) and just multiply the numbers" (Figure 8.2).

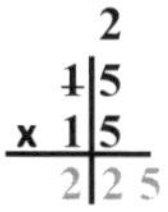

Figure 8.2. Corrine's algorithm.

Austin watched Corrine and commented "Oh, I see. Let me try it." Austin copied the original problem and demonstrated the rule to convince himself that the rule worked.

The tutor then asked the students if they might use the same "trick" to solve another problem that appeared on the worksheet, 25×25. Austin hesitated, seemingly unsure of replicating the algorithm, while Corrine replicated her algorithm to get a correct answer (Figure 8.3). Austin then used the standard algorithm to verify that Corrine's answer was correct.

[1] Italicized comments are used to indicate the actions performed by the students as observed by the researchers.

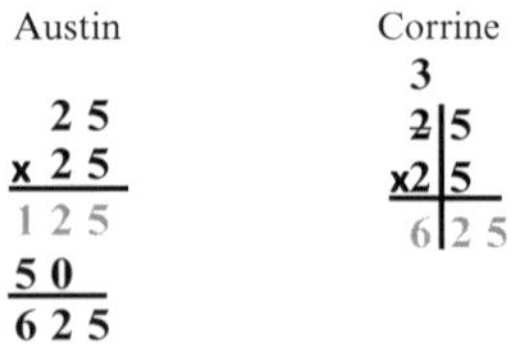

Figure 8.3. The students' solutions.

"So it works!" exclaimed Corrine to which Austin replied "Now we have an easier way to do it." The tutor noted that: (a) Adding +1 to the tens digit had to do with number place value representation and the fact that when squaring 2-digit numbers that end in 5, the result would always end in the number 25 (Figure 8.4); and (b) The algorithm would not work for numbers that do not end in 5. Although such mathematical reasoning did not come into play, as expected, the students' desire to find a better way to solve their problems and to investigate this strange new algorithm in the solution of other problems demonstrate a level of inquiry that helped them become more engaged with the situation and begin to develop new goals and purposes.

$$\begin{aligned}(10x+5)(10x+5) &= 100x^2 + 100x + 25\\ &= 100x(x+1) + 25\end{aligned}$$

Figure 8.4. A place value explanation of Corrine's algorithm.

The students then explored additional questions. These self-generated questions, which were based on the students' initial interpretations of the original task, indicate that their problem posing evolved through the applicability and efficacy of the algorithm to other problems. Specifically, the students examined several other cases, some of which were suggested by the tutor and others which they thought of themselves. As they explored multiple problems, they continuously challenged their own understandings about multiplication and how well the "new" algorithm worked for other problems. Their ability to self-generate new questions together with their ability to generate new explorations that conformed to their questions contributed to what we inferred to be their evolving understanding of the new "algorithm." For example, after they had achieved some sense of the utility of the algorithm for other problems (e.g., that it did not work for some problems that they tried such as 11×12, but that it did appear to work when the two numbers were the same and both ended in five), it occurred to Austin that they might try a "much bigger problem" than those on the worksheet. He wrote down the product 125×125 in the vertical format. The fact that he would even propose a problem was a significant step in advancing towards a solution given that the students had not yet applied the standard multiplication algorithm to multiplying pairs of three-digit numbers. Nevertheless, Austin and Corrine solved the problem both ways, with the standard algorithm and with Corrine's algorithm, with the aid of some scratch work to compute 12×13 on the side (Figure 8.5).

In the following sections, we report the actions of a graduate student who appeared to use problem posing to achieve substantial conceptual gains in her problem solving.

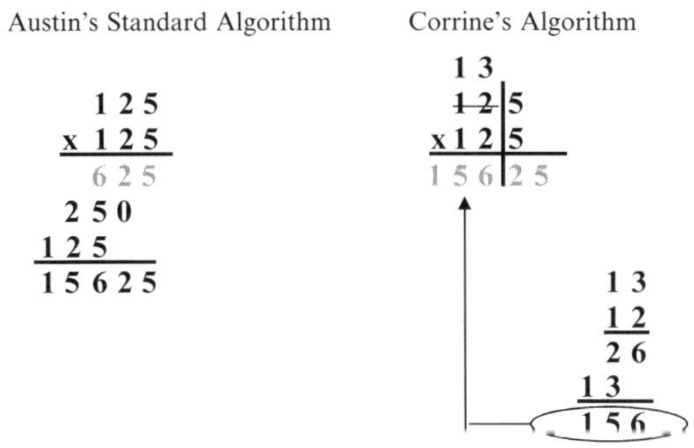

Figure 8.5. Austin and Corrine's application to a new problem.

Mathematics Education Graduate Student Solving the Number Array Task

The analysis of the episode with the graduate student, Sarah, will include her posing and solving of a problem she self-generated from the array (Figure 8.6): What is the sum of cell entries in any *N* by *N* block of numbers from the array?

Find as many relationships as possible among the numbers.

1	2	3	4	5	6	7	8	9	10
2	4	6	8	10	12	14	16	18	20
3	6	9	12	15	18	21	24	27	30
4	8	12	16	20	24	28	32	36	40
5	10	15	20	25	30	35	40	45	50
6	12	18	24	30	36	42	48	54	60
7	14	21	28	35	42	49	56	63	70
8	16	24	32	40	48	56	64	72	80
9	18	27	36	45	54	63	72	81	90
10	20	30	40	50	60	70	80	90	100

Figure 8.6. Number array task.

After exploring the array, Sarah reflected on her initial results and looked for other relationships that involved the sums of the entries in the square blocks (Figure 8.7).

Sarah: Let's see ... (*long reflection*) ... I was wondering about those square numbers on the diagonal going from left to right. They seem to relate to the dimension of the square blocks, ... I don't know, ... Maybe they relate to the sums of these blocks I had earlier (*points to the* 2×2, 3×3, 4×4 *blocks*). So, let's check it.

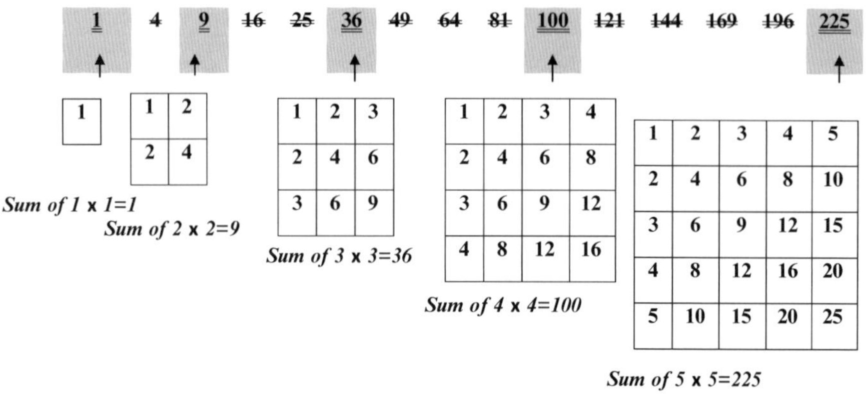

Figure 8.7. Sarah's skipping to find sums of block entries.

Sarah proceeded to examine the sum of the entries of each $N \times N$ block that contained the square numbers on the diagonal and she developed an informal method to find the sums of the entries of all $N \times N$ blocks going down the main diagonal.

> **Sarah**: So, for a 1×1, I get a sum of 1 (*points to the sequence of square numbers on the diagonal*). For a 2×2 (*points to block [1, 2:2, 4]*[2]), I get a sum of 9 … but what happened to 4? It appears to have been skipped! (*several seconds of reflection*). Okay, let me try this, I will write down the sequence of squares of all numbers, all in a row (*writes the following sequence of square numbers: 1, 4, 9, 16, 25, 36, 49, 64, 81, 100, 121, 144, 169, 196, 225*). So, the first number, 1, tells the sum of the very first matrix, a 1×1. And the first 2×2 has a sum of 9. …. So, I skipped over 4 to get the next sum (*crosses out the 4 in the sequence*), going from 1×1 to a 2×2, a sum of 9. The 4 gets skipped? Interesting!

Sarah has a new problem here—she thinks that there should be a relationship between the sequence of square numbers on the diagonal of the array and the successive sums of the entries of $N \times N$ blocks. Sarah was able to generalize her "skip" method to generate the sequence of sums of the entries of all $N \times N$ blocks.

> **Sarah**: So, for the first 3×3 (*points to [1, 2, 3: 2, 4, 6: 3, 6, 9]*), I already did this over here, so it is 36. So, in going from the 1×1 to the 2×2 to the 3×3, we go from 1, to 9, to 36—so we skipped over the 16 and the 25 (*she crosses out the 16 and 25 in the square number sequence*), a skip of 2 in this sequence!! So, okay, if this is true, then it looks like we will skip over the next 3 square numbers, and that should tell us the sum for a 4×4 should be equal to 100 (*crosses out the 49,*

[2] We use a bracket notation that lists the top-to-bottom rows of the block being considered. For example, the 2×2 block is indicated by [1, 2: 2, 4], the 3×3 block is indicated by [1, 2, 3: 2, 4, 6: 3, 6, 9].

64, 81 in the square number sequence)—that is what I have over here!! Cool! So, for a 5×5, we skip over the next 4 numbers in the sequence, (*points to the sequence 121, 144, 169, 196*) and get 225—yes, I got that one earlier for the 5×5.

Sarah then looked to make sense of her method with some further exploration (Figure 8.8).

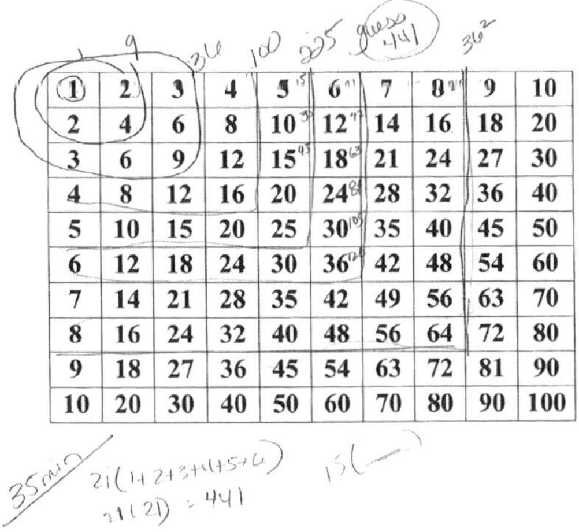

Figure 8.8. Sarah's diagram of her computation of sums in a 6×6 block.

Sarah: I wonder why this skipping works? Let's see it another way, for the 6×6, we add the entries in the rows to get 21+42+⋯+126=21(1+2+3+4+5+6)= 21×21=441. Do we get 441 by skipping the next 5 in the square sequence? (*Sarah extended her original sequence beyond 225, crossed out the corresponding "skips," and got a result of 441 as the next number in the sequence*) (Figure 8.8). But also, I notice that 21 over here (*points to the factored form 21•(1+2+3+4+5+6)*) is the sum of the first 6 numbers in that first row. Yes!

Sarah tried her idea on an 8×8 block (Figure 8.9).

Sarah: So to find the sum of these $N \times N$ blocks, I bet you just need to look at the sum of 1 to N and then square that total to get the sum. Let's try a big one, say 8×8. So, I guess that it would be .... 1+2+⋯+8=36, I don't know why I am adding these individual numbers since I know that the sum is (8×9)/2, and then I take 36^2? So that comes out to be … 1,296. And does it check with my skipping over here? Let's see, so for 8×8, I first skip 6 over 21 to get 28^2 for 7×7, and then skip 7 more to get the one for 8×8, … so 7 more is 35, and the next one is 36! So my algorithm seems to work! The algorithm is pretty efficient for larger numbers, beyond all of these. How about a 100×100 grid! But I thought that the skipping relationship was pretty cool!

1	2	3	4	5	6	7	8	36
2	4	6	8	10	12	14	16	
3	6	9	12	15	18	21	24	
4	8	12	16	20	24	28	32	
5	10	15	20	25	30	35	40	
6	12	18	24	30	36	42	48	
7	14	21	28	35	42	49	56	
8	16	24	32	40	48	56	64	

225	~~**256**~~	~~**289**~~	~~**324**~~	~~**381**~~	~~**400**~~	**441**	**Skip next 6 for 7x7 case**	**784**	**Skip next 7 for 8x8 case**	**1296**
15^2	**16^2**	**17^2**	**18^2**	**19^2**	**20^2**	**21^2**		**28^2**		**36^2**

Figure 8.9. Sarah's computation of the sum for the 8×8 block.

Conclusions

We must be careful not to infer too much from the actions of any of the students we observed. In the case of the fourth graders, they never really came up with a formal mathematical explanation for why the algorithm worked for some kinds of numbers and not others. Furthermore, we need to be vigilant whenever our students begin to see mathematics as involving tricks, rather than as an activity that is supposed to make good sense to them. But two salient points to be made with respect to the research questions are first that problems can be posed by solvers in the least expected situations, often appearing as a surprise for the students to address, and second, that solvers faced with surprising results are often motivated and even driven to seeking new explanations as a resolution to the situation. While it may be a classroom norm for teachers to challenge their students to answer questions regarding why or how well a strategy may work, the importance of within-solution problem posing is that it is self-directed and changes the solver's goals and purposes through a cycle of questioning and reflection. The students were highly motivated to answer questions that arose from their sense of surprise in their results. For Austin, the surprise came from seeing first that the strange new algorithm actually worked for the problems they were given and then that it worked for other problems as well. For Corrine, we inferred an ongoing sense of accomplishment from her purposeful actions to make sense of their ongoing applications of her algorithm to see that the algorithm would work for other problems they generated. In trying to determine why and for what kinds of multiplication problems the new algorithm might work, the students made conjectures, systematically checked out new possibilities, self-generated feedback that they then applied to their evolving ideas, and overall demonstrated mathematical thinking actions of the kind we usually associate with upper grades students.

While it is inappropriate to conclude that the fourth graders actually transformed their actions into a more sophisticated algorithm, they moved from working on a traditional task to working on authentic mathematical situations that were genuinely problematic to them. Self-generated or invoked mathematical situations, which were genuinely problematic to the students, served as opportunities for reformulation of the problem within problem solving, and thus significantly contributed to their problem solving.

In this way, the students were able to investigate the usefulness of the rule for other problems and thus developed some sense of efficacy of how well it worked. While their somewhat limited mathematical sophistication prevented them from making major conceptual gains, we were interested in seeing how a student with more sophisticated mathematical knowledge might use problem posing as a means to make major conceptual progress in their problem solving.

In contrast to the fourth-grade students, Sarah's problem posing played more of a transformative role in her solution of the number array problem: her problem about sums of entries in simple rectangular blocks evolved into more sophisticated problems about finding sums of entries in any $N \times N$ square blocks which extended beyond the actual 10×10 array. This appeared to be an example of within-problem posing in which the problem under consideration evolved in terms of scope and complexity (Silver, 1994; Silver & Cai, 1996). In addition, the evolutionary aspect of Sarah's problem posing as well as the impact on her subsequent solution of the problem provides some validation of Schoenfeld's (1985) view of the structural character of problem solving as occurring within chunks or clusters of activity.

Sarah considered several sophisticated ideas as she posed problems to solve. For example, she posed and solved the problem of finding the sum of entries in an $N \times N$ block as she "moved" the block down the diagonal. In this way, Sarah formulated a problem to solve that was quite sophisticated and which suggested a form of structuring activity that has not been reported in the literature on problem posing. Specifically, Sarah's development of her method for computing the sums appeared to be an example of metaphorical structuring (Sáenz-Ludlow, 2004). She found the sums of entries in the various blocks by "skipping" through a sequence of square numbers. In addition to her naming the method "skipping," the term also represented for Sarah the solution process involved. In this way, she invoked the use of a metaphor, "skipping," to give meaning to her subsequent solution actions. This metaphorical structuring appeared to be an example of within-solution problem posing in that it reformulated the original problem and expanded its intended scope.

With these idiosyncratic actions, she progressed from the original problem of finding as many relationships as possible to the problem of finding the sum of entries in an $N \times N$ block via a process of "skipping" or traversing through a sequence of square numbers. In this way, she had developed an informal method. In terms of our research questions, we see her problem posing as evolving from her informal reasoning of the sequence of square numbers, and progressing to the posing and investigation of more sophisticated problems. Specifically, she further developed the meaning of her idea by checking its applicability with simple cases and then drew upon the metaphor in her subsequent investigations. This finding is consistent with research that identifies informal methods as playing a prominent role in the development of formal algorithms (Cai, Moyer, & McLaughlin, 1998; Sáenz-Ludlow, 1995).

Sarah's subsequent development of the algorithm involved a subtle shifting of her attention from validation and verification activities that came with carrying out the method (demonstrating its applicability for blocks of dimension 2×2, 3×3, and 4×4), to efficacy activities (why the method appeared to work for the cases she generated). This shift provided for her an opportunity to unfold the process and relate her informal method to operations on the row and column numbers. She was able to generalize her method from skipping within a simple sequence to a formal algorithm which was more efficient for finding the sums of entries in any $N \times N$ block beyond the 10×10 array. This two-phase development demonstrates generalization that encompasses both informal and formal solution activity, a kind of generalization we did not see in the earlier episodes with the fourth graders. She generalized in two senses: First, in terms of moving from simple to more complex cases (blocks ranging from 3×3 to 100×100); and second, in terms of a transformation of an informal algorithm into a formal algorithm. These subtle shifts of reflective focus by solvers have been hypothesized by researchers (Krutetskii, 1976; Lobato, Ellis, & Munoz, 2003; Mason, 1995), but not illustrated or explained as discussed here. Sarah's extension of her informal skipping activity into a more formal method appeared indicative of a generalization that involved a conceptual jump from explanation of examples of a particular kind (i.e., relating the sums of entries in the various blocks to a sequence of square numbers) to a more mathematically sophisticated explanation of the method that involved properties of the task structure (i.e., showing how the sums of the various blocks related to the row and column dimensions). In other words, the coevolving processes of problem solving and posing enabled Sarah to move from a low level of generalization, typical of an inductive generalization of a sequential pattern based on the correspondence between the sums of entries in $N \times N$ blocks and the square number sequence, to a higher level of generalization, that captured the mathematical properties of the array. In Sarah's case, her jump to considering the row and column numbers to compute sums was based more on the mathematics of the array than on the pattern of square numbers she generated from consideration of the sequence of $N \times N$ blocks.

Sarah's use of problem posing to progress from her skipping idea to a formal rule that gave the sums of the block entries suggests a type of generalization consistent with Krutetskii's (1976) distinction between inductive and scientific generalizations. According to this view, scientific generalizations are based on more formal mathematical properties of the concept while inductive generalizations are based on less sophisticated similarities and differences from the learner's actions.

Although the two cases are somewhat disparate, we nevertheless view both of them as illustrating important ways that problem posing interacts with problem solving. First, both cases highlighted the role that informal reasoning and idiosyncratic actions can play in the formulation and subsequent solution of problems. The fourth graders, particularly the informal questions of Austin to learn more about Corrine's trick to multiply the numbers, and the graduate student's use of the metaphor of "skipping" applied to the square number sequence, indicate how useful problem formulations often arise in the least expected situations. So, the solvers' self-generated questions, often resulting from surprise and based on their ongoing reformulation of goals and purposes, triggered subtle shifts of reflective focus that helped increase their engagement within the problem situation. In the first case, these shifts of reflective focus increased the two students' overall reflective activity on the meaning of multiplication and number relations. In the second case, these shifts both helped the solver to expand the scope of the problem through metaphorical structuring and to make conceptual gains by transforming informal methods to formal explanations.

Theoretical Implications

One goal of this chapter was to document and explain the role that within-solution problem posing plays in problem-solving activities. Our analysis of the problem posing demonstrated by the fourth graders and the graduate student suggests some important roles that problem posing can play in solution activity. First, problem posing performed in the solution of a problem helps to broaden the solver's perspective of the original problem as well as expand its scope. This expansion of scope can further help students engage in unexpected generalizing activity that is rooted in students' own goals and purposes. Second, the finding that solvers make conceptual progress by posing problems that help to extend their understandings of the problem highlights the importance of self-generated activity. There is need for additional studies of how the solvers' interpretation of the problem links to planning strategies that may lead to successful solutions.

Third, while research on problem solving has highlighted the importance of planning strategies that anticipate potentially successful strategies (Schoenfeld, 1992), few studies have focused on how the solvers' ideas about potential solutions become elaborated and extrapolated as action unfolds. The results of the current study suggest that solvers actively monitor and assess the usefulness of their ideas as problem solving commences and actively pose new questions and problems when initial ideas outlive their usefulness. There is need for additional studies that shed light on different ways that solvers use results of actions to reformulate problem situations in ways that conform to their evolving ideas.

Instructional Implications

We offer some instructional implications based on the results.

Create Rich Problem-Solving and Problem-Posing Opportunities for Students

In addition to asking their students to pose problems as opportunities to apply their developing conceptions within particular mathematical situations, teachers should also look for and encourage students' problem posing to emerge naturally during their problem-solving activity. This can be accomplished by presenting them with conceptually rich problem-solving tasks that provide opportunities for them to express their knowledge and then reflect on their potential solutions.

Encourage Proactive Agency

A theme of this chapter was to view problem posing as a process that naturally coevolves with problem solving. The results demonstrate the need for teachers to recognize that problem posing can appear in varying degrees, as sense-making within problem solving, as generating new questions and problems within-solution of a problem situation, and as creating new problems to apply developing conceptions within a certain mathematical domain. Self-generated questions (while problem solving) can help students to place the current problem in a broader perspective and thus expand its scope.

In addition to teachers acknowledging the overall structure of within-solution problem posing, there is need to develop and implement teaching strategies that help students achieve a sense of self-advocacy in their problem solving. The students of the current study freely expressed and defended their ideas about the problems they faced. Students need opportunities to present and defend their ideas about problem solutions prior to carrying them out. If teachers notice, allow for, and encourage within-solution problem posing, this will help students become better problem solvers and potentially increase their problem-solving sophistication.

Emphasize Interconnection of Problem Posing and Problem Solving

Finally, problem posing has to be seen as integral to the problem-solving process and needs to be emphasized by mathematics teachers at all levels accordingly. When problem posing and solving are viewed as connected, the importance of problem posing as sense-making becomes an important goal of instructional activity.

References

Brown, S. I., & Walter, M. I. (1993). *The art of problem posing* (2nd ed.). Philadelphia, PA: Franklin Institute Press.
Cai, J. (1994). A protocol-analytic study of metacognition in mathematical problem solving. *Mathematics Education Research Journal, 6*(2), 166–183.
Cai, J. (1998). An investigation of U.S. and Chinese students' mathematical problem posing and problem solving. *Mathematics Education Research Journal, 10*, 37–50.
Cai, J., Moyer, J., Wang, N., Hwang, S., Nie, B., & Garber, T. (2012). Mathematical problem posing as a measure of curricular effect on students' learning. *Educational Studies in Mathematics, 83*(1), 57–69. doi:10.1007/s10649-012-9429-3. Online First.
Cai, J., Moyer, J. C., & McLaughlin, C. (1998). Algorithms in solving non-routine problems. In L. J. Morrow & M. J. Kenney (Eds.), *Teaching and learning of algorithms in school mathematics. 1998 Yearbook of the National Council of Teachers of Mathematics* (pp. 218–229). Reston, VA: National Council of Teachers of Mathematics.
Carlson, M., & Bloom, I. (2005). The cyclic nature of problem solving: An emergent multidimensional problem-solving framework. *Educational Studies in Mathematics, 58*(1), 45–75.
Cifarelli, V. V., & Cai, J. (2005). The evolution of mathematical explorations in open-ended problem solving situations. *The Journal of Mathematical Behavior, 24*, 302–324.
Cobb, P., & Steffe, L. P. (1983). The constructivist researcher as teacher and model builder. *Journal for Research in Mathematics Education, 14*, 83–94.
English, L. D. (1997a). Development of fifth-grade children's problem-posing abilities. *Educational Studies in Mathematics, 34*, 183–217.
English, L. D. (1997b). Development of seventh-grade students' problem posing. In *Proceedings of the 21st Annual Conference for the International Group for the Psychology of Mathematics Education* (Vol. 2, pp. 241–248). Lahti, Finland: University of Helsinki and Lahti Research and Training Center.
Ericsson, K. A., & Simon, H. A. (1993). *Protocol analysis: Verbal reports as data* (Rev. ed.). Cambridge, MA: The MIT Press.
Goldin, G. (1987). Cognitive representational systems for mathematical problem solving. In C. Janvier (Ed.), *Problems of representation in the teaching and learning of mathematics* (pp. 125–145). Hillsdale, NJ: Erlbaum.
Hall, R., Kibler, D., Wenger, E., & Truxaw, C. (1989). Exploring the episodic structure of algebra story problem solving. *Cognition and Instruction, 6*(3), 223–283.
Kilpatrick, J. (1987). Problem formulating: Where do good problems come from? In A. H. Schoenfeld (Ed.), *Mathematical thinking and problem solving* (pp. 123–148). Hillsdale, NJ: Erlbaum.
Krutetskii, V. A. (1976). *The psychology of mathematical abilities in schoolchildren*. Chicago, IL: University of Chicago Press.
Lobato, J., Ellis, A. B., & Munoz, R. (2003). How "focusing phenomena" in the instructional environment afford students' generalizations. *Mathematical Thinking and Learning, 5*(1), 1–36.
Mason, J. (1995). *Abduction at the heart of mathematical being*. Paper presented in honor of David Tall at the Centre for Mathematics Education of the Open University, Milton Keynes, United Kingdom.
National Council of Teachers of Mathematics. (2000). *Principles and standards for school mathematics*. Reston, VA: National Council of Teachers of Mathematics.
Pirie, S. E. B., & Kieran, T. (1994). Growth in mathematical understanding: How can we characterise it and how can we represent it? *Educational Studies in Mathematics, 26*, 165–190.
Roth, W.-M. (2005). *Doing qualitative research: Praxis of method*. Rotterdam, The Netherlands: Sense Publishers.
Sáenz-Ludlow, A. (1995). The emergence of the splitting metaphor on a fourth grade classroom. In *Proceedings of the 17th Annual Meeting of the American Chapter for the International Group for the Psychology of Mathematics Education* (pp. 152–157). Columbus, OH: Ohio State University.

Sáenz-Ludlow, A. (2004). Metaphor and numerical diagrams in the arithmetical activity of a fourth-grade class. *Journal for Research in Mathematics Education, 35*, 34–56.
Schoenfeld, A. H. (1985). *Mathematical problem solving*. Orlando, FL: Academic.
Schoenfeld, A. H. (1992). Learning to think mathematically: Problem solving, metacognition, and sense making in mathematics. In D. A. Grouws (Ed.), *Handbook of research on mathematics teaching and learning* (pp. 334–370). New York, NY: Macmillan.
Schoenfeld, A. H. (1994). Reflections on doing and teaching mathematics. In A. Schoenfeld (Ed.), *Mathematical thinking and problem solving* (pp. 53–70). Hillsdale, NJ: Erlbaum.
Schwarz, N. (1999). Self-reports: How the questions shape the answers. *American Psychologist, 54*(2), 93–105.
Silver, E. A. (1994). On mathematical problem posing. *For the Learning of Mathematics, 14*(1), 19–28.
Silver, E. A., & Cai, J. (1996). An analysis of arithmetic problem posing by middle school students. *Journal for Research in Mathematics Education, 27*, 521–539.
Silver, E. A., & Mamona, J. (1989). Problem posing by middle school teachers. In C. A. Maher, G. A. Goldin, & R. B. Davis (Eds.), *Proceedings of the 11th Annual Meeting of the North American Chapter for the International Group for the Psychology of Mathematics Education* (pp. 263–269). New Brunswick, NJ: Rutgers University.
Silver, E. A., & Stein, M. K. (1996). The QUASAR project: The "revolution of the possible" in mathematics instructional reform in urban middle schools. *Urban Education, 30*, 476–522.
Thompson, P. (1994). The development of the concept of speed and its relationship to concepts of rate. In G. Harel & J. Confrey (Eds.), *The development of multiplicative reasoning in the learning of mathematics* (pp. 179–234). Albany, NY: State University of New York.

Chapter 9
From Problem Posing to Posing Problems via Explicit Mediation in Grades 4 and 5

Sharada Gade and Charlotta Blomqvist

Abstract Drawing upon cultural historical activity theory (CHAT) perspectives, in this chapter we portray a classroom practice of problem posing that evolved with a cohort of students across Grades 4 and 5 in Sweden. In line with a language and literacy pedagogy, the classroom practice in which students utilised textbook vocabulary handed out on slips of paper (*lappar* in Swedish) advanced through three distinct stages namely: formulating written questions, problem posing as dyads and actively posing problems to one another. Mediated explicitly by *lappar,* such a practice provided social and public opportunities for students to attribute personal meaning and make conscious use of words in semiotic activity, as well as appropriate cultural meaning and valid norms of use. The increasing gain and display of agency by students in this practice, informed by student(s)-acting-with-*lappar*-as-mediational-means as unit of analysis, was indicative of their self-regulation, volition and independence. Developmental in approach, such classroom practice was born through teacher–researcher collaboration.

Contents

S. Gade (✉)
Department of Science and Mathematics Education, Umeå University, Umeå, Sweden

The Graduate Centre, City University of New York, New York, NY, USA
e-mail: sharada.gade@umu.se

C. Blomqvist
Berghemsskolan, Umeå, Sweden
e-mail: charlotta.blomqvist@umea.se

F.M. Singer et al. (eds.), *Mathematical Problem Posing*,
Research in Mathematics Education, DOI 10.1007/978-1-4614-6258-3_9

Introduction

As teacher and researcher we contribute to this volume by reporting a classroom practice of problem posing with a cohort of students in Grades 4 and 5, in a Grade 4–6 school, in Sweden. The establishment of this practice was conceived as part of a larger project correlating mathematics and communication for which the second author sought funding for, from The Swedish National Agency for Education (Skolverket Dnr 2009:406; http://www.skolverket.se). The practice began with Charlotta's students formulating written questions at their desks, in Grade 4 and evolved to their actively posing problems to one another as they gathered beside their teacher's desk, in Grade 5. It was also the case that although Lotta, as Charlotta is known, taught mathematics to her students at Grade 4; it was Cecelia their class teacher who taught mathematics at Grade 5, since Lotta was working with The Swedish National Agency by then. In writing this chapter, Lotta has also been able to draw on her interviews with the same cohort of students as part of interviewing them in their National Tests, at Grade 6.

The classroom practice which evolved from students formulating written questions to their actively posing problems to one another was established in a progressive manner over time. In line with the aims of correlating mathematics and communication in Lotta's project, we drew upon theory of classroom talk, taking care to train students to use talk for learning mathematics (Mercer, 2002). For example, Lotta conducted the game of *Yes and No* with her students, in which they had opportunities to guess numbers written on concealed slips of paper (*lappars* in Swedish). While Lotta responded only with *Yes* or *No* to questions posed by students, her students needed to pay attention to the place value of the digits in the numbers being guessed in the game. Although there was ample opportunity for teacher–pupil and pupil–pupil talk in playing this game, it was the numbers themselves on various *lappars* that mediated the game and/or classroom activity. After the game of *Yes and No*, we next conducted action research to rectify the faulty use of the = sign by Lotta's students (Gade, 2012b). The action cycle in this study was also *lappar*-based, wherein *lappars* now contained numbers, arithmetical operations and the = sign. As with the game of *Yes and No* and action research, Lotta's classroom practice of problem posing was *lappar*-based where textbook vocabulary or words on *lappars* mediated problem posing. While we offer details of CHAT perspectives that underpinned our approach in the next section, we describe the manner in which the classroom practice was established in the section thereafter.

Inclusive of co-authorship, it was teacher–researcher collaboration that formed a reliable backbone to our study. While the development of our teacher-researcher collaboration is detailed in the action research reported (Gade, 2012b), Lotta's co-authoring of this chapter extends this collaboration in two significant ways. First and unlike Lotta remaining silent and anonymised as Lea in the conduct of action research, Lotta is now theoriser lending voice to the scientific practice of research. By this our attempts redress the unfortunate yet recognised fact that K-12 teachers have little opportunity to contribute to theory generation, beyond merely consuming research generated by others (Cochran-Smith & Donnell, 2006). Second and unlike teacher and researcher talking across a theory/practice divide, our combined efforts are in line with recent calls in mathematics education research, for teachers to become stakeholders in university-led research, alongside researchers to become stakeholders in teachers' classroom practices (Krainer, 2011). In this way, our study provides a working model for other practitioners to emulate, towards which end we have also co-authored an expository article for a teacher journal (Blomqvist & Gade, 2013).

Prior problem-posing research in mathematics education, besides being relatively new and largely cognitive in approach, outlines many issues that our CHAT-based study can illuminate. For example, prior research draws attention to finding out who poses problems and for whom, how problems are posed around particular situations and whether problems posed by students are mathematical or not (Singer, Ellerton, Cai, & Leung, 2011). There is interest also in finding how lived experiences of students inform students' modelling of reality in the problems they pose (Greer, Verschaffel, Dooren, & Mykhopadyay, 2009) and investigating the extent to which the problems students pose are mathematical and solvable (Silver & Cai, 1996). Our study has the possibility of showing how Lotta was able to organise a problem-posing practice in her classroom (Leung, 2013) and the pedagogical strategies she was able to implement as teacher (Cai et al., 2013). It is towards the latter that our study exemplifies a CHAT-based pedagogical category that Dalton and Tharp (2002) identify as *Developing language and literacy across the curriculum.* While also detailing this category in the next section we ask: In what ways did problem-posing practice evolve across Grades 4 and 5, when students' classroom practice was explicitly mediated by vocabulary from their mathematics textbook?

Theoretical Framework

Two CHAT constructs which inform our problem-posing study in particular are *explicit mediation* and *developmental education*. These constructs build on the CHAT premise that the human mind is in a dialectical relationship with its social world, with human consciousness neither given nor produced by nature but a product of one's social interactions in the material world (Leont'ev, 1981). CHAT recognises the mediated nature of such interactions with cultural tools as well, be they spoken words of a language which are invisible, or visible tools like a pen and protractor. It is on basis of visibility that Wertsch (2007) makes a useful distinction between implicit and explicit mediation. While mediation by tools like words of a language is invisible, deemed present by inference and implicit; mediation by tools like a pen and protractor is visible, a result of conscious action and explicit. The use of the Swedish language by Lotta and her students would be an example of implicit mediation in our study, just as the use of words handed out on *lappars* by her students to pose problems, would be an example of explicit mediation. Wertsch explains how either kind of mediation transforms human action by altering the flow of human thought. Although such alteration is brought about by word meaning in implicit mediation, in explicit mediation alteration is brought about by an external agent. Lotta's handing out textbook vocabulary on *lappars* in our study was designed to alter the flow of thought and actions that her students took. While students posed problems by utilising the words given, Lotta had opportunities to guide their many attempts.

The classroom practice that Lotta established was based on explicit mediation and is in line with the principles of developmental education characterised by van Oers (2009). Exemplifying many of its features, Lotta's classroom practice provided opportunities for her students to make personal sense of words handed out on *lappars*, while at the same time accommodating personal motives and emotions. Participating in this practice was at the same time opportunity for her students to critically evaluate, as well as master meanings of words accepted as standard or the norm in wider culture. In such manner of meaningful learning (van Oers, 1996) Lotta's students first utilised words with their peers at their desks and later actively posed problems under full public scrutiny. As detailed in the next section, this practice was based upon explicit rules and expectations that Lotta spelt out, enabling her, as teacher, to reflect on the pedagogical practice she was implementing. In Lotta's practice, which allowed students to engage consciously with the words they were given, her guidance made it possible to lead her students in their zones of proximal development. Through dialogue and polylogue, both teacher and students had the opportunity to develop their respective identities. Importantly, the semiotic activity which the classroom practice promoted allowed for ontogenetic development of higher psychological functioning in each student (Vygotsky, 1987). Mediated by cultural tools in either implicit or explicit manner, the initiation of these actions was importantly social and public in origin. Following Vygotskian CHAT perspectives, such incidence is the basis for internalisation in Lotta's students of cultural word meaning, leading to self-regulation and volition that accompanied usage. The progression of Lotta's students, from initially formulating questions at their desks to actively posing problems to one another beside the teacher's desk, is indicative of the freedom and independence that accompanies human development within CHAT.

Lotta's classroom practice, which promoted development, exemplifies many features that Dalton and Tharp (2002) associated with their pedagogical category of *developing language and literacy across the curriculum*. This category primarily brings to fruition the premise that language proficiency in speaking, reading and writing is key to academic achievement, an aspect that van Oers (2009) recognised as central in his articulation of developmental education and an aspect taken into account in Lotta's classroom practice. Specifically and in line with Dalton and Tharp, Lotta's practice allowed her to listen to her students talk and to assist them as they posed mathematical problems about the world they had modelled. This included a complex of actions on her part like probing, praising, restating, making eye-contact and implementing turn-taking, as she guided the intellectual efforts of her students while they posed mathematical problems. In order to grapple with such complexity, our study focused on two units of analysis, allowing us to make claims about events that transpired in Lotta's classroom practice. The first of these is mediated action which links the actions of students to cultural, institutional and historical contexts in Lotta's classroom. The questions posed by Lotta's students with words that were handed out were mediated actions. The other and related unit is that of mediated agency which Lotta's students gained over time (Wertsch, 1998). To deal with the play that exists between the use of words by Lotta's students as they attempted to bridge personal with propositional meaning in the Brunerian (1997) sense, Wertsch forwards reformulation of agency as a composite unit of individual(s)-acting-with-mediational-means. In actively posing problems to one another by Grade 5, in line with Wertsch, the agency of Lotta's students is better conceived as student(s)-acting-with-words-on-*lappars*-as-mediational-means. Such observations extend Gade's (2006) findings in an earlier study wherein students were speaking-with-the-graph, speaking-with-a-formula and speaking-with-a-calculator. Wertsch's hyphenated expression of agency allows our study to attend to issues of power and authority imbued in the usage of words. As our data will show, Lotta's students had little hesitation in making reference to FBI agents and presidential elections on one hand, alongside tooth fairies and sausages on the other. The power perceived by her students in using the first set of words and the authority they seemed to exhibit with the second set were part and parcel of their personal meaning which in fact propelled them to participate or act-with-words-on-*lappars*-as-mediational-means.

Method

Two Vygotskin or CHAT perspectives guided our methodological arguments. First, an experimental–genetic study of classroom practice which sought prolonged observation of the students' usage of words handed out by Lotta. Such engagement enabled our study to record the qualitative transformations that accompanied students' actions in the course of their development (van Oers, 2009). Second, recognition of a process of double stimulation whereby both the Swedish language spoken and the words handed out on *lappars* mediated the many actions that Lotta and her students took in classroom practice. Following Wertsch (2007) both implicit and explicit mediation contributed to a practical–theoretic view with which CHAT perspectives both promoted and helped interpret Lotta's classroom practice. Before describing the separate methods we deployed in our progressive study, we here mention four aspects pertinent to its conduct. First, Lotta obtained permission from parents of students to conduct research. Second, the names of students we present are pseudonyms. Third, and as with action research (Gade, 2012b), we organised *lappar*-based activity for Lotta's 22 students in pairs. Fourth, the words used in the *lappars* were first selected from the textbook as in Table 9.1, to which we then supplemented those that Lotta's students considered as belonging to mathematics as in Table 9.2 (*matteord* in Swedish). The methods we now detail correspond to the

Table 9.1
Swedish Vocabulary (with Translation) Offered for Formulating Blue and Green Questions

mindre (less)	vilket (which)	många (many)	plus (plus)	störst (greatest)	minus (minus)
olika (different)	mellan (between)	mycket (many)	sist (last)	lång (long)	får (get)
jämna (even)	före (before)	först (first)	siffran (numeral)	hur (how)	före (before)
efter (after)	samma (same)	värd (worth)	kostar (costs)	är (is)	närmaste (nearest)
tal (numbers)	alla (all)	udda (odd)	bara (only)	ungefär (roughly)	siffarsum (sum of digits)
skriv (write)	pris (price)	med (with)			

Table 9.2
Swedish Vocabulary (with Translation) Offered by Students on December 8, 2009

skillnad (difference)	kvot (quotient)	grader (degrees)	addera (add)	subtraktion (subtraction)	addition (addition)	algorithm (algorithm)
summa (sum)	term (term)	multiplikation (multiplication)	tusendel (thousands)	tredjedel (one third)	fjärdedel (a quarter)	siffrorna (figures)
åttondel (one eighth)	hundratal (hundreds)	ental (units)	tiotal (tens)	hundratal (hundreds)	tiotusental (ten thousands)	mellanled (intermediate)
klockan (the clock)	micrometer (micrometer)	mil (mile)	ljusår (light years)	ekvation (equation)	svar (answer)	
division (division)	product (product)	centimeter (centimeter)	millimeter (millimeter)	siffersumma (sum of digits)	meter (meter)	
subtrahera (subtract)	lika med (equal to)	plus (plus)	minus (minus)	gånger (multiply)	udda (odd)	
jämna (even)	kvadrat meter (square meters)	omkrets (perimeter)	linjal (ruler)	avrunda (round off)	ungefär (roughly)	

three categories in which we present data; the same three categories will also be used as a framework for presenting our discussion in the section that follows.

Blue and Green Questions

The conduct of problem posing began with Lotta instructing students to formulate blue and green questions in pairs, for which we chose the words listed in Table 9.1. Lotta handed out two *lappars* to each pair of students, one having a blue number and a green word and the other having a green number and a blue word. In this chapter we denote blue writing in **bold** and green writing in an <u>underlined</u> format. The purpose of using the blue and green colour was to give a **blue word** and <u>green number</u> to one student and a <u>green word</u> and **blue number** to the other. Lotta's instructed her students to formulate a green and blue question, to ensure that they worked *as* pairs using words and numbers from either *lappar* to pose mathematical problems. Lotta's students were then asked to write the questions they formulated on a separate sheet of paper that was provided, from which we illustrate one example below:

First *lappar*		Second *lappar*	
9876	<u>alla</u> (all)	<u>8345</u>	**ungerfär** (approximately)

The blue and green questions formulated by Lotta's students with the above pair of *lappars* were **Are you approximately 9876 years old?** and <u>Are all the digits in the number 8345 odd</u> ? Five aspects in relation to our data and discussion are pertinent. First, we present an English translation of questions formulated in the Swedish language. Second, after students had formulated their questions, Lotta asked student pairs to read their questions aloud in a plenary. Third, Lotta followed their reading aloud by asking "Is this a question?" (Är det fråga?). Rhetorical in function, Lotta's question was an opportunity for her students to reflect on whether the question formulated by a peer was indeed a question or not. In line with Lotta's project aims, the object of this exercise was to focus on the societal norms of a mathematical question, relegating more realistic aspects of the content of the question such as **Are you approximately 9876 years old?** to the background. Fourth, recognising the paucity of time felt by most teachers in satisfying all the requirements of the curriculum, we conducted our study whenever we could find 10–15 minutes of time within regular everyday instruction. Finally, in the conduct of 11 sessions of explicit mediation beginning on October 13, 2009 and ending on March 31, 2011, Lotta's students formulated blue and green questions in the first three sessions and actively posed problems to one another in the final three sessions.

Problem Posing as Dyads

Sandwiched between formulating blue and green questions and actively posing problems to one another, our study witnessed problem posing by Lotta's students *as* dyads with active and brisk consultation visible in most pairs. We also observed one of the two students in some pairs to take on the role of lead writer. Commencing on December 8, 2009, Lotta first began eliciting from all her students the words they associated with mathematics. We present in Table 9.2 a list of words that Lotta collected on the whiteboard and mention three pertinent aspects. First, we observed a student, Mark, who came up to the whiteboard and added the word *farmer* (*bönde* in Swedish) to Lotta's list (Table 9.2) only to come back and erase this word. We took Mark's actions to exemplify the active nature of association that the students were making between the words belonging to mathematics and words belonging to the wider societal world. Second, we found words offered by Lotta's students (Table 9.2) to be much more extensive than those we had ourselves chosen from the textbook (Table 9.1). We therefore decided that we would no longer highlight words in blue and green colours, resorting to black print. Finally, we thought that the ease and fluency with which Lotta's students executed this stage of problem posing was critical to their posing problems in the final stage.

Two Acts of Posing Problems

Although we collected what students wrote on separate sheets of paper throughout our study, audio-recordings were made of the final three sessions during which Lotta's students made use of the whiteboard as well. Our transcribing of these audio-recordings drew on field notes taken by the first author and our combined insight as practitioners, besides Lotta's experience of conducting interviews with the same cohort of students. In response to geometry problems in National Tests at Grade 6, Lotta observed her students "to show faith in their own ability to use words and concepts that they developed" (Blomqvist, personal communication).

Data and Discussion

We present both our data and the discussion in three categories: (a) formulating blue and green questions—which corresponds to the first three sessions of problem posing; (b) problem posing as dyads—which corresponds to the next five sessions; and (c) two acts of posing problems—which correspond to the final three sessions. While we present English translation of questions originally posed in Swedish, we maintain their original flavour by retaining *kronor* as the currency.

Formulating Blue and Green Questions

Our corpus of data leads us to present students' formulations of blue and green questions in two sub-categories: questions related to properties of numbers alone and those related to their wider societal experiences. Having explained our use of blue and green colours in the methods section, we now present data without any embellishments below. We have, however, included in parentheses the hints and answers that students themselves provided in the inscriptions collected from them.

A. *Questions related to properties of numbers*

1. What is the sum of the digits in 2673?
2. How much must one add to get 7245 if one has 7000?
3. What comes before 4861?

4. Is 1008 less than 1009?
5. After 3129 comes 3130. 3130 + 1, what is that?
6. Is 6425 an odd or an even number? (Hint = same as 16)
7. Is 5000 minus 4345 an even number? (Answer: No)
8. 8282 is more than 7000. How much must one add to 7000 to obtain 8282? (1282)
9. Before 6738 comes 6737. What comes after 6738?
10. Which number is closest to 1331? (a) 1001 (b) 999 (c) 1300 (d) 1330
11. What number is it, if you add together 8345 and 5678?
12. How much is needed for 2196 to become 2200?
13. What digit comes first in the number 13312?
14. The number 4831 is odd. If one adds the same amount twice and then removes 6794, is the remaining number then odd?
15. 4861 is approximately 4860. Can you round off 4861 to the nearest thousand?

B. *Questions related to wider societal experiences*

1. Are you approximately 9876 years old?
2. 1676 is worth a lot. What is its half?
3. Have you a lottery ticket with number 8190?
4. If I get 6574 kronor as weekly allowance and 10000 kronor for monthly allowance how much money do I get in a month? (Answer: 36296 kronor)
5. If you were born in 9876, approximately how old are you now?
6. If there is a football field that is 176 meters on the short side and 2196 meters on the long side, is the perimeter an odd number or an even number?
7. If there are only 4831 football players in Umeå how many teams are there.
8. Milk costs 4831. How much do two milk cartons cost?
9. Round the prize of an internet competition, an iPhone (worth 6555) to the nearest hundred.
10. A sausage seller sells sausages. One sausage costs 4831. Nadya wants to buy two sausages. How much does she have to pay?
11. If a thing costs 8190 and I buy 8190 how much does it cost? (Over 10,000)
12. How many 50 years will one get in 2000? (4000)
13. Everyone at home in the village Normjöle is 1234 persons. About 30 are going on vacation. How many are left? (1204)
14. Ika has 6323 ice creams of which two melted. How many has she left?
15. I have 6,000 flowers and 500 water bottles and 74 postage stamps. How many do I have? (6574)

In discussing students' formulation of blue and green questions, we first shed light on Lotta's pedagogical conduct of *developing language and literacy across the curriculum*. As characterised by Dalton and Tharp (2002) Lotta's classroom practice was established to meet "students' bottom line," which was to provide students with an interactive experience which was needed to help them master academic discourse. This was initiated by having students pay attention to words and numbers in order to pose questions. We also had Lotta's students work in pairs, an aspect that alleviated Lotta's role of having to give each student individual attention (Gade, 2012b). In line with a pedagogy of developing language and literacy, Lotta's students worked at their desks and formulated written questions in consultation with their peers. Lotta's plenary in which she asked students to read out their questions and respond to her query: "Is this a question?" was an opportunity for her to gauge the questions which her students had posed. Through this approach, Lotta was able to probe, praise, seek clarification and even rephrase in some cases the question students posed in a more appropriate manner, guiding not just oral but written expression as well. Such articulation of accurately worded mathematical problems is important from a CHAT perspective, enabling students to appropriate cultural meaning and valid norms of expression.

The incidence of two sub-categories of problems in our data set is just as interesting. While the first referred to the various properties of numbers, the second offered insight into the world view Lotta's students perceived at Grade 4. We argue that the first sub-category of questions are in line with Lotta's conduct of teacher-pupil and pupil-pupil talk during instruction of the first chapter of the textbook dealing with digits, numbers and place value. During this session, Lotta drew the attention of her students to numbers she had put up on the whiteboard and asked them to reflect on number properties (e.g. whether they were odd or even, the place value of digits and the sum of digits of any number (*siffarsum* in Swedish)). The second sub-category of questions support the CHAT contention that the human mind and consciousness are dialectical products of social interactions in the material world, mediated by cultural tools and the concepts they connote. These questions evidence the vast variety of conceptual tools that Lotta's students utilised in posing mathematical problems, inclusive of age, lottery tickets, pocket money, ice cream and postage stamps. Following CHAT, Lotta's practice allowed for two aspects simultaneously. On the one hand the object or issue at stake, such as stamps or pocket money, was subjectivised in her students by their very posing of questions. Simultaneously and on the other, the subjectivities of students were objectivised in the very nature of questions they formulated (Wertsch, 1981). For example, the magnitude of the number 1676 was expressed as being worth a lot, reflecting subjective judgement about the number 1676 as object.

Problem Posing as Dyads

Sandwiched between formulating blue and green questions and actively posing problems to one another, in this category we present problems posed by Lotta's students *as dyads* during which some students took on the role of scribe as well. We mention too that Lotta's students participated with great ease in the pedagogical practice being conducted, demonstrated by the range and extent of societal reality that they were able to model in mathematical terms.

1. A farmer has 72 hectares of land. He got 27 when he won the battle against the ants. How much land has he? (Answer: 99 hectares of land)
2. 230 men in an airplane all of a sudden fell. Obama and his helpers came down from heaven. They were 50. How many were they now? (Answer: 230 + 50 = 280)
3. Peter drives 44 kilometers on his tricycle. Later he drives 55 kilometers by car. How many kilometres has he driven? (Answer: He has driven 99 kilometers)
4. Hi! I have 550 kronor and I'm going to buy a pet enclosure for 4950, how much must I have?
5. If Axel has 130 sausages and buys 60. How many sausages will Axel have then? (Answer: 190)
6. Donald is a bully. He has 32 kronor in his pocket. He takes money from someone at school. Now he has 45 kronor. How much more does he have?
7. Berta and Bert want a monocycle that costs 3265 and they have 3200. How much are they short of (missing)? (Answer: They are short of 65 kronor)
8. The length of a football field is 49 m, its breadth is 38 m. What is the perimeter? (Answer: 87 meters)
9. In Australia the temperature is 830 degrees Fahrenheit hot. In America it is 110 Fahrenheit degrees. What is the difference? (Answer: 720)
10. Santa Claus, Tooth Fairy and Easter Bunny are best friends. Santa has a 1000 Christmas gifts. And Easter Bunny has 300. Tooth fairy has 10,000 teeth and tens. How many Christmas gifts, eggs and teeth and tens have they all together? (Answer: 21300)
11. There was once a monkey who had 230 bananas, of which he had forgotten some bananas, so 50 bananas were rotten. How many bananas were not rotten? (Answer: 180 bananas)
12. Crooked Carlson had 160 kronor. He bought a robot for 50 kronor and the entire universe for 50 kronor. How much has he left?
13. There were 580 persons at a school, but 140 were under 10. How many children were over 10 years? (Answer: 440)
14. 1000 people voted for president Noel but 600 voted for President Blomqvist. How many more voted for president Noel than President Blomqvist? (Answer: 400)
15. There is a presidential election in the United States for the President. Noel was dismissed. President Ulla received 320 votes, and President Sara 165 votes. How many people voted?
16. There were 630 teachers at a really great school. One day, 200 teachers were fired for having an affair with each other. How many teachers are there left after the big scandal.

17. In 2008 Donald Duck was 74 years. How old was he 12 years ago when one counts from 2008.
18. Neils vomited 990 times in a day. The next day he threw up 500 times. How many times has he vomited?
19. Pelle, Pelle, Pelle, Pelle, Pelle, Pelle, Pelle, Pelle and Pelle are friends !! One day they share 905 cars. How many did they each get? (Answer: 100 each and 5 remaining)
20. There were 615 Gangsters. Then came five FBI agents and all FBI agents kill an equal number of gangsters. How many did each kill? (Answer: 123)

Our discussion on the present category of problems draws on van Oers' (1996) contention that meaningful learning encourages students to undertake a critical evaluation of cultural concepts and norms with their own private insights. Having had opportunity for such parley in the pedagogical practice that Lotta established, we argue that the diversity of problems posed is indicative of three aspects. First, the normal flow of thinking or actions of students was altered deliberately by words handed out on *lappars* by Lotta (Wertsch, 2007). The use of words besides expectation that they would have to read their problems aloud, made the actions Lotta's students took to be both conscious and explicit. The later ensured that the questions students posed would have to be held up for public scrutiny and acceptance. Second, Lotta's guidance in posing problems with the authority of a teacher provided her students disciplinary and cultural norms that they could appropriate and/or internalise. In addition to students' thoughts being led by peers at their desks, Lotta also led students in their zones of proximal development during which she could alter, seek clarification, extend, deem correct, or rephrase incorrectly worded problems (Vygotsky, 1987). Finally, and in line with van Oers (2009), Lotta's students had opportunity to be agents of their own learning and not merely passive participants in a rote-learning exercise. The classroom practice that Lotta established promoted a polylogue of ideas which we now examine.

In line with arguments of CHAT, the dialectic between students and their world was never in greater display than in this category of problems that Lotta's students posed (Leont'ev, 1981). The actions taken by Lotta's students and thereby their mind and/or consciousness are here evidenced as a product of their social interactions with the material world of words and cultural concepts they dealt with everyday. For example, we find reference to real-life matters such as a farmer's problems with ants, travelling to a destination with multiple modes of transport, a shortfall of money needed to buy something one really wanted or a difference in temperatures at two known places located far away from one another. With respect to cultural influences that are not surprising for students at Grade 4 to experience, we cite students alluding to Santa Claus, the Tooth Fairy, the Easter Bunny and Donald Duck. While speaking with authority on these issues, Lotta's students alluded to issues of power through reference to President Obama and FBI agents (Wertsch, 1998). Crooked Carsson could thus purchase the entire universe for a mere 50 kronor! Instances of humour, morality and human emotion were also evident in problems alluding to rotten bananas, affairs of teachers and vomiting. Pelle or many

Pelles were called upon nine times to share cars, with some left to spare as well. Yet the breadth of issues referred to by students provides evidence of the insight with which students modelled their mathematical worlds and wider reality. Far from being disengaged members of a lifeless classroom, Lotta's students exhibited diversity in their actions mediated explicitly by words handed out on *lappars* (Wertsch, 2007). We draw particular attention to meaningful learning (van Oers, 1996) exemplified with great effect in Problems 14 and 15 in our data (presented under "Problem Posing as Dyads"). Problem 14 was a problem posed by Noel and Lotta who partnered with him since his peer was absent from school that day. The content of this problem shows Noel defeating Lotta (Blomqvist) by 400 votes in a Presidential election. However, Sara and Ulla who sat at desks just behind Noel did *not* accept Noel as President and dismissed him, electing Ulla as President instead in Problem 15. We argue that multiple meanings were being mediated in these questions—the idea that Noel could be President defeating his teacher, or the idea that Noel could be summarily dismissed and another elected instead, or the widespread awareness of the 2008 Presidential elections in the USA.

Two Acts of Posing Problems

In this final category of data collected towards the end of Lotta's pedagogical practice, we offer two of many extracts that demonstrate the active posing of problems by Lotta's students to one another near the teacher's desk. The first relates to Nelly's problem set in fractional numbers, one she read aloud after being unsatisfied with the initial response she received from her peers.

Nelly	Pelle's little sister has 500 horses. 7/10 of the horses are Icelandic horses. How many horses are Icelandic horses?
Lars	Eight
Sharada	Is he right?
Nelly	No
Lars	Eighty
[Student]	190,8
Sara	Can you read the question again I did not hear that entirely
Sharada	Read again [To Nelly]
Nelly	Pelle's little sister has 500 horses. 7/10 of the horses are Icelandic horses. How many horses are Icelandic horses?
Tuva	Seven tenths of a whole
[Student]	No, but not in parts [meaning the question is asking for the *number* of horses]
Liam	150 [Possibly calculating 3/10's of the whole]
Nelly	Pelle's little sister has 500 horses. 7/10 of the horses are Icelandic horses. How many horses are Icelandic horses? [Posing the problem again]
Liam	Kind of 400 or 3
Mikael	350
Nelly	Right!
Sharada	Good!

The second relates to Jan allowing his peers to collaborate on a solution in response to an unreal problem that he posed. This was accompanied by Mikael's eagerness to pose his own problem even while Cecelia, their teacher, urged her students to move on to other events in the day.

Jan	Pelle's hot dog is thirteen thirty seven kilometers long in reality. How long is his hot dog on a scale of 1:1000? [Talking about a ridiculously long hot dog, perhaps tongue in cheek]
Noel	137! [loudly]
Leon	No!!
Anton	That will be only thirteen thirty seven, thousand times
Jan	No thirteen thirty seven kilometer [making himself clear]
Noel	The scale is what?
Jan	One is to one thousand
Leon	Aha, you mean thirteen thirty seven millimeter … centimeter …
Noel	One thousand [Loudly]
Noel	Is it like this? [Coming forward and showing a calculation on the whiteboard]
Jan	No
Cecilia	We leave it there now [Asking that students let go of these tasks for now]
Mikael	I will read another one now? [Indicating that he is willing to continue, however]
Jan	No, not yet … ok then read
Mikael	One square is …
Noel	No that is 0,001 [continuing to work at the conversion necessary]
Mikael	What?
Noel	Is it like this? [Showing his calculation again on the whiteboard]
Leon	No thirteen thirty seven kilometer
Liam	Yes, but one does not say thirteen thirty seven, one says one thousand three hundred and thirty seven
Mikael	A square is … [Goes ahead and reads with intention of posing his own problem]
Liam	One can say 1337 …
Jan	Well, yes yes 1337 km
Mikael	A square … [Wishes to pose his problem even as students loose their intention and train of thought]

The two extracts we offer in this third and final category demonstrate how Lotta's students actively *posed* problems by the end of our study. Though not instructed to act in this manner, it is important to observe how Lotta's students transformed their actions of reading their questions aloud to physical actions of *posing* problems to one another. In the two extracts above this includes Nelly reading *her* problem aloud, Jan guiding others to work on the problem *he* posed and Mikael eager to pose *his own* problem in the midst of other instructional events. In line with the genetic approach of CHAT, we argue such eventuality to be characteristic of students' volition and independence, indicative of their development. However, such development was not without their engagement with Lotta their teacher and their peers in a meaningful manner (van Oers, 1996). Embedded in a pedagogical practice geared towards language and literacy development (Dalton & Tharp, 2002), the actions of Nelly, Jan and Mikael speak also of their identities, which at least for them came to fruition within this practice (van Oers, 2009).

Although we have portrayed the manner in which we brought about explicit mediation in our study, the two extracts above make it possible to reflect also on aspects of implicit mediation (Wertsch, 2007). For example, when Nelly posed her problem by repeating herself, she implicitly recognised the inability of her peers to give her the right answer. In Tuva offering her solution but in another mathematical formulation, she implicitly acknowledged an alternative route that could be taken to arrive at a solution. These psychological aspects go beyond students' use of words handed out on *lappars* and highlight qualitative aspects that accompanied their development. In addition to the many actions of Lotta's students mediated first by textbook vocabulary and later by words that they themselves contributed, we argue that our study demonstrates progressive agency in Lotta's students which culminated in students actively posing problems to one another. In this Nelly, Jan, Mikael or Tuva were agents-acting-with-mediational-means (Wertsch, 1998) where their actions are better perceived as a composite whole, with the words they used are to be considered together and not separate from each other. Such nature of agency is evidence of two other aspects. First, that the participation of Lotta's students in semiotic activity within her pedagogical practice provided ample opportunity for students to assign personal meaning to the words they used (van Oers, 2009). The absence of this would not have had students rise from their desks and willingly pose problems to one another. Second, and by inference, Lotta's students were led in their zones of proximal development via participation in Lotta's pedagogical practice (Vygotsky, 1987). Their active posing of problems and challenging each other mathematically, we argue, lies in stark contrast to their being asked initially to work in pairs and pose blue and green questions in writing at their desks.

Conclusion

In conclusion we view our study as a case of close-to-practice research (Edwards, Gilroy, & Hartley, 2002) allowing us to adopt a theory/practice stance and deploy the theory of explicit mediation to conceive and study Lotta's classroom practice. Drawing on her research in other studies, Gade (2012a) argues such a stance to keep alive the relationship between theory-which-informs and theory-being-built, besides existing-practice and steered-practice. Yet before reflecting on this stance, we first consider wider research in problem posing in light of our study. In response to the question of whether important mathematical problems are posed by students (Singer et al., 2011), our study showed how all of Lotta's students had the opportunity to participate in problem posing, something that Lotta valued in her teaching. In thoughtfully deploying explicit mediation with a strategy of *developing language and literacy across the curriculum,* Lotta provided her students with both social as well as public opportunities by which *they* could attribute personal meaning and appropriate use of words in semiotic activity (van Oers, 1996). Given such manner of guidance, and unlike in Silver and Cai's (1996) study, most problems that Lotta's students posed were mathematical and solvable. This could also have been a result of Lotta privileging the social language of formulating questions by asking "Is this a question?" in the plenary she conducted (Wertsch, 1991).

Although the objective of our study was not to classify the mathematical nature of problems posed, our study addressed two aspects to which Leung (2013) drew attention—the feasibility of enacting an example of problem-posing practice, as well as the professional development of Lotta as teacher. In the theory-driven classroom practice we have portrayed, evidence of the strengths of our teacher-researcher collaboration is exemplified by our ability to establish a theoretically conceived pedagogical practice in Lotta's classroom. Drawing upon this and in response to Leung, our study demonstrates how via explicit mediation, we were able to treat CHAT theory on par with Lotta's classroom practice. In this study, and as with the action research conducted in an earlier study (Gade, 2012b), the professional development of Lotta remained central, something that Lotta herself described in positive terms of "doing a course in the classroom." Building upon this synergy, and taking wider concerns of practitioner inquiry into account, we have since extended our collaboration to include co-authorship (Cochran-Smith & Donnell, 2006; Krainer, 2011; van Oers, 2009). Yet, and as found desirable by Cai et al. (2013), we have adopted a qualitative rubric, in addition to our theoretic one, for designing and evaluating our study. This leads us to highlight two significant aspects. First, our study displayed the ability of Lotta's students to pose diverse problems, as was the case with students posing problems in informal contexts in English (1998). Second, Lotta's students drew on personal lived experiences, making sense not only of problems they posed but also of the mathematics they used in modelling realities, a didactical aspect that was advocated by Greer et al. (2009). These results tempt us to populate Lave's (1990) insightful dictum of "*how* math is learned in school depends on its being learned *there*" with our equivalent of "*how* problems are posed in schools depends on these being posed *there*."

The theoretical underpinnings of CHAT that our study deployed in Lotta's classroom practice, along with concerns of practitioner inquiry, allow us to reflect on our close-to-practice stance with an even hand (Edwards et al., 2002). We begin with explicit mediation (Wertsch, 2007). Extending the use of *lappars* as in our conduct of action research, our present study confirms their utility as a useful approach to intervention that can be put to many creative uses, depending on the teacher, the classroom and the school. Slips of paper are not difficult to source and use. Utilised in our study to facilitate conscious use of words or textbook vocabulary, the use of *lappars* resulted in a transformation of the classroom practice itself. We have been able to explain beyond simply describe, the development of Lotta's students from formulating written questions to their actively posing problems to one another. In such visible and increasing display of agency, insightful analysis was possible by deploying agents(s)-acting-with-mediational-means as the unit of analysis (Wertsch, 1998).

We mention, too, that the last three sessions in which students posed problems to one another were quite animated and enjoyable to watch. Early in the study, the first author observed Lotta's students take away *lappars* which contained words they personally liked and store these in the inner confines of their desks with the care given to prized possessions. By Grade 5 the use of *lappars* by students was remarkably different and transformative. In posing problems to one another Lotta's students were now challenging each other, bringing out calculators to verify and counter-challenge their peers. These actions, we argue, provide evidence of self-regulation, volition and independence, which Vygotsky (1987) argued would accompany the development of higher psychological functions in relation to cultural tools or textbook vocabulary utilised. Importantly, and following van Oers (1996), the initiative for such manner of transformation in Lotta's students was not forced upon them, but resulted from personal meaning making and engagement. Not only were there opportunities for Lotta's students to master the norms associated with formulating mathematical questions, but there were also opportunities for them to enact in an embodied sense the essence of posing problems in classroom activity. While Vygotskian and CHAT perspectives were empirically substantiated in such conduct, the aim of Lotta's project of communication for mathematics was not only met but also realised. Portraying an instance of classroom practice that realises these aims with 10–12 year olds at Grades 4 and 5 is the contribution that our study makes to this volume.

It would be possible to articulate, though not in this volume, the ways in which the germ cell of activity—inclusive of social interaction, mediating cultural tools and zone of proximal development—enabled learning by Lotta's students to lead their development (Stetsenko, 1999). An in-depth study of mathematical aspects of our corpus of data is also an exercise we leave for future research. For now we portray a case of developmental education, which rests on the Vygotskian imperative and vision that good education should always be one step ahead of our students (van Oers, 2009).

Acknowledgments Sharada wishes to thank The Graduate Center, City University of New York (CUNY), USA, for hosting her as Visiting Fellow during the academic year 2012–2013. We wish to thank Cecelia Persson for the time and space she provided for this study. We look forward to co-authoring further reporting from this study with her.

References

Blomqvist, C., & Gade, S. (2013). Att kommunicera om likamedtecknet [Communicating about the equal to sign]. *Nämnaren, 2013*(4), 39–42.

Bruner, J. (1997). Will cognitive revolutions ever stop? In D. M. Johnson & C. E. Erneling (Eds.), *The future of the cognitive revolution* (pp. 279–292). New York, NY: Oxford University Press.

Cai, J., Moyer, J. C., Wang, N., Hwang, S., Nie, B., & Garber, T. (2013). Mathematical problem posing as a measure of curricular effect on students' learning. *Educational Studies in Mathematics, 83*(1), 57–69.

Cochran-Smith, M., & Donnell, K. (2006). Practitioner inquiry: Blurring the boundaries of research and practice. In J. L. Green, G. Camilli, & P. B. Elmore (Eds.), *Handbook of complementary methods in education research* (pp. 503–518). Mahwah, NJ: Lawrence Erlbaum.

Dalton, S., & Tharp, R. (2002). Standards for pedagogy, research and practice. In G. Wells & G. Claxton (Eds.), *Learning for life in the 21st century* (pp. 181–194). Oxford, United Kingdom: Blackwell.

Edwards, A., Gilroy, P., & Hartley, D. (2002). *Rethinking teacher education: Collaborative responses to uncertainty*. New York, NY: Routledge.

English, L. D. (1998). Children's problem posing with formal and informal contexts. *Journal of Research in Mathematics Education, 29*(1), 83–116.

Gade, S. (2006). *The micro-culture of a mathematics classroom: Artefacts and activity in meaning making and problem solving* (Doctoral dissertation). Agder University College, Kristiansand, Norway.

Gade, S. (2012a). Close-to-practice classroom research by way of Vygotskian units of analysis. In S. J. Cho (Ed.), *Intellectual and attitudinal challenges: Proceedings of the 12th International Congress on Mathematics Education (ICME)* (pp. 4312–4321). New York, NY: Springer.

Gade, S. (2012b). Teacher researcher collaboration at a grade four mathematics classroom: Restoring equality to students usage of the '=' sign. *Educational Action Research, 20*(4), 551–568.

Greer, B., Verschaffel, L., Dooren, W. V., & Mykhopadyay, S. (2009). Introduction: Making sense of word problems: Past, present and future. In L. Verschaffel, B. Greer, W. V. Dooren, & S. Mukhopadhyay (Eds.) *Words and worlds* (pp. xi–xxviii). Rotterdam, The Netherlands: Springer.

Krainer, K. (2011). Teachers as stakeholders in mathematics education research. In B. Ubuz (Ed.), *Proceedings of the International Group of Psychology of Mathematics Education (PME)* (Vol. 1, pp. 47–62). Ankara, Turkey: PME.

Lave, J. (1990). The culture of acquisition and the practice of understanding. In J. W. Stigler, R. A. Shweder, & G. H. Herdt (Eds.), *Cultural psychology: Essays on comparative human development* (pp. 309–329). Cambridge, England: Cambridge University Press.

Leont'ev, A. N. (1981). The problem of activity in psychology. In J. V. Wertsch (Ed.), *The concept of activity in Soviet psychology* (pp. 37–71). Armonk, NY: M.E. Sharpe.

Leung, S. S. (2013). Teachers implementing mathematical problem posing in the classroom: Challenges and strategies. *Educational Studies in Mathematics, 83*(1), 103–116.

Mercer, N. (2002). Developing dialogues. In G. Wells & G. Claxton (Eds.), *Learning for life in the 21st century* (pp. 141–153). Cornwall, England: Blackwell.

Silver, E. A., & Cai, J. (1996). An analysis of arithmetic problem posing by middle school students. *Journal for Research in Mathematics Education, 27*(5), 521–539.

Singer, M., Ellerton, N. F., Cai, J., & Leung, E. (2011). Problem posing in mathematics learning and teaching: A research agenda. In B. Ubuz (Ed.), *Proceedings of the 35th Conference of the International Group for the Psychology of Mathematics Education* (Vol. 1, pp. 137–166). Ankara, Turkey: PME.

Stetsenko, A. P. (1999). Social interaction, cultural tools and the zone of proximal development: In search of a synthesis. In S. Chaiklin, M. Hedegaard, & U. J. Jensen (Eds.), *Activity theory and social practice: Cultural historical approaches* (pp. 235–252). Aarhus, Denmark: Aarhus University Press.

van Oers, B. (1996). Learning mathematics as a meaningful activity. In L. Steffe, P. Nesher, P. Cobb, G. Goldin, & B. Greer (Eds.), *Theories of mathematical learning* (pp. 91–113). Mahwah, NJ: Lawrence Erlbaum.

van Oers, B. (2009). Developmental education: Improving participation in cultural practices. In M. Fleer, M. Hedegaard, & J. Tudge (Eds.), *World yearbook of education 2009. Childhood studies and the impact of globalization: Policies and practices at global and local levels* (pp. 213–229). Abingdon, United Kingdom: Routledge.

Vygotsky, L. S. (1987). *The collected works of L. S. Vygotsky: Volume 1—Problems of general psychology*. New York, NY: Plenum.

Wertsch, J. V. (1981). The conception of activity in Soviet psychology: An introduction. In J. V. Wertsch (Ed.), *The concept of activity in Soviet psychology* (pp. 3–36). Armonk, NY: M.E. Sharpe.

Wertsch, J. V. (1991). *Voices of the mind: A sociocultural approach to mediated action*. Cambridge, MA: Harvard University Press.

Wertsch, J. V. (1998). *Mind as action*. New York, NY: Oxford University Press.

Wertsch, J. V. (2007). Mediation. In H. Daniels, M. Cole, & J. V. Wertsch (Eds.), *The Cambridge companion to Vygotsky* (pp. 178–192). Cambridge, United Kingdom: Cambridge University Press.

Chapter 10
Problem Posing as Providing Students with Content-Specific Motives

Kees Klaassen and Michiel Doorman

Abstract We interpret problem posing not as an end in itself, but as a means to add quality to students' process of learning content. Our basic tenet is that all along students know the purpose(s) of what they are doing. This condition is not easily and not often satisfied in education, as we illustrate with some attempts of other researchers to incorporate mathematical problem-posing activities in instruction. The emphasis of our approach lies on providing students with content-specific motives and on soliciting seeds in their existing ideas, in such a way that they are willing and able to extend their knowledge and skills in the direction intended by the course designer. This requires a detailed outlining of teaching–learning activities that support and build on each other. We illustrate and support our theoretical argument with results from two design-based studies concerning the topics of radioactivity and calculus.

Contents

K. Klaassen (✉) • M. Doorman
Freudenthal Institute for Science and Mathematics Education, Faculty of Science, Utrecht University, Utrecht, The Netherlands
e-mail: c.w.j.m.klaassen@uu.nl; m.doorman@uu.nl

F.M. Singer et al. (eds.), *Mathematical Problem Posing*,
Research in Mathematics Education, DOI 10.1007/978-1-4614-6258-3_10

Introduction

Among policy makers there is a growing appreciation of the educational relevance of mathematical problem posing. Given the central importance of problem posing in both pure and applied mathematics, it is argued that developing the ability to *pose* mathematical problems ought to be at least as important as developing the ability to *solve* them (e.g., National Council of Teachers of Mathematics [NCTM], 2000). This plea is reinforced by the recent UNESCO list of competences as challenges for basic mathematics education (UNESCO, 2012). Posing and solving mathematical problems is described as one of the eight major transverse competencies related to content acquisition in mathematics education. We are largely in sympathy with the plea that mathematical problem posing deserves a more prominent place in education. But we are cautious when it comes to the social and scientific benefits that mathematical problem posing is suggested to have according to many policy documents. For example, with reference to a statement of Einstein's, one often reads that by raising a new problem or by regarding an old problem from a new angle, many of the greatest scientists revolutionized their field of inquiry or even initiated an entirely new field of inquiry (Einstein & Infeld, 1938, p. 92). This is true, but it should be clear that problem posing at this level is and will remain the domain of exceptional genius, far beyond the reach of the vast majority. It may also be true that modern-day society requires flexible, creative, and mathematically able professionals. But we do not find it obvious that this demand will be met automatically by incorporating problem posing into mathematics education. Our aim for the incorporation of problem posing is much more humble, namely to increase the quality of students' process of learning mathematics.

Also among researchers there is a growing interest in mathematical problem posing. One area of research concerns the identification, characterization, operationalization, and framing of various aspects of mathematical problem posing (Christou, Mousoulides, Pittalis, Pitta-Pantazi, & Sriraman, 2005; English, 1997a; Silver & Cai, 1996; Stoyanova & Ellerton, 1996). Another area of research concerns the incorporation of mathematical problem-posing activities in mathematics education. This is done with a variety of partly overlapping aims. One aim is to gain insight into students' understanding of mathematical ideas and their perception of the nature of mathematics (Brown & Walter, 1983; Ellerton & Clarkson, 1996). This insight may function as a kind of formative assessment or perhaps even help the teacher to anticipate students' future understanding (Barlow & Cates, 2006; Van den Heuvel-Panhuizen, Middleton, & Streefland, 1995). Another aim is to develop students' actual problem-posing abilities by explicitly teaching them about what are considered to be key elements of mathematical problem posing (English, 1997a, 1997b, 1998). As a final goal, we mention the integration of mathematical problem posing within mathematical inquiry or modeling (Bonotto, 2010; Crespo & Sinclair, 2008; English, Fox, & Watters, 2005).

We do not pretend to have given a comprehensive overview, but it suffices to locate our own research. Our expertise is in secondary physics education (first author) and secondary mathematics education (second author). We are particularly interested in in-depth studies of teaching sequences about some physics or mathematics topics, for example mechanics or calculus. Our main concern differs from the ones mentioned above, which all relate to purposes of teachers or curriculum designers: *they* want to assess students' understanding, *they* want to establish what students like and dislike. In contrast, our aim concerns the purposes of students. The quality we want to add to *their* process of learning mathematics or physics is that all along they know the purpose of what they are doing.

The aim that students know what they are doing and why is not often satisfied in education, nor is it easily satisfiable. Gunstone (1992) writes in this respect: "This problem of students not knowing the purpose(s) of what they are doing, even when they have been told, is perfectly familiar to any of us who have spent time teaching. The real issue is why the problem is so common and why it is so very hard to avoid." As we will illustrate in the next section, this problem also applies to many attempts to incorporate mathematical problem-posing activities in instruction. Even when it is clear to us what the designer of such an activity wanted to achieve, we often feel that students will be at a loss as to why they are to engage in the activity. At best they will only in retrospect be able to appreciate what it has been good for.

Nearly two decades ago, we introduced an educational approach, the basic tenet of which was to bring students to such a position that, not only in retrospect, but already beforehand, they know the purpose(s) of what they are going to do. We have dubbed this approach *problem posing* because it would be a clear case of students knowing what they are doing and why, when they can be brought to such a position that (a) they themselves come to pose the main problems they are going to work on, and (b) in the process also come to appreciate the main means by which to tackle those problems. In this chapter, our approach will be further described, illustrated with two teaching sequences, and discussed. In the final section, we return to mathematical problem posing and reflect on it from the point of view of our problem-posing approach.

What Is the Point of Mathematical Problem Posing for Students?

In order to illustrate the problem of students not knowing the purpose(s) of what they are doing, Gunstone (1992) wrote: "In the following typical example, the student (P) has been asked by the interviewer (O) about the purpose of the activity they have just completed.

P: He [the teacher] talked about it...that's about all...

O: What have you decided it [the activity] is all about?

P: I dunno, I never really thought about it .... just doing—doing what it says ... it's 8.5 .... just got to do different numbers and the next one we have to do is this [points in text to 8.6]."

Note that it is not the case that the student has no answer at all to the question "Why are you doing this?" The student did have an answer. More fully articulated it may be something like this: "I am now working on 8.5, because I just finished 8.4; and after I finish 8.5, I am going to do 8.6; these are the numbers the teacher told us we got to do and we are supposed to do as the teacher says." Although this fragment in itself proves nothing, we hope it will strike the reader as familiar, as exemplifying the implicit didactical contract (Tiberghien, 2000) that the teacher knows what is best for students and that students simply are to follow suit. What we especially want to draw attention to is the absence of content-specific features in the student's answer. There is not even an indication of the topic or subject he or she is working on. We do not blame the student for this. Nevertheless, it is hard to suppress a feeling of disappointment. One would have hoped for more.

The problem of students not knowing the purpose(s) of what they are doing also applies to attempts to incorporate mathematical problem-posing activities in instruction. In Figure 10.1, we have collected from the literature a variety of kinds of

Write a problem to the following story so that the answer to the problem is "385 pencils."
"Alex has 180 pencils while Chris has 25 pencils more than Alex."

Write an appropriate problem for the following:
$(2300 + 1100) - 790 = n$

Last night there was a party and the host's doorbell rang 10 times. The first time the doorbell rang only one guest arrived. Each time the doorbell rang after that, three more guests arrived than had arrived on the previous ring.
Ask as many questions as you can. Try to put them in a suitable order.

Write a problem that involves use of the concept of a right-angled triangle.

Write a problem based on the following picture:

Write a problem that you would find difficult to solve.

Figure 10.1. A variety of examples of mathematical problem-posing activities.

mathematical problem-posing activities involving a variety of cognitive processes (Christou et al., 2005; Stoyanova & Ellerton, 1996).

Now, think about these examples from the point of view suggested by Gunstone (1992). What could be the point for children to engage in these activities? When asked "Why are you doing this?", would students be able to give an answer other than "Because the teacher told us so"? Would they be able to give any content-directed reasons for being involved in the activities? We are not suggesting that students will dislike such activities or that they will not learn anything from them. We are suggesting, however, that it would add quality to an activity if students had reasons for being involved that are specifically directed at topical content.

It may be said that in order to judge whether or not an activity is purposeful for students, more must be done than just to consider a single activity in isolation (such as the ones in Figure 10.1). A particular activity may rather get its point from the way it is embedded in a series of activities. We agree, but wish to make two observations. First, we know of very little research in which mathematical problem posing is embedded in a series of connected activities. Second, in the few cases we are aware of, the problem of students not knowing the purposes of what they are doing receives very little explicit or systematic attention. Let us discuss some examples.

For third-, fifth-, and seventh-grade, English (1997a, 1997b, 1998) designed and evaluated problem-posing programs comprising about 10 weeks for about 1 hour per week. The programs consisted of a sequence of main activities: exploring attitudes towards problems; classifying problems; separating problem structures from contextual features; modeling new problems on existing structures; creating new problems from given components; transforming given problems into new problems. The rationale behind this sequence seems clear enough. It was based on what in the literature were identified as key elements of mathematical problem posing. But let us now reflect on the sequence from the point of view suggested by Gunstone. What could be the point for children to engage in these activities in this order? Children may in some general sense be (made) aware that you learn more from creating and solving your own problems than from solving ones the teacher makes up. But even given this general motive, we still doubt whether it is "logical" for them subsequently to go on to classify problems or to separate contextual features from structural elements in given problems. We do not wish to underrate the efforts of English, if only because her findings show that, with some guidance from the teacher, key components of mathematical problem posing are well within reach of students. Students may also be able to tell at a later stage of the sequence, for example when modeling new problems on existing structures, why in an earlier stage they had to classify problems. That is, in retrospect they may see the reason for what they had to do earlier. Let us also stress that one need not be moved by our considerations that center on students' advance content-directed motives. But if one is, we conjecture that the sequence designed by English will not appear so "logical" any more.

Whereas mathematical problem posing is often promoted because it is part and parcel of mathematical inquiry, in educational settings the bond between mathematical problem posing and mathematical inquiry very often is broken. Crespo and Sinclair (2008) detected as symptoms of this broken bond an emphasis on de-contextualized problem posing (such as the examples in Figure 10.1), and on prescriptive problem-posing strategies that permit an almost effortless generation of new problems. Although we sympathize with this criticism, we think that in their study Crespo and Sinclair (2008) do not make real progress towards making mathematical problem posing functional for students within some worthwhile mathematical inquiry. As in the approaches they criticize, they too very much focus on mathematical problem posing per se, though in their case with an emphasis on the quality instead of the quantity of the problems posed. Again, we do not wish to underrate their efforts, in particular their finding that an easily implemented measure such as allowing students some exploration time will increase the quality of the problems they pose. Nevertheless, our point remains that students still are not provided with a purpose to pose mathematical problems in the first place.

Let us take stock. We have drawn attention to the problem of students not knowing the purpose(s) of what they are doing, and we have also illustrated that with respect to mathematical problem posing this is quite common. We agree that it is useful for a curriculum designer to have a clear idea of key elements of mathematical problem posing. We also recognize the temptation to design an educational program of which the rationale is that students first need to be trained in each of these elements as prerequisites to later mathematical problem posing. But we also urge course designers to resist this temptation if one explicitly aims to provide students with advance content-directed reasons for what they are going to do. Finally, we agree that the natural context for mathematical problem posing is mathematical inquiry. But we also note that we have found no convincing examples of weaving mathematical problem posing, in a for-students purposeful way, into an ongoing process of mathematical inquiry.

Providing Students with Content-Specific Motives as an Educational Ideal

The issue of students not knowing the purpose of what they are doing is a major concern within our problem-posing approach. Our basic tenet is that all along students know what they are doing and why, as much as possible on content-specific grounds. This ideal serves as a quality standard that as designers we aim to meet when concretely designing teaching-learning activities. Since the basic way to answer the question "Why am I doing this?" is by citing a motive (or reason or purpose), it is an essential ingredient of our approach to think of ways to induce motives in students for engaging in particular activities. In order to get a coherent sequence of activities, moreover, students' reasons for being involved in a particular activity are to be induced by preceding activities, while that particular activity in turn, together with the preceding ones, are to induce the reasons for being involved in subsequent activities. One way to achieve this coherence is by designing activities with the explicit educational function of making students pose certain content-specific problems, in particular problems that more or less coincide with the tasks they are going to work on next or that at least provide the next tasks with a clear purpose. The overall aim is to increase the quality of students' learning process by enabling students to perceive their learning process as an internally coherent one, which in important respects is driven by their own questions (either existing or induced), over which they have some control, and which point in a certain direction. Of course, there is also the traditional demand that the direction of the learning process is worthwhile from the course designer's point of view, in that it leads to specified attainment targets.

Before discussing our approach any further in general terms, we think it is illustrative to first further clarify it with concrete cases. The cases concern teaching sequences about the physics topic of radioactivity and the mathematics topic of calculus, which were developed and tested in the Ph.D. studies of Klaassen (1995) and Doorman (2005), respectively. The cases represent our efforts to meet, for the two topics at hand, the problem-posing ideal of providing students with content-specific motives. The two cases differ with respect to the vigor with which the ideal is striven for and the extent to which it is attained. But this does not matter for our main aim with presenting the cases, which is to clarify our problem-posing approach as much as possible. For this purpose, partial failures may be as illuminating as partial successes.

The details of the two cases—radioactivity and calculus—take the form of argumentative accounts rather than reports of empirical evaluations. The main steps of each teaching sequence are outlined at several intertwined levels of description:

- Descriptions of what happened in classrooms when the design was put to the test;
- Indications of why the designer expected that this would happen;
- Explanations of the cases in which the expectations did not come out; and
- Clarifying remarks and notes.

The interested reader is referred to Klaassen (1995, Chapters 6–10) and to Doorman (2005, Chapters 5 and 6) for more conventional presentations of the several cycles of small-scale in-depth developmental research involved in each case, as well as for extensive discussion of methodological issues, and for detailed information about textbooks, other materials, in-service programs, and so on.

The Case of Radioactivity

In this section, we illustrate our problem-posing approach with a teaching sequence about the topic of radioactivity. In order to better highlight the defining aspects of our approach, by way of contrast we first sketch the "traditional" way of teaching the topic. Both the traditional approach and our alternative approach are aimed at middle-ability students of about 15 years of age, and take about ten 50-minute lessons. We close with a reflection on the problem-posing features.

Traditional Treatment of the Topic of Radioactivity

Figure 10.2 represents the main structure of how the topic of radioactivity is typically taught. Given the aura of danger and mystery surrounding the topic, it is easily introduced in such a way that students are really motivated to begin with it. For this purpose it suffices to simply announce that safety measures and applications in health care will be covered.

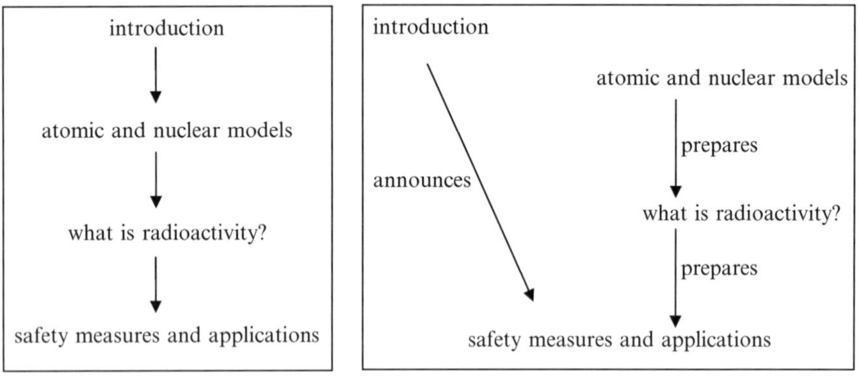

Figure 10.2. Structure of the common treatment of radioactivity. *Left*: temporal order. *Right*: rationale.

The motivating introduction is followed by a presentation of atomic and nuclear models along the following lines. Substances consist of molecules, molecules consist of atoms, atoms consist of …. At the level of middle-ability students the "models" typically take the form of pictorial representations as in Figure 10.3.

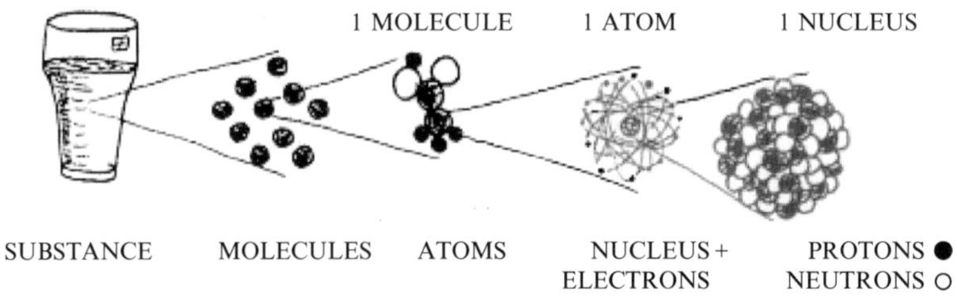

Figure 10.3. From substances to protons, neutrons, and electrons.

A subsequent step in the traditional treatment concerns the introduction of isotopes and an answer to the question "What is radioactivity?" in terms of unstable isotopes that decay while emitting radiation. Finally, safety measures and applications of radiation are treated.

The arrows on the left in Figure 10.2 represent temporal order. On the right, they represent the rationale behind the structure. The rationale seems clear enough. In order to be able to understand safety measures and applications, students should first know what radioactivity is: what a radioactive substance is, what radiation is, how it emerges, and so on. And in order to be able to understand what radioactivity is, students should first know about isotopes, helium nuclei, electrons, and so on. And in order to be able to understand that, they must first know, albeit at a simplified level, about nuclear and atomic models.

Some Comments on the Traditional Treatment

Like many traditional curricula in general, the standard treatment of the topic of radioactivity is cast in the form of a simplified rational reconstruction. Apart from a simplification appropriate to the target group, the content is sequenced in the way in which someone who has already mastered it may in hindsight conveniently reconstruct or summarize it, or build it up from first principles. For those who have not yet mastered it, however, following a simplified rational construction may not be a particularly useful route towards mastering it. What we especially want to draw attention to, in contrast to the problem-posing approach to be described later, is that following a simplified rational construction is not very suited for making students understand the purpose(s) of what they are doing. Before they are going to do what they were motivated for in the introduction (safety measures, etc.), there are five or six lessons about rather tough material (atomic models, etc.). But since middle-ability students are not familiar with (sub)microscopic models, it is not at all obvious for them to begin with such models. While observing some middle-ability classes in which the topic of radioactivity was taught in the traditional way, Klaassen (see also 1995, section 3.3) found that after 2–3 lessons on atomic models students became impatient and somewhat rebellious. The more assertive students began to complain why they were spending so much time on atoms, and when they would at last begin with radioactivity.

A further comment on the traditional treatment is that in order to arrive at a useful understanding of safety measures and applications, it is not at all necessary to first understand at a fundamental level what radioactivity is. The question if an irradiated object poses a radiation hazard to its environment, for example, is most relevantly answered by probing with a Geiger counter—which would be more relevant than by a theoretical treatment of the processes involved in the absorption of helium nuclei, electrons, and so on.

We hope to have made clear that the arrows in Figure 10.2 cannot be taken to represent motives for students to make a transition from one block to the next. Even in retrospect students may have a hard time trying to say why they have done what they did. A possible exception concerns the blocks "atomic and nuclear models" and "what is radioactivity?". While working on the block "what is radioactivity?", presumably it will be clear to students that use is made of concepts and models that were introduced in the block "atomic and nuclear models." This is why in Figure 10.4 we have drawn a backward pointing "retrospective arrow."

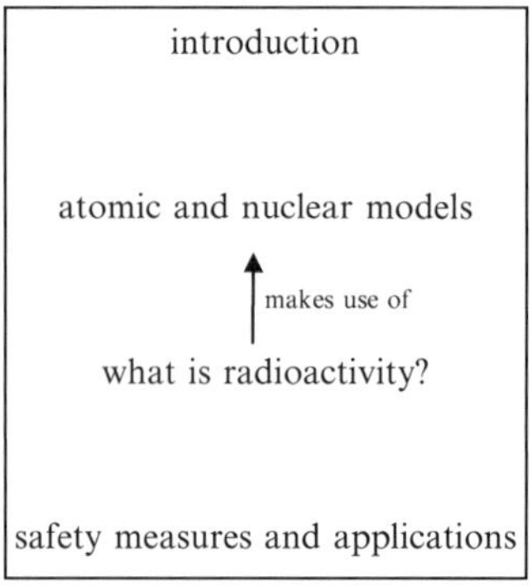

Figure 10.4. For students there is at best only a weak retrospective coherence.

Some Preliminaries About an Alternative Approach

We tried to design an alternative approach to the topic of radioactivity, such that for students there is a solid coherence and such that they do have advance motives for making a transition to the next block. We arrived at a structure that is almost a complete reversal of the traditional structure (compare Figure 10.4 with Figure 10.5, and in particular note the forward pointing arrows in Figure 10.5). Whereas the

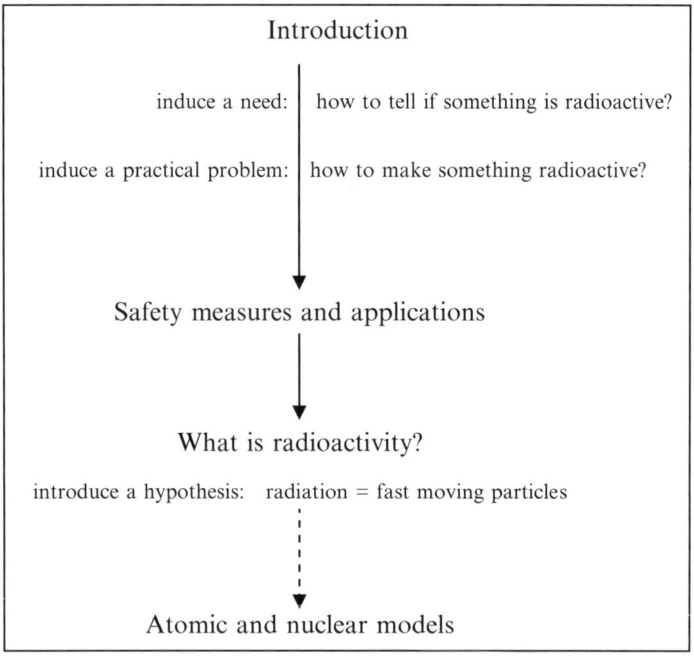

Figure 10.5. A didactical structure of radioactivity with a solid coherence.

introduction in the alternative approach is similar to the one in the traditional structure, after that the alternative approach proceeds in the direction announced in the introduction. Students are made to experience that they do already know quite a lot about radioactivity, but not enough to gain a genuine understanding of safety measures and applications.

Klaassen (1995, sections 2.3–2.5) first did some research on students' existing knowledge about radioactivity.[1] The findings were that students' existing knowledge could to a large extent be understood in terms of very basic notions concerning causation. In essence, an affector harms an object by means of an instrument.[2] In the case at hand, X-ray machines, radioactive waste, irradiated food, Chernobyl, and so on have the potential to harm something or someone because in one way or another they can make it happen that something harmful enters the thing or person. Students often call this something harmful "radiation" or "radioactivity." It functions as the instrument. In the case at hand, it is invisible, transportable, and penetrating. The Chernobyl accident was an affector because huge amounts of the instrument were released. According to many students, irradiated food is a potential affector because it contains the instrument and by eating the food we get the instrument inside. An object or person is affected as long as it contains the instrument. The effects may be reduced by applying a resistance, i.e., something that counteracts the instrument. A resistance, such as a lead wall or a special suit, prevents the instrument from entering an object or person. Furthermore, students applied semiquantitative relationships such as: the stronger the affector is, the more the object is affected; the longer the affector harms the object, the more the object is affected; the more affectors harm an object, the more the object is affected; the nearer the affector is to the object, the more the object is affected; the greater the resistance, the less the object is affected.

Sketch of an Alternative Approach

Partly based on the preceding analysis of students' existing knowledge, an alternative treatment of the topic of radioactivity was designed and tested. The structure is outlined in Figure 10.5. In the following description, we will especially focus on the way content-specific motives are induced for making a transition to the next block.

Inducing a need: How to tell if something is radioactive? After a motivating introduction, students discuss what has and what has not got to do with radioactivity. They all know that nuclear power plants and X-ray machines have got to do

[1] Klaassen's research was carried out in the late 1980s and early 1990s. At that time, Dutch students all knew about the accident that had happened just a few years earlier (in 1986) with a nuclear power plant in Chernobyl. Also in the Netherlands the accident had consequences. For example, fresh products such as milk and spinach had become radioactive and had to be withdrawn from the market. Our alternative approach also draws heavily on students' familiarity with the Chernobyl accident.

[2] We do not mean to suggest that terms such as "affector" or "instrument" are used by students. It is we who use these terms to talk about their ideas.

with it. In the terminology introduced above, students are sure that nuclear power plants and X-ray machines are affectors. But they are not so sure, or mutually disagree, about whether or not a battery has got to do with radioactivity, or a laser, or a magnet. We had foreseen these doubts and disagreements because many people find batteries, lasers, and magnets somewhat mysterious or dangerous, just as radioactivity. By referring to students' doubts and disagreements, it is relatively easy to induce a need for an objective criterion of telling when something is radioactive (as was our explicit intention).[3] This need is eventually met by a Geiger counter, which in the sequel also makes it possible for students to check their predictions and expectations experimentally.

Inducing a practical problem: How to make something radioactive? We had also foreseen that students' existing knowledge would enable them to simulate the Chernobyl accident. The teacher introduces a weak radioactive source, e.g., a mantle of a gas lamp, as the radioactive material that was stored in the power plant before the accident, and asks students to store it in such a way that it poses no radiation hazard to its immediate environment ("the people living nearby"). Students had no trouble doing this. They immediately built "walls" of lead around the mantle until a Geiger counter on the outside no longer ticked above the background rate. We expected such proposals. In the terminology introduced above, the proposals amounted to applying a resistance. Students also believed that they knew what would have to happen in order that radiation could be measured at the other side of the classroom ("the Netherlands"). They proposed that the "walls" must be broken, that there must be a wind blowing towards "the Netherlands," and that it must rain above "the Netherlands." These proposals were also expected. In the terms introduced above, the proposals all amounted to a means of transporting the instrument from the affector to the affected. As was our explicit intention, students were really surprised when it turned out that their proposals did not work. They broke down the "walls," used a fan to produce a flow of air towards "the Netherlands," sprinkled some water above a Geiger counter in "the Netherlands"—whatever they tried, the counter did not begin to tick any faster.

In another part of the simulation activity, students were asked to make an apple radioactive with the materials present in the classroom.[4] This, too, they thought they knew how to achieve. For example, they proposed to put the apple next to the mantle or to X-ray the apple for a while. Such proposals were also foreseen. In the terms introduced above, the proposals all amounted to a means to get the instrument from the affector into the apple. Students were baffled even more when these proposals

[3] Here, we have a first major example where a reason is induced in students for what they are going to do next. It is not of a general nature, such as: we are going to do this, because we want to please the teacher, get a good grade, or stay out of trouble. Instead the reason directly and specifically concerns the topical content: in order to reach mutual agreement and secure knowledge about safety measures and applications of radioactivity, we first of all need an objective criterion of telling when something is radioactive, and that is what we are going to find out now. Because this reason is specifically directed at topical content, we call it content specific or content directed.

[4] Apart from some weak radioactive sources, also a small X-ray machine was present in the classroom.

also did not work. The problem of how to make something radioactive thus thrust itself upon the students, and with quite some force given its practical relevance, as was our explicit intention.[5]

Solving the practical problem in the context of safety measures. As is often the case in situations where one oneself has framed a problem that has a clear meaning to oneself, the students were already on the way to solving the practical problem once they have framed it. For one thing, they had an open eye and mind for possible contributions to its solution. In the process that had led to their formulation of the problem, they were implicitly also provided with the conceptual equipment that was appropriate to recognize possible solutions as such. This is not to say that it was obvious to students how they might find a solution to the problem. They needed guidance. Lack of space prevents us from going into details here. We merely mention that gradually students developed what might be called a macroscopic theory of radioactivity. It consisted of relationships between the core concepts of radiation, radioactive, irradiation, and contamination. For example, objects do not get radioactive from being irradiated. Students also learned to apply the theory in the context of safety measures, for example when they thought about whether or not the prevention of irradiation required the same sort of safety measures as the prevention of contamination.

Inducing theoretical problems: What is radioactivity? The macroscopic theory answers the practical problem, as well as related questions such as how the spinach in the Netherlands did become radioactive. But the macroscopic theory also raises new questions, such as the following. Why is it that an object does not emit radiation after it has been irradiated? What, then, happens to the radiation when it enters an object and, in particular, why is it that receiving radiation *does* have harmful effects? And what is radiation anyway? We did not expect all students to raise all of these questions or to find such questions very exciting. But we did expect that at least some such questions would be raised by at least some students, and that, once raised, the other students would at least recognize that the macroscopic theory does not provide answers. This typically happened.

Note that questions such as those just mentioned do not demand an improved understanding of situations that are of practical interest, but rather require a deeper understanding than is offered by the macroscopic theory. In short, they are questions of a more theoretical nature, of the kind: what is radioactivity? Such theoretical questions were also at the forefront in the traditional treatment. But whereas in the traditional approach the questions were prematurely raised by the textbook or the

[5] This is a second major example, where a reason is induced in students for what they are going to do next. This reason is content specific: we do not yet know how to make something radioactive, but clearly this is at least one thing we need to know in order to properly understand safety measures and applications of radioactivity. So what we are going to do next is find out why all of our proposals did not work and how something *can* be made radioactive.

teacher, this time they were either raised by the students themselves or at least fell on fertile soil, as was our explicit intention.[6]

"Solving" the theoretical problems: Atomic and nuclear models. Students received some hints with which they could tackle the theoretical problems, such as the suggestion to think of radiation as consisting of very small and very fast moving particles. The challenge then was to think of some micro-level account of what happens when the particles enter an object, that explains why food is affected while it is being irradiated (e.g., the bacteria in it are killed), but no longer poses a radiation hazard after it is irradiated, also not when it is eaten. Along these lines students get a flavor of how micro-level mechanisms might enable a deeper understanding.[7]

Reflection on Problem-Posing Features

As will have become clear from the previous sketch, it is not coincidental that:

- At one stage students felt a need for an objective criterion for telling whether or not something was radioactive;
- At a later stage students came to appreciate as urgent the practical problem of how to make something radioactive; and
- At a still later stage students came to pose theoretical problems that invited an account of what radiation does in terms of what radiation is.

All of this was carefully planned and outlined, by making productive use of students' existing knowledge and by tuning activities to one another in considerable detail. The main difference with the traditional approach was that it is *not* unquestioningly assumed that students simply stand ready to absorb new knowledge, such that all one has to do is present them with this new knowledge. The main difference with conceptual-change approaches is that it is *not* deemed necessary first to delete existing knowledge in order to create a place for the knowledge to be taught to occupy. Our emphasis rather lies on providing students with content-directed motives and on soliciting seeds in their existing ideas, in such a way that they are willing and able to extend their knowledge and skills in a certain direction. This direction, moreover, from the perspective of the designer must be such that by

[6] This is a third major example, where a reason is induced in students for what they are going to do next. This reason is content specific and of a theoretical rather than practical nature: we now know a lot about safety measures and applications, but some questions are left open, especially concerning the interaction of radiation with matter and living tissue; we are going to find out more about that now. The theoretical questions invite an account of what radiation does in terms of what radiation is.

[7] It was not expected that students' theoretical questions would provide a basis that was strong enough to support the introduction of full-fledged nuclear models. The bottom arrow in Figure 10.5 is drawn dotted because it represents only a weak content-directed reason suggested by students. A rather detailed nuclear model was only included to meet the requirements of the then examination program.

following it students can be expected to get closer to the intended attainment targets. The designer must explain, for example, how in a process that is given an initial purpose and direction by the practical problem of how to make something radioactive, students can come to establish, and to value as a solution to the practical problem, what above is called the macroscopic theory of radioactivity.

Perhaps it is good to add that there is no contradiction between, on the one hand, students' bottom-up control and, on the other hand, the designer's carefully outlined plan that the process will proceed in a particular way and will lead to the attainment of certain preset targets. The students may be well aware that this was all prearranged, but still feel that they are contributing substantially to the direction taken by the process.

The Case of Calculus

This section concerns a teaching sequence about the topic of calculus. Here too we first sketch the "traditional" way of teaching the topic. Both the traditional and our alternative approach were aimed at academically streamed students in upper secondary education (Grade 10). Our approach took about ten 50-minute lessons. We again close with a reflection on the problem-posing features.

Traditional Treatment of Calculus

The traditional setup of a calculus course is presented in Figure 10.6. It builds upon an early treatment of the limit concept. The gradient of a graph is introduced as the limit of a difference quotient. This notion is extrapolated to a function that describes all gradients of the graph.

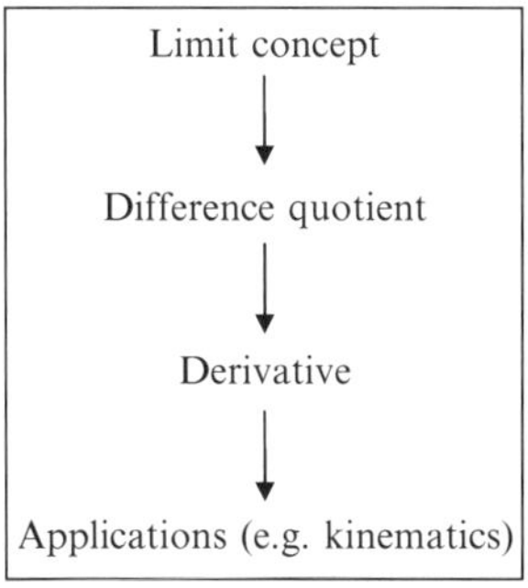

Figure 10.6. Structure of the common treatment of calculus.

The rationale is that in order to understand the derivative $f'(x)$, one has to have the concept of a limit at one's disposal because the derivative is the limit of the difference quotient $\left(f\left(x+h\right)-f\left(x\right)\right)/h$, where h tends to zero. This process is visualized with a decreasing chord on a graph, tending to the local slope of the graph. Traditionally, the variable x is initially replaced by a number or a placeholder a to define and calculate the slope of a graph at a point. The next step is to let a vary, or to replace a by an x, and to introduce the idea of the derivative of a function. Finally, this process is used to derive $f\left(x\right)=x^2$ and some other relatively simple functions, and to proceed quickly to techniques for differentiation such as product and chain rules. In this approach, the students will find at a late stage—after dealing with the concept of and techniques for integration—the connection between differentiation for grasping change and integration for finding "totals." This connection is mainly expressed as a kind of inverse relationship. The emphasis is on the techniques, and the tasks and applications are mainly meant to practice the techniques.

Some Comments on the Traditional Treatment

The late attention for applications in the traditional treatment of calculus creates difficulties for students to see connections with different notations and approaches in other disciplines. The mathematical language of functions (f, x, y, ...) and chords in graphs are hardly used in secondary school science, while the tangent method (i.e., sketching a tangent and determining its slope) is important in science but hardly treated in mathematics. The introduction of the limit concept prior to the difference quotient suddenly appears for no reason to the students. Also, the conceptual step from a limit with a fixed x to a varying x is rather difficult, since taking a limit in one point is substantially different from perceiving $f'(x)$ as a function, the values of which describe the gradient of a graph of $f(x)$.

The traditional treatment of calculus is the result of a similar rational reconstruction as in the case of radioactivity. Tall (1991) suggested that it was no wonder that mathematicians especially tended to make this typical error when they designed instructional sequences. The general approach of a mathematician is to try to simplify a complex mathematical topic by breaking it up into smaller parts which can be ordered in a sequence that is logical from a mathematical point of view. From the expert's viewpoint the components may be seen as part of a whole. But the student may see the pieces as they are presented, in isolation, like separate pieces of a jigsaw puzzle for which no total picture is available (Tall, 1991, p. 17). It may be even worse if the student does not realize that there is a total picture.

Freudenthal's interest in mathematics education started with his critique of such rational reconstructions. He was fiercely opposed to what he called an anti-didactical inversion (Freudenthal, 1973), where the end results of the work of mathematicians are taken as starting points for mathematics education. Mach (1976) had already pointed out this inversion in the presentation of mathematical theorems: "mathematicians more than others tend to eliminate all trace of development as soon as they present their findings. The perfectly clear recognition of mathematical propositions is by no means attained all at once, but is preceded and prepared by incidental observations, surmises, thought-experiments and physical experiments with counters and geometrical constructions" (pp. 182–183).

As an alternative for this inversion, Freudenthal advocated that mathematics education should take its starting point in mathematics as an activity, and not in mathematics as a ready-made system (Freudenthal, 1973, 1991). For him the core mathematical activity was mathematizing, i.e., organizing from a mathematical perspective. Mathematizing involves both mathematizing everyday-life subject matter, and mathematizing the mathematical activity itself. The main idea is to allow students to come to regard the knowledge they acquire as their own knowledge.

Some Preliminaries About an Alternative Approach

In order to realize our problem-posing ideal, we looked for problems that students would recognize as relevant and real, and that would evoke solution strategies that have the potential of being mathematized towards the desired concepts and skills. Our emphasis was on students developing a thorough understanding of basic principles rather than on the training of techniques. In particular, we aimed at genuine understanding of the relationship between taking differences and adding them up, and of the difference quotient as a means for grasping and quantifying changing quantities.

Historically, the basic principles of calculus originated from thought experiments about falling objects and from grasping the relationship between velocity and distance traveled (Sawyer, 1961). In addition, graphs and other mathematical symbols such as tables and algebraic notations play key roles. Traditionally, these are presented as ready-made symbols to students. However, for students it is not at all obvious how to interpret graphs. Terms such as "high," "steep," "quick," and "constant," which have specific meanings in interpreting graphs, are very quickly mingled with the situations that are represented by the graphs, especially in the case of motion (Doorman & Gravemeijer, 2009).

It seems that learning calculus and learning kinematics are intertwined, and it is difficult, maybe even impossible, to say what must be taught first. In the historical development of calculus (starting before Leibniz and Newton), clues can be found for how graphical representations of motion emerged and supported the understanding of the relation between velocity and distance traveled (Doorman & van Maanen, 2008).

A starting point for reasoning about changing quantities is students' common-sense understanding that when you travel at high speed, you will cover more distance in equal time intervals than when you travel slower. Intervals of distances traveled have proven to be basic structuring elements for reasoning about motion (Boyd & Rubin, 1996). Often this reasoning with intervals is sufficient, but it does not always lead to precise predictions. In order to meet a demand for more precision, our idea is to connect reasoning with intervals to reasoning about change with two-dimensional graphs that represent motion. This connection has the potential to be mathematized into reasoning with difference quotients.

Sketch of an Alternative Approach

The principal theme of our alternative approach to calculus is grasping change in order to make predictions. The structure is outlined in Figure 10.7.

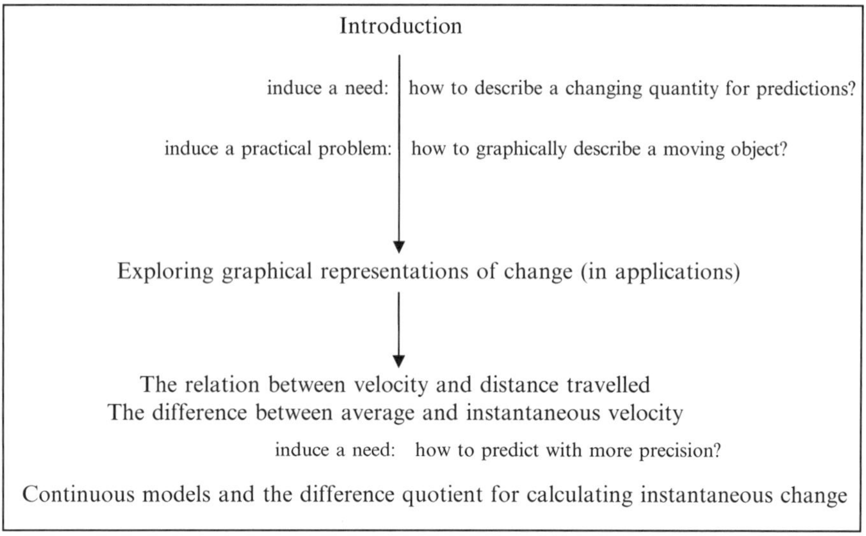

Figure 10.7. A didactical structure with coherence between modeling motion and grasping change.

Inducing a need: How to describe a changing quantity for predictions? The overarching question of the sequence is how to describe changing quantities in order to better predict. This question is initially posed in the context of motion by considering weather forecasts (moving clouds and hurricanes). Change and predictions are well-known notions in this context and we expected that this context would provide students with content-specific reasons to make predictions. During the sequence, the perspective on this overarching question changes from situation specific, to generalizing over different kinds of quantities in various contexts, and finally to context-independent concepts and skills expressed in a formal mathematical language. The overarching question supports coherence between the successive lessons by evoking contributions from students to the problems that have to be solved in order to improve conceptual understanding and tackle the global overarching question.

Inducing a practical problem: How to describe a moving object graphically? The sequence started with two satellite photos taken with 3 hours between them. The aim was to predict whether the clouds, which clearly changed position, would reach the Netherlands in the next 6 hours. This was important to know for the organizers of a pop concert that evening. The context was expected to provide a need-to-know for students and to offer opportunities for an initial orientation on the main theme. As expected students measured displacements, and extrapolated these in making predictions. Next students were shown successive positions of an accelerating hurricane on a map. They were asked to predict when and where it would hit the coastline. These questions led to opportunities for discussing patterns and for using changes in successive positions as a basis for predictions. As Boyd and Rubin (1996) have found, students naturally think of intervals as a measure of change of velocity. They were therefore expected to realize that it made sense to display the measurements graphically for investigating and extrapolating patterns in intervals.

Exploring graphical representations of change. After working with the hurricane and the stroboscopic photographs, two types of two-dimensional graphs emerged: discrete graphs of intervals between successive positions, and discrete graphs of total distances traveled. The classroom discussion led to consensus about the use of, and the relationship between, these two-dimensional graphs for describing and predicting motional phenomena. It also became clear that drawing such graphs was a sensible way to proceed.

In the graphs distances are represented, not as the height of a dot, but as lengths of vertical bars. The discrete case of the main theorem of calculus was implicitly touched on in this kinematic context. The sum of intervals was equal to the total distance traveled, and the difference between two successive values of the distance traveled was equal to the interval (see Figure 10.8).

Inducing a need: How to predict with more precision? The newly developed tools were evaluated with respect to the overarching question: do the tools enable us

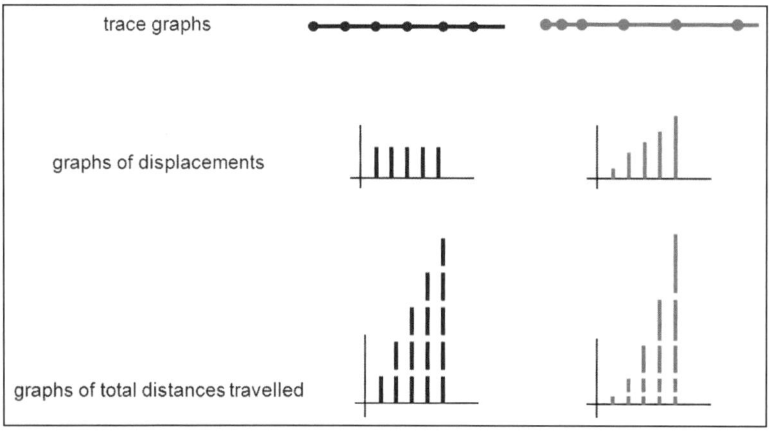

Figure 10.8. From trace graphs to discrete two-dimensional graphs of motion.

to make better predictions? We expected students to suggest measuring the successive positions of a hurricane at shorter time intervals in order to gain a better view of the pattern in the displacements. Furthermore, they were expected to differentiate between changes in average velocities based upon the measurements and the actual velocity after the last measurement. Subsequently, this was to be used by the teacher to induce a content-directed motive for introducing hypothetical continuous models for predictions. However, the next lesson dealt with a new (historical) context, which turned out to hinder rather than facilitate the teacher in tapping, emphasizing, and using the required conceptual connections.

Introducing a hypothesis: A continuous model for free fall? The transition to continuous models was introduced in the context of a narrative about Galileo's work. Students were asked to interpret Galileo's hypothesis that the velocity of a falling object increases in proportion to the time it falls, and to compare this hypothesis with other ideas in that period. We chose the story about Galileo because we thought that it would be a relevant problem for the students. Moreover, it offered opportunities for students to connect discrete approximations and discrete graphs with continuous models. Finally, it gave students a view on a milestone in science.

With intervals of distances traveled in specific time intervals students could calculate constant average velocities for the chosen time intervals. The graph of the average velocities will also increase linearly. The multiplication of a time interval and a constant velocity resulted in a displacement in the corresponding time interval. From there on, as expected, students saw the connection with the discrete case. Adding intervals traveled (areas in the velocity graph) resulted in total distances traveled (inspired by Kindt, 1996; Polya, 1963). These procedures used an informal limit concept.

By approximating changing velocities with bars (representing constant velocities in specific time intervals), the first step was made towards creating an experiential base for the process of describing motion leading up to integrating functions.

Improving prediction by using continuous models. A situation about a Dutch comic character who drove his car through a village (inspired by Kindt, 1979) was presented together with a continuous time graph of his distance traveled. The question was: Do you think he broke the speed limit? We expected students to reason about velocity with discrete approximations of time and distance (Δt and Δs) in this graph. Students managed to reach consensus on how to calculate instantaneous velocity approximately. After discussing the activity, students used a computer program for drawing a difference quotient on a graph as a chord, and for zooming in on part of the graph (inspired by Tall, 1996). As a result of this exploration, students developed a strong graphic and dynamic image to support the formalization in mathematical language of the relationship between the slope of a chord and the approximation of instantaneous change. During subsequent lessons, the teacher and the students regularly referred to this dynamic image.

The unit closed with a reflection on the successive steps that had been taken, from the perspective of the overarching question: To what extent are we now capable of describing and predicting change? The connection between the successive representations in the context of modeling motion supported students in reconstructing the meaning of the difference quotient, its power (you can be precise), and its limitations (you need a function, a continuous model, which is not always at hand).

Reflection on Problem-Posing Features

When we look back at our teaching sequence from the point of view of our problem-posing ideal, we have mixed feelings. Throughout, our aim has been to introduce situations that evoke the need for new tools or concepts by problematizing students' understandings and experiences in the context of the overarching goal of grasping change in order to make more appropriate predictions. We provided the teacher with information on students' reasoning about changing quantities, and on how this reasoning could be used to elicit productive questions and suggestions. In the first part of the teaching sequence, this worked out rather well. The teacher was indeed able to regulate classroom discussions in such a way that students understood that displaying and investigating patterns in displacements was a sensible way to proceed for describing and predicting motion. Suggestions by students for using two-dimensional graphs were welcomed by the teacher as a valuable way of reasoning. Moreover, this way of reasoning was accepted by all students, as we concluded from their contributions and questions while discussing the graphs.

The transition from discrete motion graphs to continuous models, however, did not proceed so smoothly. Although in the end students managed to reason adequately with continuous models, we did not succeed in providing students with advance content-specific motives for the transition itself. Above we indicated that the historical context of Galileo's work somehow hindered adequate scaffolding for students in building their reasoning with formula-based graphs upon their reasoning with data-based graphs. In retrospect, we now have a clearer view of the cause of the observed "friction." The transition to continuous models simply was not functional for students in view of the overarching goal of making better predictions. Up to the transition, students had made predictions on the basis of available data by linear extrapolation. In order to improve the predictions, there was a sudden switch to making predictions on the basis of imagined data or hypothesized models. But instead, it may have been more "logical" for students to use readily available data and to improve their predictions by extending the method of extrapolation beyond *linear* continuation. Furthermore, it was possible to do so, even if no hypothetical continuous models were available. In retrospect, this reinterpretation of the friction we observed during the transition was so obvious that one may wonder why we did not see it before, when we designed the relevant teaching–learning activities. We will not address this question here. Our point merely is to illustrate how our problem-posing ideal at least in retrospect has guided us to understand more fully what may have caused the observed friction.

We conclude that, as designers, we are faced with a choice. Either we retain our original aim of reasoning with continuous models, in which case there still is the need to provide students with a motive for making the transition. This seems to demand a change of overarching goal. Or, we retain the original overarching goal, and then the natural course rather seems to be towards the idea of Taylor-expansion as a controlled step-by-step improvement of prediction. We will not argue here for either option, and there may be more. Our point merely is to indicate how our problem-posing ideal has oriented us towards possible resolutions.

Reflection and Extension

Our problem-posing approach is not a general theory of learning or teaching, but a programmatic view of the possibilities for improving educational practice at a content-specific level which can be further explored and empirically realized by educational research. It is not easy to achieve the goal that all along students know, on content-specific grounds, what they are doing and why. It is not just a matter of asking students what they would want to learn. In order to appreciate the difficulty, it is useful to distinguish between the content-specific purposes of students (their goals) and the aims of the course designer (the attainment targets). The student goals should become worthwhile to them in advance of comprehending the attainment targets. Students should also come to experience the work they are going to do as instrumental to reaching their goals. From the perspective of the course designer, moreover, students' work should bring them closer to their attainment targets. It is a difficult challenge to meet all of these requirements at the same time. Hence, the reason why the problem pointed out by Gunstone (1992)—students not knowing the purpose(s) of what they are doing—is difficult to avoid. But to the extent that one manages to meet these requirements, it will contribute to having students regard the knowledge they acquire as their own. First, because the knowledge is then acquired on a need-to-know basis. Second, because the knowledge is then acquired by continually tapping their own conceptual resources, thus helping to avoid alienation and compartmentalization.

From the two cases discussed above, it should be apparent that meeting our problem-posing ideal involves a detailed analysis of students' existing knowledge and abilities, as well as a careful and detailed outlining of teaching–learning activities that support and build on each other. There are no general procedures for how to achieve this. It is a matter of finding local solutions to local problems, and in many cases critical details such as the actual wording of tasks are of vital importance (Viennot, 2003). It typically takes several cycles of design, testing, and redesign, before the ideal is just beginning to come in sight. In this respect, we feel we have made more progress in the case of radioactivity than in the case of calculus. In part, this will have to do with the nature and complexity of the topics at hand. It is much easier to involve students in the practical concerns associated with radioactivity than to set and keep them in the right kind of theoretical mood that is required for calculus.

Of course this does not imply that the ideal must be abandoned for the case of calculus, though it may make one wonder if one values the ideal strongly enough to go through the amount of trouble that apparently is needed to attain it. As far as we are concerned, we have not yet reached the stage that we would rather leave our teaching sequence on calculus as it is. Instead, our tendency is to analyze the weak points of our approach and to try harder, perhaps by exploring other avenues. In our discussion, we indicated what we see as weak points, in particular that the guiding theme of how to describe change for predictions was not always functional for the students. Could the weak points be addressed by changing some of the examples, or is a more drastic modification needed such as a replacement of the guiding theme itself? An alternative avenue may be to explore the educational usefulness of one of Zeno's paradoxes. Achilles and a turtle are involved in a running contest. The turtle has a head start on Achilles. Zeno reasons that Achilles will never overtake the turtle because when Achilles reaches the spot where the turtle started, the turtle will already have moved on, and so on ad infinitum. It will be obvious for students that Zeno's conclusion is false (of course Achilles will overtake the turtle), but it will not be obvious at all for them to pinpoint the flaw in Zeno's reasoning. The potentially useful element of this example is that it naturally sets students to think about change *within a theoretical context*, that is, within the context of sound reasoning. We have not sufficiently worked out this line of thought though. Clearly, clever ideas are needed here. Of course, it cannot be enforced that one gets good ideas, but at least we are more receptive now. We do hope that some readers, after having been sensitized to our ideal, will come up with useful suggestions. In our opinion, it is an essential aspect of educational research to thus engage the broader research community.

Several other attempts have been made, with more or less success, to realize the ideal that all along students know what they are doing and why. Vollebregt (1998) designed a teaching sequence on particle models, in which conceptual progress on particle models drives and is driven by issues of a metaphysical, ontological, and epistemological nature (e.g., What does it mean to explain something? Do particles really exist? How do we know which properties they have?). Kortland (2001) designed a teaching sequence in environmental education. In a process structured by students' existing decision-making skills and basic knowledge about life cycles of materials, students eventually arrive at well-argued decisions in the context of dealing with household package waste. Another attempt concerns an introductory mechanics course. By tapping core causal knowledge and epistemic resources, students eventually arrive at theoretical insights in explanations of motion and a justified preference of Newton's to Kepler's theory of planetary motion (Emmett, Klaassen, & Eijkelhof, 2009; Klaassen, Westra, Emmett, Eijkelhof, & Lijnse, 2008). Other attempts have been based on the idea of adapting an established professional practice, e.g., the chemistry-related practice of monitoring water quality (Westbroek, Klaassen, Bulte, & Pilot, 2010).). A professional practice can be thought of as an organized system of activities, the coordinated execution of which leads to the attainment of some goal. The basic idea is to "transform" this hierarchy of means-to-end relations in the context of professional practice into a hierarchy of content-specific motives for students to engage in learning activities.

Mathematical Problem Posing from the Point of View of Our Problem-Posing Approach

When we think about mathematical problem posing from the perspective of our problem-posing approach, our main message is *not* to view mathematical problem posing as an optional activity alongside, or over and above, students' learning about some mathematical topic (long division, calculus, statistics, or whatever). If one thinks about organizing a teaching sequence about a particular topic in such a way that all along students know on content-specific grounds what they are doing and why, one cannot but think about appropriate contexts to make students raise the right sort of problems. What makes the problems of the right sort is that they are clearly connected to a worthwhile goal (for students) and also suggest a direction for a solution. Following that direction, moreover, is to lead students eventually to the attainment targets, perhaps via some redirections engendered by newly raised problems or reformulated old problems, and so on. Just like in Vollebregt's (1998) approach, students' learning about the nature of science is not something added on to their learning of science, but naturally integrated within their learning of science, so we think of mathematical problem posing as something to be naturally integrated within students' learning of mathematics, and the same goes for mathematical modeling.

We have one final reflection. We have argued against the tendency of structuring a teaching sequence along the lines of a rational reconstruction. But this does not rule out the possibility, within a problem-posing approach, of inviting students to make a rational reconstruction, namely towards the end of the teaching sequence, in order to summarize what they have learned. Such a rational reconstruction may also concern the role played by mathematical problem posing in the teaching sequence. It may even be given a useful point within an educational setting, as a preparation for the test. That is, in order to prepare well for the test students can be challenged to design good test items for each other and to reflect on why they think these are good problems. The aim for students would then be to make explicit the sorts of elements that were also central to the programs of English (1997a, 1997b, 1998) discussed earlier in the chapter, by classifying the types of problems that have been treated, separating problem structures from contextual features, and so on.

Acknowledgments We would like to thank Carolina Kühn, two anonymous reviewers, and the editors of this book, for their help in improving previous versions of this chapter.

References

Barlow, A. T., & Cates, J. M. (2006). The impact of problem posing on elementary teachers' beliefs about mathematics and mathematics teaching. *School Science and Mathematics, 106*(2), 64–73.

Bonotto, C. (2010). Engaging students in mathematical modelling and problem-posing activities. *Journal of Mathematical Modelling and Application, 1*, 18–32.

Boyd, A., & Rubin, A. (1996). Interactive video: A bridge between motion and math. *International Journal of Computers for Mathematical Learning, 1*, 57–93.

Brown, S. I., & Walter, M. I. (1983). *The art of problem posing*. Hillsdale, NJ: Lawrence Erlbaum.

Christou, C., Mousoulides, N., Pittalis, M., Pitta-Pantazi, D., & Sriraman, B. (2005). An empirical taxonomy of problem-posing processes. *Zentralblatt für Didaktik der Mathematik, 37*, 149–158.

Crespo, S., & Sinclair, N. (2008). What makes a problem mathematically interesting? Inviting prospective teachers to pose better problems. *Journal of Mathematics Teacher Education, 11*, 395–415.

Doorman, L. M. (2005). *Modelling motion: From trace graphs to instantaneous change*. Utrecht, The Netherlands: CD-Beta.

Doorman, L. M., & Gravemeijer, K. P. E. (2009). Emergent modeling: Discrete graphs to support the understanding of change and velocity. *ZDM Mathematics Education, 41*(1), 199–211.

Doorman, L. M., & van Maanen, J. (2008). A historical perspective on teaching and learning calculus. *Australian Senior Mathematics Teacher, 22*(2), 4–14.

Einstein, A., & Infeld, L. (1938). *The evolution of physics*. Cambridge, United Kingdom: University Press.

Ellerton, N. F., & Clarkson, P. C. (1996). Language factors in mathematics teaching and learning. In A. J. Bishop, M. A. Clements, C. Keitel, J. Kilpatrick, & C. Laborde (Eds.), *International handbook of mathematics education* (pp. 987–1033). Dordrecht, The Netherlands: Kluwer.

Emmett, K., Klaassen, K., & Eijkelhof, H. (2009). Introducing mechanics: II. Towards a justified choice between alternative theories of motion. *Physics Education, 44*, 492–498.

English, L. D. (1997a). Seventh-grade students problem posing from open-ended situations. In F. Biddulph & K. Carr (Eds.), *People in mathematics education* (pp. 39–49). Waikato, New Zealand: Mathematics Education Research Group of Australasia.

English, L. D. (1997b). The development of fifth-grade children's problem-posing abilities. *Educational Studies in Mathematics, 34*, 183–217.

English, L. D. (1998). Children's problem posing within formal and informal contexts. *Journal for Research in Mathematics Education, 29*, 83–106.

English, L. D., Fox, J. L., & Watters, J. J. (2005). Problem posing and solving with mathematical modelling. *Teaching Children Mathematics, 12*(3), 156–163.

Freudenthal, H. (1973). *Mathematics as an educational task*. Dordrecht, The Netherlands: Kluwer.

Freudenthal, H. (1991). *Revisiting mathematics education: China lectures*. Dordrecht, The Netherlands: Kluwer.

Gunstone, R. (1992). Constructivism and metacognition: Theoretical issues and classroom studies. In R. Duit, F. Goldberg, & H. Niedderer (Eds.), *Research in physics learning: Theoretical and empirical studies* (pp. 129–140). Kiel, Germany: IPN.

Kindt, M. (1979). *Differentiëren 1. Een manier om veranderingen bij te houden*. Utrecht, The Netherlands: IOWO.

Kindt, M. (1996). *Som & verschil, afstand & snelheid. Differentiaal- en Integraalrekening deel 1*. Utrecht, The Netherlands: Freudenthal Institute.

Klaassen, C. W. J. M. (1995). *A problem-posing approach to teaching the topic of radioactivity*. Utrecht, The Netherlands: CD-Beta. Retrieved from http://igitur-archive.library.uu.nl/dissertations/01873016/inhoud.htm

Klaassen, K., Westra, A., Emmett, K., Eijkelhof, H., & Lijnse, P. (2008). Introducing mechanics by tapping core causal knowledge. *Physics Education, 43*, 433–439.

Kortland, J. (2001). *A problem-posing approach to teaching decision making about the waste issue*. Utrecht, The Netherlands: CD-Beta.

Mach, E. (1976). Knowledge and error. Dordrecht, The Netherlands: Reidel. (Erkenntnis und Irrtum; first published in 1905)

National Council of Teachers of Mathematics. (2000). *Principles and standards for school mathematics*. Reston, VA: Author.

Polya, G. (1963). *Studies in mathematics, volume XI. Mathematical methods in science*. Stanford, CA: School Mathematics Group.

Sawyer, W. W. (1961). *What is calculus about?* New York, NY: Random House.
Silver, E. A., & Cai, J. (1996). An analysis of arithmetic problem posing by middle school students. *Journal for Research in Mathematics Education, 27*, 521–539.
Stoyanova, E., & Ellerton, N. F. (1996). A framework for research into students' problem posing in school mathematics. In P. Clarkson (Ed.), *Technology in mathematics education* (pp. 518–525). Melbourne, Australia: Mathematics Education Research Group of Australasia.
Tall, D. (Ed.). (1991). *Advanced mathematical thinking*. Dordrecht, The Netherlands: Kluwer.
Tall, D. (1996). Functions and calculus. In A. J. Bishop, M. A. Clements, C. Keitel, J. Kilpatrick, & C. Laborde (Eds.), *International handbook of mathematics education* (pp. 289–325). Dordrecht, The Netherlands: Kluwer.
Tiberghien, A. (2000). Designing teaching situations in the secondary school. In R. Millar, J. Leach, & J. Osborne (Eds.), *Improving science education: The contribution of research* (pp. 27–47). Buckingham, United Kingdom: Open University Press.
UNESCO (2012). Challenges in basic mathematics education. Paris, France: UNESCO. Retrieved October 29, 2012, from http://unesdoc.unesco.org/images/0019/001917/191776e.pdf
Van den Heuvel-Panhuizen, M., Middleton, J. A., & Streefland, L. (1995). Student-generated problems: Easy and difficult problems on percentage. *For the Learning of Mathematics, 15*(3), 21–27.
Viennot, L. (2003). Relating research in didactics and actual teaching practice: Impact and virtues of critical details. In D. Psillos, P. Kariotoglou, V. Tselfes, E. Hatzikraniotis, G. Fassoulopoulos, & M. Kallery (Eds.), *Science education research in the knowledge-based society* (pp. 383–394). Dordrecht, The Netherlands: Kluwer.
Vollebregt, M. J. (1998). *A problem-posing approach to teaching an initial particle model*. Utrecht, The Netherlands: CDβ.
Westbroek, H. B., Klaassen, K., Bulte, A., & Pilot, A. (2010). Providing students with a sense of purpose by adapting a professional practice. *International Journal of Science Education, 32*, 603–627.

Chapter 11
Statistical Literacy in the Elementary School: Opportunities for Problem Posing

Lyn D. English and Jane M. Watson

Abstract This chapter addresses opportunities for problem posing in developing young children's statistical literacy, with a focus on student-directed investigations. Although the notion of problem posing has broadened in recent years, there nevertheless remains limited research on how problem posing can be integrated within the regular mathematics curriculum, especially in the areas of statistics and probability. The chapter first reviews briefly aspects of problem posing that have featured in the literature over the years. Consideration is next given to the importance of developing children's statistical literacy in which problem posing is an inherent feature. Some findings from a school playground investigation conducted in four, fourth-grade classes illustrate the different ways in which children posed investigative questions, how they made predictions about their outcomes and compared these with their findings, and the ways in which they chose to represent their findings.

Contents

L.D. English (✉)
School of Curriculum, Queensland University of Technology, Brisbane, Australia
e-mail: l.english@qut.edu.au

J.M. Watson
Faculty of Education, University of Tasmania, Hobart, Australia
e-mail: Jane.Watson@utas.edu.au

F.M. Singer et al. (eds.), *Mathematical Problem Posing*,
Research in Mathematics Education, DOI 10.1007/978-1-4614-6258-3_11

Introduction

Problem posing has featured in the literature for several decades now, as indicated in Stoyanova and Ellerton's (1996) review of earlier studies on the topic. As far back as 1945, Duncker considered problem posing as the creation of a new problem or a reformulation of an existing problem, a perspective that has been foundational in subsequent studies (e.g., Silver, 1994). Over time, the notion of problem posing has broadened to include its relationship with problem solving, students' strategies in posing problems, how teachers might facilitate a problem-posing classroom, and how problem posing can contribute to students' conceptual development (e.g., Cai et al., 2012; Ellerton, 1986; English, 1997, 1998; English, Fox, & Watters, 2005; Kilpatrick, 1987; Silver & Cai, 1996; Stoyanova & Ellerton, 1996). This chapter begins with a brief overview of some of these perspectives, and then considers how problem posing can play an important role in developing children's statistical literacy.

Perspectives on Problem Posing

In reviewing the literature on problem posing, both Stoyanova and Ellerton (1996) and English (1997) lamented that the potential of problem posing for developing students' understanding of mathematics had been hindered by the lack of suitable frameworks, ones that link problem posing and problem solving within the regular curriculum. In addressing this still timely concern, Stoyanova and Ellerton (1996) developed a framework comprising three forms of problem-posing situations, namely, *free*, *semi-structured*, and *structured*. In the first situation, students generate a problem from a contrived or naturalistic situation presented to them. In the semi-structured form, students explore the structure of an open situation and complete it by applying existing mathematical knowledge, concepts, and relationships. The last category, which appears less broad, involves problem-posing activities based on a specific problem.

The core assumptions of Stoyanova and Ellerton's (1996) framework are still pertinent and warrant further attention in research on problem posing. These include the importance of problem-posing situations corresponding to, and arising naturally out of, students' classroom mathematics activities, and problem posing being a part of students' problem-solving experiences. The framework of English (1997) adopted a complementary set of features for facilitating problem posing, namely, the importance of children recognizing and utilizing problem structures, the need to consider students' perceptions of and preferences for different problem types, and a focus on developing their diverse mathematical thinking.

Although different, both frameworks highlight the importance of linking problem solving and problem posing within the course of conceptual development. There have been several studies investigating relationships between problem solving and problem posing, with findings suggesting a strong correlation between the two; the focus, however, has mainly been on the nature and complexity of the mathematical problems generated by problem solvers of varying capabilities (e.g., Cai et al., 2012; Cai & Hwang, 2002; Ellerton, 1986; Kilpatrick, 1987; Silver & Cai, 1996). Furthermore, a good deal of the research to date has targeted the posing of problems in response to a specific goal or stimulus (e.g., cultural artefacts; Bonotto, 2012), the posing of real-world scenarios upon solving given problems (e.g., Cai et al., 2012), and the reformulating/extending of existing problems (e.g., Brown & Walter, 2005). As we indicate in this chapter, our approach to problem posing differs from these studies and aligns with Stoyanova and Ellerton's (1996) definition of problem posing, namely, "the process by which, on the basis of mathematical experience, students construct personal interpretations of concrete situations and formulate them as meaningful mathematical problems" (p. 518). They deliberately broadened their definition of problem posing to enable it to fall within the goals of mathematics curricula.

Prior to addressing our approach to problem posing from this perspective, it is worth noting the work of Pittalis, Constantinos, Mousoulides, and Pitta-Pantazi (2004). They produced a structural model for cognitive processes incorporating the major constructs of:

> filtering quantitative information, translating quantitative information from one form to another, comprehending and organizing quantitative information by giving it meaning or creating relations between provided information, and editing quantitative information from given stimulus (pp. 51–52).

Pittalis et al. restricted their cognitive processes model to specific types of problem-posing tasks. Editing quantitative information is primarily concerned with tasks that require students to pose a problem without restriction from the provided stimulus. In tasks that require the filtering of quantitative information, students pose problems or questions that are appropriate to given, specified answers. Such answers provide a restriction, which makes filtering more demanding than editing. Posing problems from mathematical equations or computations requires comprehending the structural context of the problem and the relations between the given information. Translating requires students to pose problems or questions from diagrams, graphs, or tables. The findings of their study suggested that, although all four processes contributed to problem-posing competence, the filtering of important and critical information and problem editing play a stronger role than comprehending structural relations in quantitative information and translating this from one mode to another.

The authors thus concluded that students' competence in filtering and editing problems is strongly related to posing problems.

Despite the work that has been undertaken, problem posing remains a complex learning issue requiring a good deal more research (Kontorovich, Koichu, Leikin, & Berman, 2012), especially in underrepresented domains such as statistical literacy. In this chapter we expand the interest in problem posing in yet another direction by considering the opportunities afforded by the needs of statistical literacy in society and a school curriculum that now includes Statistics and Probability (e.g., Australian Curriculum, Assessment and Reporting Authority [ACARA], 2012; Common Core State Standards for Mathematics, 2010; http://www.corestandards.org/the-standards/mathematics/).

Statistical Literacy

Young students are very much a part of our data-driven society. They have daily exposure to the mass media where various displays of data and related reports can easily mystify or misinform, rather than inform, their minds. For students to become statistically literate citizens, they need to be introduced early to the powerful mathematical and scientific ideas and processes that underlie this literacy (e.g., Langrall, Mooney, Nisbet, & Jones, 2008; Whitin & Whitin, 2011). Numerous curriculum and policy documents have highlighted the importance of children working mathematically and scientifically in dealing with real-world data in the elementary school years (e.g., Curriculum Corporation, 2006; Franklin et al., 2007; National Council of Teachers of Mathematics, 2006). Limited attention however, is given to the statistical literacy that children need generally for decision-making in the twenty-first century. This is of substantial concern given that students need to make both personal and public decisions based on data when entering society beyond school (Watson, 2009).

Numerous definitions of statistical literacy abound (e.g., Gal, 2002; Watson, 2006). As used in this chapter, statistical literacy is viewed as "the meeting point" of statistics and probability and "the everyday world, where encounters involve unrehearsed contexts and spontaneous decision-making based on the ability to apply statistical tools, general contextual knowledge, and critical literacy skills" (Watson, 2006, p. 11).

Gal (2002) identified core requirements for statistical literacy in the wider society, the rudiments of which we argue need to commence in the younger school grades. These include the ability to "interpret and critically evaluate statistical information… and data-based arguments encountered in diverse contexts," and the ability to "discuss or communicate their reactions to such statistical information." Being able to communicate an understanding of what the information means and one's opinions on this, together with concerns about the acceptability of conclusions drawn are important aspects here.

An important, yet underrepresented component of statistical literacy is beginning inference, which includes the foundational components of variation, prediction, hypothesizing, and criticizing (Garfield & Ben-Zvi, 2007; Makar, Bakker, & Ben-Zvi, 2011; Shaughnessy, 2006; Watson, 2006). There has been little research on these components, including children's abilities to make predictions based on data. Children need experiences in drawing inferences from a range of statistical situations and representations including everyday events, raw data sets, graphs, and tables.

Although it is not expected that young students develop a sophisticated understanding of the components of informal inference, it is important that they gain an appreciation of the nature of the statistical process as they answer questions of relevance to them (Watson, 2009). In answering their questions, however, adequate evidence (data) needs to be collected and conclusions drawn with a stated degree of uncertainty that reflects the nature of the investigation and the evidence. If children are not exposed to these various facets of statistical literacy in the elementary school, the introduction of formal statistical tests in the late secondary school can become a meaningless experience because students will not have developed an intuition about the stories conveyed by the data.

Problem Posing and Statistical Investigations

As noted above, a key aspect of these early experiences in statistical literacy is undertaking investigations (Curcio, 2010). Problem posing is an inherent feature of such experiences, especially given that any situation involving problem posing incorporates a certain degree of uncertainty (Kontorovich et al., 2012). Accompanying this uncertainty is the need for critical analysis, a core component of problem posing that requires further attention in the classroom. Indeed, statistically literate citizens faced with making their own judgements on media and other reports need to be critical consumers of information.

The end point of statistical investigations is an inference, or decision, based on the evidence analyzed using available tools. Statistical literacy requires consumers of public reports based on these investigations to judge the inferences claimed within them. Without gaining experience through conducting their own investigations and understanding the uncertainty with which they reach their conclusions (informal as they will be), students will not gain the understanding to judge other claims they meet later. To illustrate further the role of problem posing in statistical investigations, we give consideration to the model presented by Watson (2009).

The model in Figure 11.1 begins with a statistical question. Such questions must be more refined and succinct than the general inquiries students initially pose. That is, their statistical questions must be unambiguous and enable the collection of manageable data to answer their queries of interest. As in all problem posing, there needs to be a context within which a question or problem can be posed. A context might entail, for example, global warming, or road fatalities, or national testing of school students. Whatever the context, skill is needed in posing the question to be

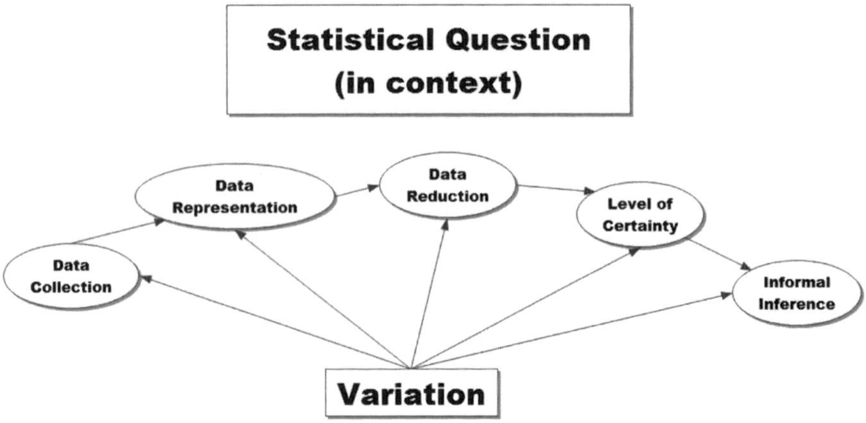

Figure 11.1. Stages of a statistical investigation (adapted from Watson, 2009).

asked through the statistical investigation (English, 2013a; Whitin & Whitin, 2011). Often, however, students are *given* the question to explore in an investigation. In contrast, statistically literate adults must be able to pose their own questions or query those of others. Understanding the relationship of samples and populations, and the uncertainty associated with any conclusion drawn about a population from a limited sample, is essential in posing the question to investigate. Giving school students the opportunity to create their own questions to explore and answer is a stepping stone to statistical literacy.

Following the posing of a statistical question, the stages of a typical investigation, as shown in Figure 11.1, provide other opportunities to pose "problems" in the sense of posing methods from which to choose to collect data, represent data, and reduce data to summary form. The presence of underlying variation in all statistical investigations influences the posing of methods appropriate to the context and to the type of data that are collected. This is likely to be a very open-ended process. The inference that is made at the end of a statistical investigation answering the posed question is constrained to some extent by the choices made during the investigation and is usually stated with a level of confidence in the inference, which may be stated as a chance.

In the present case, we were concerned with elementary school children posing their own questions and response options in undertaking an investigation within a meaningful and appealing context. Although all stages of a statistical investigation shown in Figure 11.1 are essential to students' development of statistical literacy, we concentrate our discussion here on beginning inference. In particular, we consider students' responses in predicting the outcomes of the questions they posed and comparing their predictions with their findings. An open-ended process in which the students chose their own forms of representations to display their findings completed the activity.

We focus specifically on prediction given that one of the core goals of statistics education is to assist students in making predictions that have a high probability of being correct or at least judging what this likelihood might be (Watson, 2007). As Watson emphasizes, many aspects of the mathematics curriculum caution against making predictions without certainty. Beyond the classroom, however, students are often confronted with problematic situations where decisions regarding several alternatives may appear reasonable. Again, critical thinking comes to the fore when students are asked to make a prediction because they must take into account all of the perspectives available in the statistical context.

Investigating the School Playground

The investigative activity addressed here was a replication of one conducted the previous year with third-grade students in the final year of a 3-year, longitudinal project on data modelling (e.g., English, 2013b). The present activity, implemented with fourth-grade students at a different school, served as a benchmark for the impending implementation of a new 3-year longitudinal project on beginning inference.

Although four, fourth-grade classes and one combined fourth/fifth-grade class completed the investigative activity, we report on the responses of the fourth-grade classes only ($n=81$, students with permission for their work to be included).

The activity comprised two components, namely, creating a survey and then implementing it within the classroom. The context of the investigation was exploring playgrounds, with a focus on the school's playing area. The class teachers implemented the first component in one lesson of 1.5 hours duration, with the second component in one, 2 hour lesson. The lessons were conducted on consecutive days, using lesson plan booklets developed to guide the implementation. A highly experienced senior research assistant met with the teachers initially, and subsequently monitored the activity implementation across all classes. The core learnings targeted included: posing and refining of questions; identifying, deciding on, and measuring attributes; developing and conducting a survey; collecting and recording data; organizing, interpreting, analyzing and representing data; and developing data-based explanations, arguments, inferences, and predictions.

The activity was introduced by discussing photographs of varied playgrounds from around the world, including those in underdeveloped countries as well as in Australian cities and outback regions. Students then reflected on their favorite neighborhood playground or park and offered reasons for their enjoyment in these areas. Within this context, the students were invited to find out more about their peers' thoughts on playgrounds, in particular, their own school playing area. The creation of a survey was subsequently discussed, with students offering some possible questions they might ask their peers. These suggestions served as examples only, with the students to pose their own survey questions.

The challenging aspect for students (of all ages) in designing a survey is turning their general inquiries into statistical questions. As noted previously, these questions must be clear to the respondent and enable the collection of manageable data to answer their queries. For fourth-grade students, the discussion focused on multiple-choice questions with four options for responses. Students worked in groups of about four, each first posing a question with corresponding response options. All students in each group then answered their group's survey, choosing one response option for each question, an example of which appears in Figure 11.2.

	Answer	Answer	Answer	Answer
Q1 Where is your favorite part of the whole of the seinor playgrond?	the stuck up poles to us people on to climb to the top	the ladder For playing off ground taggy	The part where it's kind of like a seat	The part of the wobbley bridge
Q2 How Long do you play for On second break?	Whole break	half of the second break	A third Of the break	A quarter of the break
Q3 What is your favorite thing to play in the Play ground?	Tag	Ofground Tag	Hide and seek.	ghost Hunter
Q4 Why do you like the Play ground?	because We can play games	because there is equipment	there is my friend go there	close to the eatting area

Figure 11.2. Example of one group's initial survey.

Each group then selected one of their group's questions to be the focus question for the remaining groups in the class to answer and also made a prediction of how this question would be answered. The groups' focus questions were copied, collated as a booklet, and distributed to all groups. Each student in the class then responded to all the focus questions within the class. A whole class discussion on ways in which the students might deal with their data then took place, but no specific direction was provided. The student groups subsequently collated their data, examined their findings, and compared these with their predictions. Finally, each group developed its own representation for displaying its findings; if time permitted, the groups were encouraged to complete more than one representation. A range of recording materials was provided for the representations, including blank chart sheets, lined paper, 2.5 cm squared grid paper, a sheet displaying a circle, and another with unmarked axes. The students were not directed to use any one recording format, however.

All student responses were recorded in their booklets (20 groups), and combined with their representations, served as the basis of our data analysis. We report here some findings from across the four classes, with a focus on (a) the types of question posed, (b) the basis on which students made their predictions for their focus question outcomes, (c) students' comparisons of their predictions with their findings, and (d) the representational forms students chose to convey their findings.

Posing Questions About the School Playground

The initial questions posed by students in each group were analyzed in terms of the type of query and the variation in the number of types suggested by groups. The question types included asking *why*, *what*, *how*, *which*, *where*, *when*, and *if*. Of the 20 groups, 9 posed three or four different types of questions, 10 created two types, and 1 group, just one type. Examples of these question types, with corresponding multiple-choice response options, appear in Table 11.1. It is interesting to note the inclusion of the conditional, *if*, which appeared in eight groups' questions, with three of these groups including two different question types.

Table 11.1
Examples of Survey Questions Posed

Question type	Examples of questions and response options
Why	*Why do you like to play in the playground?* Fun; Cheerful; Exciting; Amazing
What	*What can we add to make the playground a better place to play?* Giant Slide; Monkey Bars; Huge Rock Climbing Wall; Small Water Sprays
How	*How would you rate the playground?* OK; Bad; Good; Excellent
Which	*Which playground equipment would you spend the most time on?* The Spinner; The Nest; The Twisted Spider Web; The Ladder
When	*How do you feel when you go back to class?* Exhausted; Sad; Sweaty; Hot
If/what/where	*If the P&C had $1,000 to spend, what would you do to the playground?* Improve safety for the playground; make the slide 4.5 meters; get noughts and crosses; Get a longer flying fox *If the school built a new playground, where would you put it?* Oval; Grassy Patch; G block (year 1 area); Behind B block (bottom of back stairs)

Making Predictions

In analyzing the students' predictions of how their peers might answer their chosen focus question, we first noted that all student groups in one of the classes made a quantitative prediction that encompassed all four multiple-choice options for the question. For example, in predicting their peers' response to the question listed in Table 11.1, "If the P&C had $1,000, what would you do to the playground?" one group stated, "3 will improve safety, 7 will make the slide 4.5 meters, 3 will get noughts and crosses, 13 will get a longer flying fox." A response of this nature contrasted with the group predictions in the remaining classes, where 12 groups chose only one "most popular" option and three groups selected two possibilities.

Our data collection did not reveal whether the teacher whose students gave quantitative predictions directed the students to do so; such a direction was not provided in the teacher guidelines. This particular teacher, however, strongly encouraged detailed responses in all her students' work, and these did provide greater insights into the students' predictions of the option outcomes.

Analysis of the bases for the students' predictions yielded five categories, namely, those based on: (a) students' personal preferences and assumptions; (b) their observations of what takes place in the playground, such as how long peers spend on items of equipment or the most popular item; (c) students' factual, personal knowledge of the playground equipment/structure; (d) students' assumptions about their peers' preferences/perspectives; and (e) students' informal notion of probability, such as the unlikelihood of an event occurring or of a particular piece of equipment appearing in a school playground. Six groups made predictions that combined two of these categories.

The most common basis for prediction was the students' personal preferences and assumptions, with over half of all the groups displaying this reason (category (a)). For example, one group reasoned that their focus question, "What is your favorite part of the playground?" would be answered as follows: "The flying fox is quick to get around, while the monkey bars need strength, which some children don't have and you can get blisters on your hands which gets annoying. The spider tunnel we think will be the second most popular because it's cool, has a great view, and [you] can quickly escape from someone when playing tiggy."

The next most common prediction base, offered by five groups, was observation of what happens in the playground area (category (b)). One such group based their prediction for their focus question, "Which game would you play in the playground?" on their observation that, "We see that [children] play lots of off-ground tiggy. We do not see anyone play hide and seek; only a few people play hand games, and quite a lot of people play tiggy."

Four groups based their predictions on factual, personal knowledge of the playground (category (c)). One of these groups also included assumptions about their peers' preferences (category (d)). For their focus question, "If the P&C had $1,000, what would you do to the playground?" (Table 11.1), the group justified their prediction on the basis that, "People like to play on flying foxes. There is a lot of safety in our playground. Our slide is quite short so people want to make it much longer so people can have a longer ride and have more fun."

Only two groups used an informal notion of probability as a basis of their predictions (category (e)). One of these groups, who asked, "If you could add one thing, what would you add?" predicted that "The most popular would be the trampoline," because "We have never heard of a school with trampolines."

Comparing Predictions with Findings

The students' comparisons of their predictions with their findings mostly stated that they matched or otherwise, with the predictions and outcomes cited in their written responses. For example, the group that asked, "What is your favorite part of the playground?" stated that "The flying fox was quite correct, whereas the monkey bars got exactly 5, and rock wall we got 5, said it was 1–2, and spider tunnel was 6, and we said 6–7." It was disappointing that only one group offered a substantial justification in their comparisons. The group that posed the focus question, "What would you improve in the playground if you had $50,000?" predicted that "10 will choose the oval, 7 the spider web, 3 the slide, 3 the flying fox." In making their comparison, the group explained, "We found that a few of our predictions were off track and unfortunately made our predictions wrong. Here are our results: flying fox 8, slide 2, spider web 6, oval upgrade 10. A couple were off track because our decisions were based on what we see at play time but on the day of the vote, people [students] had not voted according to what we saw, making our predictions incorrect."

Posing Ways of Representing Findings

As noted, the students were to decide on their own ways of representing their findings and had no specific instructions. Three of the four classes completed the second activity component in the allotted 2 hours, while the teacher of the remaining class chose to allow her class extra time. Consequently, her students created a greater range of representations. One of the other classes who did not have extended time, however, also generated multiple, varied representations. Of the 50 completed representations generated by all groups, vertical bar graphs were the most common, with 50% of the representations being of this nature. There was considerable variation in the nature of these graphs, with some groups using the 2.5 cm grid paper, others creating their own grid paper, and a few drawing their own set of labelled axes.

The popular use of these bar graphs suggests that the students were not given adequate opportunities to pose their own forms of representations in their earlier school years, reflecting standard curriculum recommendations. For example, in the Australian Curriculum: Mathematics, "column" graphs, the equivalent term for vertical bar graphs, are mentioned at all grades from third to sixth (ACARA, 2012, pp. 29, 33, 38, 44). This finding contrasts with that of the previous longitudinal study across grades 1–3, where children displayed a rich repertoire of representational forms following experiences in generating their own means to display their data (e.g., English, 2013a, 2013b). Nevertheless, there were creative representations from the remaining groups, with seven vertical dot plots, three horizontal bar graphs, and three line graphs produced. The vertical dot plots, an example of which appears in Figure 11.3, were interesting variations of the traditional bar graph and foreshadowed, in part, our subsequent implementation of the *TinkerPlots* software program in the new project (Konold & Miller, 2011).

On one hand, the creation of only three horizontal bar graphs and three line graphs again seems to reflect the apparent limited opportunities for posing diverse representations in the early school years. On the other hand, the use of lists and tallies by 24% of the groups did display interesting and diverse approaches to collating

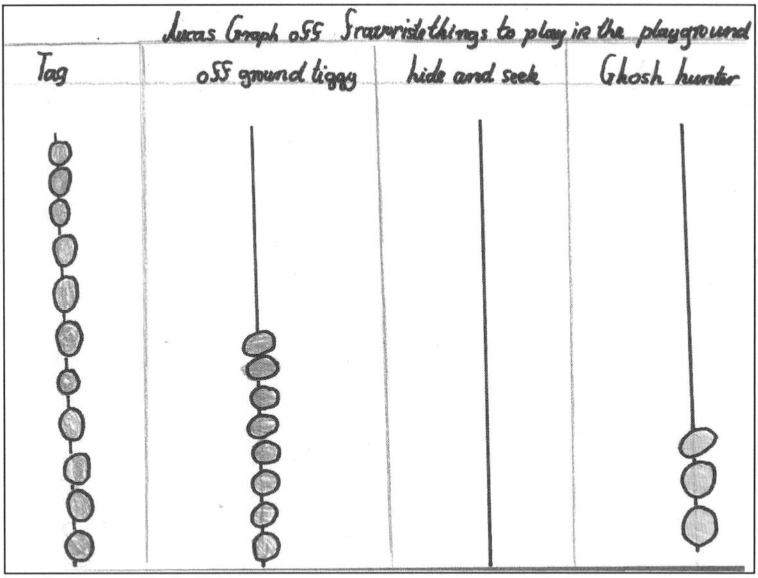

Figure 11.3. A vertical dot [Graph of favourite things to play in the playground: Tag, Off-ground tiggy, Hide and seek, Ghost hunter].

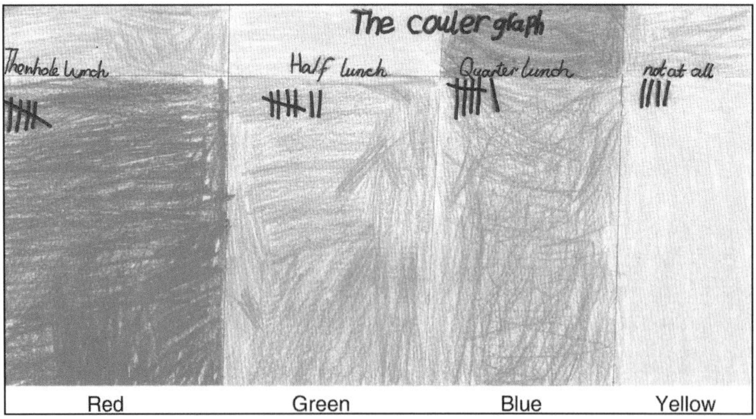

Figure 11.4. Structured tallies representation.

and representing data. An example of one group's creation appears in Figure 11.4. This group's representation displayed their findings for their focus question, "How long do you play in the playground?" predicting that "Most will play for a quarter of the lunch break," because "Most people get a bit bored because there [*sic*] so use [*sic*] to the playground," the group's representation supported their conclusion, namely, "We thought that most people would spend a quarter of there [*sic*] time in the play ground... BUT most people spend half of there [*sic*] time."

Discussion and Concluding Points

In relation to the previous research on problem posing, this study illustrates the rich potential of statistical literacy as a context for the extended development of problem-posing skills. As well, it provides a motivation for those who have been working within more narrow definitions of problem posing, to extend their research into this emerging area of the mathematics curriculum.

With respect to Stoyanova and Ellerton's (1996) framework with three forms of problem-posing situations, this study fits the semi-structured form in that students explored the structure of an open situation related to their school's playground and completed it by posing questions and applying their existing knowledge of survey construction, administration, analysis, and representation of outcomes. The examples of semi-structured situations given by Stoyanova and Ellerton were related to incorporating unfinished structures while posing problems or posing sequences of interconnected problems, using content in geometry, arithmetic, and algebra. In this study, the unfinished structures arose from the need to pose questions and alternative response options that would make up a reasonable survey for the class. Going further, the students were required to connect the survey construction with the other aspects of the investigation, including prediction of results, analysis of data collected, and presentation of their results visually.

Turning to the complementary framework of English (1997), all three of her features for facilitating problem posing were present in this study. The importance of children recognizing and utilizing problem structures was clearly seen as students worked creatively and critically with the format of writing a meaningful and understandable question with four multiple-choice options. The need to consider students' perceptions of and preferences for different problem types was shown in the various types of questions posed for the survey, recognizing the potential difference in responses to "how" and "when" questions and the power of "if" to set a conditional question. Different types of graphs were also considered when suggesting a method of displaying the groups' results. Finally, the opportunity to develop diverse statistical thinking arose throughout the investigation, but particularly in the reporting of the results and visual justification.

The work of Pittalis et al. (2004) in developing a structural model for cognitive processes contributing to problem-posing competence is also relevant to the outcomes in this study, although their research focused primarily on quantitative examples. Their four major constructs of filtering, translating, comprehending, and editing were all observed during the investigations carried out by students in this study. Although we did not quantify the impact of the four constructs, constant reminders of all were present throughout. Comprehending was required and took place from the start when students had the context set with pictures of playgrounds from around the world; further at each stage of the investigation it was necessary to comprehend the posing task required. Translating took place, not only from the beginning in interpreting the questions of others in the group, but also in the tasks of representing the results of the analysis in pictorial/graphical form. Filtering was particularly evident when students were choosing the focus questions for each group to contribute to the class survey. Much discussion took place on clarity. Editing was related not only to filtering in the development of the survey items, but also to the production of the final representation of results. As these constructs of Pittalis et al. appear foundational to statistical investigations as well as to problem posing, they support the contention of the potential to make an explicit link between the two important aspects of mathematical learning.

Besides presenting content under the three headings of Number and Algebra, Measurement and Geometry, and Statistics and Probability, *The Australian Curriculum: Mathematics* (ACARA, 2012) requires four proficiencies to be developed alongside the content. These are *Understanding*, *Fluency*, *Problem Solving*, and *Reasoning*. The four constructs of Pittalis et al. (2004) make contributions to each of these proficiencies and problem posing itself is a part of *Problem Solving* (called "problem formulating"). Comprehending links directly to *Understanding*, and translating, filtering, and editing are necessary for success with all proficiencies. Referring to Figure 11.1, any complete statistical investigation also utilizes the four proficiencies, as illustrated by the work of the students in this study. Again the strength of the association between problem posing and carrying out a statistical investigation is shown through the proficiencies in the Australian curriculum and should encourage both researchers and teachers to explore further the power to move students to creative and critical thinking. This is the type of experience that will give students the confidence to pose questions of statistical literacy when they meet suspicious data-driven claims when they leave school (or even before).

Acknowledgements This research study is supported by a Discovery Grant (DP120100158) from the Australian Research Council (ARC). Any opinions, findings, and conclusions or recommendations expressed in this chapter are those of the authors and do not necessarily reflect the views of the ARC. We wish to acknowledge the enthusiastic participation of all the classroom teachers and their students, as well as the excellent support provided by our senior research assistant, Jo Macri.

References

Australian Curriculum, Assessment and Reporting Authority. (2012). *The Australian Curriculum: Mathematics, Version 3.0, 23 January 2012*. Sydney, Australia: Author.

Bonotto, C. (2012). Artifacts as sources for problem-posing activities. *Educational Studies in Mathematics, 83*(1), 37–55.

Brown, S. I., & Walter, M. (2005). *The art of problem posing*. Mahwah, NJ: Lawrence Erlbaum.

Cai, J., & Hwang, S. (2002). Generalized and generative thinking in US and Chinese students' mathematical problem solving and problem posing. *The Journal of Mathematical Behavior, 21*(4), 401–421.

Cai, J., Moyer, J. C., Wang, N., Hwang, S., Nie, B., & Garber, T. (2012). Mathematical problem posing as a measure of curricular effect on students' learning. *Educational Studies in Mathematics, 83*(1), 57–69.

Common Core State Standards for Mathematics. (2010). Retrieved December 27, 2012, from http://www.corestandards.org/the-standards/mathematics/

Curcio, F. R. (2010). *Developing data-graph comprehension in grades K-8*. Reston, VA: National Council of Teachers of Mathematics.

Curriculum Corporation. (2006). *Statements of learning for mathematics/Statements of learning for science.* Carlton South, VIC: Author. Retrieved October 7, 2010, from http://www.curriculum.edu.au/verve/_resources/StmntLearning_Maths_2008.pdf and http://www.curriculum.edu.au/verve/_resources/StmntLearning_Science_2008.pdf

Duncker, K. (1945). On problem solving. *Psychological Monographs, 58* (5, Whole No. 270).

Ellerton, N. F. (1986). Children's made-up mathematics problems—A new perspective on talented mathematicians. *Educational Studies in Mathematics, 17*(3), 261–271.

English, L. D. (1997). The development of fifth-grade children's problem-posing abilities. *Educational Studies in Mathematics, 34*, 183–217.

English, L. D. (1998). Children's problem posing within formal and informal contexts. *Journal for Research in Mathematics Education, 29*(1), 83–106.

English, L. D. (2013a). Surviving an avalanche of data. *Teaching Children Mathematics, 19*, 364–372.

English, L. D. (2013b). Reconceptualising statistical learning in the early years. In L. D. English & J. Mulligan (Eds.), *Reconceptualising early mathematics learning* (pp. 67–82). Dordrecht, The Netherlands: Springer.

English, L. D., Fox, J. L., & Watters, J. J. (2005). Problem posing and solving with mathematical modelling. *Teaching Children Mathematics, 12*(3), 156–163.

Franklin, C., Kader, G., Mewborn, D., Moreno, J., Peck, R., Perry, M., & Scheaffer, R. (2007). *Guidelines for assessment and instruction in statistics education (GAISE) report: A preK-12 curriculum framework*. Alexandria, VA: American Statistical Association. Retrieved July 3, 2009, from http://www.amstat.org/education/gaise/

Gal, I. (2002). Adults' statistical literacy: Meanings, components, responsibilities. *International Statistical Review, 70*, 1–51.

Garfield, J., & Ben-Zvi, D. (2007). How students learn statistics revisited: A current review of research on teaching and learning statistics. *International Statistical Review, 75*, 372–396.

Kilpatrick, J. (1987). Problem formulating: Where do good problems come from? In A. H. Schoenfeld (Ed.), *Cognitive science and mathematics education* (pp. 123–147). Hillsdale, NJ: Lawrence Erlbaum.

Konold, C., & Miller, C. D. (2011). *TinkerPlots: Dynamic data exploration* [computer software, Version 2.0]. Emeryville, CA: Key Curriculum Press.

Kontorovich, I., Koichu, B., Leikin, R., & Berman, A. (2012). An exploratory framework for handling the complexity of mathematical problem posing in small groups. *The Journal of Mathematical Behavior, 31*, 149–161.

Langrall, C., Mooney, E., Nisbet, S., & Jones, G. (2008). Elementary students' access to powerful mathematical ideas. In L. D. English (Ed.), *Handbook of international research in mathematics education* (2nd ed., pp. 109–135). New York, NY: Routledge.

Makar, K., Bakker, A., & Ben-Zvi, D. (2011). The reasoning behind informal statistical inference. *Mathematical Thinking and Learning, 13*(1/2), 152–173.

National Council of Teachers of Mathematics. (2006). *Curriculum focal points for prekindergarten through Grade 8 mathematics: A quest for coherence*. Reston, VA: Author.

Pittalis, M., Constantinos, C., Mousoulides, N., & Pitta-Pantazi, D. (2004). A structural model for problem posing. In M. J. H. Hoines & A. B. Fuglestad (Eds.), *Proceedings of the 28th Conference of the International Group for the Psychology of Mathematics Education* (Vol. 4, pp. 41–49). Bergen, Norway: Bergen University College.

Shaughnessy, J. M. (2006). Research on students' understanding of some big concepts in statistics. In G. Burrill & P. Elliott (Eds.), *Thinking and reasoning with data and chance* (pp. 77–98). Reston, VA: National Council of Teachers of Mathematics.

Silver, E. A. (1994). On mathematical problem posing. *For the Learning of Mathematics, 14*(1), 19–28.
Silver, E. A., & Cai, J. (1996). An analysis of arithmetic problem posing by middle school students. *Journal for Research in Mathematics Education, 27*(5), 521–539.
Stoyanova, E., & Ellerton, N. F. (1996). A framework for research into students' problem posing in school mathematics. In P. C. Clarkson (Ed.), *Technology in mathematics education* (pp. 518–525). Melbourne, Australia: Mathematics Education Research Group of Australasia.
Watson, J. M. (2006). *Statistical literacy at school: Growth and goals*. Mahwah, NJ: Lawrence Erlbaum.
Watson, J. M. (2007). Inference as prediction. *Australian Mathematics Teacher, 63*(1), 6–11.
Watson, J. M. (2009). The development of statistical understanding at the elementary school level. *Mediterranean Journal for Research in Mathematics Education, 8*(1), 89–109.
Whitin, D. J., & Whitin, P. E. (2011). *Learning to read the numbers: Integrating critical literacy and critical numeracy in K-8 classrooms*. New York, NY: Routledge.

Chapter 12
Problem Posing in the Upper Grades Using Computers

Mitsunori Imaoka, Tetsu Shimomura, and Eikoh Kanno

Abstract Problem-posing activities in mathematics classrooms have been found to have rich outcomes in helping students to develop profound understandings of mathematics and in fostering their problem-solving abilities and creative dispositions. We describe in this chapter practical ways of introducing problem posing using computers, based on our previous studies on problem-posing activities for university students (prospective teachers) and high school students. Studies on problem-posing activities have been rare for students in the upper grades (i.e., high-school and university-level students), and classroom practices involving such activities are less known. We first identify aspects associated with problem posing in the upper grades using computers and introduce practical activities. We report surveys on some of our concrete problem-posing activities and demonstrate their validity. We also present the results concerning the effects of computer use for problem posing in our setting.

Contents

M. Imaoka (✉) • T. Shimomura
Hiroshima University, Hiroshima, Japan
e-mail: imaoka@hiroshima-u.ac.jp; m.imaoka.mx@it-hiroshima.ac.jp

E. Kanno
Department of Mathematics Education, Aichi University of Education, Kariya, Japan

F.M. Singer et al. (eds.), *Mathematical Problem Posing*,
Research in Mathematics Education, DOI 10.1007/978-1-4614-6258-3_12

Problem Posing in the Upper Grades Using Computers

One day, in response to an assignment, a university student brought to his teacher a mathematical problem which he thought up by himself. The problem was: "Represent the figure given by the equation $|z-|z-1||=1$ for the complex number z." The student expressed excitement and surprise at how interesting he found the figure. The teacher, one of the authors of this chapter, was also amazed with the student's work since he had not seen such an equation taking twofold magnitudes but with a clear solution. Whereas the student was usually inconspicuous in the class, he seemed to have been intensely stimulated by working out this mathematical problem on his own. He is now a high school mathematics teacher.

Problem posing in mathematics is a discipline in which students create mathematical problems. Problem-posing activities in classrooms appear to offer potential for students' mathematical growth: to help students to deepen their conceptual understanding of mathematical content, to foster their ability to solve problems, and to cultivate students' creativity. Problem-posing activities also serve to build mathematical communication between students mediated by the posed problems. During the past three decades, various studies which support the authenticity of problem posing have been presented (e.g., Brown & Walter, 1983; Ellerton, 1986; Gonzales, 1994; Hashimoto & Sakai, 1983; Lavy & Bershadsky, 2003; Pelzer & Gamboa, 2009; Perrin, 2007; Saito, 1986; Silver, 1994; Silver & Cai, 1996; Silver, Mamona-Downs, Leung, & Kenney, 1996; Singer, Ellerton, & Cai, 2011).

The importance of problem posing in high school mathematics classrooms was recognized by the National Council of Teachers of Mathematics (1989): "Students in grades 9–12 should also have some experience recognizing and formulating their own problems, an activity that is at the heart of doing mathematics" (p. 138). In spite of statements like this, the practice of problem posing by upper-grade students has not been common, and only a few studies about the practical side of problem posing for the upper grades could be found. It may be the case that some consider that inculcation is so necessary for the sufficient understanding of mathematics in the upper grades that self-generated learning like problem posing is unsuitable for effective study. Varying levels or degrees of competency or interest on the part of the students may be another difficulty encountered when implementing the activity. But, it is also conceivable that many upper-grade class teachers do not realize that it is possible to incorporate problem posing effectively into their daily instruction.

In this chapter, we will present some practical ways of introducing problem posing in the upper grades, based on our studies of its use among high school students (Imaoka, 2001; Kanno, Shimomura, & Imaoka, 2007, 2008) and university students (Imaoka, 2001; Shimomura, Imaoka, & Mukaidani, 2002, 2003a, 2004; Shimomura, Imaoka, Mukaidani, & Kanno, 2003b; Shimomura & Imaoka, 2007, 2009, 2011). We believe that problem posing should be closely connected with everyday class activities through appropriate practical activities. Without this, problem posing may be regarded as peripheral to the tight curricula adopted in the upper grades, or as an activity that only expert teachers could manage. Despite such potential deterrents, we believe that problem-posing activities create significant opportunities for upper-grade students to learn from their classmates.

Computer environments provide rich contexts for visualizing objects and for allowing students to experiment on various cases as well as to explore problems developmentally. Although the use of a computer does not always ensure that students can pose problems to their satisfaction, it is, nevertheless a powerful auxiliary for problem-posing activities. Isoda, Okubo, and Ijima (1992) described a method for supporting problem posing using self-developed software and presented data on students' enhanced heuristic learning.

In this chapter we place emphasis on enriching problem-posing activities with the aid of existing computer software. We will present feasible ways for the introduction of problem posing through actual classroom practice, and we aim to demonstrate the effectiveness and significance of problem posing in the upper grades. Our analysis of a survey we conducted with participants will also be presented. We hope that this chapter will encourage upper-grade mathematics teachers to include problem posing in their own mathematics classrooms.

The Study and Framework for Analysis

The problem-posing activities used in the mathematics classrooms involved in this study, and the analysis of data on problem posing which incorporated the use of computers, focused on the following questions:

1. How did teachers organize the problem-posing activity?
2. Which student-created problems were considered adequate?
3. What types of problems were generated by students?
4. In what ways did students utilize computers?
5. How did students evaluate the activity?

We shall describe the organization of the study and criteria for analysis (Questions 1 and 2) in this section; Questions 3 and 4 will be addressed in subsequent sections by referring to our practices; and Question 5 will be discussed in a later section that presents the results of questionnaires.

In designing this study, we applied traditional Japanese methods developed through our practice of using open-ended approaches to mathematics teaching. In Japan, the practice of problem posing was attempted in the 1920s. Teachers of Nara Female Teachers College affiliate elementary school carried out what was described as "arithmetic problem making classroom." This groundbreaking practice was taken up by many teachers, but became controversial at that time, and declined around 1925 because of methodological insufficiency. The basic principle of the practice was to let children freely make "close-at-home" problems, and teachers who adopted the approach were considered to have been influenced by Dewey pragmatism (cf., Hirabayashi, 1958). Inheriting this tradition, the practice of problem posing was re-introduced extensively in the 1970s and was based on studies about the developmental treatment of problems (e.g., Hashimoto, 1997; Nakano, Tsubota, & Takii, 1999; Sawada & Sakai, 1995; Takeuchi & Sawada, 1984). Assessment of students' understanding of mathematics adopted open-ended approaches, and practitioners worked towards establishing a common way of problem posing. Originally devised in elementary schools, the method is better suited for lower grade students. Although we have employed a fundamentally similar approach to problem posing, we have made adjustments appropriate for upper-grade students.

Our approach adopted the following steps:

Step 1: The teacher introduced a problem-posing activity by explaining the process, showing original problems, or by giving remarks as the occasion demanded.

Step 2: Students were assigned the task of posing problems with the aid of computers and were required to give answers; they also needed to show how they contrived their problems and how they used the computers.

Step 3: The teacher checked each student's posed problems individually, and then exhibited them to all students.

Step 4: Each student was assigned several classmate-posed problems to solve and to comment on.

Step 5: As a final step, the teacher chose several posed problems and asked students to solve and develop them in front of the whole class.

In this study, an original problem is defined as a problem prepared by the teacher at the first stage of the activity. It serves as an example for students to refer to later on, and it sets the tone for the later stages. Teachers in Matsubara Elementary School (1984) indicated the following requisites for choosing an original problem: (a) "generalization" for easy consideration; (b) some "analogy" for applicability, and if possible, an "opposite" construction; and (c) "combinations" of these requisites for easy new constructions. Taking into consideration these factors, as well as knowing that the upper-grade students would be asked to add elements for the activity through the use of their computers, we believe that, in addition to the requisites indicated by Matsubara Elementary School, the following additional elements should be present in the original problems.

1. The original problems should have some characteristics that students can target in their problem posing. For instance, the problems should include multiple representations (such as graphs), or should involve measuring of figures, like areas.
2. The original problems should include some elements which would be particularly suitable for involving computer use. For instance, the solutions can be inferred by experimentally examining particular values using computers.

The software used by high school students in this study was the free software *Grapes* (Ver.6.50c) developed in Japan around 2000, and that used by university students was Wolfram Research's *Mathematica*. The high school students used *Grapes* in their school computer room and on their home computers after downloading it from the Internet. The university students installed *Mathematica* on their own computers using the university's group license.

Activities

Activity I (for High-School Students)

One author (Kanno) of this chapter studied 320 second-year high school students in eight classes as they explored various equations through problem posing using their computers. The mathematical focus of the activity was to reinforce the understanding of relationships between equations and graphs of functions. The five steps outlined in the preceding section were followed in the activity. The teacher prepared original problems as summarized below and assigned students the task of posing problems. After completing the task, students were instructed to offer comments on their classmates' problems. Each class invested three class hours to discuss several posed problems, allowing time for the students who posed the problems to explain their work, and for their classmates to solve them.

For the first original problem, the teacher displayed the corresponding graphs for several concrete values of k.

Original Problem 1

Find the number of distinct real solutions for the quadratic equation $x^2 - 3x - k = 0$.

Students were familiar with this sort of problem and solved it immediately using the sign of the discriminant $D = 9 + 4k$. Then, the teacher posed the next problem.

Original Problem 2

Find the number of distinct real solutions for the cubic equation $x^3 - 3x - k = 0$.

In this case, many students could not come up with the solution since they had not been acquainted with this sort of problem. The teacher reminded the students that they could get the solution of the first problem by observing the move of the corresponding graph with varying k. Students then noticed that the solution of the latter problem could also be detected from the move of the corresponding graph. The teacher gave a further suggestion that it is better to fix the graph $y = x^3 - 3x$ and move the graph $y = k$.

The most popularly posed problems were polynomial equations of higher degrees, such as the following: "What are the distinct real solutions for $x^6 - 14x^4 + 49x^2 - 36 - k = 0$ and $x^6 - 6x^5 + 9x^4 - 10 + k = 0$, respectively?" Figure 12.1 presents two examples of problems (and graphical solutions) posed by students.

It is noteworthy in Figure 12.1 that the students employed the last hint given by the teacher and made use of fixed polynomial graphs. Although Japanese high school textbooks usually treat at most fourth degree polynomial equations, students

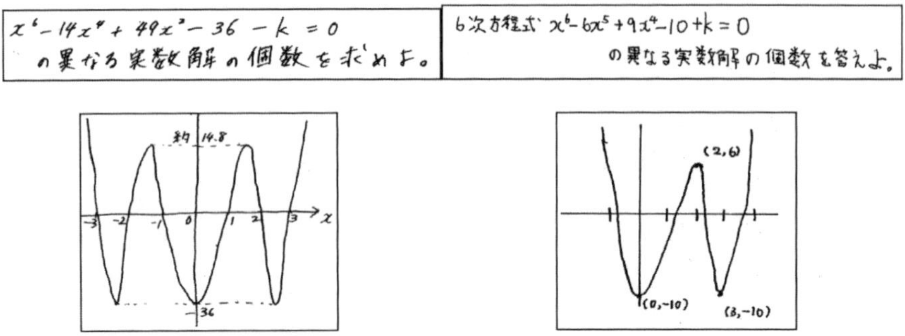

Figure 12.1. Two polynomial problems posed by students.

were free from such a constraint because they were able to use appropriate computer software. In subsequent class discussion, some students noticed that the graphs of polynomial functions associated with posed problems with no odd degree terms were generally symmetric with respect to the y-axis, and other students noticed that there exists at least one solution for any equation of an odd degree. Observations such as these were, in fact, not isolated occurrences in which significant mathematical ideas were discussed. Through the problem-posing activity, students were often able to make significant mathematical observations and connections.

Figure 12.2 shows a typical example of a student-posed problem that involved absolute values: "What are the distinct solutions of the equation $|x^3 - 6x^2 + 9x - 2| - k = 0$?".

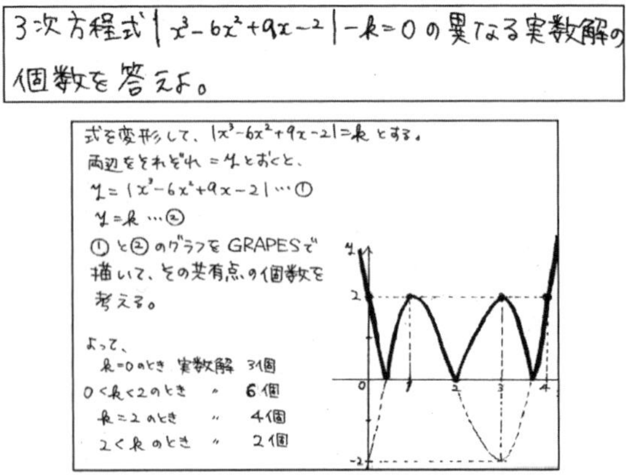

Figure 12.2. An absolute-value problem posed by a student.

Finding solutions to functions which involve absolute values is usually challenging to Japanese high school students. In spite of such challenges, we found students who did create absolute value problems. The ease of finding computer solutions might have made exploring such challenging problems more accessible to students. However, we have observed in other contexts that students are willing to create problems that involve content that they want to master (see, e.g., Imaoka, 2001). This tendency represents a special mentality which appears when students embrace the challenge to think through their work.

Activity II (for Preservice Teacher-Education Students)

Problem-posing activities that incorporate the use of computers have also been studied with preservice teacher-education students (Shimomura & Imaoka, 2007, 2009, 2011, Shimomura et al., 2002, 2003a, 2003b, 2004), and a case will be described here. Although studies on problem posing by preservice or in-service teachers have been presented (e.g., Gonzales, 1994; Lavy & Bershadsky, 2003; Silver et al., 1996), reports of the activity using computers were not found at the time of our research. Our study looks at the practices taking place in preservice teacher-education classes where students are preparing to become secondary school mathematics teachers. The number of students in each class was limited to less than 50. Prior to each time the activity was introduced, both the instructor and the students had learned to solve various problems using *Mathematica*. When activities were introduced, the five steps outlined earlier in this chapter were followed.

The objective of our activities was to enhance students' capacity for devising instruction in their future profession and to let them recognize the possible benefits of using computers for mathematical activities. Being university students, they had considerable competency both in mathematics and in the use of computers, and therefore we could expect them to try to work out quite sophisticated mathematical problems.

We introduced the original problem (referred to as the OPA problem) from Shimomura and Imaoka (2007), as shown in Figure 12.3.

Let P be a point on the curve C: $y=e^x$, and Q the point where the tangent line of C at P intersects with the x-axis. We draw a rectangle whose diagonal is PQ and one edge is on the x-axis. Then, the rectangle is divided into two parts by the curve C. Explore whether or not the ratio of the areas of these two parts is constant for any point P.

Figure 12.3. The OPA problem.

In the class, many students anticipated that the ratio might be constant, and conjectured the ratio as 1:2, 1:3, 2:3, and so on. The teacher urged students to examine the ratio using their computers. The diagram in Figure 12.4 represents the displayed graphs in the cases that the *x*-coordinates of *P*'s are –.5, .5, and 3, respectively. Students experimented with the graphs and began to determine the ratio using the computer. Surprisingly enough for the students, the ratio became the constant 1:(*e*–1). After that, the teacher explained the answer using both a computer and the blackboard and assigned the following task as homework: "Referring to OPA, make two developmental problems using the computer effectively. Also, prepare answers for the problems, and describe how you used computers to work out the problems."

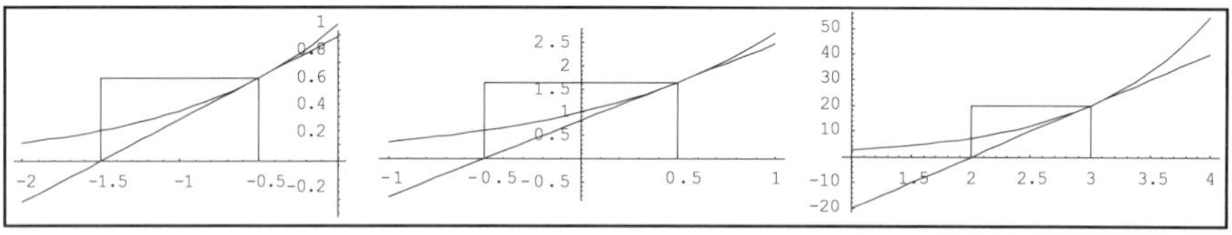

Figure 12.4. Graphs produced for *P* values of –.5, .5, and 3, respectively, in the OPA problem.

Some students posed problems by making simple changes from e^x to e^{ax}, ae^x or be^{ax}, but many students elaborated their problems by making extensive changes to the OPA problem. In Figures 12.5, 12.6, and 12.7, we present three problems posed by Students 1, 2, and 3, respectively.

Let *P* be a point on the curve *C*: y = log *kx* (*k* > 0), and *Q* the point where the tangent line of *C* at *P* intersects with the *y*-axis. We draw a rectangle whose diagonal is *PQ* and one edge is on the *y*-axis. Then, the rectangle is divided into two parts by the curve *C*. Is the ratio of the areas of these two parts constant for any point *P*?

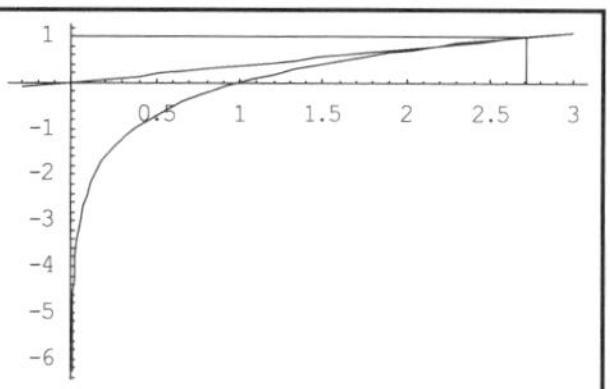

Figure 12.5. Problem posed by Student 1.

Let P be a point on the hyperbola $\frac{x^2}{9} - \frac{y^2}{4} = 1$, and Q and R the two intersections of the tangent line at P and the two asymptotes of the hyperbola. Then, is the area of the triangle △ OQR constant for any point P on the hyperbola? Here, O is the origin.

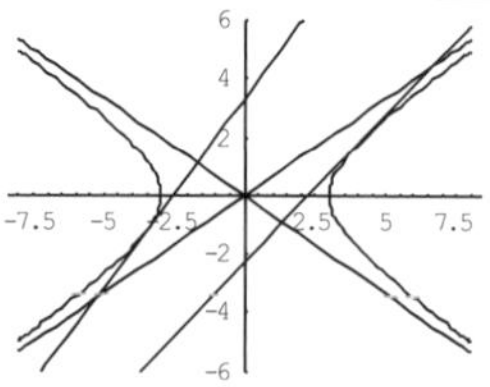

Figure 12.6. Problem posed by Student 2.

Let $z = f(x,y) = x^n + y^n$ and $z = g(x,y) = x + y$ be two surfaces over the region $0 \le x \le 1$ and $0 \le y \le 1$. Does the ratio of the volume of the solid surrounded by the surfaces $z = f(x,y)$ and $z = g(x,y)$ and the volume of the solid surrounded by $z = f(x,y)$ and $z = 0$ have some principle with respect to n?

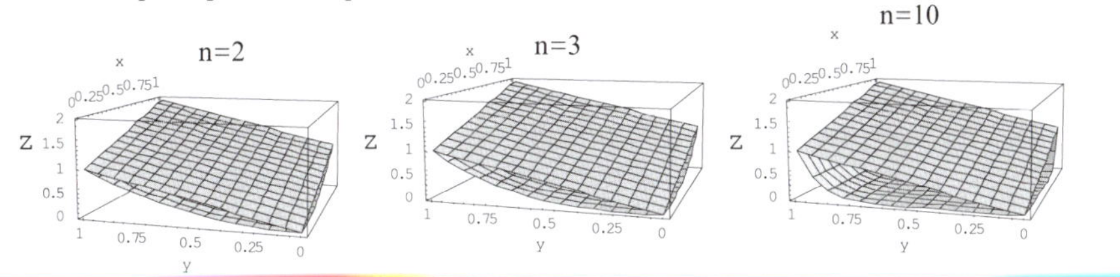

Figure 12.7. Problem posed by Student 3.

Student 1 wrote in his note that he first examined the case of $k = 1$ using a computer, which is a direct analogy to the OPA problem since it changes the original function to its inverse function, and he generalized it with $k > 0$.

Student 2 changed the setting of the OPA problem a little differently, but kept the characteristic concerning invariance of the area. The student reported that she created various quadratic curves using a computer and took a chance on coming upon this problem when she recollected her study on quadratic curves in the class.

Student 3 created the problem shown in Figure 12.7 and changed the ratio of areas in the OPA problem to that of volumes. This student wrote that, when drawing the graphs in the cases of $n = 2$, 3, 4, and 10, he found the graphs of $z = f(x, y)$ and $z = g(x, y)$ intersect only at the points (0,0,0), (1,0,1), (0,1,1), (1,1,2). He examined the volumes in each case using the computer, which led him to formulate this problem.

With these examples, we can observe that students tried appropriate experimental methods using computers. In a later class, the teacher chose the problem posed by Student 3 and invited Student 3 to present the problem to the class. After being shown the above computer diagrams, many students arrived at the answer in about 30 minutes.

Although the OPA problem might appear to be advanced, the student opinions summarized in Figure 12.8 suggest that many students were able to cope with the level of mathematics involved. We conclude that the OPA problem satisfies the requisites outlined earlier in this chapter, particularly those concerned with including the use of graphical properties that represent some invariance of an object. Based on the student-posed problems and on students' opinions about the activity, we are convinced that properties associated with graphs are a good fit for problem-posing activities using computers since various explorations about invariance become possible. Students can explore a range of properties of mathematical objects, thus opening creative approaches problem posing.

OPA is adequate to develop the problems. (8 students expressed similar opinions)
OPA was helpful for my problem posing. (3)
OPA is a helpful first step which makes problem posing easy to understand (3)
OPA can also be solved by paper-pencil computation, which is more familiar to me. (2)
Since I have never created a problem like OPA before, it was a significant occasion. (2)
I challenged myself to pose a problem equal to the OPA.
OPA fits the graphical function of the software. (6)
I was impressed by the constant ratio of areas in the OPA. (5)
I stuck to using the ratios in my problem posing and could not devise different kinds of problems. (2)
I associated OPA with problems related to volumes or using the Monte Carlo method.

Figure 12.8. Summary of students' comments about the OPA problem. The number given in parentheses indicates the number of students who expressed a similar opinion.

Survey Results

After Activity I, we asked the following two questions to the high school students who posed mathematics problems based on Original Problem 1 and 2:

1. Was the use of computer valid in your problem posing?
2. Did you find some unintentional findings or interesting results?

In response to the first question, "Was the use of computer valid in your problem posing," 65.5% replied "Yes, very valid" and 31.6% "Yes, rather valid." The responses show that most students believed that the computer had been useful for the activity. We conclude that students appreciated the convenience of the computer which enabled them to treat complicated equations easily through displayed graphs. In response to the second question, "Did you find some unintentional findings or interesting results," 47.7% answered either "Yes, many" or "Yes, some." Nearly half of the students made unexpected discoveries during their problem-posing attempts. Table 12.1 shows a further breakdown of students' responses to Question 2, with responses grouped under one of three levels—high, average, and low—based on the students' terminal mathematics examination results.

Table 12.1
High School Students' Responses to Survey Question 2 Grouped Under Three Performance Levels

	Response to Question 2				
Ranking	Many	Some	Few	None	Total
High	18	32	40	6	96
Average	15	42	54	7	115
Low	12	29	51	7	99
Total	45	103	145	17	310

A χ^2 test of the correlation between the groups and findings was found to be 4.998, and the Cramer V was .090. Hence, we did not find a clear relationship between the responses to Question 2 about whether students made unexpected discoveries in their problem-posing attempts, and their results in their terminal mathematics examination results. We interpret this in the following way: Every student has an equal opportunity to come across some unexpected findings through problem posing, that is, to experience real mathematical activity. Such a low correlation was also reported by Saito (1986) in a different situation of problem posing. In fact, such equality spread across a wide range of student achievement is an excellent feature of problem posing.

Next, we shall present the results of the questionnaires to university students in three studies:

- *Study I*: Our first study on the activity of problem posing using computers.
- *Study II*: Our focus on the reflections of students on their posed problems.
- *Study III*: Our use of the OPA problem, and our subsequent inquiries into the requisites for the original problem.

In Tables 12.2, 12.3, 12.4, 12.5, and 12.6, the notation $a(b)$ has been used to represent the number of answers, a, with a's percentage in each category represented by (b).

Since more than 80% of students answered "yes" in every study, we can say that our way of implementing the activity was found to be appropriate and acceptable for the students.

Table 12.2
Preservice Students' Responses to Questionnaire Question 1: Did You Find the Problem Posing Activity Beneficial and Satisfactory?

	Beneficial			Satisfied		
Activity	Study I	Study II	Study III	Study I	Study II	Study III
Yes, absolutely	9 (25.7)	8 (16.3)	16 (38.1)	11 (31.45)	13 (26.5)	14 (33.3)
Yes, mostly	23 (65.7)	36 (73.5)	23 (54.8)	19 (54.3)	27 (55.1)	23 (54.8)
Neither yes or no	3 (8.6)	5 (10.2)	3 (7.1)	4 (11.4)	9 (18.4)	3 (7.1)
Mostly no	0 (0)	0 (0)	0 (0)	1 (2.9)	0 (0)	0 (0)
Absolutely no	0 (0)	0 (0)	0 (0)	0 (0)	0 (0)	1 (2.4)

Table 12.3
Preservice Students' Responses to Questionnaire Question 2: How Difficult Was it for You to Think out the Problems?

Difficulty	Study I	Study II	Study III
Very difficult	15 (42.8)	14 (28.6)	11 (26.2)
Somewhat difficult	17 (48.6)	29 (59.2)	24 (57.1)
Difficult	3 (8.6)	6 (12.2)	7 (16.7)
Slightly difficult	0 (0)	0 (0)	0 (0)
Easy	0 (0)	0 (0)	0 (0)

Table 12.4
Preservice Students' Responses to Questionnaire Question 3: What Was the Most Difficult Step in the Problem-Posing Process?

The most difficult process	Study II	Study III
Problem design	41 (83.7)	33 (78.6)
Showing methods, computations, etc.	5 (10.2)	2 (4.8)
Careful examinations	3 (6.1)	7 (16.7)

Table 12.5
Preservice Students' Responses to Questionnaire Question 4: Was the Use of Computers Helpful?

Useful	Study I	Study II	Study III
Yes, very	21 (60.0)	26 (53.1)	26 (61.9)
Yes, somewhat	11 (31.4)	21 (42.9)	11 (26.2)
No opinion	2 (5.7)	2 (4.1)	5 (11.9)
Not so useful	1 (2.9)	0 (0)	0 (0)
No	0 (0)	0 (0)	0 (0)

Table 12.6
Preservice Students' Responses to Questionnaire Question 5: How Was a Computer Useful for Anticipating, Solving Problems, and Acquiring New Knowledge?

Items		Study I	Study II	Study III
A computer is a useful tool in anticipating problems	Yes, very	18 (51.4)	26 (53.1)	24 (53.1)
	Yes, somewhat	13 (37.1)	17 (34.7)	14 (33.3)
	No opinion	1 (2.9)	4 (8.2)	3 (7.1)
	Not so useful	3 (8.6)	2 (4.1)	0 (0)
	No	0 (0)	0 (0)	1 (2.4)
A computer is a useful tool for solving the anticipated problem	Yes, very	24 (68.5)	30 (61.2)	27 (64.3)
	Yes, somewhat	9 (25.7)	13 (26.5)	10 (23.8)
	No opinion	1 (2.9)	3 (6.1)	4 (9.5)
	Not so useful	1 (2.9)	3 (6.1)	1 (2.4)
	No	0 (0)	0 (0)	0 (0)
A computer is a useful tool in acquiring the new knowledge	Yes, very	18 (51.4)	23 (46.9)	25 (59.5)
	Yes, somewhat	16 (45.7)	19 (38.8)	12 (28.6)
	No opinion	1 (2.9)	7 (14.3)	5 (11.9)
	Not so useful	0 (0)	0 (0)	0 (0)
	No	0 (0)	0 (0)	0 (0)

More than 80% of students answered "difficult" ("Very difficult" to "Difficult") to Questionnaire Question 2: "How difficult was it for you to think out the problems?" The results in Table 12.3, along with those for Questionnaire Question 1 in Table 12.2, support our conclusion that problem posing in our setting established sufficient challenge to provide our students with satisfaction after they had completed the activity. This suggests that there may be an important difference between the outcomes of problem-posing activities in elementary schools and in our setting for the upper-grade students.

The types of responses shown in Table 12.4 were anticipated since our target was for students not to make simple modifications to the given problem, but rather for them to engage in more creative problem posing.

Similar rates of approval as in Survey Question 1 in Activity I can be seen in the responses to Questionnaire Question 4 (Table 12.5), but we note that the university students used computers in various ways according to their respective purposes as in Activity II.

Rates of responses for the three items under Questionnaire Question 5 are shown in Table 12.6. Although the number of students was limited in each activity, we observed that students could use computers effectively for every step of the problem-posing process.

Discussion

In the upper grades, giving students sufficient time and opportunity to devise and elaborate mathematics problems is a realistic, and indeed appropriate, mathematical activity. In textbooks or in surveys of elementary or junior high school students in Japan, we find tasks about problem posing related to events or phenomena in the real world (cf., NIER, 2003). Also, in lower grades, problem posing usually took place during class hours. Upper-grade students are generally expected to acquire a certain level of knowledge and skills in mathematics and to enhance their competency through additional mathematical activities. Naturally, problem posing should fit with such circumstances. Activity I (or equivalent activities) would complement students' mathematics classroom experiences and would enable them to explore various equations or functions. Activity II (or equivalent activities) would serve prospective mathematics teachers (and subsequently their students) both by acknowledging the importance of creative thinking and by the effective use of computers. We do not propose that problem posing be the central activity in mathematics learning of upper-grade students, but that the activity is an appropriate supplement for enhancing the study of mathematics.

Our approach to problem posing is also fully applicable without using computers (e.g. Imaoka, 2001; Kanno et al., 2007, 2008). For instance, Kanno, Shimomura, and Imaoka described an activity that took place in senior high school mathematics classrooms, when students were preparing for university entrance examinations. We utilized past problems of entrance examinations as original problems and practiced problem posing without computers. The activity intensified students' attention since it enabled them to experience many kinds of entrance examination problems in a creative way by collaborating with their classmates, exchanging useful information, and discussing solutions with each other.

It is also possible to begin without the stimulus of an original problem. Instead, students can be given only a theme for problem posing and given time to devise their own problems. In particular, problem posing without an original problem seems to be effective in the upper grades when students review what they have been studying in a broader context (cf. Imaoka, 2001; Shimomura et al., 2003a, 2003b).

The exchange of comments, questions, or opinions among students about posed problems increases the effect of problem posing. However, such exchanges might increase teacher involvement since he/she may need to organize the posed problems and may need to facilitate or at minimum initiate discussion about the posed problems. In other words, Step 3 ("The teacher checked each student's posed problems individually, and then exhibited them to all students") needs input from the teacher. In Activity I, the teacher converted all posed problems into PDF files and placed them on his server. Furthermore, he utilized the local area network in the school for distributing posed problems to each student. In this way, if a teacher stores the files of the posed problems in his/her computer, he/she can use them efficiently, for instance, some good posed problems can later be used as original problems.

Figure 12.9 summarizes preservice teacher-education students' comments about Activity II, as a whole, and such opinions have often been observed in other applications of problem posing in our classrooms. Based on student opinions such as those in Figure 12.9, we are convinced that many students feel that our approach to problem posing provided a significant learning activity which supplemented their studies, albeit with some useful suggestions for improvement.

- I found interest and pleasure in examining concrete values first and then making problems by generalizing them while anticipating results.
- I was pleased to be able to make good problems by thinking various patterns and by drawing graphs using computers.
- I tried to make some understandable problems using *Mathematica* , and I was successful.
- I felt that my posed problem was successful, since friends asked me how to solve my problem.
- I almost made problems without a computer. But, I felt that it was very convenient to *Mathematica* to check the result.
- I had only solved given problems before, and I felt strongly that one cannot pose a problem without profound understanding of mathematics.
- It was a good experience, but I felt the shortage of time since I could not use computers freely and quickly.
- It was very difficult, because I could not operate computers well. But, it was a wonderful experience to make problems using *Mathematica*.

Figure 12.9. Opinions expressed by preservice teacher-education students about Activity II.

In conclusion, we would repeat our main findings. Our approach to problem posing was found to encourage our students by inspiring them to create their own work. The experience of problem posing had a positive effect on upper-grade students' overall mathematical experiences. If students are encouraged to exchange comments on posed problems with each other, they were found to gain many ideas through mathematical communication with their peers.

References

Brown, S. I., & Walter, M. I. (1983). *The art of problem posing*. Philadelphia, PA: Franklin Institute Press.

Ellerton, N. F. (1986). Children's made-up mathematics problems—A new perspective on talented mathematicians. *Educational Studies in Mathematics, 17*, 261–271.

Gonzales, N. A. (1994). Problem posing: A neglected component in mathematics courses for prospective elementary and middle school teachers. *School Science and Mathematics, 94*, 78–84.

Hashimoto, Y. (1997). Learning and Teaching of the developmental treatment of mathematical problems. In *Japanese Mathematical Society, Year book 3* (pp. 211–222). Tokyo, Japan: Japanese Mathematical Society.

Hashimoto, Y., & Sakai, H. (1983). A study on developmental approach to mathematical problems. *Journal of Japan Society of Mathematical Education, 65*, 9–16.

Hirabayashi, I. (1958). One process of Japanese arithmetic education history: Arithmetic education centered on making problems. *Journal of Japan Society of Mathematical Education, 40*, 2–12.

Imaoka, M. (2001). Problem posing by students in high school or university. *Journal of JASME: Research in Mathematics Education, 7*, 125–131.

Isoda, M., Okubo, K., & Ijima, Y. (1992). *Mathematics study using media: Improvement of classroom by problem posing from scene*. Tokyo, Japan: Meijitosho.

Kanno, E., Shimomura, T., & Imaoka, M. (2007). High school lessons on problem posing to extend mathematical inquiry: Effective use of problems in university entrance examinations. *Journal of Japan Society of Mathematical Education, 89*, 2–9.

Kanno, E., Shimomura, T., & Imaoka, M. (2008). Incorporating problem posing in high school instruction on differential calculus aiming to active improvements over mechanical drills. *Journal of JASME: Research in Mathematics Education, 14*, 111–117.

Lavy, I., & Bershadsky, I. (2003). Problem posing via "what if not?" strategy in solid geometry—A case study. *Journal of Mathematical Behavior, 22*, 369–387.

Matsubara Elementary School. (1984). *Problem posing in arithmetic course*. Tokyo, Japan: Toyokan.

Nakano, Y., Tsubota, K., & Takii, A. (1999). *Children make problems: Problem posing classroom in the elementary school arithmetic*. Tokyo, Japan: Toyokan.

National Council of Teachers of Mathematics. (1989). *Curriculum and evaluation standards for school mathematics*. Reston, VA: Author.

National Institute for Educational Policy Research. (2003). *Report of survey on curriculum implementation of the school year 2001*. Tokyo, Japan: Gyosei.

Pelzer, I., & Gamboa, F. (2009). Problem posing: Comparison between experts and novices. In M. Tzekaki, M. Kaldrimidou, & H. Sakonidis (Eds.), *Proceedings of the 33rd Conference of the International Group for the Psychology of Mathematics Education* (Vol. 4, pp. 353–360). Thessaloniki, Greece: International Group for the Psychology of Mathematics Education.

Perrin, J. R. (2007). Problem posing at all levels in the calculus classroom. *School Science and Mathematics, 107*, 182–192.

Saito, N. (1986). Interests of the learner in the learning to create problem. *Journal of Japan Society of Mathematical Education, 68*, 24–33.

Sawada, T., & Sakai, H. (1995). *Problem posing classroom*. Tokyo, Japan: Toyokan.

Shimomura, T., & Imaoka, M. (2007). Problem posing using computer (IV): Focusing on how to pose original problems. *Journal of JASME: Research in Mathematics Education, 13*, 225–234.

Shimomura, T., & Imaoka, M. (2009). Problem posing using computer (V): Focusing on problems posed by university students. *Journal of JASME: Research in Mathematics Education, 15*, 137–146.

Shimomura, T., & Imaoka, M. (2011). Problem posing using computer (VI): Focusing on problem posing process by university students. *Journal of JASME: Research in Mathematics Education, 17*, 1–12.

Shimomura, T., Imaoka, M., Mukaidani, H., & Kanno, E. (2003a). Mathematical problem posing in high school: For high school (grade 3) students. *Journal of JASME: Research in Mathematics Education, 13*, 225–234.

Shimomura, T., Imaoka, M., & Mukaidani, H. (2002). Problem posing using computer (I): Through the practice at university. *Journal of JASME: Research in Mathematics Education, 8*, 235–242.

Shimomura, T., Imaoka, M., & Mukaidani, H. (2003b). Problem posing using computer (II): On planning problems freely for students in university. *Journal of JASME: Research in Mathematics Education, 9*, 235–241.

Shimomura, T., Imaoka, M., & Mukaidani, H. (2004). Problem posing using computer (III): Focusing on the effective use of problems posed by university students. *Journal of JASME: Research in Mathematics Education, 10*, 207–217.

Silver, E. A. (1994). On mathematical problem posing. *For the Learning of Mathematics, 14*, 19–28.

Silver, E. A., & Cai, J. (1996). An analysis of arithmetic problem posing by middle school students. *Journal for Research in Mathematics Education, 27*, 521–539.

Silver, E. A., Mamona-Downs, J., Leung, A. S., & Kenney, P. A. (1996). Posing mathematical problems: An exploring study. *Journal for Research in Mathematics Education, 27*, 293–309.

Singer, F. M., Ellerton, N. F., & Cai, J. (2011). Problem posing in mathematics learning and teaching: a research agenda. *Proceedings of the 35th Conference of International Group for the Psychology of Mathematics Education, 1*, 137–166

Takeuchi, Y., & Sawada, T. (1984). *From problems to problems*. Tokyo, Japan: Toyokan.

Chapter 13
Using Problem Posing as a Formative Assessment Tool

Meek Lin Kwek

Abstract To become an effective mathematics problem solver, students must go beyond problem solving to pose problems and finally to create mathematical problems. It is only at this highest level of creation that students will begin to realize their true potential and experience the excitement of mathematical discovery and research. Considering the pedagogical benefits of problem posing, should the tasks not also be embedded into the classroom assessment for learning? This chapter explores the use of problem-posing tasks as a formative assessment tool to examine students' thinking processes, understandings, and competencies. Based on the analytic scheme developed by Silver and Cai (*Journal for Research in Mathematics Education 27*: 521–539, 1996), this chapter describes how a team of teacher–researchers implemented problem-posing tasks and analyzed the problems posed by 75 high-ability secondary school students. The students' performances were analyzed and evaluated in the light of problem complexity. Through this lens, patterns in students' mathematical learning and thinking processes were interpreted in terms of their mathematical knowledge and how that knowledge is applied.

Contents

M.L. Kwek (✉)
Department of Mathematics, Raffles Girls' School (Secondary), Singapore
e-mail: meeklin.chew@rgs.edu.sg

F.M. Singer et al. (eds.), *Mathematical Problem Posing*,
Research in Mathematics Education, DOI 10.1007/978-1-4614-6258-3_13

Introduction

Problem Solving as an Instructional Goal

For most school mathematics curricula, the primary goal of mathematics education is for their students to become mathematically literate. This implies that each student is capable of critical thinking, problem solving, communicating, and reasoning mathematically, as well as using mathematics confidently. In order to achieve these goals, mathematics educators and researchers have invested much thought and effort into the strategies that are aimed to help students develop desirable mathematical concepts, thinking processes, and attitudes towards mathematics. Sheffield (2003) suggested that students be directed along a mathematical continuum from novice to expert through the following stages: innumerates, doers, computers, consumers, problem solvers, problem posers, and creators.

The essential knowledge and skills required of a learner to move from a lower level to a higher level in this continuum are featured in the Singapore's Mathematics Curriculum Framework (Figure 13.1), developed by the Ministry of Education, Singapore (2006). When teachers place problem solving at the core of their instructional goal, teaching via a wide range of problems that are situated in both familiar and unfamiliar contexts can be expected to serve as a means for the students to acquire mathematical concepts and develop mathematical skills.

As an introduction to the design of problem-posing activities, it is helpful to reflect on problem-solving tasks, and ask: What constitutes a rich problem-solving task? Lappan, Fey, Fitzgerald, Friel, and Phillips (1996) proposed the following elements:

- Has important, useful mathematics embedded in it;
- May have different solutions or allow for different decisions or positions to be taken and defended;
- Can be approached by students in multiple ways using different solution strategies;

- Encourages student engagement and discourse;
- Requires higher level thinking and problem solving;
- Contributes to the conceptual development of students;
- Promotes the skilful use of mathematics; and
- Creates opportunities for teachers to assess what their students are learning and where they are having difficulty.

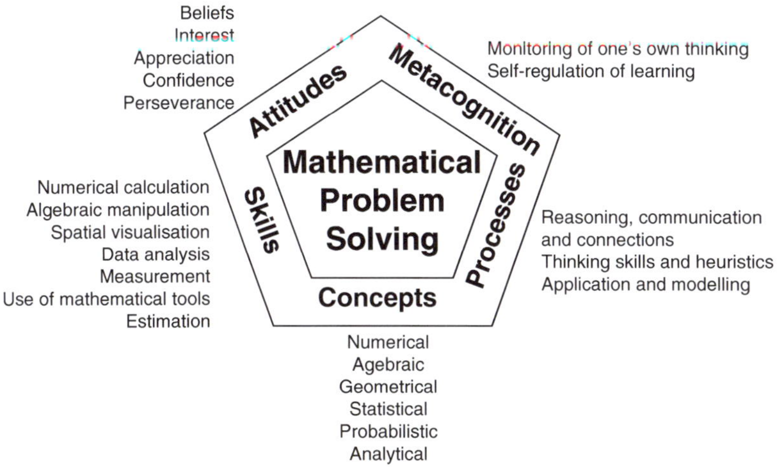

Figure 13.1. Singapore's Mathematics Curriculum Framework, developed by the Ministry of Education, Singapore (2006).

Problem Posing Enhances Problem Solving

Problem posing can refer both to the generation of new problems from a mathematical context and to the reformulation of a given problem (Silver, 1994). When generating new problems, the problem poser would usually consider the nature of the context and possible solution paths to the problem posed. Reformulation of problems, however, often occurs when the problem poser interacts with the problems by asking "How to begin?" and "How to continue from here?" The problem solver then reacts by selecting suitable strategies or creating an easier problem or modifying conditions in the problem so as to be able to continue. This process of problem generation and consideration for multiple solution paths provides excellent opportunities for fostering divergent and flexible thinking. We found that thinking habits such as these not only enhanced problem-solving skills, but also helped to reinforce and enrich basic mathematical concepts. Hence from a pedagogical perspective, problem-posing activities potentially present themselves as powerful assessment tools (English, 1997; Lowrie, 1999).

Problem Posing as a Formative Assessment Tool

With purposeful and consistent planning for problem-posing activities, opportunities for students to reason, reflect on their own thinking, and make connections between mathematics and the real-world often spring from such creative experiences (Brown & Walter, 1993; English, 1996). More important for the purpose of this study, problem posing can potentially fill the gaps between what students know and what is not known to them; and inform teachers about their understandings, knowledge, skills, and dispositions as they interact with the situations presented to them.

In order to use problem posing effectively in the classroom as a generative activity from which information about the students' engagement, competencies, and areas of improvement can be drawn, it is essential for teachers to understand its value as a formative assessment tool. In the context of this study, problem-posing tasks were developed to help both teachers and students find out how much learning had taken place, and to what extent students' mathematical knowledge could be applied creatively from a designer's perspective. Such information allows the teacher to identify students' strengths, as well as areas that need improvement, and provides data for teachers to chart students' mathematical growth. When "feed-back" from problem-posing tasks is effectively used to "feed-forward," then the intent of good formative assessment can be achieved. It is this interest in the process of gathering data about students' growth in learning mathematics for improving classroom instruction and learning that drives this study.

Theoretical Background

A problem-posing task in this study is defined as a task designed by teachers that requires students to generate one or more word problems. The quality of the problems posed by students in these tasks can be evaluated in several different ways. Silver and Cai (2005) noted that

> Because of the open-ended nature of such tasks, there is often considerable variability in the responses that students generate. Although this aspect is desirable from an instructional perspective, it can often present challenges from an assessment perspective (p. 131).

Based on the principles of assessment, teachers must make decisions about assessment tasks by first considering their instructional goals and the form of the tasks that can potentially provide evidence of the attainment of these goals. The particular features of the tasks related to the instructional goals contribute largely to the set of criteria teachers used for evaluating students' performance in the assessment. In their study on assessing students' mathematical problem posing, Silver and Cai (2005) identified three criteria that are commonly applicable to most problem-posing tasks: *quantity*, *originality,* and *complexity*.

Quantity refers to the number of correct responses generated from the problem-posing task. Counting the number of (correct) responses may be deemed by many as a trivial way of evaluating students' responses to generative activities such as problem posing. Nevertheless, the fluent generation of responses can potentially inform the teacher about students' characteristics such as creativity. *Originality* of the problems posed is another feature of responses that can be used as a criterion to measure students' creativity.

Problem *complexity* can be examined from various perspectives. Silver and Cai (2005) identified four facets of problem complexity—sophistication of the mathematical relationships embedded in problems, problem difficulty, linguistic complexity, and mathematical complexity. Of particular interest to this study is mathematical complexity. Mathematical complexity refers to the cognitive demands of the task. It can be categorized as low, moderate, or high. Each level of complexity includes aspects of knowing and doing mathematics, such as reasoning, performing procedures, understanding concepts, or solving problems.

The levels of complexity form an ordered description of the demands that a problem may make on the problem solver. Problems that are categorized with low levels of complexity are usually solved by recalling and recognizing facts or having a one-step solution. Moderate complexity problems would require a solver to move beyond simple recall and involve more thought and decision-making points. Such problems usually demand a combination of mathematics skills and knowledge, and involve reasoning, problem-solving strategies, application of theories, or multiple-step solutions. High complexity problems make demands on solvers' thinking by engaging them in reasoning, analyzing, generalizing, synthesizing, or making connections in multiple-step solutions.

Method

In examining the potential of problem posing for the formative assessment of learning, the range of problems posed by students as they responded to a semi-structured situation was first explored. Samples of students' work were analyzed in the light of the mathematical complexity of the problem posed. These findings, together with the teachers' observations of student behavior during the problem-posing task, were used to identify patterns in the students' mathematical learning and thinking.

Participants

Thirty-two Grade 7 students and 43 Grade 9 students from an all-girls school participated in this study. From their consistently high performance and interest in mathematics, these mathematically promising students were identified and placed in advanced classes. However, none had had prior exposure with problem-posing activities.

Procedure

The classroom preparation for using problem-posing tasks as formative assessment began with two researchers modelling problem-posing behavior during the students' regular classroom instruction. Several problem-generating strategies were presented to the students during class periods dealing with Arithmetic and Algebra, respectively, for the Grade 7 and Grade 9 students. These strategies included the techniques of changing a problem to create new ones by changing the numbers and operations, or by removing and adding conditions in a given problem. This exposure to problem-posing techniques was embedded within each of the 55-minute regular lessons, three times a week, and took place over a period of 6 weeks.

To help align each problem-posing task with the goals of formative assessment, a rubric explaining the different levels of mathematical complexity in the problems was provided for the students' reference throughout this teaching and assessment period. It served as a guide for students to pose problems with high mathematical complexity; and the descriptors provided some directions for them to move their problems from a lower to a higher level of complexity.

At the end of the study, a new problem-posing task was administered. The students first worked individually to pose one or more problems to a task involving semi-structured situations, and assessed its quality using the rubric. Then they worked collaboratively in their assigned groups to select, refine, and evaluate the problems posed by one another, as part of the process of self- and peer-assessment, and to help them internalize the characteristics of a quality problem. Similar to the study by Pittalis, Christou, Mousoulides, and Pitta-Pantazi (2004), this study aimed to have students use their prior mathematical experiences to "complete a situation and the structure of this situation." During the problem-posing process, the teacher–researchers made field notes of their observations about the student behaviors in one another's classes, followed by a brief discussion about those observations.

After all of the groups had discussed and formulated the word problems for their group, students exchanged problems and challenged one another to solve them. The solutions from the problem posers and problem solvers were then compared, checked, and verified. Finally, when a group had successfully attempted a problem posed by another group, students in the group indicated whether or not they had found the problem interesting and challenging.

At the end of the problem-posing activity, a three-item reflection survey was administered to the students. They were given 1 day to think carefully about their learning experiences with problem posing, and then respond to the open-ended questions. Students spent a total of 3 hours (over three class periods) to complete this problem-posing task.

Instruments

Problem-posing tasks. The format of the problem-posing task administered to each grade in this study was similar to that developed by Silver and Cai (2005). The stimulus used involved a real-life scenario with an incomplete problem. Some quantitative information was given with which the students could use creatively to generate problems related to the stipulated topic. Figure 13.2 shows the problem-posing task for Grade 7 students. Students were required to pose word problems that demonstrated their ability to apply their pre-knowledge of proportions and percentages set within a realistic and meaningful context.

Grade 9 students' knowledge of inequalities was evaluated based on their ability to pose and complete the problem by modifying or adding conditions for the given scenario. Students were encouraged to make multiple linkages with other mathematical concepts in this optimization task, as shown in Figure 13.3.

> Pose mathematical problems, with solutions, to demonstrate your competency in constructing real-word problems involving one or a combination of the following aspects: percentages, hire purchase, simple and compound interest, money exchange and taxation. Solve the problem you have posed, reformulating it where necessary.

Figure 13.2. Problem-posing task objective for Grade 7 students.

> Use the information below to pose a mathematical problem, and solve the problem to demonstrate your competency in
> - using the basic rules of manipulating inequalities
> - simplifying inequalities involving linear, quadratic or modulus functions
> - solving a pair of simultaneous inequalities
>
> A gardener is planting a new orchard. The young trees are arranged in the rectangular plot, which has its longer side measuring 100m.

Figure 13.3. Problem-posing task objective for Grade 9 students.

Rubric

Students' responses to the problem-posing task were evaluated for the mathematical complexity of the problems posed. Mathematical complexity is an important attribute of the posed problems, and reflects students' mathematical understandings and cognitive processes. The rubric presented in Figure 13.4 describes three different levels of mathematical complexity. The descriptors were adapted from the Mathematics Framework section of *The Nation's Report Card Mathematics 2005*, produced by the National Assessment of Educational Progress (NAEP), in the United States. The rubric was used to evaluate the complexity of the anticipated solutions to the problems posed by students.

	Low complexity	Moderate complexity	High complexity
Description	This category relies heavily on the recall and recognition of previously–learned concepts. Items typically specify what the solver is to do, which is often to carry out some procedure that can be performed mechanically. It leaves little room for creative solutions. The following are some, but not all, of the demands that items in the low–complexity category might make:	Items in the moderate–complexity category involve more flexibility of thinking and choice among alternatives than do those in the low–complexity category. They require responses that may go beyond the conventional approach, or require multiple steps. The solver is expected to decide what to do, using informal methods of reasoning and problem–solving strategies. The following illustrate some of the demands that items of moderate complexity might make:	High–complexity items make heavy demands on solver, who must engage in more abstract reasoning, planning, analysis, judgment, and creative thought. A satisfactory response to the item requires that the solver think in an abstract and sophisticated way. The following illustrate some of the demands that items of high complexity might make:
Cognitive demand	• Recall or recognize a fact, term, or property • Compute a sum, difference, product, or quotient • Perform a specified procedure • Solve a one–step word problem • Retrieve information from a graph, table, or figure	• Represent a situation mathematically in more than one way • Provide a justification for steps in a solution process • Interpret a visual representation • Solve a multiple-step problem • Extend a pattern • Retrieve information from a graph, table, or figure and use it to solve a problem • Interpret a simple argument	• Describe how different representations can be used to solve the problem • Perform a procedure having multiple • steps and multiple decision points • Generalize a pattern • Solve a problem in more than one way • Explain and justify a solution to a problem • Describe, compare, and contrast solution methods • Analyze the assumptions made in solution • Provide a mathematical justification

Figure 13.4. Rubric for evaluating the complexity of problems posed by students, adapted from the Mathematics Framework included in National Assessment of Educational Progress (2005).

Teacher's Observation Checklist

The acquisition and application of mathematical knowledge and skills are common goals of most school mathematics curricula and programs. The achievement of these learning goals is supported by key mathematical processes in the programs.

For the purpose of this study, four of the most relevant processes (representing, reflecting, connecting, and problem solving) were chosen for the Teachers' Observation Checklist.

Figure 13.5 shows the Teachers' Observation Checklist which enabled teachers to record observable student behaviors associated with each process. A box was checked when a group (labelled as U1, U2, etc. for Grade 9 students and L1, L2, etc. for Grade 7 students) demonstrated a behavior more than once; and it was left empty if the behavior was not exhibited.

Mathematical Processes	**Observable behaviours**	U1	U2	U3	U4	U5	U6	U7	U8	U9	U10
Representing											
Reflecting											
Connecting											
Problem solving											

Figure 13.5. Classroom observation checklist.

Student Reflections

Students' reflections on their experiences with the problem-posing task were prompted by the following survey questions:

- How is the problem(s) posed by your peers interesting/not interesting?
- How is the problem(s) posed by your peers challenging/not challenging?
- What is the most important thing you learned in this problem-posing activity?

Students were asked to write several sentences in response to these prompts.

Data Collection and Analysis

This study was concerned with students' responses to problem-posing tasks and activities. Three main sources of data informed the study: (a) the problems posed by the students; (b) the students' reflection on their problem-posing experiences; and (c) the teachers' observations of student behaviors in the generative activity. When the data collection phase was completed, the empirical data were reviewed and analyzed to help the researcher identify and describe possible patterns in students' mathematical learning and thinking.

The problems posed by the students were first examined to determine the range and commonalities among them, so that they could be classified accordingly. In each category, the problems were further compared in order to refine the search for similar or distinguishing features. These underlying patterns potentially served to provide supporting evidence for the way in which students perceived mathematical problems, and of their disposition towards learning through problem posing and creative thinking.

The students' reflections on their problem-posing experiences were reviewed in order to identify possible themes related to their interests, motivation, beliefs, and self-knowledge about the generative activity, particularly in the context of high-ability learners. The findings of this study have the potential to add to the body of knowledge about the connection between affective and cognitive factors as students adapt to "new" learning environments.

The teachers' observation of student behavior served to triangulate the data about students' cognitive processes obtained from the students' work samples and their reflections. It involved interpretation and making correspondences between observable behaviors and the thinking processes triggered by the problem-posing activities and problem-solving tasks.

Findings and Discussion

This section focuses on the data collected from the problems posed by the Grade 7 and Grade 9 students, their perceptions of the activity, and their observable behaviors in the context of a problem-posing activity.

Problems Posed by Students: Classification of Problems

In examining the problems posed by middle school students, Silver and Cai (1996) proposed an analytic scheme for classifying the problems, as shown in Figure 13.6.

According to the schema, posed problems that are statements or nonmathematical questions were first sieved out, before focusing on the remaining mathematical

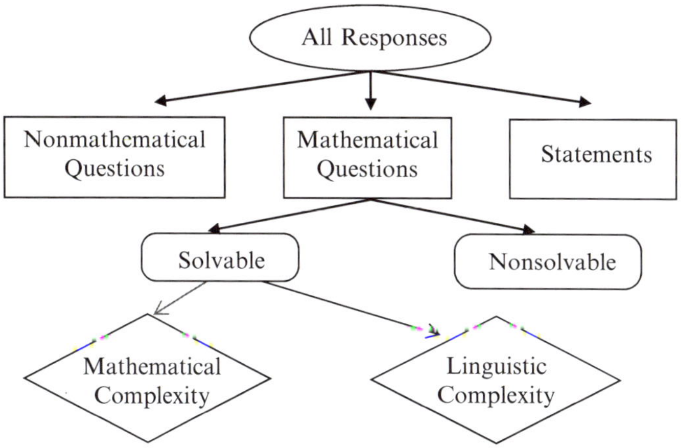

Figure 13.6. Analytic scheme for classifying students' posed problems (from Silver and Cai (1996)).

problems. Within the set of mathematical problems posed by students, those that were solvable were further examined for their nature and mathematical complexity.

In this study, all Grade 7 and Grade 9 students successfully generated at least one mathematical problem, the majority of which were solvable. This set of solvable mathematical problems was of particular interest to the study. The problems generated showed wide variation in terms of mathematical complexity, challenge, and the potential to arouse student interest. Table 13.1 presents a summary of the proportion of the problems which were solvable, those which were classified to have low, moderate, and high levels of mathematical complexity, and those which were perceived by the students to be interesting and challenging.

Table 13.1
Percentage of Posed Problems for Different Classifications

Classification of posed problems	Grade 7 (%)	Grade 9 (%)
Mathematical questions	100	100
Solvable	81	78
Low mathematical complexity	81	67
Moderate mathematical complexity	13	30
High mathematical complexity	6	3
Interesting	81	58
Challenging	19	50

The students' ability to pose solvable mathematical problems successfully can be attributed to their previous exposure to problem-posing strategies during the 6-week intervention study. As the teachers modeled problem-posing strategies, classroom discussions began with problems that were amenable to being analyzed, and possibly solved using mathematical approaches with which students were already familiar. The definition of a mathematical problem was further reinforced when the students made frequent reference to the rubric for assessing the quality of the problems posed.

In classifying the problems posed by students, it was observed that many of the unsolvable problems were due to one or more of the following: unclear wording, important assumptions not being stated, and the use of overly complex algebraic expressions. Examples of these unsolvable problems are shown in Figure 13.7a, b, c, respectively.

a

Grade 7: L6

To buy a TV in Singapore, John has to pay $1048 and a tax of 7%. If he buys it from China, he will pay $843, a shipping fee of 10% and another 20% duty fee. If the size of the TV is a ratio to its cost, then which is a better deal?

b

Grade 7: L1

Grace decides to settle down permanently either in Singapore or Hong Kong. She intends to buy a 5-room flat with her current salary of S$5000. Use the following information to find out in which country should she choose to stay.

	Singapore	**Hong Kong**
Cost of flat	S$700000	HK$3300000
Down payment	S$50000	HK$
Instalment	S$2000 per month	HK$8250 per month
Bank interest	5%p.a	8%p.a

c

Grade 9: U2

The gardener decides to enclose the rectangle such that its four vertices touching a circular enclosure. He wants to plant trees within the circular enclosure but outside the rectangular plot. If the trees were to occupy 30% to 60% of the land area, what is the range of values of the rectangle's width?

Figure 13.7. (**a**) Example of an unsolvable problem (posed by a Group 6 Grade 7 student). (**b**) Example of an unsolvable problem (posed by a Group 1 Grade 7 student). (**c**) Example of an unsolvable problem (posed by a Group 2 Grade 9 student).

Unsolvable mathematical problems. Compared to their seniors, the Grade 7 students faced greater challenges in crafting problem statements. Most of the Grade 7 students' English language competency was regarded as above average in comparison with many of their peers in the national examination. It seems reasonable to conclude, therefore, that any lack of clarity in the problem statements could be attributed to factors related to mathematical communication rather than to their facility with the English language. In this respect, students were generally unaware of the distinction between mathematical writing for a problem and for its solution. In developing problem solvers, it was observed that most mathematics programs have placed greater emphasis on the clarity of solutions and less on the problems themselves.

Both Grade 7 and Grade 9 students generated unsolvable problems that were lacking in important assumptions. Analysis of the problems posed by the Grade 7 students suggested that these students had an inadequate understanding of the meaning of assumptions in a mathematical problem. In contrast, unsolvable problems posed by Grade 9 students tended to reflect these students' limited exposure to real-life mathematical modeling experiences.

Problems with low mathematical complexity. Of those mathematical problems that were solvable, more than half of them were further categorized as having a low mathematical complexity (see Table 13.1). For both Grade 7 and Grade 9 students, problems with low mathematical complexity may appear at first reading to be somewhat difficult to solve, yet the demand on one's mathematical skills could in fact be fairly low. Many of the problems posed by the Grade 7 students were identified as belonging to this category, with all of these problems being strongly reflective of the routine questions commonly found in mathematics textbooks. This trend, although not as prevalent as among the Grade 9 students, was observed in more than half of the solvable mathematical problems posed. Figure 13.8 shows two mathematical problems with low mathematical complexity.

a

Grade 7: L4

Mrs Jill wants to buy a car for $120 000. She is offered loans by Bank A and Bank B. Bank A allows her to pay in 20 monthly instalments, with 15% p.a. simple interest. For Bank B, Mrs Jill has to pay $12 000 as deposit, and the rest in 16 monthly instalments, at an interest rate of 10% p.a., compounded annually. Which bank offers a better deal? How much does she save?

b

Grade 9: U1

If a fence around the orchard measures more than 330m, and the area of the orchard is not more than 7000m^2, find the range of values of the shorter side.

Figure 13.8. (**a**) Example of a low complexity mathematical problem (posed by a Group 4 Grade 7 student). (**b**) Example of a low complexity mathematical problem (posed by a Group 1 Grade 9 student).

Problems with moderate mathematical complexity. Problems which were classified as being of moderate mathematical complexity generally reflected real-world contexts. Although the solutions to these problems were not necessarily more demanding on the solver in terms of mathematical skills, the context often required some consideration in order to take account of the underlying assumptions. In this respect, the Grade 9 students were more successful than their juniors, given their earlier exposure to real-life experiences and their mathematical maturity.

It was observed that the majority of the Grade 7 students posed problems by applying the strategy of imitation, producing problem structures that bore close semblance to textbook questions of a more challenging nature. An example of such a problem posed by Grade 7 students is shown in Figure 13.9a. By comparison,

a

Grade 7: L3

Ms X had S$83264, which she wanted to change to Z$ to buy an item in Zork. The exchange rate is at Z$1 = S$4.1632 (selling rate) and Z$1 = S$4.0071 (buying rate). She bought the item using hire purchase, paying a 20% deposit and was charged a 3.2% simple interest rate per annum over a period of 36 months. She had S$36.993.55 leftover when she converted the money she had left after paying the total balance. What was the selling price of the item in Z$?

b

Grade 9: U8

The gardener decides to divide the plot of land into three sections for growing three different types of plants. It is given that section C is representative of a quadrant and the area of section C is bigger than that of section A. The various sections require different types of soils of different prices. If the gardener has a budget of $400 for buying soil for the orchard, what is the maximum value of X ?

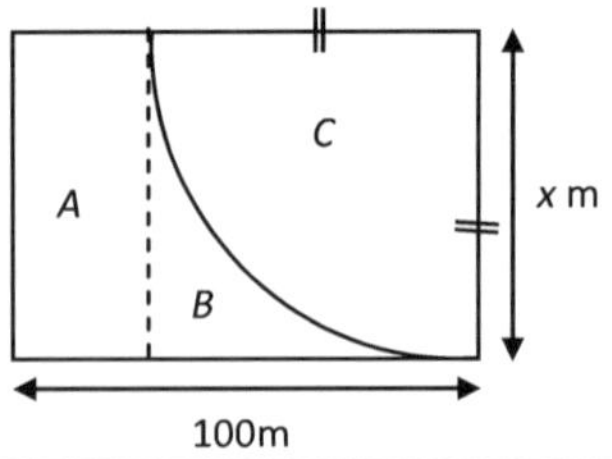

Types of soil	Price
Soil A	$2 /m^2
Soil B	$3 /m^2
Soil C	$8 /m^2

Figure 13.9. (**a**) Example of a problem with moderate mathematical complexity (posed by a Group 3 Grade 7 student). (**b**) Example of a problem with moderate mathematical complexity (posed by a Group 8 Grade 9 student).

Grade 9 students tended to be more adventurous in generating new problems. Problems posed by Grade 9 students generally showed the application of a wider range of strategies such as using a combination of mathematical concepts, adding conditions, or requiring different representations. This resulted in a greater variety of problem types. An example of such a problem posed by a Grade 9 student is shown in Figure 13.9b.

Teachers' observational notes on the Grade 9 students indicated that the groups working on those problems with real-world contexts generally did not confine their discussions to mathematics per se; but also concerned themselves with realistic views about the world. These groups were observed to be more engaged in the problem-posing activity and some of them willingly spent time outside the curriculum to refine and complete their tasks.

Problems with high mathematical complexity. Problems posed by students have varying cognitive demands. In this study, those problems that were categorized as having a high mathematical complexity were a minority, and showed one feature that set them apart from the rest—the opportunity for multiple solution paths to the

Grade 9: U3

It is given that the width of the orchard is 50m. Starting from point B, a worker P walked along the edge in a clockwise direction and back to B at a speed of 2m/s. Another worker, Q, started from point A and walked along the edge in the clockwise direction and back to point A at a speed of 1 m/s. What is largest possible area of triangle BPQ?

Figure 13.10. Example of a problem with high mathematical complexity (posed by a Group 3 Grade 9 student).

problem posed. An example of such a problem posed by Grade 9 students from group U3 is shown in Figure 13.10.

The teacher also observed that the students who posed problems with high mathematical complexity were fully engaged in group and class discussions, had good conceptual knowledge, and demonstrated a wide repertoire of mathematical skills. In fact, for students to craft problems that require abstract reasoning, judgment, and analysis, it is not unreasonable to expect them to demonstrate these traits.

Challenge and student interest. In their roles as problem solvers, students in each group were asked to discuss problems created by other groups, and to decide whether these problems were interesting and challenging. The Grade 7 and Grade 9 students differed in their responses.

Although 81% of the problems posed by Grade 7 students were classified as those with low mathematical complexity, most of them were rated by their peers as interesting problems. When students were asked in the survey: "*How is the problem(s) posed by your peers interesting/not interesting*?" four types of responses were identified. The following four responses were typical of these four types:

"*The questions are interesting because they are not taken from the worksheets.*"

"*I like to work on the questions we created together.*"

"*It is like breaking a code. We are racing to break one another's code. It is real fun.*"

"*They are not too hard.*"

These responses strongly suggest that the Grade 7 students placed emphasis on the learning environment from which the problems were derived, rather than on the problems themselves. Hence problems posed by their friends were more likely to be considered as interesting. When the problems were original creations, their creative products could be used for competing against one another, and the problems were not overly demanding.

When students were asked: "*How is the problem(s) posed by your peers challenging/not challenging*?" two types of responses were obtained. Examples of each type of response follow:

"*They are not as difficult as the examination questions, but they can be kind of hard to understand sometimes. Luckily Mrs D. allowed us to ask the other group.*"

> "*Working with my group is a greater challenge for me. At least we solved the problem correctly.*"

The responses suggested that the students had more difficulty in understanding the problem posed by their peers and managing group dynamics, than in applying their mathematical knowledge to solve the problems.

The responses by the Grade 9 students suggested that they had a different view of what they felt constituted an interesting problem posed by their peers, compared with Grade 7 students. Grade 9 students tended to be more appreciative of the mathematical structure than the surface structure of the problems posed. In this study, the term "mathematical structure" was taken to refer to the underlying mathematical relationships between the quantities occurring in the problem. Three types of comments were made by Grade 9 students in response to the survey question about how the problems posed by their peers were interesting (or not interesting):

> "[Group 8] … *has an interesting question. They managed to put together the idea of geometrical shapes, inequalities, and rate to form an optimization problem. It was a little contrived though.*"

> "*There did not seem to be a unique answer to the problem. Interesting to think about the possible scenarios.*"

> "*The farmer can do so many different interesting things with his piece of land! Can we use with other shapes?*"

The Grade 9 teacher noted that students were generally more tolerant towards ambiguities in the questions posed by their peers. Grade 9 students tended to focus on the mathematics involved in the problem, and this focus accounted for their high engagement and productive "arguments" with one another during the activity. These productive "arguments" mostly occurred when important assumptions about real-world situations were not stated clearly, and the students would have to draw upon their personal knowledge and experiences in order to solve the problem. Such problems were deemed to be the more challenging ones as shown in the responses below:

> "[Group 5] … *did not state the size or type of trees to be planted! Our group spent more time on talking about the trees than solving for x, because we could not set up the initial inequality.*"

Overall, the Grade 7 students paid more attention to the novelty of creating the problems rather than dealing with details of the mathematics of the problem, and they were less tolerant towards ambiguities in real-world problems. In contrast, by focusing on the mathematical structure of the problems, the Grade 9 students were more able to generate problems with moderate to high mathematical complexity. The Grade 9 students also showed their potential for generating problems that exceeded the degree of difficulty and novelty of those found in textbooks and school worksheets.

Students' Perceptions of Problem Posing

Reflections about problem posing. With no prior experiences with generative activities, it is reasonable to assume that students' perceptions about mathematical problem posing were mainly derived from their involvement in this 6-week program. In order to gather information about their problem-posing experiences, all students were asked to reflect on the most important thing that they had learnt about problem posing and the related activities. Based on data from their reflections, 83% of the Grade 7 students felt positively about their learning experiences. Three types of responses to the survey were identified:

> "[problem posing] … *is harder than it looks, especially if the problem has to have important mathematics, is interesting and yet challenging. Maybe it is worth a try again next time.*"

> "*It is like making my favourite muffin—more chocolate chips and less sugar but still delicious.*"

> "*My friends enjoyed the problem posed by my group, though we are not exactly the strongest group around. Anyway, I am glad it turned out okay.*"

These responses reflected different aspects of problem posing that students considered important, including:

- Problem-posing skills can be developed over time;
- The amount of "control" the poser has when generating a problem. Such control includes being able to establish the given conditions for the problem, or being able to create the real-world stimulus that can be related to specific mathematical concepts; and
- Posing problems collaboratively can possibly increase one's confidence in learning and applying mathematical knowledge.

Of the Grade 9 students in the study, 67% felt that the problem-posing activities had a significant impact on their views on mathematical learning. The following responses illustrate the three types of responses obtained from the survey:

> "*Problem posing is itself a problem-solving activity, where the final answer is the problem itself. It has to satisfy all the conditions in the given situation.*"

> "*I learnt that* $x^a > y^b$ *does not necessarily imply that* $a > b$."

> "*I can use SCAMPER* [*from Design & Technology class*] *to pose/create the problems.*"

These responses differed from those identified in the Grade 7 students' reflections, and suggested other important aspects of problem posing, including:

- Problem posing involved similar skills required in solving a mathematical problem, and are therefore transferrable;
- Learning mathematics is important in problem-posing activities; and
- Problem-posing strategies are not unique, and, as a consequence, creative thinking skills are possibly transferrable from other disciplines.

Teacher Observations

Mathematical processes such as representing, reflecting, connecting, and problem solving are mental activities which may or may not be accompanied by observable behaviors. Behaviors of students during the problem-posing activities were noted by the teacher, and these may help serve as indicators for making inferences about students' mathematical thinking and learning related to problem posing. Figure 13.11 shows a summary of the behaviors observed, and the mathematical thinking that was likely to be associated with those behaviors during the problem-posing tasks. Implications for students' possible ways of learning mathematics are proposed.

Mathematical Process	Observable behaviours	Likely Thinking	Learning
Representing	Use a variety of representations of mathematical ideas	Imagine, reason or work flexibly with multiple representations	Use a "visual" approach to learn about mathematical concepts
	Make connections between different representations	Compare and select representations for solving problems	Focus on the big ideas first before attending to details of a concept
Reflecting	Consider the reasonableness of problem posed	Examine problems and their solution from multiple perspectives	Verify outcomes by considering underlying assumptions
	Assess the effectiveness of strategies used	Relate concepts and procedures applied in the strategies	Experiment with ideas for the purposes of validation or extension
Connecting	Apply knowledge of solving one problem to another	Identify commonalities in the mathematical structure of problems	Imitate procedures as a way to assimilate new ideas and knowledge
	Relate mathematical ideas to real life situations	Evaluate the relevance to have broader perspective of a concept	Use a "kinaesthetic" approach to learning about concepts
Problem solving	Provide multiple responses (problems) to a specific solution	Reverse the thinking in the first solution to look for "new" problems	Persevere by focusing on establishing fluency in generating ideas
	Provide multiple solution paths to a problem	Reformulate the first problem to look for "new" solutions	Deconstruct problems to look for alternative perspectives

Figure 13.11. Behaviors of students during the problem-posing activities.

Implications for Further Research

This study offers information about how teachers can plan systematically for the introduction of problem-posing activities into the classroom. The results of the study suggest how these activities might provide insight for teachers about students' understandings of mathematical concepts in the problem posed and competence in problem solving. This informal way of gathering information about students' thinking processes, their strategies, and their developing mathematical understandings are present in most formative assessments. In addition, the problem-posing activities presented the students with a full array of tasks that required them to be effective performers as they applied their knowledge and skills. From the affective point of view, the students' beliefs and other attributes—such as their desire to take risks and to be open to constructive feedback—were revealed. Hence, problem posing can potentially be used as an assessment tool for the effective teaching and learning of mathematics.

Conclusions

The objective of this study was to identify patterns in students' mathematical learning and thinking during classroom-based problem-posing tasks. Students' attempts to construct well-defined, solvable problems within given contexts incorporating suitable objects and events, while at the same time utilizing specific mathematical concepts, proved to be a demanding experience. The study identified several cognitive factors: the students' ability to identify the mathematical structure of problem, their familiarity with creative thinking strategies, reverse engineering techniques, mathematically modeling with real-life situations, tolerance towards ambiguities, and productive thinking and communication through writing mathematical problems. Cognitive skills can be developed and acquired over time and most importantly, evidence from the study suggests that these skills are transferrable, both from within mathematics (via problem-solving activities), and from other fields. Hence, there are important implications for teachers to include both problem-solving *and* problem-posing activities in their instructional goals for mathematics teaching and learning.

Affective factors are closely related to cognitive factors, particularly in social settings like those presented in this study. The findings indicated that student motivation, perseverance, and risk-taking are positive dispositions which students can develop which assist them, and their teachers, to harness benefits that problem posing can bring to the learning environment. When given the space and resources for students to have "control" over their own creations, including the opportunity to use the problems they pose to compete with their peers, the students' self-confidence and learning can be enhanced. In this context, the teacher plays an important mediating role in establishing an effective collaborative, interactive, and safe environment for the students.

Merely asking students to pose problems may not be enough to detect the quality of their mathematical understanding, thinking, and learning. When regarded as a feedback mechanism, problems posed individually, and then shared collectively, can inform teachers about what teaching and learning activities might enable their students to move from their current level of understanding to the next. This study is not merely about the potential of using problem posing as an assessment tool, but rather about raising awareness of the importance of planned approaches for the use of problem-posing activities in the classroom.

The data presented on the levels of mathematical complexity of problems posed by Grades 7 and 9 students, and on their thinking processes during problem posing, should be seen as a good starting point for exploring and building a taxonomy of thinking processes related to problem posing. Such a taxonomy has the potential to give teachers a more complete account of students' mathematical competencies and understandings. With its potential to foster more diverse and flexible thinking, the use of problem-posing activities can not only enhance students' problem-solving skills, but can also help to broaden their perceptions of mathematics and enrich and consolidate their understanding of basic mathematical concepts. In such ways, problem posing can become a highly viable tool for evaluating students' learning of mathematics.

References

Brown, S. I., & Walter, M. I. (1993). *Problem posing: Reflections and applications*. Hillsdale, NJ: Lawrence Erlbaum.

English, L. D. (1996). Children's problem posing and problem-solving preferences. In J. Mulligan & M. Mitchelmore (Eds.), *Children's number learning: A research monograph* (pp. 227–242). Adelaide, Australia: Australian Association of Mathematics Teachers.

English, L. D. (1997). Promoting a problem-posing classroom. *Teaching Children Mathematics, 4*(3), 172–179.

Lappan, G., Fey, J. T., Fitzgerald, W. M., Friel, S. N., & Phillips, E. D. (1996). *Getting to know CMP: An introduction to the Connected Mathematics Project*. Palo Alto, CA: Dale Seymour.

Lowrie, T. (1999). Free problem posing: Year 3/4 students constructing problems for friends to solve. In J. M. Truran & K. M. Truran (Eds.), *Making the difference* (pp. 328–335). Sydney, Australia: Mathematics Education Research Group of Australasia.

Ministry of Education. (2006). *Revised syllabus for primary mathematics*. Singapore: Curriculum Planning and Development Division, Ministry of Education.

National Assessment of Educational Progress. (2005). *The nation's report card: Mathematics 2005*. Washington, DC: Author.

Pittalis, M., Christou, C., Mousoulides, N., & Pitta-Pantazi, D. (2004). A structural model for problem posing. In M. J. Hoines & A. B. Fuglestad (Eds.), *Proceedings of the 28th Conference of the International Group for the Psychology of Mathematics Education* (Vol. 4, pp. 49–56). Bergen, Norway: PME.

Sheffield, L. J. (2003). *Extending the challenge in mathematics: Developing mathematical promise in K–8 students*. Thousand Oaks, CA: Corwin Press.

Silver, E. A. (1994). On mathematical problem solving. *For the Learning of Mathematics, 14*(1), 19–28.

Silver, E. A., & Cai, J. (1996). An analysis of arithmetic problem posing by middle school students. *Journal for Research in Mathematics Education, 27*(5), 521–539.

Silver, E. A., & Cai, J. (2005). Assessing students' mathematical problem posing. *Teaching Children Mathematics, 12*(3), 129–135.

Chapter 14
An Investigation of High School Students' Mathematical Problem Posing in the United States and China

Xianwei Van Harpen and Norma Presmeg

Abstract In the literature, problem posing is claimed to be important in learning mathematics. This study investigated US and Chinese high school students' attitudes and abilities in posing mathematical problems. All of the participants were taking advanced mathematics in high school. A mathematics content test and a mathematical problem-posing test were administered to the students. The mathematical content test was adapted from the National Assessment of Educational Progress for 12th graders. The problem-posing test included three situations, namely a free problem-posing situation, a semi-structured problem-posing situation, and a structured problem-posing situation. Students who scored 39 or above out of 50 points were interviewed. During the interviews, the majority of the students reported that they had not had any prior experience in posing mathematical problems. Many students did not have a specific strategy for posing problems, and many had difficulty explaining their problem-posing processes. Most of the US students for various reasons said that problem posing was important in mathematics. Most Chinese students said that problem posing was not important in high school learning because of college entrance examinations.

Contents

X. Van Harpen (✉)
Department of Mathematical Sciences, University of Wisconsin-Milwaukee, Milwaukee, WI, USA
e-mail: vanharpe@uwm.edu

N. Presmeg
Department of Mathematics, Illinois State University, Normal, IL, USA
e-mail: npresmeg@msn.com

F.M. Singer et al. (eds.), *Mathematical Problem Posing*,
Research in Mathematics Education, DOI 10.1007/978-1-4614-6258-3_14

Literature Review

In 1957, Polya stated that "the mathematical experience of the student is incomplete if he never had an opportunity to solve a problem invented by himself" (p. 68). Silver (1997) concluded that inquiry-oriented mathematics instruction, which includes problem-solving and problem-posing tasks and activities, can assist students to develop more creative approaches to mathematics. It is claimed that through the use of such tasks and activities, teachers can increase their students' capacity with respect to the core dimensions of creativity, namely, fluency, flexibility, and originality (Presmeg, 1981; Torrance, 1988). According to Barlow and Cates (2006), problem posing gave students a sense of ownership of mathematics. Brown and Walter (2005) observed that problem posing promotes a sense of mathematical autonomy. English claimed that problem posing helped students to develop skills in recognizing problem structure (English, 1997, 2003; English & Halford, 1995).

The importance of the role of problem posing in mathematics teaching and learning has also been documented in official documents in different parts of the world. According to the National Council of Teachers of Mathematics (1989), students should be given opportunities to solve mathematical problems using multiple solution strategies and to formulate and create their own problems from given situations. Similarly, in China, in a document entitled the *Interpretation of Mathematics Curriculum* (Trial Version) (Mathematics Curriculum Development Group of Basic Education of Education Department, 2002), it was pointed out that students' abilities in problem solving and problem posing should be emphasized and that students should learn to find problems and pose problems in and out of the context of mathematics.

Much research has been conducted on students' problem posing in mathematics. For example, Cai's (1998, 2000) studies, and Cai and Hwang's (2002) study indicated that, although Chinese students were superior in tasks involving computation skills and were more efficient in routine problem solving than US students, the latter performed as well as or better than their Chinese counterparts on more open, creative problem-solving tasks, and problem-posing tasks. Li and Wang (2006) studied 20 eighth-grade students and found a positive correlation between their problem posing and problem solving, which means that students of higher ability in problem posing also had higher ability in problem solving and vice versa. However, most studies on mathematical problem posing chose primary school students as participants (e.g., Cai, 1998, 2000; Cai & Hwang, 2002; Zeng, Lu, & Wang, 2006). Not much has been reported about high school students' mathematical problem-posing abilities. Also, despite the emphasis on problem posing by educators, little is known about students' perception of the importance of problem posing in their learning of mathematics. Thus, this study investigated US and Chinese high school students' attitudes and abilities in posing mathematical problems. The following three research questions were addressed in this study:

1. How much problem posing is involved in students' learning of mathematics?
2. How do students pose mathematical problems?
3. What are students' perceptions of the role of problem posing in their mathematics learning?

Conceptual Framework

Different terms are used for problem posing, such as problem finding, problem sensing, problem formulating, creative problem discovering, problemizing, problem creating, and problem envisaging (Dillon, 1982; Jay & Perkins, 1997). In the present study, mathematical problem posing will be defined as the process by which, on the basis of mathematical experience, students construct personal interpretations of concrete situations and from these situations formulate meaningful mathematical problems (Stoyanova & Ellerton, 1996). Mathematical problem-posing abilities will be measured by means of a mathematical problem-posing test. More details about the test are discussed in the "Methodology" section. Also, the literature review will explicate different terms, such as problem finding, problem formulating, and creative problem discovering.

The present study uses the framework proposed by Stoyanova and Ellerton (1996) who classified a problem-posing situation as free, semi-structured, or structured. According to this framework, a problem-posing situation is referred to as *free* when students are asked to generate a problem from a given, contrived, or naturalistic situation (see Example 1 below). A problem-posing situation is referred to as *semi-structured* when students are given an open situation and are invited to explore the structure of that situation, and to complete it by applying knowledge, skills, concepts, and relationships from their previous mathematical experiences (see Example 2 below). A problem-posing situation is referred to as *structured* when problem-posing activities are based on a specific problem (see Example 3 below). The following three tasks were used in the problem-posing test in this study. Three pilot studies were conducted with both Chinese students and US students before this study to make sure that the content of the tasks were appropriate for high school students and would encourage them to pose a variety of mathematical problems. The three tasks had also been validated in previous studies. Specifically, the first two tasks were adapted from Stoyanova's (1997) dissertation and the third task was adapted from Stoyanova's (1997) dissertation and Cai's (2000) research.

Task 1 (Free problem-posing situation): There are ten girls and ten boys standing in a line. Make up as many problems as you can that use the information in some way.

Task 2 (Semi-structured problem-posing situation): In the following picture (Figure 14.1), there is a triangle and its inscribed circle. Make up as many problems as you can that are in some way related to this picture.

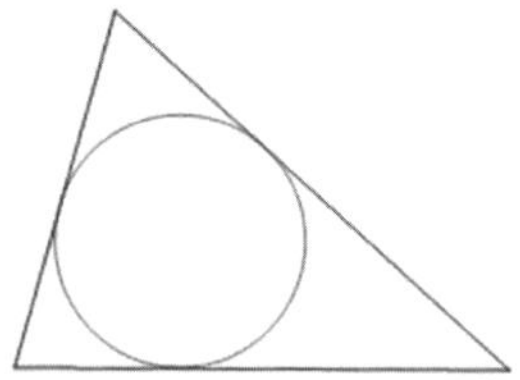

Figure 14.1. Diagram for the semi-structured problem-posing situation (Task 2).

Task 3 (Structured problem-posing situation): Last night, there was a party at your cousin's house and the doorbell rang ten times. The first time the doorbell rang only one guest arrived. Each time the doorbell rang, three more guests arrived than had arrived on the previous ring.

(a) How many guests will enter on the tenth ring? Explain how you found your answer.

(b) Ask as many questions as you can that are in some way related to this problem.

Methodology

Participants

According to Peverly (2005), even within one country, different locations in China can vary greatly in terms of culture. Thus, this study selected students from a large city (Shanghai) in the south of China and a small city (Jiaozhou) in the north of China. Because of Shanghai's cultural and economic status in East Asia, it represents everything considered modern in China, including education. At the time of the study, Shanghai had a population of about 23 million. Jiaozhou is located in Shandong Province, where Confucian culture has a significant influence. Jiaozhou had a population of about 843,000. In China, high school students are divided into two strands, namely, a science strand and an arts strand. Science students take more advanced mathematics courses in high school than arts students. The 44 Shanghai participants in this study are from two 11th-grade science-strand classes. The 55 Jiaozhou participants were from one 12th-grade science-strand class. Unfortunately, there was only one group of US participants in this study because at the time they were the only available participants. Among the 30 US students in this study, there were 13 students from a Pre-calculus class (11th grade) and 17 students from an Advanced Placement Calculus class (12th grade) from a mid-western town which had a population of about 120,000. All of the participants in this study were taking advanced topics in mathematics in high school and were 18 years old. Since the samples were convenience samples, the findings cannot be generalized to other students in the two countries.

Instruments

Two tests were administered to the students, namely, a mathematical problem-posing test and a mathematics content test. The three tasks mentioned earlier in the "Conceptual Framework" section were included in the mathematical problem-posing test. The purpose of the mathematics content test was to measure the participants' mathematical content knowledge. Instead of developing a test for this study, the researchers adapted the National Assessment of Educational Progress (NAEP) 12th-grade Mathematics Assessment for the study since the items on NAEP fitted the purpose of this study. According to the National Center for Educational Statistics (2009), NAEP is the only nationally representative and continuing assessment of what America's students know and can do in various subject areas. The 2005 NAEP Mathematics Assessment results for the 12th grade were released and available to be used by the time this study started. NAEP assesses an appropriate balance of content along with a variety of ways of knowing and doing mathematics. The tasks in the 12th-grade assessment involved 4 mathematics content areas: number properties and operations, measurement and geometry, data analysis and probability, and algebra. Since the participants in this study were from 3 different locations with different curricula, and since the sample included some 11th-grade students, only tasks based on content studied by all 3 groups were included. Decisions on the questions to include were made only after examining the textbooks used by the students in the three locations, and after talking to the mathematics teachers of the three participating schools. Of the 50 questions in the mathematics content test, 9 were concerned with Numbers and Operations, 16 with Measurement and Geometry, 16 with Algebra, and 9 with Data Analysis and Probability. Thus, since the participants of this study were from three different locations with different mathematics curricula and instruction, the modified version of the NAEP assessment instrument allowed a relatively comprehensive and fair examination of the participants' mathematical knowledge base.

Interviews

Twelve US students, 12 Shanghai students, and 8 Jiaozhou students were interviewed. The interviewees were chosen based on their scores on the mathematical content test (more details are provided in the "Results" section). The interviews were audio taped and transcribed. All interviews were conducted by the first author.

Results

The mathematics content test was graded according to the instructions provided by the NAEP assessment developers. The mathematics problem-posing test was graded according to the rubrics developed for this study. During the development of the rubrics, the 2 authors examined all of the problems posed by the 30 US students and discussed any differences in their opinions. Eventually, both researchers were able to reach agreement on all aspects of the rubrics. This chapter focuses on reporting the results of interviews conducted with selected students. Detailed results of the mathematical content test and of the mathematical problem-posing test have been reported in Yuan and Sriraman (2010a, b), Van Harpen and Presmeg (2013), and Van Harpen and Sriraman (2013).

Mathematical Content Test Results

After eliminating the items that had translation or cultural issues from the mathematics content test, 50 items were left in the test. Each item was assigned one point. US students, Shanghai students, and Jiaozhou students scored means of 36.5, 36.2, and 45.8, respectively. All of the 32 students who were interviewed achieved 39 or more points out of 50 in the mathematics content test.

Mathematical Problem-Posing Test Results

Nonviable problems versus viable problems. In analyzing the problems generated by the students, the problems that were non-appropriate (e.g., How old are the children?) and problems that lacked the information needed to determine solutions (e.g., How many girls and how many boys are there at the party?) were defined as nonviable problems and were excluded from further analysis. Since the numbers of students in each of the three groups were different, the average percentage of nonviable problems generated by the students in each group was calculated and listed in

Table 14.1. The percentages were calculated by finding the ratio of the number of nonviable problems to the total number of posed problems (i.e., nonviable + viable problems).

It should be pointed out that the criteria classifying problems as viable or nonviable were not tightly defined but were based on the researchers' judgment. Therefore, the numbers in Table 14.1 might not be rigorously precise.

Table 14.1
Average Percentage of Nonviable Problems Generated by Students

	Nonviable problems in Task 1	Nonviable problems in Task 2	Nonviable problems in Task 3
US students (%)	9	31	8
Shanghai students (%)	12	42	13
Jiaozhou students (%)	12	15	3

Number of posed problems. After any nonviable problems were eliminated, the problems posed by each student were counted. US students, Shanghai students, and Jiaozhou students scored a mean of 13.1, 7.7, and 14.1, respectively, and a standard deviation of 6.45, 5.25, and 6.54.

Trivial problems versus nontrivial problems. After any nonviable problems had been eliminated, the remaining problems were analyzed for their triviality. Problems that required at most elementary mathematics to solve were defined as trivial problems. For example, the following problems were considered to be trivial problems.

Problem for Task 1: *How many children are there in all?*

Problem for Task 2: *How many geometric shapes are formed in the given image?*

Problem for Task 3: *Half of the guests wear blue shirts, how many wear blue shirts?*

Table 14.2 shows the percentage of trivial problems posed by students in the three groups. The ratio of the number of trivial problems to the number of viable problems was found for each task and for each group of students.

Table 14.2
Average Percentage of Trivial Problems Generated by Students

	Trivial problems in Task 1	Trivial problems in Task 2	Trivial problems in Task 3
US students (%)	16	9	19
Shanghai students (%)	17	8	7
Jiaozhou students (%)	14	6	6

Variety of posed tasks. Students of the three groups all posed a variety of problems for each task. For example, Table 14.3 lists eight different categories of problems posed for Task 1. Table 14.4 shows the percentage of each category of problems posed by the different groups of students.

Table 14.3
Variety of Posed Problems for Problem-Posing Test Task 1

Category name	Examples
1. Combination and/or permutation	How many different ways are there of arranging the 20 children in a line if a girl has to be in the first place?
2. Probability	If the teacher wants to pair the 20 children up, what is the probability that student A and student B are paired up?
3. Arithmetic	One child eats one chicken egg. Two children eat one duck egg. Four children eat one goose egg. Five children eat one other egg. How many eggs do these 20 children eat?
4. Data analysis	The heart rates of the 20 children are as following. Graph the data: 82, 85, 69, 70, 83, 71, 90, 77, 76, 69, 81, 77, 88, 69, 75, 82, 84, 78, 70, 68 (number of heart beats per minute)
5. Geometry	If the 20 children are to form a rectangle, how many different rectangles can they form?
6. Sequence	The first person has one candy. The second person has two candies. The third person has three candies. The fourth person has five candies. The fifth person has eight candies. How many candies does the 20th person have?
7. Algebra	Among the 20 children, every 4 boys share 1 book A. Every girl has one book A. Every boy has a book B. Every two girls share a book B. the total number of book B is five more than the total number of book A. Find out the number of boys and girls
8. Others	Twenty people pass a ball. Boys can pass to the third person next to them. Girls can pass to the second person next to them. In total, nine people touched the ball. What is the arrangement of the line like?

Table 14.4
Categories of Posed Problems for Task 1

Category name	US students (30)	Shanghai students (44)	Jiaozhou students (55)
1. Combination and/or permutation	27 (24.1%)	27 (30.0%)	143 (50%)
2. Probability	13 (12.6%)	3 (2.4%)	74 (25.9%)
3. Arithmetic	42 (37.5%)	35 (28.5%)	30 (10.5%)
4. Data analysis	8 (7.1%)	3 (2.4%)	5 (1.7%)
5. Geometry	3 (2.7%)	1 (0.8%)	5 (1.7%)
6. Sequence	10 (8.9%)	25 (20.3%)	14 (4.9%)
7. Algebra	3 (2.7%)	16 (13.0%)	4 (1.4%)
8. Others	6 (5.4%)	13 (10.6%)	11 (3.8%)
Total	112	123	286

Interviews

Literature has suggested that students with higher ability in mathematics tended to be better problem posers (e.g., Ellerton, 1986; Krutetskii, 1976). Van Harpen and Presmeg (2013), in their study with the same groups of students as this present study, also found that students who achieved high scores on the mathematical content test posed more problems, and a greater diversity of problems than their peers who did not score as well on the mathematics content test. The problem-posing test was conducted with paper and pencil and does not reveal how students went about posing their problems. Interviews were therefore conducted with students who achieved higher scores on the mathematical content test.

The following questions were used as a guide for the interview. Examples of student responses will be presented in the remainder of this chapter.

1. What are some of the questions or difficulties you had when you were working on the tests?
2. As to the problems or questions that you posed in the problem-posing test, have you ever seen or heard the same or similar problems before? If yes, where and when did you see or hear them? Can you describe the situation to me?
3. If you have not seen or heard any problems that are similar to those posed by you in the problem-posing test, how did you come up with those problems or questions?
4. When and how were you asked to pose problems in mathematics before?
5. Do you think problem-posing abilities are important in learning mathematics? Why or why not?
6. Further clarification of the responses in the test.

Students' prior experience in posing mathematical problems. Despite the emphasis on problem posing in official curriculum documents, interviews with the students suggested that teachers rarely used problem posing in mathematics instruction.

Students' mathematical problem-posing processes. During the interview, students were given plenty of time to look through their responses to the problem-posing test to help them recall writing the problems and to think about how they had gone about the problem-posing process for each of the three tasks. Despite the fact that students in this study reported little previous experience in posing problems in mathematics, they were all able to pose problems in the test. It is important to note that these students were all taking advanced mathematical courses in their schools. When asked to pose problems within a limited amount of time, students posed a variety of problems in very different ways.

U.S. students. When asked how they posed the problems on the test, US students frequently used the terms "thinking outside of the box," "something fun," and "something interesting." US students also tended to make comments about freeing their minds and letting ideas come to them. When asked how they posed those problems, many of them did not know how to explain what they had done, and said "I don't know" or "I am not sure." For example,

Mulnon: Some of them [ideas] came and I kind of took them and sat back for a second just stared up the space and looked down again and then something else appeared.
Kyle: I was just putting down the first things that came into my mind. Things just popped into there.

Shanghai students. Most Shanghai students said that they posed problems which were very similar to those that they usually did in class. For example, Lijun commented:

> *I just related to some knowledge from the textbook. ... For example, we have been learning sequences recently. This one is a sequence problem. As to the geometric problem, I thought of geometry I did before, such as finding the minimum and maximum area.*

Some Shanghai students reported that it was very difficult for them to pose new problems. For example, Xueying observed:

> *We have done so many problems and therefore there is not much to do ... I just thought of the problems I have done before when I saw the information in the test.*

Jiaozhou students. When asked how they posed problems on the test, Jiaozhou students tended to focus more on the mathematical content of the problems they were posing. Also, all Jiaozhou students who were interviewed explained clearly how they went about posing the problems. Two examples of these explanations are presented below.

> *Xiang: I first started with a mathematical idea and then I tried to connect it to real life. For example, in the third task, the doorbell problem, it is obviously an arithmetic sequence, so I just made up a problem related to it.*

> *Yanan: When I saw the circle, I thought of radius, area, circumference, etc. Then when I saw the triangle, I immediately thought of area, perimeter, altitude, etc. Then I just tried to connect all of them to make the problems harder.*

Students' perceptions of the role of problem posing in mathematics learning. In the interview, students were asked if problem posing was important in their mathematics learning.

U.S. students. In the U.S. student group, 5 out of the 12 students did not seem to understand the question. Instead of talking about the importance of problem posing in mathematics learning, they thought that the question was asking them how they thought the teaching and learning of mathematics could be improved. Of the students who did give relevant responses, Ramona said that the act of posing problems herself helped her to see the structure of the problems. Scarlett explained that posing problems would help students see what is important in their mathematics learning. Kurt indicated that he thought problem posing was helpful because one could work backwards to find the answer and that helped him to see what worked and what did not. Two students explained that posing problems would help them solve problems, in the sense that they learned how to read problems more thoroughly. For example, Iris noted:

> *Yes, I think that ... you have to know both sides to have a full understanding. If you only ...if you get so focused on finding the answer, you forget how to read the question completely.*

Shanghai students. US students mostly said that they thought it was important to be able to pose problems in their mathematics learning. When it came to Chinese students, however, the answers were not as consistent. Four of the 12 Shanghai students reported that they did not think problem-posing ability was important in math ematics learning. For example,

> *Lijun: If it is for the college entrance exam in Shanghai, I would say that it is not important. The reason is that, in the exam, you should have seen most of the problems. If you did not get it the first time, then the second time you should understand the way of solving it, then you should know how to do it and you don't really need any creativity. Once you've got it, you will always get it.*

The other eight students reported that they thought problem posing was important in mathematics learning, but the reasons varied. Examples from two students follow.

> *Lulu: I think so, because I think to pose a problem, one has to have the basic knowledge. It is like you need to prepare for it, which makes you think better. Instead of simply solving problems in certain ways, you need to ask things. That is very helpful to organize my own ideas.*

> *Feng: Yes, it is important, because if you can think of some more problems similar to your teachers' thinking, then it will be easier for you to solve problems later on. In other words, you can think of what your teacher will test you on, what is the most important knowledge, and you will learn to summarize your knowledge better.*

Jiaozhou students. Among the eight Jiaozhou students interviewed, two students said that it was not important in their mathematics learning in high school because of the college entrance examination. For example, Yuan said:

> *I think it is in fact pretty important in mathematics learning, but it is not that important in the college entrance examination. I think it is important in mathematics learning because it can help you enhance your thinking ability, so it is very important. But the fact is that in high school, especially the senior year, it is more important to solve a lot of problems.*

Two students said that posing problems would help them see the way the tests are designed so that they would be able to do better in solving problems. Yanan noted:

> *I think it is very important. I think problem posing helps us see how our teachers make up test problems to test us. It will help us think what they will put in the tests.*

Three students stated that it was important for them to be able to understand mathematical concepts before they could pose problems. Two of these comments have been included here.

> *Xiang: I think it is pretty important. First of all you have to understand the methods of working on the problem before you can pose a problem yourself.*

Ning: I think it is very important. It is an ability of summarizing and thinking. For example, if you are to pose a problem about an inscribed circle in a triangle, you have to know all about triangle and circle. That involves a great amount of thinking.

A summary of key points arising from the interviews with students from the three groups (U.S., Shanghai, and Jiaozhou students) can be found in Table 14.5.

Table 14.5
Interview Results

	U.S. students	Shanghai students	Jiaozhou students
Experience in problem posing	Little or none	Little or none	Little or none
How problems were posed	Think outside of the box	Recall problems solved in the past	Start with the mathematical concepts and relate to real life
	Think of something fun	Difficult to think beyond the problem seen before	Come up with relevant concepts and connect them
	Let ideas come to mind		
Role of problem posing	Important	Important but not tested in college entrance examinations	Important but not tested in college entrance examinations
	Helps see structure of mathematics	Help summarize organize knowledge	Enhance thinking
	Helps see important mathematics	Help see how teachers make up test problems	Help understand methods
	Helps to read and fully understand other problems		

Conclusion and Implications

This chapter reports the findings of a study which investigated how much students used problem posing in their mathematics learning, how they posed problems, and how much they valued problem posing. Three groups of students from two locations in China and one location in the United States participated in the study. Although the findings of this study should not be generalized to student populations of the two countries, they shed some light on problem posing from students' perspectives, and suggest some ideas about how more problem-posing activities might be integrated into classroom instruction.

Exposure to Problem Posing

The findings of this study suggest that, despite any emphasis on problem posing in curriculum documents, problem posing is, in fact, not often practiced. During the interviews, the majority of the students reported that they had had little or no experience in posing mathematical problems. But when talking about the role of problem posing in their mathematics learning, most mentioned that it might have many benefits in helping them learn mathematics. Those findings suggest that students were open to the idea of using more problem-posing activities in mathematics learning and tended to have a positive attitude towards problem posing.

Problem-Posing Process and Products

On the one hand, many students did not have a specific strategy for posing new problems and had difficulty explaining how they went about posing problems. This could explain why the students posed a significant number of trivial problems and nonviable problems. On the other hand, some students were easily able to see through the context of the problems and focus on the mathematics in the scenarios. These findings suggest that even students who were more advanced in mathematics are not necessarily good problem posers. Without teacher modeling and guiding, problem posing may not always be an efficient tool for learning.

In his study of mathematical "giftedness" Krutetskii (1976) included problem-posing tasks in which there were unstated questions (e.g., "A pupil bought $2x$ notebooks in 1 store, and in another bought 1.5 times as many"). For tasks like this, students were required to pose and then answer questions on the basis of the given information. Krutetskii reported that high-ability students were able to take account of the given information and to pose relevant problems, whereas students of lesser ability either required hints or were unable to pose matching problems. In Ellerton's (1986) study, 11- to 13-year-old students were asked to pose mathematics problems that would be difficult for friends (who had been absent from class) to solve. She found that the "more able" students posed problems of greater computational difficulty (i.e., more complex numbers and requiring more operations for solution) than did their "less able" peers.

For the same groups of students described in this chapter, Van Harpen and Presmeg (2013) reported correlations between students' mathematical content test scores and the number of problems they each posed and also with the different categories of problems they each posed ($p<.01$ for both correlations). Specifically, the correlations reported suggest that mathematics content knowledge predicts 32.3% of the number of posed problems and 38.3% of the number of different categories of posed problems. Given that all students in this study were in advanced mathematics course in high school, this result is surprising, as it seemed reasonable to expect there to be higher correlations. It is possible that the nature of the tasks, as well as the fact that the tests were in addition to their normal classroom work, may have contributed to these lower-than-expected correlations.

Another possible reason might be the fact that, as some Shanghai students reported, students had solved so many problems that it was hard for them to pose new problems without recalling those they had already seen. In addition, many students in the interview mentioned the purpose of problem posing that they had experienced was in the preparation for tests or examinations, leading them to think of a more limited range of problems. Further research is clearly needed in this area.

The Role of Problem Posing in School Mathematics

Li and Wang (2006) described three factors—the learning environment, personality, and the ability to reflect on learning—which might explain why high-ability students in mathematics sometimes posed poor problems. The findings reported in this chapter are consistent with the first of these three factors (the learning environment).

Most of the U.S. students who were interviewed in this study said that problem posing was important in mathematics. Most Chinese students said that problem posing would be useful in college, but not important in high school where they focused on preparing for the college entrance examination. In addition, many Chinese mentioned valid benefits of posing problems in mathematics learning. For example, they said that posing problems helped them to organize ideas, summarize knowledge, and enhance understanding of mathematical methods, etc. Such comments are consistent with Li and Wang's (2006) finding that problem posing relies on one's ability to reflect on learning, and that problem posing may help enhance such reflection.

An important area for further research would be to explore possible links between students' ability to reflect on their own learning and the quality of the mathematics problems they pose.

References

Barlow, A. T., & Cates, J. M. (2006). The impact of problem posing on elementary teachers' beliefs about mathematics and mathematics teaching. *School Science and Mathematics, 106*(2), 64–73.

Brown, S. I., & Walter, M. I. (2005). *The art of problem posing* (3rd ed.). New York, NY: Routledge.

Cai, J. (1998). An investigation of U.S. and Chinese students' mathematical problem posing and problem solving. *Mathematics Education Research Journal, 10*(1), 37–50.

Cai, J. (2000). Mathematical thinking involved in U.S. and Chinese students' solving process-constrained and process-open problems. *Mathematical Thinking and Learning, 2*, 309–340.

Cai, J., & Hwang, S. (2002). Generalized and generative thinking in U.S. and Chinese students' mathematical problem solving and problem posing. *The Journal of Mathematical Behavior, 21*(4), 401–421.

Dillon, J. T. (1982). Problem finding and solving. *The Journal of Creative Behavior, 16*, 97–111.

Ellerton, N. F. (1986). Children's made-up mathematics problems: A new perspective on talented mathematicians. *Educational Studies in Mathematics, 17*(3), 261–271.

English, L. D. (1997). Development of seventh-grade students' problem posing. In E. Pehkonen (Ed.), *Proceedings of the 21st Conference of the International Group for the Psychology of Mathematics Education* (Vol. 2, pp. 241–248). Lahti, Finland: PME.

English, L. D. (2003). Engaging students in problem posing in an inquiry-oriented mathematics classroom. In F. K. Lester Jr. & I. C. Randall (Eds.), *Teaching mathematics through problem solving: Prekindergarten-grade 6* (pp. 187–269). Reston, VA: National Council of Teachers of Mathematics.

English, L. D., & Halford, G. S. (1995). *Mathematics education: Models and processes*. Mahwah, NJ: Lawrence Erlbaum.

Jay, E. S., & Perkins, D. N. (1997). Problem finding: The search for mechanism. In M. A. Runco (Ed.), *The creativity research handbook* (Vol. 1, pp. 257–293). Cresskill, NJ: Hampton Press.

Krutetskii, V. A. (1976). *The psychology of mathematical abilities in school children*. Chicago, IL: The University of Chicago Press.

Li, X., & Wang, J. (2006). *Problem-posing-based mathematical learning* [基于问题提出的数学学习] (Doctoral dissertation). East China Normal University, Shanghai, China.

Mathematics Curriculum Development Group of Basic Education of Education Department [教育部基础教育司数学课程标准研制组]. (2002). *The interpretation of mathematics curriculum* (Trial Version) [数学课程标准(实验稿)解读]. Beijing, China: Beijing Normal University Press.

National Center for Educational Statistics. (2009). *The National Assessment of Educational Progress Overview.* Retrieved August 28, 2009, from http://nces.ed.gov/nationsreportcard/mathematics/

National Council of Teachers of Mathematics. (1989). *Curriculum and evaluation standards for school mathematics*. Reston, VA: National Council of Teachers of Mathematics.

Peverly, S. (2005). Moving past cultural homogeneity: Suggestions for comparisons of students' educational outcomes in the United States and China. *Psychology in the Schools, 42*(3), 241–249.

Presmeg, N. C. (1981). *Parallel threads in the development of Albert Einstein's thought and current ideas on creativity: What are the implications for the teaching of school mathematics* (Unpublished M.Ed. dissertation in Educational Psychology). University of Natal.

Silver, E. A. (1997). *Fostering creativity through instruction rich in mathematical problem solving and problem posing*. Retrieved July 29, 2007, from http://www.emis.de/journals/ZDM/zdm973a3.pdf

Stoyanova, E. (1997). *Extending and exploring students' problem solving via problem posing: A study of years 8 and 9 students involved in Mathematics Challenge and Enrichment Stages of Euler Enrichment Program for Young Australians* (Unpublished doctoral dissertation). Edith Cowan University, Perth, Australia.

Stoyanova, E., & Ellerton, N. F. (1996). A framework for research into students' problem posing in school mathematics. In P. C. Clarkson (Ed.), *Technology in mathematics education* (pp. 518–525). Melbourne, Australia: Mathematics Education Research Group of Australasia.

Torrance, E. P. (1988). The nature of creativity as manifest in its testing. In R. J. Sternberg (Ed.), *The nature of creativity: Contemporary psychological perspectives* (pp. 43–75). New York, NY: Cambridge University Press.

Van Harpen, X. Y., & Presmeg, N. (2013). An investigation of relationships between students' mathematical problem-posing abilities and their mathematical content knowledge. *Educational Studies in Mathematics, 83*(1), 117–132.

Van Harpen, X. Y., & Sriraman, B. (2013). Creativity and mathematical problem posing: An analysis of high school students' mathematical problem posing in China and the United States. *Educational Studies in Mathematics, 82*(2), 201–221.

Yuan, X., & Sriraman, B. (2010). An exploratory study of relationships between students' creativity and mathematical problem posing abilities—Comparing Chinese and U.S students. In B. Sriraman, & K. Lee (Eds.), *The elements of creativity and giftedness in mathematics* (pp. 5–28). Rotterdam, The Netherlands: Sense Publishers.

Yuan, X., & Sriraman, B. (2010b). An exploratory study of relationships between students' creativity and mathematical problem posing abilities—Comparing Chinese and U.S students. In B. Sriraman & K. Lee (Eds.), *The elements of creativity and giftedness in mathematics* (pp. 5–28). Rotterdam, The Netherlands: Sense Publishers.

Zeng, X., Lu, C., & Wang, B. (2006). Actuality and countermeasure of "posing mathematical problem" on junior high school students [初中生“提出数学问题”的现状与对策]. *Journal of Mathematics Education, 15*(3), 51–53.

Chapter 15
Enhancing the Development of Chinese Fifth-Graders' Problem-Posing and Problem-Solving Abilities, Beliefs, and Attitudes: A Design Experiment

Limin Chen, Wim Van Dooren, and Lieven Verschaffel

Abstract The present study reports the design, implementation, and evaluation of a training program aimed at developing Chinese students' problem-posing abilities, problem-solving abilities, and their beliefs about, and attitudes toward, mathematical problem posing and problem solving. In this study, a framework for teaching and assessing problem posing was developed. Results revealed that the training program had a significant positive effect on the originality of the problems posed by the students (but not on the appropriateness, complexity, and diversity of the problems posed), as well as on their problem-solving abilities and on their problem-posing and problem-solving beliefs and attitudes.

Contents

L. Chen (✉)
Liaoning Research and Training Center for Basic Education,
Shenyang Normal University, Shenyang 110034, China
e-mail: chenlimin_leuven@126.com

W. Van Dooren • L. Verschaffel
Faculty of Psychology and Educational Sciences,
University of Leuven, Leuven, 3000, Belgium
e-mail: Wim.VanDooren@ppw.kuleuven.be; Lieven.Verschaffel@ppw.kuleuven.be

F.M. Singer et al. (eds.), *Mathematical Problem Posing*,
Research in Mathematics Education, DOI 10.1007/978-1-4614-6258-3_15

Introduction

Worldwide recommendations for the reform of school mathematics suggest an important role for problem posing. For example, the *Principles and Standards for School Mathematics* in the United States (National Council of Teachers of Mathematics, 2000) calls for students to "formulate interesting problems based on a wide variety of situations, both within and outside mathematics" (p. 258). In addition, that document recommends that students should make and investigate mathematical conjectures and learn how to generalize and extend problems by posing follow-up questions. Likewise, *Compulsory Education Mathematics Curriculum Standards* (Ministry of Education of The People's Republic of China, 2012) pays attention to students' acquisition of problem-posing abilities, emphasizing that students should learn to discover and pose problems from the perspective of mathematics (p. 9). So, according to these reform documents, the development of problem-posing competency is an important goal of mathematics teaching and learning that lies at the heart of mathematical activity. Moreover, the potential value of problem posing in developing students' problem-solving abilities, creativity, and mathematical understanding has been recognized by several researchers (Brown & Walter, 1990; English, 1997a, 1997b, 1998; Kilpatrick, 1987; Lavy & Bershadsky, 2003; Lowrie, 2002; Silver, 1994; Yuan & Sriraman, 2011).

Theoretical and Empirical Background

Since the late eighties, there has been growing interest in problem posing among researchers. First, some studies revealed that many students suffer from some difficulties in posing problems (Cai & Hwang, 2002; Chen, Van Dooren, Chen, & Verschaffel, 2005, 2007; Ellerton, 1986; English, 1997a, 1997b, 1998; Silver & Cai, 1996; Verschaffel, Van Dooren, Chen, & Stessens, 2009). Other studies revealed that some teachers also face difficulties in posing problems (Chen, Van Dooren, Chen, & Verschaffel, 2011; Leung & Silver, 1997; Silver, Mamona-Downs, Leung, & Kenney, 1996). Researchers have found that there is a close relationship between students' abilities to pose and solve problems (Cai & Hwang, 2002; Chen et al., 2005, 2007; Ellerton, 1986; Silver & Cai, 1996; Verschaffel et al., 2009). Second, several design experiments aimed at implementing and testing new instructional approaches that incorporate problem-posing activities into the mathematics curriculum have been carried out. These experiments were designed to improve students' mathematical understanding, problem-posing and problem-solving abilities, as well as their beliefs about and attitudes toward problem posing and problem solving (Bonotto & Baroni, 2008; English, 1997a, 1997b, 1998; Lavy & Bershadsky, 2003; Rudnitsky, Etheredge, Freeman, & Gilbert, 1995; Verschaffel, De Corte, Lowyck, Dhert, & Vandeput, 2000; Winograd, 1997).

Rudnitsky et al. (1995) implemented a "structure-plus-writing" instruction with third-grade and fourth-grade students to test whether the instruction, intended to help students construct knowledge about addition and subtraction story problems, could be transferred to helping them to solve problems. Children were instructed with the concept of a mathematics story (i.e., any story, happening, or event that has to do with quantities or amounts) and its relationship to a mathematics problem, and were engaged in creating their own mathematics stories, categorizing their own stories, and making up mathematics problems from these mathematics stories. It was found that children with structure-plus-writing instruction outperformed children who only received a problem-solving treatment based on practice and provision of explicit heuristics, and children who received no explicit instruction in arithmetic word problem solving. Winograd (1997) implemented a problem-posing training program with fifth-grade students, wherein different ways of sharing student-authored word problems (i.e., posing and solving mathematics problems like a mathematician, publishing their problems on worksheets) were attempted. Classroom observations revealed that students were highly motivated to pose problems that their classmates would find interesting or difficult, and that their personal interest was sustained during the process of sharing posed problems.

In a study by Verschaffel et al. (2000), problem posing was integrated into a computer-supported learning environment in which upper elementary school children were guided and supported in becoming more strategic, motivated, communicative, mindful, and self-regulated mathematical problem solvers. Various problem posing and solving activities were integrated, such as solving mathematical application problems and putting them on a networked knowledge forum, learning to pose and solve mathematical application problems, and so on. It was found that learning environments in which problem posing played an important role, had a positive effect on the problem-solving competency of the sixth-graders, but not on that of the fifth-graders. It also yielded a positive influence on all pupils' beliefs about, and attitudes toward, collaborative learning in general. In the study of Lavy and Bershadsky (2003), a *What-if-not* strategy was adapted into two learning workshops for preservice teachers on complex solid geometry. The results showed that the preservice teachers strengthened their understanding of geometrical concepts and the connections between the given and new concepts while creating new problems. In a study involving problem-posing and problem-critiquing activities (Bonotto & Baroni, 2008), children were able to create problem situations that were more original, complex, and realistic in their content than traditional word problems after the training.

In many studies, problem posing is not only considered as a vehicle to develop students' problem-solving abilities, but also as one of the central aims of mathematics teaching in itself, and, consequently, is treated as a critical, if not the most important, dependent variable in the evaluation. English (1997a, 1997b, 1998) carried out a 3-year study in which various (related) problem-posing programs were implemented with third-, fifth-, and seventh-grade students who displayed different profiles of achievement in number sense and mathematical problem solving. English (1998) found that third-grade students had difficulties in posing a range of problems in informal contexts (e.g., a picture or a piece of literature) and even more difficulties in formal contexts (e.g., a standard addition and subtraction number sentence). Furthermore, the program was effective in increasing the number of problems generated in general and the number of multi-step problems in particular, but not effective in increasing the diversity of the third-grade students' self-generated problem types. In problem-posing training programs with fifth- and seventh-grade students (English, 1997a, 1997b), it was found that, compared to children in a control group, students who followed the programs displayed an increase in their abilities to generate more diverse and more semantically and computationally complex problems, to identify problem structures, and to model new problems on the structure of a given problem. English also found an increase in the range of problems that students indicated they would like to solve.

Taken as a whole, the intervention studies reviewed above suggest that engaging students in instructional activities related to problem posing has a positive influence on their mathematical understanding (e.g., Lavy & Bershadsky, 2003), word problem-posing abilities (e.g., English, 1997a, 1997b, 1998), and problem-posing motivation (e.g., Winograd, 1997), as well as on their word problem-solving abilities (e.g., Rudnitsky et al., 1995; Verschaffel et al., 2000) and beliefs (e.g., Verschaffel et al., 2000). However, these intervention studies have some limitations. First, the ecological validity of some studies can be questioned because (a) the intervention involved only selected subgroups of children and not intact classes (e.g., English, 1997a, 1997b, 1998); (b) the training program was conducted separately from normal mathematics lessons (e.g., Verschaffel et al., 2000); or (c) the participants were selected only from preservice teachers (Lavy & Bershadsky, 2003). Second, some studies do not allow strong conclusions because of the lack of an appropriate control group (e.g., Verschaffel et al., 2000; Winograd, 1997). Third, some studies only address one type of problem-posing activity, for example, making up mathematics problems from mathematics stories (e.g., Rudnitsky et al., 1995), when in fact many forms of problem-posing activities are available—such as posing problems from a symbolic expression, or from verbal statements (English, 1997a, 1997b, 1998).

The present study tries to overcome the shortcomings described above. First, apart from an initial set of training units that was separated from normal mathematics lessons and was given by the researcher, the training program involved a second series of experimental lessons—taught by the regular classroom teacher—in which problem-posing activities were integrated. Second, we worked with intact classes instead of specifically chosen subgroups of students. Third, rather than doing only one kind of problem-posing activity, various problem-posing situations and activities were used. Finally, we developed and used a systematic assessment battery to examine students' problem-posing and problem-solving capacities, as well as problem-posing and problem-solving beliefs and attitudes. More particularly, as far as problem-posing capacity is concerned, we made use of an assessment tool that evaluated the problems posed along four dimensions: appropriateness, complexity, originality, and diversity.

Description of the Intervention Program

Aims of the Intervention Program

The first aim of the intervention program was that students would acquire problem-posing skills, and positive beliefs about and attitudes toward problem posing. Given the claimed close relationship between problem posing and problem solving, a second aim of the program was to develop students' problem-solving abilities and positive beliefs about and attitudes toward problem solving.

With respect to problem-posing skills, we intended that students would acquire metacognitive strategies for generating problems from a given situation or by reformulating a given problem, which consists of four steps, namely: (a) understanding the problem-posing task presented; (b) identifying the category of the problem-posing task presented; (c) applying appropriate strategies to pose problems; and (d) evaluating the posed problems (for more details, see Figure 15.2). With respect to the development of positive beliefs and attitudes toward problem posing, we intended that students would be more explicitly aware of their erroneous beliefs about and their negative attitudes toward problem posing (e.g., "I will give up immediately if I can't pose a mathematical problem in a given situation" or "I don't like communicating my problem-posing strategies with peers"), and that, after the intervention, they would be more inclined to change them into more positive beliefs and attitudes.

Major Design Principles of the Intervention Program

The intervention program incorporated three design principles—drawn from the above aims—related to the learning tasks, instructional techniques, and socio-mathematical norms (English, 1997a, 1997b, 1998; Rudnitsky et al., 1995; Verschaffel et al., 2000; Winograd, 1997). These three principles, which are depicted in Figure 15.1, were the basic pillars of the training program that, together and in close mutual interaction, guided the activities of and interactions between the teacher and students.

- In a typical lesson, the teacher presented a meaningful and realistic task and asked students to pose mathematics problems starting from that task (principle 1).
- During this problem-posing task, the teacher encouraged students to pose appropriate problems using powerful instructional techniques. For example, if students posed a nonmathematical problem, the instructor would scaffold the students by means of a series of focused questions to help them realize that this problem was not a good word problem although it was a meaningful one (e.g., "Is it a mathematical problem?", "What do we need to have a mathematical problem?", or "What are the givens and the requirements?") (principle 2).

- Meanwhile, the teacher created a classroom climate conducive to the development of students' appropriate dispositions toward mathematical problem posing (principle 3).

Below, we discuss and illustrate these three design principles in greater detail.

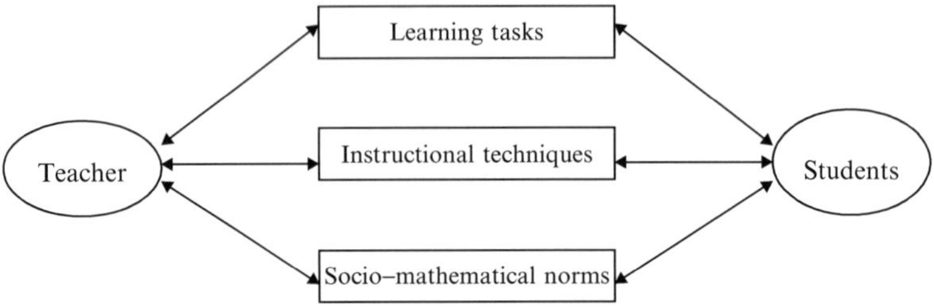

Figure 15.1. Three design principles of the training program.

First, a varied set of meaningful and realistic learning tasks (i.e., problem-posing situations) was used. The problem-posing tasks were presented in various formats including stories, formulae, pictures, tables, and games. Problems were generated in various semantic structures, different problem-posing strategies were applied to pose problems, and attention was paid to the meaningful and realistic nature of the problem-posing situations.

Some examples of problem-posing tasks are:

- Writing appropriate problems for the following symbolic expressions and equations

  $76+28$, $96-24$, 11×3, and $24\div 3$

  $100\div 8=12.5$, $100\div 8=12$, and $100\div 8=13$

- Writing a problem based on the following story, "Teddy Bear Sells Fish." "Teddy bear's mother was ill, so he must earn money to cure his mother's disease by selling fish. One day, a fox, a dog, and a wolf wanted to buy fish from Teddy bear. They asked: 'How much is the fish per kilo? … so he sold the fish to the fox, dog, and wolf. 35 kilos of bellies were sold for 70 yuan, 15 kilos of heads were sold for 15 yuan, and 10 kilos of tails were sold for 10 yuan";
- Solving the following word problem and posing some new problems based on the given problem using the what-if-not strategy, modifying the attributes of the given problem by replacing them with more general or more restricted ones (Brown & Walter, 1993);
- "Calculate the area of a rectangle given that its width is 2 m and its length is 3 m."

Second, a varied set of instructional techniques was used (Collins, Brown, & Newman, 1989; Verschaffel et al., 2000). Most of the experimental lessons/training units followed an instructional model consisting of the following sequence of classroom activities: (a) a short whole-class introduction; (b) posing problems in fixed heterogeneous groups; (c) solving problems generated by other groups; and (d) an individual problem-posing task, followed by a final whole-class discussion. During all of these activities, the instructor's[1] role was to stimulate and scaffold students in the problem-posing and problem-solving activities. We relied heavily on the list of six instructional techniques distinguished in the cognitive apprenticeship model of Collins et al. (1989) to help ensure that the problem-posing instruction would have the features of a *powerful* instructional environment: modeling, coaching, scaffolding, fading, articulation, and reflection. For example, assuming that initially students do not know how to pose problems, modeling was used by the instructor at the outset to show how a problem-posing process unfolds and explains why it happens that way (Collins et al., 1989) to allow the students to follow and see what and how an expert problem poser thinks and to pay special attention to the overall strategy of posing problems. During the process of posing a problem, the students were given an instruction card, as shown in Figure 15.2, with scaffolding instructions that they

Steps of problem posing
1. Understand the problem posing task
2. Identify the category of the problem posing task presented
◇ Category 1: Generating new problems from a problem posing situation
◇ Category 2: Transforming a given problem into new problems
3. Pose new problems by applying appropriate problem posing strategies
◇ Category 1: Generating new problems from a problem posing situation
Think of a question that you would ask yourself if you were actually in that situation
Think about different types of additive or multiplicative word problems that you have learnt
◇ Category 2: Transforming a given problem into new problems
Try reversing knowns and unknowns
Try adding more knowns and/or more constraints
Try applying the "what-if-not" strategy
4. Evaluate the posed problems
◇ Is the problem a solvable math problem?
◇ Is the wording of the problem sufficiently clear?
◇ Is the problem sufficiently interesting?
◇ Is the problem sufficiently original?
◇ Is the problem sufficiently complex?
◇ Is the problem sufficiently realistic?

Figure 15.2. Problem-posing instruction card.

[1] The term "instructor" refers to the researcher who was acting as the teacher in the first series of special problem posing training units and to the regular classroom teacher in the second series of lessons wherein problem-posing activities were integrated into the regular mathematics lessons.

were to follow sequentially. This card was initially used intensively and systematically and was gradually removed as students began to internalize its contents.

The third principle of the intervention program was the establishment of sociomathematical norms concerning mathematical problem posing aimed at creating a classroom climate conducive to the development of students' appropriate dispositions toward mathematical problem posing (English, 1997a, 1997b; Silver, 1997; Verschaffel et al., 2000; Winograd, 1997; Yackel & Cobb, 1996). Norms about problem posing included (a) thinking of problem posing as a genuine and valuable mathematical activity; (b) agreements about what makes a problem (sufficiently) different from another one, why more challenging and/or more realistic problems are better, how problem posing and problem solving are related, etc. and (c) expectations of the role that students and teachers should play in the problem-posing activities. Examples of such norms are: "Just increasing the size of the given numbers is not the best way to increase the complexity of a problem" or "There is not a single best problem for a given problem-posing task."

Content and Organization of the Intervention Program

The training program consisted of eleven 90-minute training units taught by the first author (LC) with one training unit per week, and twenty-four 45-minute lessons taught by the regular classroom teacher of the experimental class wherein problem-posing activities were integrated into the regular mathematics lessons, with two lessons per week. An overview of the 11 training units is presented in Table 15.1.

Table 15.1
Overview of the Intervention Program

Training unit	Topic
1	Exploration of the concept of problem posing, i.e., generating new problems from a problem-posing situation and transforming a given problem into new problems
2	Exploration of the assessment criteria of problem posing (e.g., Is it solvable? Is it clear? Is it interesting? Is it complex?)
3–6	Generating new problems starting from a numerical answer, a symbolic expression, a mathematics story, verbal statements,[a] or a table
7–8	Generating new problems from a mathematical game
9–10	Transforming a given problem into new problems by reversing knowns and unknowns, adding more knowns and/or more constraints, or using a what-if-not strategy
11	Mixed practice on problem posing

[a]Verbal statements refer to one or two verbal descriptions with data information in them, while a mathematics story refers to a more-or-less longer text with some plots and data information.

According to the influential instructional theory of Kaiipob (Ma, 2003), there are five steps in a typical mathematics lesson in a Chinese classroom: (a) introduction; (b) new knowledge introduction; (c) new knowledge exploration; (d) practice and consolidation; and (e) summary. In the experimental program, problem posing was integrated into three of these five instructional steps in the regular lessons taught by the classroom teacher, namely, steps (b), (d), and (e).

Teacher Support

Because the second series of experimental lessons of the training program was not taught by the researcher but by the experimental teacher, the experimental teacher was prepared for and supported in implementing the program. The model of teacher development used was inspired by Verschaffel et al. (2000) and emphasized the creation of a social context wherein the teacher and researcher learn from each other, rather than a model whereby the researcher directly transmits knowledge to the teacher. The teacher support involved three elements: (a) provision of a general teacher guide containing an extensive description of the experimental program; (b) provision of a description of one exemplified lesson showing what each lesson looks like and how it differs (precisely) in terms of the problem posing tasks between the experimental and control class; and (c) the presence of the researcher during one lesson per week, and feedback to the teacher with suggestions for possible improvements.

The experimental teacher's preparation was implemented during the months that preceded the actual intervention and consisted of three meetings—each lasting 1 hour—attended by the teacher and the researcher, wherein (a) the theory of problem posing; (b) different instructional techniques of integrating problem posing into the three instructional steps of the mathematics lessons; (c) an extensive description of the training program; and (d) a description of one exemplified lesson for the experimental and the control group was introduced to the experimental teacher, and wherein a try-out lesson on problem posing (with the researcher being the only audience) was presented and feedback was given. During a fourth meeting, held shortly after the end of the intervention, the researcher obtained some feedback and suggestions from the teacher about the experimental program and the way she had been coached.

Method

Participants

The training program took place in two mixed-gender 4th-grade classes with 69 students (average age = 12.2 years) of a primary school located in the countryside near Shenyang City, China. One of the 2 classes, with 33 students, was designated to be the experimental class, and the other class, with 36 students, acted as the control group. The socioeconomic and educational level of most students' parents was relatively low in both groups. An experimental class teacher with about 12 years of teaching experience and a control class teacher with about 34 years of teaching experience participated in the program. Each of them was in charge of one class, and their duty included teaching mathematics, teaching Chinese, and some daily managerial tasks. Before participating in this study, the students had had some occasional experiences in problem-posing activities since a few problem-posing situations appear in the regular textbooks to meet the goal of problem posing described in the *Compulsory Education Mathematics Curriculum Standards in China* (Ministry of Education of The People's Republic of China, 2012).

Instruments

Before and after the intervention, five instruments—a problem-posing test (PPT), a problem-solving test (PST), a problem-posing questionnaire (PPQ), a problem-solving questionnaire (PSQ), and a standard achievement test (SAT)—were collectively administered in the two participating classes. The first four instruments were administered in two sessions on two successive days, shortly before and after the intervention, and each session lasted for about 1 hour. In the first session, the experimental and control classes were administered the PPT, and in the next session, they were administered the PST, PPQ, and PSQ. The SAT was administered to the students as the final exam in the first and the second term of the academic year in which the experiment was implemented, respectively.

Problem-posing test. Two parallel PPTs were designed, consisting of 12 problem-posing items aimed to assess students' problem-posing abilities. They were administered before and after the intervention. The problem-posing items were selected from different curricular subfields (arithmetic, geometry, and statistics). In each item students were asked to pose two problems. When administering the PPT, one half of the experimental and control classes were administered PPT 1 and the other half of each class was administered PPT 2. Before administering the actual PPT, all students were introduced to the test by means of one example of a problem-posing item.

Problems posed in the PPT were evaluated along four dimensions, i.e., appropriateness,[2] complexity, originality, and diversity. Appropriateness refers to

---

[2] To be considered appropriate, a problem, first, should involve a quantity which is not given in the situation, but which can be computed by means of one or more mathematical operations with the given numbers. Second, the problem should satisfy the requirements of the problem situation (e.g., posing two different word problems was required for each item) or relate to the given problem situation (i.e., using at least one of the knowns, or the goal provided in the situation). Third, the problem should be solvable, i.e., the problem should provide sufficient information to obtain its answer or its goal should be compatible with the given information. Finally, the problem should accord with real-world constraints. (For more details, see Chen, Verschaffel, & Van Dooren, 2011.)

the number of appropriate mathematics problems posed. A posed problem was awarded 1 point if it was scored as appropriate or 0 points if scored as inappropriate. Since two problems were required to be posed in each item, each item was awarded a maximum of 2 points, resulting in a total score for the dimension of appropriateness of 0–24 (2 × 12) points. All the appropriate problems were also scored along the other three dimensions (i.e., complexity, originality, and diversity) with a higher score reflecting a higher level of problem-posing ability.

Complexity refers to the linguistic complexity—whether the word problem involved propositions with an assignment, a relational and/or a conditional structure (Silver & Cai, 1996)—and the semantic complexity—combine, change, compare, and equalize structure for addition and subtraction word problems (Fuson, 1992) and equal group, multiplicative comparison, rectangular pattern, and Cartesian product for multiplication and division word problems (Verschaffel & De Corte, 1996)—of an appropriately posed mathematics problem. More specifically, in line with Silver and Cai (1996), a problem with conditional and/or relational propositions was considered to be more complex than a problem containing only assignment propositions (e.g., a conditional problem "The price for 1 scarf is 20 yuan. If Xiaoming bought 3 scarves and gave the seller 70 yuan, how much was returned?" was considered more complex than an assignment problem "The price for 1 scarf is 20 yuan, and for 1 pair of gloves is 10 yuan. How much is 2 scarves and 1 pair of gloves?"). A problem involving a greater variety of semantic relationships was considered to be more complex than a problem involving fewer semantic relationships (e.g., the posed problem "The price for 1 scarf is 20 yuan, and for 1 pair of gloves is 10 yuan. How much is 2 scarves and 3 pairs of gloves?" was scored as more complex than "The price for 1 scarf is 20 yuan. How much is 2 scarves?").

Originality refers to the uncommon or rare nature of the appropriate mathematics problems being posed. More specifically, a problem belonging to a problem type (defined and operationalized in terms of its linguistic, semantic, and mathematical structure) that occurred with a smaller frequency in our data set was considered to be more original than a problem that occurred with a larger frequency.

For the dimensions of complexity and originality, each self-generated problem was awarded from 1 to 5 points, and so each item (consisting of two problems) was awarded from 2 to 10 points. So the total score for the dimension of complexity and originality ranged from 24 (2 × 12) to 120 (10 × 12) points.

The first three criteria can be applied to each individual self-generated problem, whereas the fourth criterion, diversity, addresses the relationship between the two problems that had to be generated in a given problem-posing item. More specifically, it assesses how much variation there is for the two posed problems in terms of their semantic, linguistic, and mathematical features. For the dimension of diversity, each item (except for one[3]) was awarded from 1 to 5 points, so the total score for the dimension of diversity was from 11 (1 × 11) to 55 (5 × 11) points.

[3] Since its requirements state "Pose one mathematical problem whose solution would require only addition or subtraction, and one mathematical problem whose solution would require at least one multiplication or division," it does not make sense to evaluate the diversity of the posed problems with these specific requirements.

To assess the reliability of the scoring method, ten students were randomly selected and their posed problems in the pre-test and post-test were independently scored by two researchers based on the scoring system described above (complemented with a note with more detailed scoring instructions and examples). Inter-rater agreement for the dimension of appropriateness, complexity, originality, and diversity was 1.00, 0.86, 0.93, and 0.93, respectively. The two researchers then met, jointly examined the posed problems that had yielded different scores, and reached an agreement on the final scores for those problems. Finally, 1 researcher scored all of the problems posed by the remaining 59 students based on the assessment criteria and asked for advice if any uncertainties occurred during this coding process. As another test of the reliability of the scoring system, we also computed the correlation between the control group students' total scores on the two parallel versions of the PPT for each of the four scoring dimensions. This correlation analysis showed that the PPT has a sufficiently high positive and statistically significant parallel forms reliability (Nunnally & Bernstein, 1994) for all four dimensions: appropriateness ($r = .51$, $p = .00$), complexity ($r = .32$, $p < .001$), originality ($r = .31$, $p < .001$), and diversity ($r = .29$, $p < .001$).

Problem-solving test. Two parallel PSTs were designed, consisting of ten problem-solving items aimed at assessing students' problem-solving abilities. They were administered before and after the intervention. They were also selected from three different curricular subfields (arithmetic, geometry, and statistics). In each item, students were required to answer one or two questions. A similar procedure to the PPT was used for the administration of the PST. Each answer was scored either as a correct answer, a wrong answer (i.e., an answer using one or more faulty arithmetic operations), a technical error (i.e., an answer with a purely technical mistake in the execution of the arithmetic operation), or no answer. However, because the intervention especially aimed at the improvement of students' problem-solving abilities (rather than at students' computational proficiency), purely technical errors were ultimately also considered correct. So, items consisting of 2 questions were awarded 2 points if the 2 questions were answered correctly, 1 point if only 1 question was answered correctly, and 0 points when neither of the questions was answered correctly, whereas items consisting of only 1 question were awarded 2 points if that question was answered correctly, and 0 points in all other cases. This resulted in a maximum total score of 20 points for the PST. The PST had relatively high parallel forms reliability (Nunnally & Bernstein, 1994); the correlation between the control group students' total score on the two parallel versions of the PST was $r = .76$, $p < .001$.

Problem-posing and problem-solving questionnaires (PPQ and PSQ). The PPQ and PSQ were designed to assess students' beliefs about and attitudes toward problem posing and solving, and were administered before and after the intervention.[4]

[4]The PPQ and PSQ with different item order were used before and after the intervention.

The PPQ consisted of twenty 5-point Likert-scale items dealing with students' values about, preference for, perseverance in, and confidence in mathematical problem posing (e.g., "I think pupils can learn a lot from posing mathematical problems," "I like to pose mathematical problems similar to those in textbooks," or "I don't have the confidence that I can improve my problem-posing ability by effort"). With respect to each item of the PPQ students had to respond by indicating whether they strongly agreed, agreed, were uncertain, disagreed, or strongly disagreed with the statement. The PSQ had a similar content and design to the PPQ, except that the statements were about problem solving instead of problem posing. A similar procedure to the PPT was used for the administration of the PPQ and PSQ. Each response to the problem-posing/solving questionnaire was awarded 1–5 points with a higher score reflecting a more positive belief about or attitude toward problem posing/solving. For a positively formulated item like "In most cases, I can pose/solve mathematical problems successfully in a given situation," the option "strongly disagree" was awarded 1 point, "disagree" 2 points, "uncertain" 3 points, "agree" 4 points, and "strongly agree" 5 points. In case of a negatively formulated item like "I am not very sure whether I can pose mathematical problems in a given situation," or "I don't like solving mathematical problems," the scores were reversed. This resulted in a total score from 20 to 100 points for the PPQ and for the PSQ. Cronbach's (1951) α for the PPQ and PSQ was 0.81 and 0.87, respectively, which is considered to be a sufficient level of internal consistency (Nunnally & Bernstein, 1994).

Standard achievement test. To assess students' general mathematical knowledge and skills, two SATs developed by the Shenyang Municipal Educational Committee were used to assess students' general mathematical knowledge and skills before and after the intervention. As stated above, the two SATs were administered as the final exams in the first and second terms. The items on the final exam administered in each term related to the various curricular subfields being covered in the program, such as number, addition and subtraction of fractions, solving equations, area of plane or solid figures, word problem solving, probability, and statistics. The two SATs collected from the experimental and control groups were scored by the experimental teacher and control group teacher with each teacher being responsible for her own class. The maximum score for each SAT was 100 points.

Hypotheses and Research Questions

A first hypothesis was that the experimental program would result in a positive effect on students' problem-posing abilities based on the results of some intervention studies (English, 1997a, 1997b, 1998; Winograd, 1997). We predicted in the experimental group—as compared to the control group—that there would be a significantly larger increase from pre-test to post-test of the global score on the PPT in the four dimensions, appropriateness, complexity, originality, and diversity.

A second hypothesis was that the experimental program would result in a positive effect on students' problem-solving abilities because of the close relationship between problem posing and problem solving revealed by some investigations (Cai & Hwang, 2002; Chen et al., 2005, 2007; Ellerton, 1986; Silver & Cai, 1996; Verschaffel et al., 2009). Therefore, a significantly larger increase of the global score on the PST from pre-test to post-test was predicted for the experimental group than for the control group.

Third, based on the results of previous intervention studies (English, 1997a, 1997b, 1998; Verschaffel et al., 2000; Winograd, 1997), we hypothesized that the experimental program would result in a positive effect on students' problem-posing/solving beliefs and attitudes. More specifically, we expected a significantly larger increase of the global score on the problem-posing/solving questionnaire from pre-test to post-test for the experimental group than for the control group.

Finally, for the same reasons as argued by Verschaffel et al. (1999), no prediction was formulated for the results of the SAT after the intervention.

Results

The impact of the training program on the students' results on the five assessment instruments was analyzed by means of independent sample t-tests and an alpha level of .05 for all statistical tests was used. The outcomes of these analyses are presented below.

Table 15.2

Mean Score (and Standard Deviation) on the Problem-Posing Pre-test for the Dimensions of Appropriateness, Complexity, Originality, and Diversity in the Experimental and Control Group

	Experimental group ($n=33$)		Control group ($n=36$)	
	Mean	SD	Mean	SD
Appropriateness	0.83	0.38	0.80	0.40
Complexity	1.99	1.15	1.92	1.12
Originality	1.75	1.17	1.73	1.23
Diversity	3.00	1.38	2.93	1.51

Note: The minimum and maximum mean score on the PPT for the dimension of appropriateness is 0 and 1 point, respectively, and for the other three dimensions (i.e., complexity, originality, and diversity) is 1 and 5 points, respectively

First, the results of the problem-posing pre-test (see Table 15.2) revealed that there was no significant difference between the experimental and control groups in the four dimensions, i.e., appropriateness (*t*-test, two-tailed, $t(1{,}652.29)=1.37$, $p=.70$), complexity (*t*-test, two-tailed, $t(1{,}220)=0.68$, $p=.50$), originality (*t*-test, two-tailed, $t(1{,}220)=0.23$, $p=.82$), and diversity (*t*-test, two-tailed, $t(563.58)=0.51$, $p=.61$), which indicates that the two groups were comparable for the PPT before the intervention.

Furthermore, there was significantly different progress from pre-test to post-test between the experimental and control group in the dimension of originality (*t*-test, one-tailed, $t(1{,}198.77)=1.99$, $p=.02$) in favor of the experimental group, but not in the dimensions of appropriateness (*t*-test, one-tailed, $t(1{,}654)=-0.44$, $p=.33$), complexity (*t*-test, one-tailed, $t(1{,}187.64)=1.13$, $p=.13$), and diversity (*t*-test, one-tailed, $t(479)=1.15$, $p=.13$). The effect size for the dimension of originality was 0.114, which is considered small (Cohen, 1988). So, the first hypothesis was confirmed only for one of the four problem-posing dimensions and only to some extent. The progress from the problem-posing pre-test to post-test in the four dimensions is provided in Table 15.3.

Table 15.3
Progress (and Standard Deviation) from the Problem-Posing Pre-test to Post-test for the Dimensions of Appropriateness, Complexity, Originality, and Diversity in the Experimental and Control Group

	Experimental group ($n=33$)		Control group ($n=36$)	
	Progress	SD	Progress	SD
Appropriateness	0.00	0.42	0.01	0.39
Complexity	0.13	1.45	0.04	1.34
Originality	0.18	1.52	0.01	1.44
Diversity	0.05	1.74	−0.13	1.67

Second, the results of the problem-solving pre-test revealed there was no significant difference between the experimental and control groups (*t*-test, two-tailed, $t(67)=-0.28$, $p=.78$). The mean score for the experimental group was 14.39 (SD=3.86) and 14.64 (SD=3.33) for the control group, which indicates that the two groups were comparable for the PST before the intervention. Results further revealed that there was significantly different progress from the problem-solving pre-test to the post-test between the experimental and control groups in favor of the experimental group (*t*-test, one-tailed, $t(67)=2.46$, $p=.01$). The effect size was 0.57, which is relatively large (Cohen, 1988). The change of the mean score for the

Table 15.4
Mean Score (and Standard Deviation) on the PPQ and PSQ in the Experimental and Control Group at the Pre-test

	Experimental group ($n=33$)		Control group ($n=36$)	
	Mean	SD	Mean	SD
PPQ	72.61	9.63	77.36	10.02
PSQ	73.85	11.92	77.14	9.63

Note: The minimum and maximum mean scores on the PPQ and PSQ are 20 points and 100 points, respectively

experimental and for the control groups was 1.26 (SD = 2.51) and −0.18 (SD = 2.54), respectively. So, the second hypothesis was confirmed.

Third, before the start of the experimental intervention, for both PPQ and PSQ, the mean score for the control group tended to be higher than that for the experimental group, and it was significantly different for the PPQ (t-test, two-tailed, $t(67) = -2.00$, $p = .049$), but not for the PSQ (t-test, two-tailed, $t(67) = -1.27$, $p = .21$) (see Table 15.4).

Table 15.5
Progress (and Standard Deviation) from the Problem Posing Pre-test to Post-test on the PPQ and PSQ in the Experimental and Control Group

	Experimental group ($n = 33$)		Control group ($n = 36$)	
	Progress	SD	Progress	SD
PPQ	2.81	8.55	−5.94	8.69
PSQ	2.18	8.53	−8.78	8.70

After the intervention, the control group declined in its scores quite strongly, whereas the experimental group made limited progress which led to a significant difference for the PPQ from the pre-test to the post-test between the two groups (t-test, one-tailed, $t(67) = 4.21$, $p < .001$) and for the PSQ (t-test, one-tailed, $t(67) = 5.28$, $p < .001$) in favor of the experimental group (see Table 15.5). The effect size was 1.02 and 1.27, respectively, each of which is very large (Cohen, 1988).

Fourth, the results of the two SATs revealed that there was no significant difference either on the mean score of the two SATs between the experimental and control group before the intervention (t-test, two-tailed, $t(67) = -0.35$, $p = .73$), nor on the gain from pre-test to post-test (t-test, one-tailed, $t(67) = -0.33$, $p = .37$).

Discussion

In the present study, a training program aimed at developing Chinese students' problem-posing abilities and indirectly developing their problem-solving abilities and beliefs about and attitudes toward mathematical problem posing and problem solving, given the claimed close relationship between problem posing and problem solving, was designed, implemented, and evaluated. The study focused on the impact of the program on students' problem-posing and problem-solving abilities and beliefs, rather than on the interaction processes between teacher and students and/or between the researcher and the teacher, or on the impact of the involvement in the program on the teachers' professional knowledge and beliefs about mathematical problem posing and problem solving. First, we found that, compared to students from the control group, students who followed the program demonstrated more improvement in their abilities in posing original problems, but not in posing appropriate, complex, or diverse problems after the training. Second, the students in the experimental group also showed better performance on a problem-solving test. Finally, they also improved more in their beliefs about and attitudes toward problem posing and problem solving.

We end this contribution with a reflection on some restrictions of the present study and some theoretical, methodological, and educational issues that need to be addressed in further research. First, the present study sheds some light on the complex relationship between students' problem-posing and problem-solving abilities. More specifically, it confirms the close relationship between students' problem-posing and problem-solving abilities using an intervention study since we found that experiences with problem posing had a positive effect on students' problem-solving abilities. In other words, even if the experimental group students were not explicitly and systematically instructed with any problem-solving strategies, they still made more progress in problem solving than the students from the control group. However, the students from the experimental group were also frequently asked to solve the problems they posed in some training units, and some lessons given by the regular classroom teacher might have had a positive impact on their problem-solving abilities. So, in order to detect the relationship between problem posing and problem solving more accurately, in future research, it might be necessary to involve an experimental group with only problem-posing activities in addition to one experimental group with both problem-posing and problem-solving activities and one control group with only regular lessons.

Second, we developed a self-made PPT together with a problem-posing coding system to assess students' problem-posing abilities. This assessment tool evaluates the problems that the students posed along four dimensions, namely appropriateness, complexity, originality, and diversity. However, some intriguing questions remain, such as: Are these four dimensions sufficient to assess the quintessence of students' problem-posing abilities? And how should the (meta) cognitive processes underlying students' problem-posing performance be assessed? Indeed, the four dimensions described in the assessment tool only focused on evaluating the students' *performance* in the problem-posing tasks, but the assessment tool was unable to assess students' underlying (meta) cognitive *processes*. The four-step problem-posing model (see Figure 15.2) that was developed for and used in the intervention program could be taken as a starting point for developing a more process-oriented measure.

Third, the students who participated in our study were all selected from one particular, relatively small region in China. Moreover, the sample size was small and involved only one experimental and one control class. Both elements evidently jeopardize the external validity of the results. So, follow-up studies should involve a larger sample of classes randomly selected from both the countryside and inner cities from different regions in China.

Fourth, while we made a detailed lesson plan for the first series of lessons taught by the researcher and prepared a detailed teacher guide for the teacher for the second series of lessons, together with an individual preparation and coaching program, we have to acknowledge that a detailed picture of what actually occurred in the experimental class in terms of the realization of the three design principles, is largely lacking. Therefore, future studies need to analyze the specific effects of these various design principles and the relative contribution of more specific instructional features within each principle. This requires the unraveling of the "black box" of the experimental treatment by means of videotaped lessons and/or systematic observations of what happens during these lessons.

Fifth, the training program was evaluated with only a pre-test and a post-test, but without a retention test. Therefore, it is impossible to know whether the observed positive effects of the experimental program on the development of students' problem-posing and problem-solving abilities, beliefs, and attitudes would last after the program had stopped. So, in future research, a repeated measurement design should be used to allow assessment of lasting effect. Moreover, only paper-and-pencil tests and questionnaires were used to assess students' abilities in and beliefs about problem posing and problem solving. Interviews could have allowed us to know more about students' problem-posing and problem-solving abilities and beliefs. So, it might also be interesting in future research to supplement the tests and questionnaires with interviews. In particular, there is a specific challenge in measuring students' problem-posing skills by means of collective tests which is different from measuring problem-solving skills.

In an exploration of individual student profiles, we were surprised to find that some experimental group students posed quite complex problems on the pre-test and easier ones on the post-test. This might be due to the fact that problem-posing tasks are—according to the students' beliefs—a kind of activity with more openness and freedom since it typically elicits multiple possible results and more divergent thinking processes (Çildir & Sezen, 2011). So, given the rather "open" nature of most problem-posing tasks (as compared to typical word problem-solving tasks), students may not always try to pose difficult or original problems and tend to be satisfied with easy and familiar ones. Therefore, we recommend including in future problem-posing tasks and tests, more instructions like "Pose complex problems" or warnings like "Complex problems will get a higher score." It may also be interesting to supplement the paper-and-pencil test with an interview with a carefully selected subgroup sample in order to explore why students who received training in problem posing may not do their very best to come up with the most complex and unfamiliar problems they can think of.

Sixth, some other recent intervention studies on problem posing revealed that a training program on problem posing can improve students' problem-posing abilities, problem-solving abilities, and students' standardized mathematics achievement test performance to a greater extent than found in our study (e.g., Chen & Ye, 2007; Xia, Lü, Wang, & Song, 2007). As noted above, our training program, first, only had a significant positive effect on the originality of the problems posed by the students (but not on the appropriateness, complexity, and diversity of the problems posed), second, the decreased performance of the control group between pre-test and post-test resulted in augmenting the gains for the experimental group, so the positive effect on students' problem-solving ability was more due to the negative effect for the control group than to a significant increased performance in the experimental group, and, third, the program did not have a significant positive effect on the experimental students' standardized mathematics achievement test performance. Therefore, to conclude, we list some factors that may help to explain the rather modest effects of our intervention. First, there is the small sample size of the experimental and control groups. As a result, the regular absence of two students during the intervention (who were kept in the analysis because of the small sample size), might have had a negative effect on the post-test results of the experimental group. Second, as a consequence of the new Chinese mathematics curriculum, most teachers already pay some attention to problem posing. Because of the lack of systematic control over the experimental teacher's actual implementation of the three design principles during the problem-posing moments, the actual instructional difference between the two classes with respect to the intensity and quality of the problem-posing moments may have been less extreme than intended by the researchers. Finally, even though much attention was paid to the selection and construction of the assessment instruments, it is possible that some instruments were unable to detect possible (positive) learning and transfer effects in the students of the experimental class.

References

Bonotto, C., & Baroni, M. (2008). Using maths in a daily context: Experiences in Italian compulsory education. In H. W. Henn & S. Meier (Eds.), *Planting mathematics. First Annual Publication of the Comenius-Network Developing Quality in Mathematics Education II—DQME II* (pp. 19–47). Dortmund, Germany: TU Dortmund.

Brown, S. I., & Walter, M. I. (1990). *The art of problem posing*. Hillsdale, NJ: Lawrence Erlbaum.

Brown, S. I., & Walter, M. I. (1993). Problem posing in mathematics education. In S. I. Brown & M. I. Walter (Eds.), *Problem posing: Reflections and applications* (pp. 16–27). Hillsdale, NJ: Lawrence Erlbaum.

Cai, J., & Hwang, S. (2002). Generalized and generative thinking in U.S. and Chinese students' mathematical problem solving and problem posing. *Journal of Mathematical Behaviour, 21*, 401–421. doi:10.1016/j.bbr.2011.03.031.

Chen, L., Van Dooren, W., Chen, Q., & Verschaffel, L. (2005). The relationship between posing and solving division with remainder problems among Chinese elementary school children. *Mediterranean Journal for Research in Mathematics Education, 4*(2), 85–109.

Chen, L., Van Dooren, W., Chen, Q., & Verschaffel, L. (2007). The relationship between posing and solving arithmetic word problems among Chinese elementary school children. *Journal of the Korea Society of Mathematical Education Series D: Research in Mathematical Education, 11*(1), 1–31.

Chen, L., Van Dooren, W., Chen, Q., & Verschaffel, L. (2011). An investigation on Chinese teachers' realistic problem posing and problem solving ability and beliefs. *International Journal of Science and Mathematics Education, 9*, 919–948. doi:10.1007/s10763-010-9259-7.

Chen, L., Verschaffel, L., & Van Dooren, W. (2011). *The relationship between posing and solving (realistic) mathematical problems in elementary school in China* (Doctoral dissertation).

Center for Instructional Psychology and Technology, Katholieke Universiteit Leuven, Leuven, Belgium.

Chen, Z., & Ye, X. (2007). 数学课堂教学中学生问题提出能力培养的研究 *[The research on developing students' ability to put forward questions in math class]* (Master's thesis). Fujian Normal University, Fuzhou, China.

Çildir, S., & Sezen, N. (2011). Skill levels of prospective physics teachers on problem posing. *Hacettepe Üniversitesi Eğitim Fakültesi Dergisi [Hacettepe University Journal of Education], 40*, 105–116.

Cohen, J. (1988). *Statistical power analysis for the behavioral sciences*. Hillsdale, NJ: Erlbaum.

Collins, A., Brown, J. S., & Newman, S. E. (1989). Cognitive apprenticeship: Teaching the craft of reading, writing and mathematics. In L. B. Resnick (Ed.), *Knowing, learning and instruction: Essays in honor of Robert Glaser* (pp. 453–494). Hillsdale, NJ: Erlbaum.

Cronbach, L. J. (1951). Coefficient alpha and the internal structure of tests. *Psychometrika, 16*(3), 297–334. doi:10.1007/BF02310555.

Ellerton, N. F. (1986). Children's made-up mathematical problems: A new perspective on talented mathematicians. *Educational Studies in Mathematics, 17*, 261–271.

English, L. D. (1997a). The development of fifth-grade children's problem-posing abilities. *Educational Studies in Mathematics, 34*, 183–217. doi:10.1023/A:1002963618035.

English, L. D. (1997b). Development of seven-grade students' problem posing. In E. Pehkonen (Ed.), *21st Conference of the International Group for the Psychology of Mathematics Education* (Vol. 2, pp. 241–248). Lahti, Finland.

English, L. D. (1998). Children's problem posing within formal and informal contexts. *Journal for Research in Mathematics Education, 29*(1), 83–106.

Fuson, K. C. (1992). Research on whole number addition and subtraction. In D. A. Grouws (Ed.), *Handbook of research on mathematics teaching and learning* (pp. 243–275). New York, NY: Macmillan.

Kilpatrick, J. (1987). Problem formulating: Where do good problems come from? In A. H. Schoenfeld (Ed.), *Cognitive science and mathematics education* (pp. 123–147). Hillsdale, NJ: Lawrence Erlbaum.

Lavy, I., & Bershadsky, I. (2003). Problem posing via "What if not?" strategy in solid geometry—A case study. *Journal of Mathematics Behavior, 22*, 369–387. doi:10.1016/j.jmathb.2003.09.007.

Leung, S. S., & Silver, E. A. (1997). The role of task format, mathematics knowledge, and creative thinking on the arithmetic problem posing of prospective elementary school teachers. *Mathematics Education Research Journal, 9*(1), 5–24. doi:10.1007/BF03217299.

Lowrie, T. (2002). Young children posing problems: The influence of teacher intervention on the type of problems children pose. *Mathematics Education Research Journal, 14*(2), 87–98. doi:10.1007/BF03217355.

Ma, Y. (2003). 小学数学教学论 *[Instructional theory on elementary school mathematics]*. Beijing, China: People's Education Press.

Ministry of Education of The People's Republic of China. (2012). 义务教育数学课程标准 [*Compulsory education mathematics curriculum standards*]. 北京, 中国: 北京师范大学出版社 [Beijing, China: Beijing Normal University Publishing House].

National Council of Teachers of Mathematics. (2000). *Principles and standards for school mathematics*. Reston, VA: Author.

Nunnally, J. C., & Bernstein, I. H. (1994). *Psychometric theory* (3rd ed.). New York, NY: McGraw-Hill.

Rudnitsky, A., Etheredge, S., Freeman, S., & Gilbert, T. (1995). Learning to solve addition and subtraction word problems through a structure-plus-writing approach. *Journal for Research in Mathematics Education, 26*, 467–486. doi:10.2307/749433.

Silver, E. A. (1994). On mathematical problem posing. *For the Learning of Mathematics, 14*(1), 19–28.

Silver, E. A. (1997). Fostering creativity through instruction rich in mathematical problem solving and posing. *ZDM: International Reviews on Mathematical Education, 29*(3), 75–80. doi:10.1007/s11858-997-0003-x.

Silver, E. A., & Cai, J. (1996). An analysis of arithmetic problem posing by middle school students. *Journal for Research in Mathematics Education, 27*, 521–539.

Silver, E. A., Mamona-Downs, J., Leung, S. S., & Kenney, P. A. (1996). Posing mathematical problems: An exploratory study. *Journal for Research in Mathematics Education, 27*, 293–309.

Verschaffel, L., & De Corte, E. (1996). Number and arithmetic. In A. J. Bishop, K. Clements, C. Keitel, J. Kilpatrick, & C. Laborde (Eds.), *International handbook of mathematics education. Part I* (pp. 99–138). Dordrecht, The Netherlands: Kluwer.

Verschaffel, L., De Corte, E., Lasure, S., Van Vaerenbergh, G., Bogaerts, H., & Ratinckx, E. (1999). Learning to solve mathematical application problems: A design experiment with fifth graders. *Mathematical Thinking and Learning, 1*, 195–229. doi:10.1207/s15327833mtl0103_2.

Verschaffel, L., De Corte, E., Lowyck, J., Dhert, S., & Vandeput, L. (2000). Supporting mathematical problem solving and posing in upper elementary school children by means of knowledge forum (deliverable of project no. 2017 CL-Net: *Computer supported collaborative learning networks in primary and secondary education*). Leuven, Belgium: Center for Instructional Psychology and Technology, Katholieke Universiteit Leuven.

Verschaffel, L., Van Dooren, W., Chen, L., & Stessens, K. (2009). The relationship between posing and solving division-with-remainder problems among Flemish upper elementary school children. In L. Verschaffel, B. Greer, W. Van Dooren, & S. Mukhopadhyay (Eds.), *Words and worlds: Modelling verbal descriptions of situations* (pp. 143–160). Rotterdam, The Netherlands: Sense Publishers.

Winograd, K. (1997). Ways of sharing student-authored story problems. *Teaching Children Mathematics, 4*(1), 40–47.

Xia, X., Lü, C., Wang, B., & Song, Y. (2007). Experimental research on mathematics teaching of "situated creation and problem-based instruction" in Chinese primary and secondary schools. *Frontiers of Education in China, 2*, 366–377.

Yackel, E., & Cobb, P. (1996). Sociomathematical norms, argumentation, and autonomy in mathematics. *Journal for Research in Mathematics Education, 27*, 458–477.

Yuan, X., & Sriraman, B. (2011). An exploratory study of relationships between students' creativity and mathematical problem posing abilities. In B. Sriraman & K. H. Lee (Eds.), *The elements of creativity and giftedness in mathematics* (pp. 5–28). Rotterdam, The Netherlands: Sense Publishers. doi:10.1007/978-94-6091-439-3_2.

Part III
Mathematics Problem Posing in Teacher Education Programs and Teacher Professional Development

Chapter 16
Middle-Grade Preservice Teachers' Mathematical Problem Solving and Problem Posing

Roslinda Rosli, Mary Margaret Capraro, Dianne Goldsby, Elsa Gonzalez y Gonzalez, Anthony J. Onwuegbuzie, and Robert M. Capraro

Abstract Empirical data were gathered from 51 middle-grade preservice teachers who were randomly assigned into one of two groups. The first group solved a task and then posed new problems based on the given figures, and the second group completed these activities in reverse order. Rubrics were developed to assess the written responses, and then thoughts and concerns related to problem-posing experiences were collected to understand their practices. Results revealed that the preservice teachers were proficient in solving simpler arithmetic tasks but had difficulty generalizing and interpreting numerals in an algebraic form. They were able to pose some basic and reasonable problems and to consider important aspects of mathematical problem solving when generating new tasks. Thus, teacher educators should provide substantial educational experiences by incorporating both problem-solving and problem-posing activities into engaging instruction for preservice teachers.

Contents

R. Rosli (✉)
The National University of Malaysia, Bangi, Malaysia
e-mail: roslinda@ukm.edu.my

M.M. Capraro • D. Goldsby • E.G. y Gonzalez • R.M. Capraro
Texas A&M University, College Station, TX, USA

A.J. Onwuegbuzie
Sam Houston State University, Huntsville, TX, USA

F.M. Singer et al. (eds.), *Mathematical Problem Posing*,
Research in Mathematics Education, DOI 10.1007/978-1-4614-6258-3_16

Introduction

Problem solving has a long history within the mathematics education community. Major interest originally began in the 1940s with George Polya's (1945) work and has been part of classroom instruction since the 1980s (Schoenfeld, 1992). According to Lester (1994), problem solving is a complex task that requires much more than memorized facts and procedures involving an individual's reflective mathematical thinking when resolving problems. In developing individuals' abilities to think mathematically, Polya (1945) articulated four phases of the problem-solving process: (a) understanding the problem; (b) making a plan; (c) carrying out the plan; and (d) looking back. This approach has been well recognized as a central part of mathematics instruction at all levels of education.

Along with the emphasis placed on problem solving, subsequent researchers (Cai, 1998; Crespo, 2003; Ellerton, 1986; Silver, 1994; Silver & Cai, 1996) have claimed that problem posing has become an integral part of mathematics education reform. When viewed from the process of problem solving, it is generally recognized that problem posing is a useful cognitive activity centering on a constructivist perspective. Problem posing has received more focused attention since 1989 when the National Council of Teachers of Mathematics' (NCTM) *Curriculum and Evaluation Standard for School Mathematics* acknowledged problem posing and problem solving in the mathematics classroom (NCTM, 1989). Many scholars believe that problem solving and problem posing are closely interrelated; thus, students can pose problems before, during, or after finding solutions to mathematical problems (Cai, 1998; English, 1997; Silver, 1994).

There is scant research within the teacher education literature emphasizing the link between mathematical problem solving and problem posing (e.g., Chen, Van Dooren, Chen, & Verschaffel, 2010; Silver, Mamona-Downs, Leung, & Kenney, 1996). For example, Silver et al. (1996) investigated 53 middle-school and 28 pre-service secondary-school teachers' problems posed prior to and after solving problems utilizing a similar methodology to a previous study by Silver and Mamona (1989). Results showed these teachers posed many low-quality problems in both phases suggesting a complex association between these activities. To date, research is limited to exploring *how* preservice teachers solve mathematical tasks and reformulate the given tasks sequentially during the same time setting. Thus, exploring the relationship between teachers' mathematical problem-solving and problem-posing activities would likely be fruitful.

The focus of our study was to understand a complex circumstance (Newman, Ridenour, Newman, & Paul DeMarco, 2003) through the exploration and description (Johnson & Christensen, 2012) of preservice teachers' mathematical problem-solving and problem-posing performance. A block pattern task (Appendix A), adapted from Cai and Lester (2005), was used to examine preservice teachers' abilities to solve and pose problems. Specifically, the present study attempted to answer the following research questions:

1. How do select middle-grade preservice teachers solve the multipart block pattern task before or after posing mathematical problems?
2. How do select middle-grade preservice teachers pose mathematical problems before or after solving the multipart block pattern task?
3. What is the relationship between select middle-grade preservice teachers' abilities in solving and posing problems?
4. What are select middle-grade preservice teachers' perceptions and concerns when posing mathematical problems?

We utilized a mixed-methods design that integrated qualitative and quantitative approaches into a seamlessly intertwined single research design (Onwuegbuzie, Johnson, & Collins, 2009).

The Link Between Problem Solving and Problem Posing

The NCTM's *Principles and Standards* (2000) stated, "Problem solving is an integral part of all mathematics learning" (p. 52); thus, it can be considered the central focus of mathematics education. Students learn to be competent problem solvers through solving worthwhile problems and mathematical tasks that can lead to the development of their knowledge. When students are involved in the problem-solving process, they often pose problems based on situations they see. Through problem-solving and problem-posing activities, students draw on their prior knowledge, then discover and inquire about related subject-matter knowledge in the given situations or initial problems (NCTM, 2000).

As problem-solving and problem-posing activities have emerged as important components for learning mathematics (NCTM, 1991, 2000), teachers have a crucial role in helping students develop a repertoire of associations for proficient problem posing and effective problem solving (Moses, Bjork, & Goldenberg, 1990). If teachers are to become autonomous problem solvers and problem posers, they should have substantial educational experiences (Kilpatrick, 1987; Silver et al., 1996). Nevertheless, a body of literature documents that many preservice and in-service teachers lack the skills and confidence to go beyond solving a problem (e.g., Ball, 1990; Cai, 1998; Cai & Hwang, 2002; Crespo & Sinclair, 2008; Gonzales, 1994; Silver & Cai, 1996).

According to Kilpatrick (1987), there is a positive link between the ability to solve and to pose a problem. However, the results from previous studies have been mixed, suggesting a complex relationship between problem-posing and problem-solving success (Cai, 1998; Cai & Hwang, 2002; Chen, Van Dooren, Chen, & Verschaffel, 2007; Chen et al., 2010; Silver & Cai, 1996; Silver & Mamona, 1989). For example, a cross-national study conducted by Cai (1998) revealed a positive relationship between students' performance in mathematical problem posing and problem solving. Cai and Hwang (2002) have suggested a close link between these two components, but the results from a study by Silver and Mamona (1989) demonstrated no direct link. Further research is needed in this area to examine the nature of the relationship between these two activities.

Theoretical Framework

The shift in learning theory toward constructivism has had an enormous impact on the mathematics education community (Hatfield, Edwards, Bitter, & Morrow, 2003). According to von Glasersfeld (1989), students learn by constructing and restructuring their own knowledge. Similarly, Van de Walle, Karp, and Bay-Williams (2009) stated that students often unpack and discover mathematical concepts during the teaching and learning processes for knowledge construction. Therefore, effective teaching requires teachers to be able to facilitate an environment that encourages students to think mathematically and be reflective in order to become good problem solvers and to pose new problems (Schoenfeld, 1992).

A number of scholars have developed frameworks that classified problem posing into a variety of components: situations of problems posed (Stoyanova, 1999), stages of problem-posing activity (Silver, 1994), and processes of problem posing (Christou, Mousoulides, Pittalis, Pitta-Pantazi, & Sriraman, 2005). Our study involved the use of Silver's (1994) problem-posing framework, which refers to the generation of new problems and reformulation of given ones (see Figure 16.1) before, within, and after problem solving. Silver (1994) stated the generation of new problems can occur before the problem-solving process based on a given context or situation in which finding the solution is not the objective of a task. Instead, the focus is to create new mathematical problems. Students are given a set of information or context that can be used in creating new problems (Silver, 1994). Also, problem posing can occur after solving a given problem similar to Polya's fourth phase of problem solving—looking back (Silver, 1994). After solving the problem, students are encouraged to generate new related problems, which can be extended, modified, and varied from the original ones depending on students' mathematical knowledge and creativity (Silver, 1994).

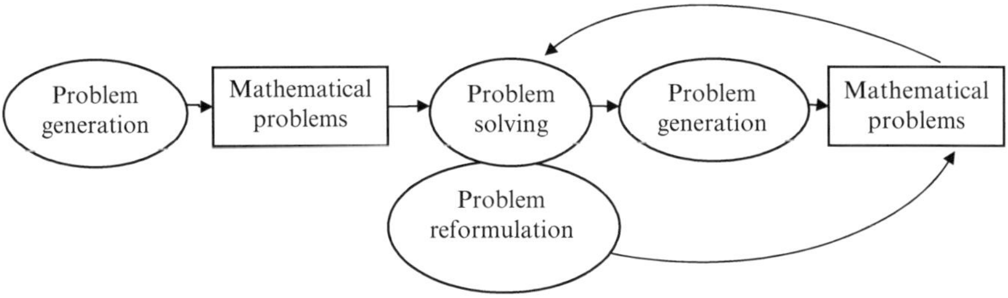

Figure 16.1. Silver's (1994) problem-posing model.

Additionally, problem posing can be categorized as problem reformulation when the given mathematical problem is formulated or transformed into a new version within the process of problem solving (Silver, 1994). For this kind of problem posing, the main goal is first to solve a problem and then to encourage students to think about any related problems in order to make the given one easier to solve (Silver, 1994). Throughout the process of solving and posing problems, students can make connections among mathematical ideas and then construct and restructure their profound knowledge based on prior ones (Moses et al., 1990). In the present study, the focus was problem generation and problem reformulation before and after a problem-solving activity.

Method

Using Teddlie and Tashakkori's (2006) general typology of mixed-methods research, this study represented a conversion mixed research because one type of data was collected and analyzed quantitatively and qualitatively. The mixing of the qualitative and quantitative methods occurred during both the data analysis and data interpretation stages (Nastasi, Hitchcock, & Brown, 2010). Specifically, written responses were collected and transformed using scoring rubrics, analyzed, and then qualitized through narrative discussion.

Participants and Setting

The qualitative data were gathered through a convenience-sampling scheme (Collins, Onwuegbuzie, & Jiao, 2007) of 51 middle-school preservice teachers enrolled in a problem-solving course in the Fall 2011 academic semester. This course was required for the completion of an undergraduate degree in interdisciplinary studies for middle-school certification at a public university in Texas and consisted of 75 minutes of face-to-face meetings for 16 weeks in conjunction with an online module via the university's Blackboard Learning System (eLearning). The professor for the course emphasized Polya's four-step problem-solving process and occasionally integrated problem-posing activities during class instruction. In regard to ethical considerations, permission to conduct the study was granted through the university's Institutional Review Board (IRB) and participants were provided with an information sheet informing participants that the study was part of regular class activities.

Instrument and Procedures

The data were mainly gathered using two multipart problem-solving and problem-posing tasks that were adapted from Cai and Lester (2005). Each task consisted of three parts displayed in Appendix A. The tasks were revised with some minor wording changes and were content validated by three mathematicians and mathematics education experts. On the problem-posing worksheet, we also included two open-ended questions asking about preservice teachers' thinking and concerns when they had to pose their own mathematical problems.

A pilot study was conducted in Spring 2011 and several changes were made to the tasks and procedures. The tasks were administered in week 12 during Fall 2011 in two parts. One half of the participants (the first group) were randomly selected and asked to solve the first part of the block pattern problem-solving task individually. The other half (the second group) was asked to reformulate three new problems (problem-posing task) with a variety of difficulty levels based on the given block pattern situation. Each individual participant was requested to write his or her responses on an answer sheet provided that was collected upon the completion of the task. In the second part of the study, the participants completed the problem-solving or problem-posing task that they had not yet completed. They were given another 15 minutes to complete both tasks and multilink cubes were provided to assist them in solving and posing problems.

Additional data were collected from class presentations of weekly homework assignments. As part of the course assessment, preservice teachers were involved in solving 11 mathematical homework problems individually every week and submitting their solutions online. The weekly assignment was graded using a standardized problem-solving rubric. Then, homework problems were randomly selected and assigned to a group of 4–6 students for a class presentation. Preservice teachers were required to present the strategies used in solving the problem. In addition, they were asked to create a problem similar to the given ones. The examination of class presentations was based on the PowerPoint slides of homework problems that each group solved and posed.

Data Analysis

Initially, we utilized the written responses of the block pattern task to develop the scoring rubrics for problem solving and problem posing, which were validated by the course professor and an experienced researcher in the field. The rubrics were designed to codify the response with guidance from prior published research studies. The content validation process was necessary because if the problems lacked content validity, then little could be said about problem solving/posing. The problem-solving rubric (Appendix B) covered preservice teachers' understanding of the problem, strategies/procedures, and clarity and completeness of presentation. Based on this rubric, their written responses were assessed according to performance indicators 1 (unsatisfactory) through 4 points (extended). Then, the responses were classified into problem-solving strategies: (a) look for a pattern; (b) make a table/chart/organized list; (c) draw a picture/diagram; (d) use direct/logical reasoning; (e) write an equation; and (f) guess and check (Polya, 1945).

The problem-posing rubric (Appendix C) was created to examine problem structure/context, understanding, mathematical expression, and appropriateness of problem-posing design with 1 (unsatisfactory) through 4 (extended) points. In order to minimize researcher bias, 10% of the responses were randomly selected and coded independently by another coder. The raters achieved 73–87% consistency. Disagreements were discussed and resolved resulting in a 100% agreement before the first researcher analyzed the remaining written responses. A sequential mixed analysis was utilized to analyze the data (Onwuegbuzie & Teddlie, 2003); the qualitative data were transformed into numerical form (Tashakkori & Teddlie, 1998). Descriptive statistics pertaining to preservice teachers' mathematical problem-solving and posing-performance was computed (Research Questions 1 and 2) using Statistical Package for Social Science (SPSS) version 17.0 (SPSS 2008). At the final stage, the total points for each task were used to examine the performance differences (Mann–Whitney test) between groups and correlation (Spearman's rho) between preservice teachers' mathematical problem-solving and problem-posing abilities (Research Questions 1, 2, and 3). Then, utilizing constant comparison analysis (Glaser & Strauss, 1967) and classical content analysis (Berelson, 1952), we explored qualitatively, preservice teachers' thinking and concerns while posing their own mathematical problems. The data were coded using QDA Miner 4.0.6 (Provalis Research, 2011). The units of data became the underlying themes, which were identified throughout the coding process. Data transformation was applied in which the qualitative data were analyzed descriptively (Onwuegbuzie & Combs, 2010). Categories/patterns found from the problem-solving and problem-posing rubrics were used to examine the group PowerPoint slides. Findings were presented through narrative discussion to support the results from the quantitative analysis.

Results

Group A (25 preservice teachers) solved the block pattern problem first, and then posed new problems. At the same time, group B (26 preservice teachers) posed the problems based on the given figures and then completed the problem-solving set. We hypothesized that the participants in group A, who solved the block pattern might have had some ideas about the problems they wanted to pose as compared to the ones who initially posed the new problems. We believed the participants in group A had a greater capacity than did those in group B to generate excellent mathematical problems based on their experiences solving the block pattern task. On the other hand, we wanted to determine if the participants in group B performed better on the problem-solving task after posing new problems as compared to participants in group A.

The following section provides evidence that supported the four research questions previously posed. In general, based on the data analysis conducted quantitatively and qualitatively, we observed some interesting findings on the select middle-school preservice teachers' mathematical problem-solving and problem-posing activities. First, we described the abilities of middle-school preservice teachers in solving and posing the block pattern problem according to the rubric performance indicators. Next, we explored the processes used in solving the problems based on the selected solution strategies, which revealed some important insights into their problem-solving abilities. Then, we presented results of the correlation analysis among performance indicators in finding the relationship between preservice teachers' mathematical problem solving and problem posing. Also, through narrative discussion, emerging themes of preservice teachers' perceptions and concerns while posing their own mathematical problems were revealed. Finally, we elaborated the findings from group presentations to support the quantitative results.

Solving the Block Pattern Task

The participants in the study solved three parts of the block pattern task (Appendix A) with 12–36 possible total points. In general, more than 70% of the select preservice teachers were able to demonstrate their understanding of the mathematical concepts used when finding the number of blocks to build a staircase of 6 and 20 steps (part i and ii). They applied appropriate strategies, provided clear explanations, and stated complete reasoning to support their working solutions. In contrast, even though a majority of preservice teachers understood the problem statement, most (53–63%) were not able to write an equation or describe in words the generalized solution for any number of steps (part iii). Only 6 of 51 preservice teachers scored a maximum of 12 points because they fulfilled each element on the problem-solving rubric. The remaining preservice teachers were not able to recognize nor understand the pattern as the number of steps increased to n, which could possibly indicate an insufficient ability to make the transition from arithmetic to algebraic form. They had difficulty seeing and expressing mathematical relationships between the number of n steps and the corresponding number of blocks using an algebraic equation.

When we made comparisons between groups, the right-skewed distribution of scores revealed participants in group B received higher points on all three parts of the problem-solving task, showing that their performance was slightly better with higher mean ranks as compared to group A. However, the Mann–Whitney test

($U = 297.50$, $p = .60$, $r = .07$) showed no statistically significant difference between group A and group B on their problem-solving abilities. We can conclude that their performance was similar whether they solved the block pattern problem before or after posing new problems (Table 16.1).

Table 16.1
Performance on Problem-Solving Task

	Min score	Median	Max score	Mean rank
Group A Solve first ($n = 25$)	15.00	27.00	36.00	24.90
Group B Pose first ($n = 26$)	22.00	27.00	36.00	27.06

Further examination of the problem-solving strategies used revealed that many preservice teachers utilized a variety of approaches in finding the solution for the block pattern problem. A majority used combinations of 2–3 problem-solving strategies to solve the first part of the task (building a staircase of six steps) in order to recognize the pattern. We observed that many preservice teachers continued drawing the figures for five and six steps, made a table/chart/organized list of numerical values, and then looked for a pattern. They successfully identified the number of blocks needed when building the staircase of six steps through this arithmetic sequence. When dealing with a staircase of 20 steps (part ii), many preservice teachers were not able to apply efficient or elegant strategies to solve the problem. They continued listing the numerical values from the sequence previously found to find the number of blocks needed. For the last part (iii), six preservice teachers provided clear and complete explanations and reasoning in generalizing the block pattern. They were able to recognize the relationship of the sequence and wrote an algebraic equation to find the blocks with 20 steps. Meanwhile, others tended to use a direct reasoning strategy when generalizing the solution for any number of steps. For example, one preservice teacher mentioned, "I found out the number of squares for n steps by adding n to the previous number of squares, e.g., for 7 steps, we know 6 has 21 blocks, so $21 + 7 = 28$ so there are 28 blocks in step 7." Many preservice teachers were able to describe the problem-solving process in words based on the pattern they observed from the previous parts (i, ii). Thus, it did not result in an equation or rule showing the mathematical relationship algebraically between the number of steps and the number of blocks.

Posing New Problems

There were some interesting similarities and differences on the problem-posing performance between groups, which might be due to the influence of the intervening experiences with the prior task. All except one participant in group A (solve first) successfully posed three new problems. We analyzed the data by summing the points the participants scored for each posed problem according to performance indicators (1–4 points) on problem structure/context, understanding of problem, mathematical expression, and appropriateness of design. Sixty four percent of the participants in group A scored a total of 8–12 points on problem structure, most of them obtained three or four points for each problem they posed. This group was able to pose new problems by modifying the structure or context of the given figures. We observed that 44–56% of the preservice teachers scored in the low range (1 or 2 points) on understanding of the problem. The majority created problems that had similar mathematical concepts to the ones they solved. Only 2–4 preservice teachers used advanced mathematical concepts when they posed new problems. For example, a preservice teacher reformulated the following:

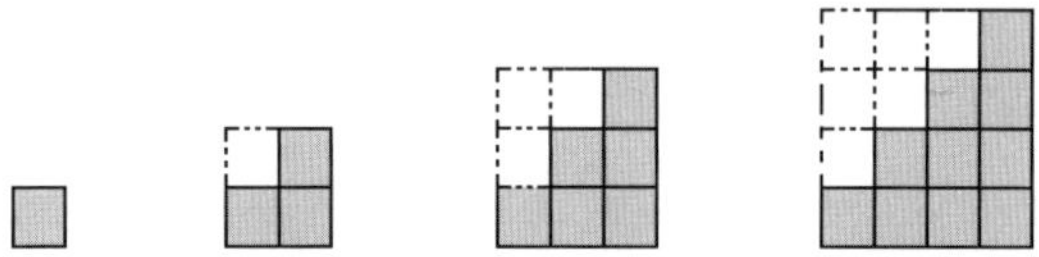

i. If the figures above are all that's left of a square, then how many cubes are missing (to make a complete square) if there are eight steps?
ii. How many would be missing if there were 32 steps?
iii. What is the formula used to find n (where n is the number of cubes missing to complete a square)?

We noticed that this preservice teacher was able to make major structural changes by modifying the original context of the block pattern problem. She posed three new problems related to missing cubes making a complete square that would be mathematically appropriate, feasible, realistic, and engaging for middle-school students, and modeled her new problems similar to the ones in the problem-solving set. Based on our problem-posing rubric, about 62–72% of the preservice teachers scored 3–4 points on the mathematical expression element demonstrating their ability to pose new problems with appropriate language, and effectively use mathematical terms. However, for the criteria appropriateness of design, most of the problems were scored as two points because the posed problems were much less workable, realistic, and engaging. We observed that the preservice teachers were not able to make sense of the problems they had posed possibly because of a time constraint or lack of experience and knowledge in posing their own mathematical problems.

On the other hand, each of the 26 participants in group B (pose first) was able to generate three new problems based on the illustrated block pattern figures. In terms of the problem structure of their first mathematical problems, a majority obtained 1–2 points. We noticed that 81% of the problems they posed seemed trivial and required a minimal level of thinking to solve. The following are samples of the lowest rated problems:

1. In step 1, there is 1 square, 3 in step 2, and 7 in step 3. How many squares are there in step number 4?
2. Using the first four steps in this sequence, draw the fifth figure.
3. If the process continues, how many blocks will be in five steps?

Also, these problems utilized very basic mathematical concepts (85%); fairly clear language with somewhat accurate use of mathematical expression (60%); and less workable, realistic, and engaging tasks (65%). However, as they became more experienced posing problems on their second and third problems, a majority of the participants in group B (50–60%) scored 3–4 points on problem structure, understanding the problem, mathematical expression, and appropriateness of problem-posing design. Five participants scored the maximum (16 points) on their third attempt at creating a problem satisfying the requirements of each element on the rubric. The posed problems created were of a higher level, mathematically appropriate, realistic, feasible, and engaging for middle-school students. They presented a clear and precise statement of the problem and used effective and accurate mathematical expressions. The following are some examples of the highest rated problems from group B:

1. If you were told how many blocks were in 1,000th step, how would you go about finding the number of blocks in the 999th step? What about the 1,001th step? Explain what you would need to do, then solve for the number of blocks in both cases.

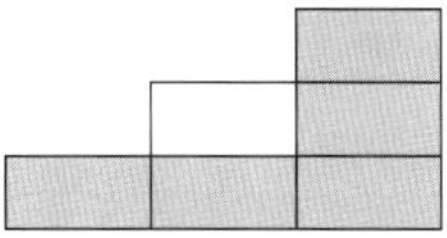

2. The outside of each figure shows above an L shape as shown above with the shaded region. The square not shaded is in the "interior" of the "L." How many "interior" squares will there be for the figure with 20 steps, with 25 steps, and with 100 steps?
3. Which steps from 1 to 100 would have a number of blocks that is a prime?

The descriptive statistics showed that the mean rank performance of group A on the problem-posing task was higher than group B; however, the Mann–Whitney test analysis ($U=295.00$, $p=.57$, $r=.08$) showed no statistically significant difference on problem-posing performance between group A and B. These results suggested that the participants' abilities to pose new problems based on the given figures were similar between these two groups (Table 16.2).

Table 16.2
Performance on Problem-Posing Task

	Min score	Median	Max score	Mean rank
Group A	23.00	32.00	47.00	27.20
Solve first ($n=25$)				
Group B	25.00	31.50	40.00	24.85
Pose first ($n=26$)				

Relationship in Solving and Posing for the Block Pattern

Spearman's rho was computed based on the total scores of preservice teachers' performance on problem solving and problem posing. The correlation coefficients showed the magnitude and direction of the linear relationship between the ability to solve and pose problems for each group. For participants in group A, the results demonstrated a statistically significant association ($r = .44$, $p < .05$) between problem solving and problem posing. The relationship was moderately positive—as the participants' scores on the problem-solving task increased, their scores on problem posing also increased. Meanwhile, there was a positive relationship between problem-posing and problem-solving scores for participants in group B, but it was not statistically significant ($r = .37$, $p = .062$).

Perceptions and Concerns when Posing New Problems

After the generation of new problems, preservice teachers completed two open-ended questions stating their perceptions and concerns with problem posing. Their responses were coded into meaningful unit words, phrases, sentences, or paragraphs (units of data) and were grouped into similar emergent themes (Johnson & Christensen, 2012). For the first open-ended task on preservice teachers' thoughts while posing problems, 14 themes were developed from the coding process. Through a constant comparison analysis, these themes were classified into two emergent meta-themes: characteristics of problems posed (11 themes) and problem-posing experiences (three themes).

Preservice teachers considered the characteristics of the problem as being an important feature while they were posing their own mathematical problems wherein 106 units of data were created and categorized (Figure 16.2). Among these 11 characteristics of problems, a *variety of difficulty level* (18 units of data), *realistic, made sense, and understandable* (15 units of data), and *age appropriateness for middle grades* (15 units of data) were frequently used indicating these might be the most important concepts noted by the participants in both groups while posing their block pattern problems. Participants in group B (pose first) believed posing *real-life situation problems*, *challenging problems*, *similar problems they had seen/worked in the past*, and *problems that fit the required mathematical concepts of arithmetical sequence* were important characteristics as compared to group A (solve first). In contrast, group A greatly emphasized *originality*, *creativity*, and using *the given information of the block pattern figures*. The least used unit of data by the participants was posing an *engaging* problem (group B).

On the other hand, participants also talked about their experiences during the problem-posing activity. Many preservice teachers, particularly in group B (pose first), mentioned that the problem-posing task was difficult for them (19 units of data). This situation possibly happened because the participants were not fully

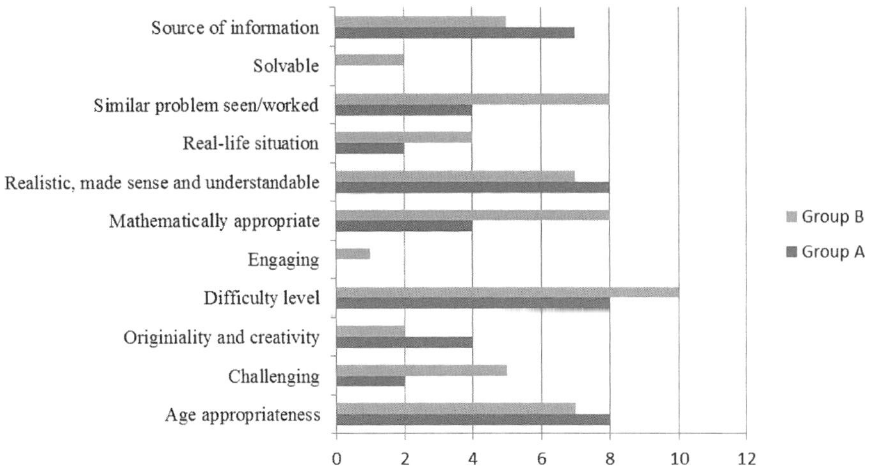

Figure 16.2. Themes under characteristics of problems.

acquainted with problem posing but were required to generate problems for the block pattern task. A participant from group B described her experience:

> At first my brain kind of froze because I am not used to having to create my own problems to be solved, but after the first shock, I began to think of problems I have seen before that deal with a pattern similar to the one given, and I took that similar problem and altered it to fit three different age levels.

Also, another participant said, "I had a difficult time coming up with more than one problem. I felt a little unsure of my questions, but I feel that as time goes on and I get more practice, I will get better at posing problems." Even though the problem posing was difficult, the participants were able to rationalize the activity. They tried to pose problems similar to ones they had seen/worked on the past and believed that more practice was needed to be able to pose better problems. Some expressed adverse feeling about problem posing and tried to avoid it when possible. However, we identified two preservice teachers who believed problem posing was a good opportunity to help them think outside the box and test their mathematical understanding.

We found slightly similar patterns with the preceding section in relation to preservice teachers' concerns after they posed their own mathematical problems. A total of 120 units of data were extracted from the responses mainly focusing on the features of problems posed (117 units of data). The analysis showed that participants in both groups worried whether the problems they posed *made sense and were understandable* (35 units of data) to middle-grade students. An example of this is shown by one preservice teacher who discussed, "I was concerned that the students wouldn't understand the problems. They would have difficulties trying to get what I was trying to say and wouldn't be able to complete the math problem." Additionally, many preservice teachers were not sure the problems they generated were *too easy or too difficult*, or *grade-level appropriate*, *mathematically accurate*, and *solvable*. Group A showed their concerns for making the problems *challenging* and *not repetitive* to the ones they posed and solved. In contrast, participants in group B greatly emphasized other features of the problems such as *difficulty level*, *student's prior knowledge*, and *originality and creativity*.

The results from this section might indicate that select preservice teachers conveyed their thoughts and concerns related to the importance of generating and reformulation of mathematics problems with a variety of characteristics/features. Of course, it could be argued that their lack of experience in problem posing might have resulted in the limited quality of problems posed. However, the emergent themes developed from their written responses showed that they were able to identify important pedagogical aspects generally considered by teachers when they generate mathematical problems for middle-grade students.

Additional Results

Nine groups of 4–6 students worked collaboratively and presented their homework assignment each week throughout the semester. The findings revealed their outstanding performance on using Polya's steps in solving problems: understanding the problem, devising a plan, carrying out the plan, and looking back. They showed deep understandings of mathematical concepts used and were able to apply a variety of appropriate strategies when solving their homework problems. The working solution was clear and supported by accurate mathematical explanation/reasoning.

In regard to creating a similar problem, most of the groups were able to come up with at least one problem. They reformulated new problems that had the same pattern as the given ones by changing the information in the problem. Most of new problems posed were replications, utilized from identical problem contexts, and used basic mathematical concepts. For instance, a group was assigned "Don't Fence Me" and then posed a new problem "Birdie Birdie":

> Farmer Nolan is separating her prized cattle collection. 8 haystacks surround one cow, 10 haystacks surround two cows, and 12 haystacks surround three cows. How many haystacks would be required to surround 100 cows? (given problem)

> Jane, the bird collector, collects all different types of birds. Currently, Jane has 2 bird feeders surrounding 1 bird. She has 5 bird feeders surrounding 2 birds, and 8 bird feeders surrounding 3 birds. How many bird feeders will Jane have surrounding 66 birds? (generated problem)

In addition, there was a group that came up with two new problems with different levels of difficulty that were not connected to any middle-grade level mathematics content. Even though the problems were workable, realistic, and engaging, we noticed that they were taken directly from websites that contained games and mind puzzles. No modifications were made to problem context in order to match the required mathematical concepts.

Discussion and Implications

The results from the present study deepen our understanding as researchers and practitioners concerning the complex circumstances of the select middle-grade preservice teachers' problem solving and problem posing. It is important to note that any generalization of the present study is limited because the results are based on the data we gathered from a small sample of middle-grade preservice teachers from a particular university. Nevertheless, we believe that the outcomes from this investigation provide educational value and a potential contribution of problem solving and problem posing to student learning of mathematics. Further research on this topic is significant and needed.

In this study, preservice teachers were randomly assigned to group A or group B. Preservice teachers in group A solved the task first and then posed new problems based on the given figures. At the same time, group B performed the tasks in reverse order. The results from the descriptive statistics showed group A performed better on the second task (problem posing) whereas group B performed better on the first task (problem solving) as assessed by the rubric. It could be argued that their performances might have been highly influenced by the particular time the tasks were taken. However, factors such as prior knowledge and other related task experiences also might have potentially affected their performances on both tasks. How did preservice teachers' performance on problem solving affect their ability to pose problem or vice versa? Unfortunately, the data that we collected do not allow us to know for certain the extent to which these variables might have had an effect on each task; this is a potential area for future research.

In addition, we noticed that, when the participants' scores on the first task increased, their scores on the second task also increased for both groups. This might indicate that good problem solvers generated more good mathematical problems or vice versa (Silver & Cai, 1996). However, the relationship was statistically significant for preservice teachers who solved first (group A) but not for the ones who posed first (group B). These results suggest a complex relationship between preservice teachers' problem solving and problem posing supported by the results from Chen et al. (2010) and Silver et al. (1996). A further reflection is warranted for exploring and illuminating the close links between problem solving and problem posing (Cai & Hwang, 2002) that was claimed as "an important companion" by Kilpatrick (1987, p. 123) and was evidenced in previous research. These findings have implications for mathematics teacher educators.

Mathematics teacher educators are faced with making tough content-coverage decisions. Those choices are precipitated because there is not enough time in one course to teach all of the essential mathematical concepts a new teacher will need. Therefore, it is necessary to utilize reliable information about scope and sequence. This study indicates that problem solving should precede problem posing. Structuring instruction in this way could make the most effective use of the short time mathematics teacher educators have and sets a foundation on which to build problem-posing skills—a necessary skill for *every* new mathematics teacher.

Given that the emphasis of the methods course was on problem solving, preservice teachers frequently solved mathematical tasks throughout the semester and they became proficient problem solvers by the end of the course. The block pattern task that we adapted in this study is similar to the ones they regularly solved during class. When finding the number of blocks to build a staircase with six steps, they selected appropriate solving strategies such as a diagram, picture, table, and listing of numbers. Even though preservice teachers were able to determine the total blocks for 20 steps, the strategies they used were inefficient. Most of them still could not identify the number pattern as an arithmetic sequence that possibly limited their capability to find the *n-th* formula. They preferred using narrative descriptions for making generalizations for any number of steps rather than translating them into algebraic expressions. The majority of the preservice teachers had difficulty interpreting arithmetic as algebra; thus, they failed to represent arithmetic generalizations with variables (Warren, 2003).

It was interesting to observe that preservice teachers who posed first (group B) had certain capabilities when creating their own mathematical problems without seeing exemplars of similar problems. Even though most of the problems they created were basic and repetitive as compared to the ones on the problem-solving task, they were able to generate some mathematically effective problems for middle-grade students. These findings were similar to Silver et al. (1996). In contrast, preservice teachers who solved the task first (group A) performed better when reformulating new problems—still many made only trivial changes to problem context and used basic mathematical concepts. From an instructional perspective, we strongly believe instructors should emphasize problem-posing activities in class and provide scaffolding for preservice teachers to experience the process of generating and reformulating mathematical problems. When they gain some background knowledge and experience posing their own mathematical problems, they hopefully will not hesitate incorporating them into their repertoire of teaching strategies. Similarly, Silver et al. (1996) suggested that opportunities should be provided for preservice teachers and in-service teachers to engage in problem-posing activities and to analyze their own posed problems in term of their practicability and standards.

The findings obtained from examining the responses of the two open-ended questions, "What did you think about when you had to pose your own mathematical problems?" and "What were your concerns when you posed your own mathematical problems?" showed some similarities and differences in preservice teachers' insights about problem posing. They emphasized significant characteristics of problems when creating new ones and were concerned about the quality of mathematical tasks they posed. While posing new problems, preservice teachers were thinking about diverse difficulty levels that were appropriate for middle-grade students. At the same time, they were trying to create realistic problems that made sense and were understandable for students to solve. However, after they posed the problems, they were concerned whether they achieved their targets in meeting the standards and characteristics that they wanted and needed for middle-school students.

Another important finding was reflected in their attitudes toward problem posing in general. We found that a number of preservice teachers thought that the activity was an eye opener for them, and they deemed it beneficial for their future teaching. There were some preservice teachers who responded adversely to their experiences; however, most of them were able to rationalize the situation, and they tried to pose new problems. They believed more educational practice was needed to help them pose suitable and effective mathematical problems. This was supported by Kilpatrick's (1987) argument about the need and time allocation for individuals to be involved in problem posing during their educational experiences. In contrast, a few preservice teachers stated they would not involve themselves in problem posing because it was too difficult a task to accomplish. We strongly believed their lack of knowledge and experiences with problem posing might have influenced their feelings toward this activity. As prospective teachers, they should discover and create mathematical problems because they will regularly pose mathematical problems in their future classrooms and should not rely solely on commercially created problems in textbooks that may or may not be relevant for their students. By experiencing learning through exploration and reflection, these preservice teachers will be able to construct and restructure their own knowledge profoundly (Van de Walle et al., 2009; von Glasersfeld, 1989).

After examining the outcomes from the group presentations, we found similar results to those described by Silver et al. (1996) suggesting that cooperative work during the problem-posing task might not have been successful. The reformulated problems from group work were either ill structured or identical to the given ones and others were taken directly from websites. The collaborative groups were not able to function and work effectively as we expected in order to produce effective mathematical problems. It becomes a norm that curriculum materials such as textbooks, workbooks, educational websites, and practice problems are sources of mathematical problems and routinely used during class instruction (Kilpatrick, 1987). However, we strongly believe it is necessary for teachers and educators to assume more responsibility in making themselves aware of mathematics education reforms so that traditional teaching practices will soon be changed. With continuous nurturing from teachers and educators, it is hoped that students will become engaged and active in constructing their own learning when they are regularly involved in posing and solving their own mathematical problems.

Appendix A

Problem-Solving Task

Please work *individually* on the task given below.
Look at the figures below.

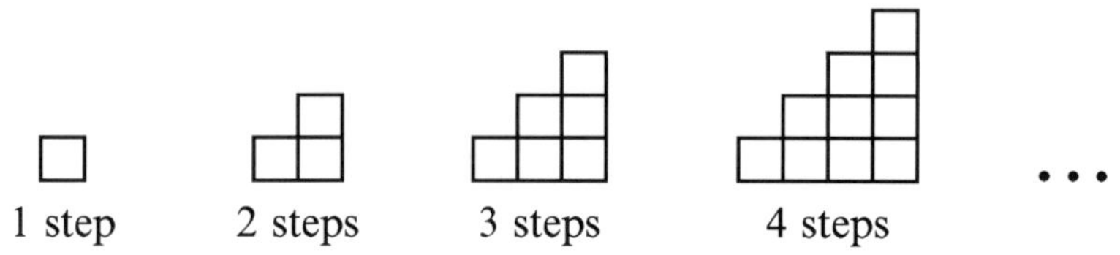

i. How many blocks are needed to build a staircase of five steps? Explain how you found your answer.
ii. How many blocks are needed to build a staircase of 20 steps? Explain how you found your answer.
iii. Based on part i. and ii. write a rule to generalize the solution for any number of steps or describe in words how to find the numbers of blocks used in each step.

Problem-Posing Task

Please work *individually* on the task given below.

i. Look at the figures below.

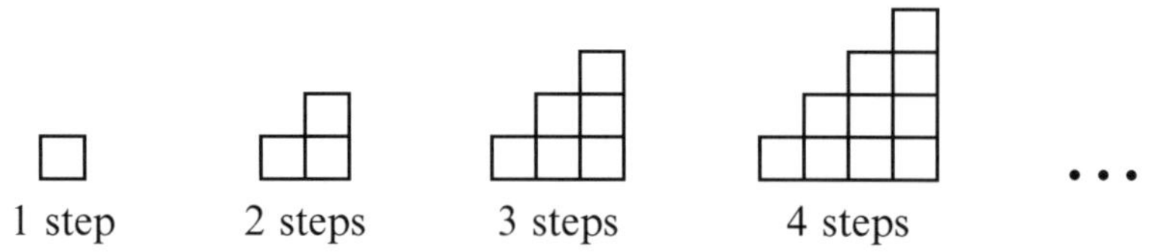

Pose three (3) mathematically appropriate problems for middle-school students in the space provided that are based on the above figures. *Use your creativity and originality when you pose your new problems*. Do not solve them.

ii. What did you think about when you had to pose your own mathematical problems?
iii. What were your concerns when you posed your own mathematical problems?

Appendix B

Problem-Solving Task Rubric

	Performance Indicators			
Elements	Unsatisfactory (one point)	Minimal (two points)	Satisfactory (three points)	Extended (four points)
Understanding of the problem	• Limited understanding of the mathematical concepts used to solve the problems	• Some understanding of the mathematical concepts used to solve the problems	• Substantial understanding of the mathematical concepts used to solve the problems	• Deep understanding of the mathematical concepts used to solve the problems
Strategies/procedures	• Applies incorrect strategies/procedures	• Applies inefficient strategies/procedures	• Applies appropriate strategies/procedures	• Applies efficient and effective strategies/procedures
Clarity and completeness of presentation	• Little/unclear explanation of attempts/solutions	• Presentation/description/diagram is not completely clear/complete	• Clear explanations	• Clear and complete explanations and reasoning

Appendix C

Problem-Posing Task Rubric

	Performance indicators			
Elements	Unsatisfactory (one point)	Minimal (two points)	Satisfactory (three points)	Extended (four points)
Problem structure/ context	Replication/ trivial changes to problem structure/context or trivial problem structure/context (for G2)	Moderate trivial changes to problem structure/context or minimal level of thinking required (for G2)	Some modifications/ extensions to problem structure/context or good problem structure/context (for G2)	Major structural/ contextual changes or higher level problem (for G2)
Understanding of problem	Does not fit the required mathematical concepts	Uses similar/ basic mathematical concepts	Uses other basic mathematical concepts	Uses more advanced mathematical concepts
Mathematical expression	Unclear statement of problem (language)	Fairly clear statement of problem (language)	Moderately clear statement of problem (language)	Clear and precise statement of problem (language)
	Ineffective/ inaccurate use of mathematical expressions	Somewhat effective/ accurate use of mathematical expressions	Moderately effective/ accurate use of mathematical expressions	Effective and accurate use of mathematical expressions
Appropriateness of problem-posing design	No consideration of whether situation is feasible (workable), realistic, and engaging	Somewhat feasible (workable), realistic, and engaging	Moderately feasible (workable), realistic, and engaging	Feasible (workable), realistic, and engaging

Note: G2 = Group B

References

Ball, D. L. (1990). The mathematical understandings that prospective teachers bring to teacher education. *Elementary School Journal, 90*, 449–466. doi:10.1086/461626.

Berelson, B. (1952). *Content analysis in communicative research*. New York, NY: Free Press.

Cai, J. (1998). An investigation of U.S. and Chinese students' mathematical problem posing and problem solving. *Mathematics Education Research Journal, 10*(7), 37–50. doi:10.1007/BF03217121.

Cai, J., & Hwang, S. (2002). Generalized and generative thinking in U.S. and Chinese students' mathematical problem solving and problem posing. *The Journal of Mathematical Behavior, 21*, 401–421. doi:10.1016/S0732-3123(02)00142-6.

Cai, J., & Lester, J. F. K. (2005). Solution representations and pedagogical representations in Chinese and U.S. classrooms. *The Journal of Mathematical Behavior, 24*(3–4), 221–237. doi:10.1016/j.jmathb.2005.09.003.

Chen, L., Van Dooren, W., Chen, Q., & Verschaffel, L. (2007). The relationship between posing and solving arithmetic word problems among Chinese elementary school children. *Journal of the Korea Society of Mathematical Education Series D: Research in Mathematical Education, 11*(1), 1–31.

Chen, L., Van Dooren, W., Chen, Q., & Verschaffel, L. (2010). An investigation on Chinese teachers' realistic problem posing and problem solving abilities and beliefs. *International Journal of Science and Mathematics Education, 9*, 1–30. doi:10.1007/s10763-010-9259-7.

Christou, C., Mousoulides, N., Pittalis, M., Pitta-Pantazi, D., & Sriraman, B. (2005). An empirical taxonomy of problem posing processes. *Zentralblatt für Didaktik der Mathematik, 37*, 149–158. doi:10.1007/s11858-005-0004-6.

Collins, K. M. T., Onwuegbuzie, A. J., & Jiao, Q. G. (2007). A mixed methods investigation of mixed methods sampling designs in social and health science research. *Journal of Mixed Methods Research, 1*, 267–294.

Crespo, S. (2003). Learning to pose mathematical problems: Exploring changes in preservice teachers' practices. *Educational Studies in Mathematics, 52*, 243–270. doi:10.1023/A:1024364304664.

Crespo, S., & Sinclair, N. (2008). What makes problem mathematically interesting? Inviting prospective teachers to pose better problems. *Journal of Mathematics Teacher Education, 11*, 395–415. doi:10.1007/s10857-008-9081-0.

Ellerton, N. F. (1986). Children's made-up mathematics problems-A new perspective on talented mathematicians. *Educational Studies in Mathematics, 17*, 261–271. doi:10.1007/BF00305073.

English, L. D. (1997). Promoting a problem posing classroom. *Teaching Children Mathematics, 4*, 172–179.

Glaser, B. G., & Strauss, A. L. (1967). *The discovery of grounded theory: Strategies for qualitative research*. Chicago, IL: Aldine.

Gonzales, N. A. (1994). Problem posing: A neglected component in mathematics courses for prospective elementary and middle school teachers. *School Science and Mathematics, 94*, 78–84. doi:10.1111/j.1949-8594.1994.tb12295.x.

Hatfield, M. M., Edwards, N. T., Bitter, G. G., & Morrow, J. (2003). *Mathematics methods for elementary and middle school teachers* (4th ed.). New York, NY: Wiley.

Johnson, B., & Christensen, L. (2012). *Educational research: Quantitative, qualitative, and mixed approaches* (4th ed.). Boston, MA: Sage.

Kilpatrick, J. (1987). Problem formulating: Where do good problems come from? In A. H. Schoenfeld (Ed.), *Cognitive science and mathematics education* (pp. 123–147). Hillsdale, NJ: Erlbaum.

Lester, F. (1994). Musing about mathematical problem solving: 1970–1994. *Journal for Research in Mathematics Education, 25*, 660–675.

Moses, B. E., Bjork, E., & Goldenberg, P. E. (1990). Beyond problem solving: Problem posing. In T. J. Cooney & C. R. Hirsch (Eds.), *Teaching and learning mathematics in the 1990's* (pp. 82–91). Reston, VA: National Council of Teachers of Mathematics.

Nastasi, B. K., Hitchcock, J. H., & Brown, L. M. (2010). An inclusive framework for conceptualizing mixed methods design typologies. In A. Tashakkori & C. Teddlie (Eds.), *Handbook of mixed methods in social and behavioral research* (2nd ed., pp. 305–338). Thousand Oaks, CA: Sage.

National Council of Teachers of Mathematics. (1989). *Curriculum and evaluation standards for school mathematics*. Reston, VA: Author.

National Council of Teachers of Mathematics. (2000). *Principles and standards for school mathematics*. Reston, VA: Author.

National Council of Teachers of Mathematics, (1991). *Professional standards for teaching mathematics*. Reston, VA: Author.

Newman, S., Ridenour, C. S., Newman, C., & Paul DeMarco, G. M. (2003). A typology of research purposes and its relationship to mixed methods. In A. Tashakkori & C. Teddlie (Eds.), *Handbook of mixed methods in social and behavioral research* (pp. 167–188). Thousand Oaks, CA: Sage.

Onwuegbuzie, A. J., & Combs, J. P. (2010). Emergent data analysis techniques in mixed methods research. In A. Tashakkori & C. Teddlie (Eds.), *Handbook of mixed methods in social and behavioral research* (2nd ed., pp. 397–430). Thousand Oaks, CA: Sage.

Onwuegbuzie, A. J., Johnson, R. B., & Collins, K. M. T. (2009). Call for mixed analysis: A philosophical framework for combining qualitative and quantitative. *International Journal of Multiple Research Approaches, 3*, 114–139. Retrieved from http://search.proquest.com/publication/76073.

Onwuegbuzie, A. J., & Teddlie, C. (2003). A framework for analyzing data in mixed methods research. In A. Tashakkori & C. Teddlie (Eds.), *Handbook of mixed methods in social and behavioral research* (pp. 351–383). Thousand Oaks, CA: Sage.

Polya, G. (1945). *How to solve it*. Princeton, NJ: Princeton University Press.

Provalis Research. (2011). *QDA Miner 4.0.6. User's guide*. Montreal, QC, Canada: Author.

Schoenfeld, A. H. (1992). Learning to think mathematically: Problem solving, metacognition, and sense making in mathematics. In D. Grouws (Ed.), *Handbook for research on mathematics teaching and learning* (pp. 334–370). New York, NY: Macmillan.

Silver, E. A. (1994). On mathematical problem posing. *For the Learning of Mathematics, 14*(1), 19–28.

Silver, E. A., & Cai, J. (1996). An analysis of arithmetic problem posing by middle school students. *Journal for Research in Mathematics Education, 27*, 521–539. doi:10.2307/749846.

Silver, E. A., & Mamona, J. (1989). Problem posing by middle school mathematics teachers. In C. A. Maher, G. A. Goldin, & R. B. Davis (Eds.), *Proceedings of the eleventh annual meeting of the North American Chapter of the International Group for the Psychology of Mathematics Education* (pp. 263–269). New Brunswick, NJ: Center for Mathematics, Science, and Computer Education, Rutgers-The State University of New Jersey.

Silver, E. A., Mamona-Downs, J., Leung, S. S., & Kenney, P. A. (1996). Posing mathematical problems: An exploratory study. *Journal for Research in Mathematics Education, 27*, 293–309. doi:10.2307/749366.

SPSS. (2008). *SPSS 17.0 for Windows. [Computer software]*. Chicago, IL: Author.

Stoyanova, E. (1999). Extending students' problem solving via problem posing. *Australian Mathematics Teacher, 55*(3), 29–35.

Tashakkori, A., & Teddlie, C. (1998). *Mixed methodology: Combining qualitative and quantitative approaches (Applied Social Research Methods Series, No. 46)*. Thousand Oaks, CA: Sage.

Teddlie, C., & Tashakkori, A. (2006). A general typology of research designs featuring mixed methods. *Research in the Schools, 13*(1), 12–28. Retrieved from http://search.proquest.com/publication/10235.

Van de Walle, J. A., Karp, K. S., & Bay-Williams, J. M. (2009). *Elementary and middle school mathematics: Teaching developmentally* (7th ed.). Boston, MA: Allyn & Bacon/Merrill.

von Glasersfeld, E. (1989). Constructivism in education. In: T. Husen, & T. N. Postlethwaite, (Eds.), *The international encyclopedia of education* (supplementary vol., pp. 162–163). Oxford, England: Pergamon.

Warren, E. (2003). The role of arithmetic structure in the transition from arithmetic to algebra. *Mathematics Education Research Journal, 15*, 122–137.

Chapter 17
Problem-Posing/Problem-Solving Dynamics in the Context of a Teaching-Research and Discovery Method

Vrunda Prabhu and Bronislaw Czarnocha

Abstract Problem posing is practiced in the context of an integrated teaching/research methodology which has become known as TR/NYCity methodology (Teaching-Research/New York City methodology) (*Dydaktyka Matematyki,* 2006, 29: 251–272). This approach has been utilized in mathematics classrooms in the New York area for a decade. Problem solving turned out to be an essential teaching strategy for developmental mathematics classrooms of Arithmetic and Algebra, where motivation in learning, interest in mathematics, and the relevance of the subject is unclear to adult learners. Problem posing and problem solving are brought into play together so that moments of understanding occur, and a pattern of these moments of understanding can lead to self-directed discovery, becoming the natural mode of learning. Facilitation of student moments of understanding as manifestations of their creative capacity emerges from classroom teaching-research practice and its relationship with the theory of the act of creation (*The Act of Creation*. 1964. Macmillan) as the integrative element leading to discovery. Discovery returns to the remedial mathematics classroom, jumpstarting reform. This teaching-research report is based on the collaborative teaching experiment (*C3IRG 7 Problem Solving in Remedial Arithmetic: Jumpstart to Reform*. 2010. City University of New York) supported by C^{3}IRG grant of CUNY.

Contents

V. Prabhu (✉)
Formerly at Bronx Community College, City University of New York, New York, NY, USA
e-mail: vrunda.prabhu@bcc.cuny.edu

B. Czarnocha
Hostos Community College, City University of New York, New York, NY, USA

F.M. Singer et al. (eds.), *Mathematical Problem Posing*,
Research in Mathematics Education, DOI 10.1007/978-1-4614-6258-3_17

Introduction: Posing the General Problem

Enquiry is the path to discovery along which the central problem decomposes into a series of posed questions (see Figure 17.1).

Problem-posing decomposition is the essential link for reaching discovery; its absence derails success by denying access to that discovery. Transformation of the process of enquiry into a series of smaller posed problems generated by the participants allows every student to reach, and to discover, a sought-after solution. Duncker (1945) thought deeply about the psychological processes involved in problem solving, and Silver, Mamona-Downs, Leung, and Kenney (1996) asserted that "problem solving consists of successive reformulations of an initial problem" (p. 294). This view became increasingly common among researchers studying problem solving. Moreover, Brown and Walter (1983), in *The Art of Problem Posing*, posed and answered the question:

> Why, however, would anyone be interested in problem *posing* in the first place? A partial answer is that problem posing can help students to see a standard topic in a sharper light and enable them to acquire a deeper understanding of it as well. It can also encourage the creation of new ideas derived from any given topic—whether a part of the standard curriculum or otherwise (p.169).

The central problem faced by mathematics teachers teaching within an urban community has dimensions that are of both global and local scales. Both ends of the scale can generate the solution of the problem if appropriate questions are posed to reformulate it to the needed precision for the scale at hand. Such a problem is the Achievement Gap. Thus, the central problem addressed in this chapter is how to

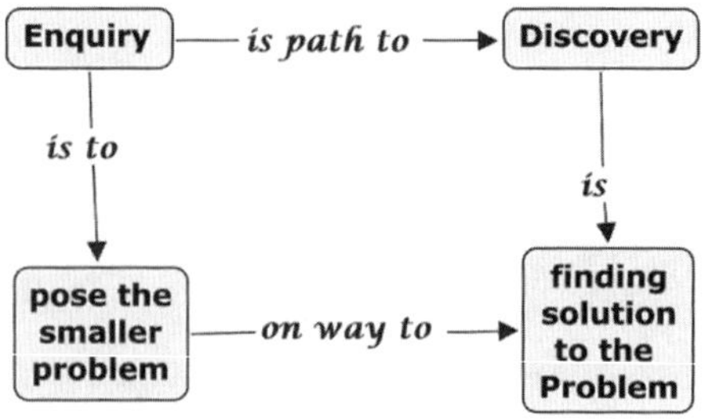

Figure 17.1. Enquiry method of teaching and the decomposition into posed questions/problems.

bridge the Achievement Gap and the role of problem-posing/problem-solving dynamics in this process. Its two scales are, on the one hand, that which drives political machinery: funding initiatives at the National Science Foundation, Department of Education, and other funding agencies, and on the other hand, the situation in a community college mathematics classroom—talent, capacity for deep thinking, yet its clarity disturbed, so grades awarded are not high. The gap for both scales is just a gap; so that the solution to the common posed problem at one end of the scale, of how to fill/bridge/eliminate the gap, can lead to a flow between the local and the national problem, in that the solution at the local scale informs the problem posed at the global national scale. The posed problem has multiple dimensions including:

1. Student voices with the actual classroom difficulties, such as: "what is $-3+5$, why is it not -2," or "why must I take a long answer test, when the final exam is multiple choice," or "why don't you teach, you just make us solve problems";
2. Teachers' voices with the curricular fixes that they think will/has definitely eliminated the gap in their own classroom, of say fractions; and who through that discovery/solved problem, wish to let the secret be available to all students to fix the fraction gap on a broader scale; and
3. Administration obsessed with standardized exams measuring student skills development but not their understanding.

The problems posed by the different constituents are sub-probes to the challenge of closing the Achievement Gap and each of these sub-problems fall into mutually affecting strands. In the classroom, these fall under the categories discussed by Barbatis, Prabhu, and Watson (2012): (a) Cognition; (b) Affect; and (c) Self-Regulated Learning Practices.

In this chapter, we will illustrate our classrooms' problem-posing possibilities. Mathematics is thinking technology through which posing problems, attempting to solve them, and solving them to the extent possible with the available thinking strategies represent the foundational core of the discipline. By repeatedly posing questions to solve the problem in its broad scope, we have discovered that creativity, and in particular, mathematical creativity, can jumpstart remedial reform, thus confirming the assertions of Silver et al. (1996), and Singer, Pelcher, and Voica (2011). Mathematics answers questions—"why?" and "how?" as it uses minimal building blocks on which its edifice is constructed. Thus at any level of the study of mathematics, problem posing and problem solving are inextricable pieces of the endeavor.

TR/NYCity Model is the classroom investigation of students learning conducted simultaneously with teaching by the classroom teacher, whose aim is the improvement of learning in their classroom, and beyond (Czarnocha & Prabhu, 2006). The Teaching-Research, NYCity (TR/NYCity) Model has been used effectively in mathematics classrooms of Bronx Community College and Hostos Community College, the Bronx community colleges of the City University of New York, for more than a decade. The investigation of student learning, as well as related mathematical thinking, necessitates the design of questions and tasks that reveal its nature to the classroom teacher–researcher. That is the original source of problem posing to facilitate student thinking employed by TR/NYCity. This method of teaching naturally connects with the discovery method proposed originally by Dewey and Moore. Utilization of TR/NYCity in conjunction with the discovery method let us, as teacher–researchers, to discover that repeatedly posing questions to students facilitates student creativity, and as such it can jumpstart remedial reform in our classrooms (Czarnocha, Prabhu, Baker, & Dias, 2010). That realization is consistent with the work of Silver et al. (1996), Singer et al. (2011) and others in the field who assert that problem posing is directly related to the facilitation of student creativity.

The *Act of Creation* by Koestler (1964) allows us to extend our understanding of classroom creativity to the methodology of TR/NYCity itself. The *Act of Creation* asserts that bisociation—the moment of creative understanding—is facilitated and can take place only when two or more different frames of discourse or action are present in the activity. Since teaching-research is the integration of two significantly different professional activities, teaching and research, TR/NYCity with its constant probing questions to reveal student thinking presents itself as the natural facilitator of teacher's creativity as well. The TR cycle shown in Figure 17.2 shows the theoretical framework within which problem-posing/problem-solving dynamics as the terrain of student and teacher classroom creativity is being iterated through consecutive semesters. The process of iteration produces new knowledge about learning and problem-posing/problem-solving instructional materials.

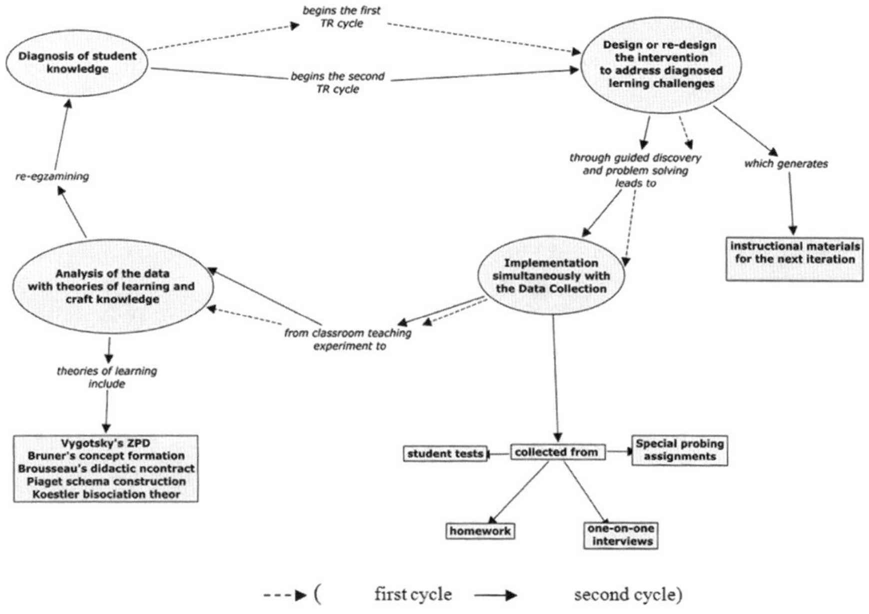

Figure 17.2. Teaching-research cycle with two iterations.

TR cycle iteration is the consecutive run of the investigation or intervention through several subsequent cycles of days, semesters or years. During each semester, student difficulties are cycled over at least twice so that the diagnosed difficulty can be addressed and its success assessed in agreement with the principles of adaptive instruction (Daro, Mosher, & Corcoran, 2011). Over the span of several semesters, the methodology creates an increasing set of materials which are refined over succeeding cycles and acquire characteristics of use to all students studying the mathematical topics under consideration. The learning environment itself develops into a translatable syllabus for the course from several perspectives. Learning environments developed in (TR/NYC) classrooms can be replicated for other instructors facing similar difficulties related to the Achievement Gap in their own classrooms, and for instructors who are interested in becoming teacher–researchers looking for solutions to larger problems in their classrooms.

In classes of Remedial Mathematics (i.e., classes of Arithmetic and Elementary Algebra) at the community college, Teaching-Research Experiments have been carried out since 2006. In the period from 2006 to 2012, success began to be evidenced in 2010 following a broader teaching-research team approach described later in this section.

The initiative in Remedial Mathematics followed the successful use of the methodology in calculus classes under the NSF-ROLE#0126141 award, entitled, Introducing Indivisibles in Calculus Instruction. In the calculus classes (NSF-ROLE#0126141), when the appropriate scaffolding dynamic had been embedded in the Learning Environment, students who were underprepared in, to name the main difficulties, fractions on the line, logic of if-then, algebra of functions and limit (essential for definite integral conception as the limit of the sequence of partial Riemann sums), were nonetheless able to perform at an introductory analysis level (as distinct from the level of standard calculus course). Discovery was the "natural" means of exploration in calculus classes and enquiry leading to discovery through problem-posing/problem-solving dynamics was able to take place without student resistance.

In classes of Remedial Mathematics, however, the situation is markedly different. Student resistance to learning is prompted by years of not succeeding in the subject, and the general attitude is of "just tell me how to do it." Discovery and enquiry are not welcome means. In the period 2006–2010, development of the mathematical materials was continued, and the learning trajectory for fractions described later in this chapter was also investigated. However, the success was not in student learning. In 2007–2008, as part of a CUNY-funded teaching experiment, Investigating Effectiveness of Fraction Grid, Fraction Domino in mathematics classrooms of community colleges of the Bronx, it was found that a satisfactory student partnership in learning, a didactic contract (Brousseau & Balacheff, 1997) or in classroom language, a mutual "handshake" confirming the commitment to student learning, was essential in confirming the role of problem posing on the affect and self-regulatory learning (Akay & Boz, 2010). In 2010, following a Bronx Community College consultancy to Further Education and Training colleges in

South Africa, a new direction to address the problem was established. The situation in classrooms, whether in South Africa or in the Bronx, needed simultaneous attention to student affect as well as to student learning.

Development of Learning Environments

The relationship between cognitive and affective components of learning has recently received increased attention (see, for example, Araujo et al., 2003; Gomez-Chacon, 2000). According to Goldin (2002) "When individuals are doing mathematics, the affective system is not merely auxiliary to cognition—it is central" (p. 60). Furinghetti and Morselli (2004), in the context of the discussion of mathematical proof, asserted that "the cognitive pathway toward the final proof presents stops, dead ends, impasses, steps forward. The causes of these diversions reside only partially in the domain of cognition; they are also in the domain of the affect" (p. 217). There is a need, in addition to attention being paid to possible cognitive pathways, to consider—and find the impact of—affective pathways. DeBellis and Goldin (1997) described affective pathways as "the sequence of (local) states and feelings, possibly quite complex, that interact with cognitive representation" (p. 211).

A learning environment began to develop under iterative loops of the TR cycle, and the components of this learning environment are captured in the concept map below. At that time, the teaching-research team constituted a counselor (also the Vice President for Student Development), a librarian, and the mathematics instructor.

A brief explanation on how to read the concept map shown in Figure 17.3, with its emphasis on the improvement of classroom performance as a function of motivation, self-regulated learning, and cognitive development, is given in Appendix 2.

In the period 2010–2012, during the process of developing the conducive learning environment, three factors emerged as anchoring the learning environment (Barbatis, Prabhu & Watson, 2012). These authors advocated simultaneous attention to:

1. Cognition (materials and classroom discourse well scaffolded, paying attention to the development of the zone of proximal development via meaningful questioning in the classroom and via instructional materials designed in accordance with Bruner's (1978) theoretical position on concept development with concrete, iconic, and symbolic stages).
2. Affect (classroom discourse and independent learning guided by the development of positive attitudes toward mathematics through instances and moments of understanding of enjoyment of problems at hand, extended by self-directed means of keeping up with students' changing attitudes toward mathematics and its learning).

3. Self-Regulated Learning Practices (learning how to learn, usefulness of careful note-taking, daily attention to homework, asking questions, paying attention to metacognition and independent work).

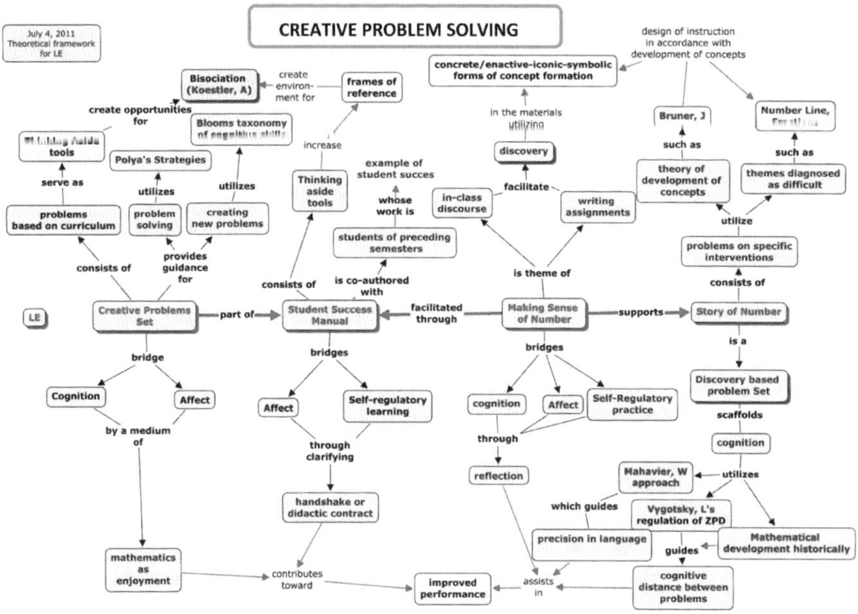

Figure 17.3. The components of a learning environment centered on creative problem solving.

Two simultaneous developments took place during the construction of the learning environment anchored in these three aspects. The craft knowledge of the teaching-research team had a common focus of employment—the development and viewing of the mathematical material on several planes of reference (Koestler, 1964). For example, with a problem such as ½+ 1/3, the counselor of the mathematician–counselor pair would keep the mathematical focus constant while alternating between concrete examples of cookies, pizzas, etc. an approach which exposed students to the process of generalization. This was then extended by the mathematics instructor in removing the monotony of "not remembering" the rules for operations on fractions by using the rules for operations on fractions in more complex problems such as those involving rules of exponents. It was found that the novelty and intrigue of decoding problems that involved exponents made the rules for fractions "easier" to remember or look up. Creativity had emerged as an organic development from the craft knowledge of the instructor. However, it was the support of Arthur Koestler's (1964) *The Act of Creation* that provided a theoretical base in which to anchor thinking and the development of creativity.

Theory of the Act of Creation

Koestler (1964) sketched the theory of the act of creation, or the creative act and coined the term bisociation to indicate the creative act. Bisociation refers to the "flash of insight" resulting from "perceiving reality on several planes at once" and hence, not just associating two familiar frames, but seeing a new one through them, which had not been possible before. This moment of understanding or bisociation is facilitated in the teaching-research classroom through problem posing which can lead to a pattern that changes habit to originality. Mathematics is no longer the "old and boring stuff that needs to be done," but is a source of enjoyment, so that even when the class period ends, students are still interested in continuing to puzzle over problems. Then, when enjoyment translates into performance, the Achievement Gap begins to close, one student at a time.

Koestler's (1964) theory of creativity was based on making connections of the concept in question across three domains or shades of creativity: humor, discovery, and art. Note that our Creative Learning Environment was anchored in Cognition, Affect, and Self-Regulated Learning Practices and assumes overlapping and mutually conducive roles. Humor addresses affect, discovery addresses cognition and learning how to learn when refined so that it is natural, the learner can transform his or her discoveries to deeper levels, or art. A quick glimpse of Koestler's theory is encapsulated in the concept maps shown in Figures 17.4 and 17.5. The habit and originality concept map provides the workings of the transformation involved in the creative process. The Habit + Matrix = Discovery concept map probes more deeply into this transformative process, showing the important role of affect/humor in the creative process. Both become directly usable in the development of the Creative Learning Environment in the classroom.

Mathematics Teaching-Research though the TR cycle clearly lends itself to creating a problem-posing/problem-solving dynamics. How does it do so? In the next section, we provide several classroom instances where problem posing has helped to bring discovery and enquiry "back on track." The concept map shown in Figure 17.5 links creativity with the problem-posing/problem-solving dynamics.

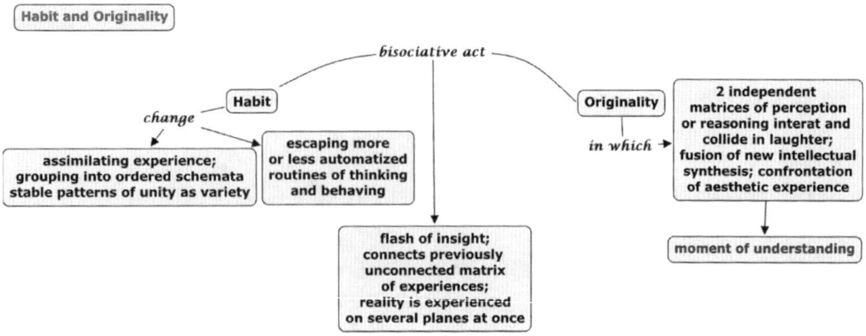

Figure 17.4. The role of the bisociative act in transforming the habit into originality.

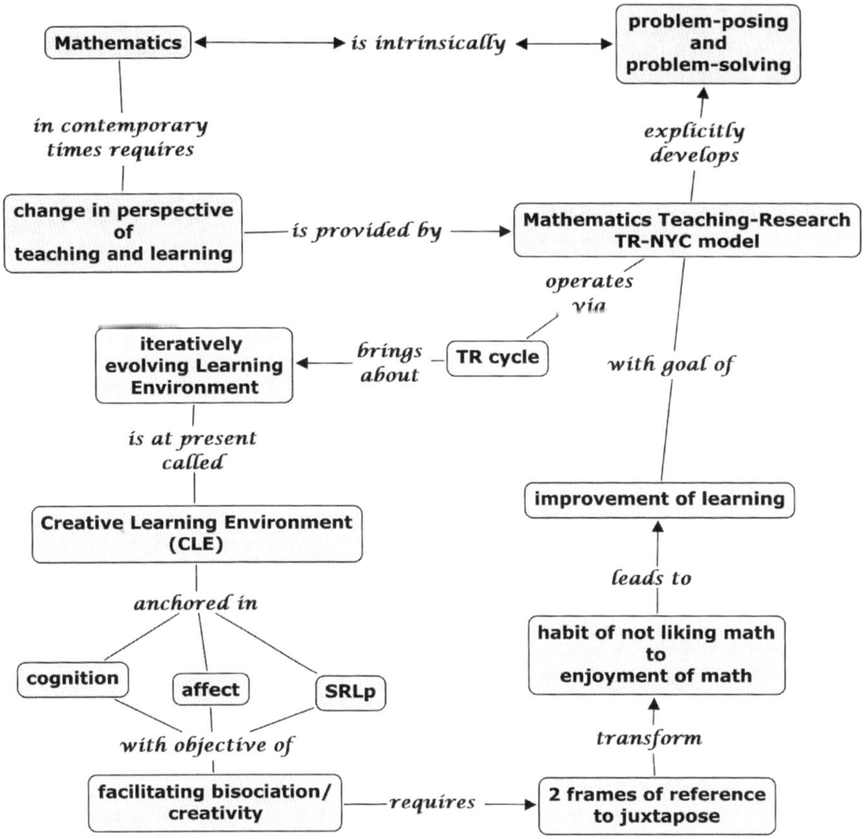

Figure 17.5. The role of mathematical creativity for the improvement of learning.

Problem-Posing/Problem-Solving Dynamics

Problem-Posing Illustration 1

This particular example is from an Elementary Algebra class. The time was just after the first exam, about a month into the semester. Students had had shorter quizzes before. On the day from which this example is taken, almost the entire class staged a rebellion. They stated that the instructor did not teach, that they solved problems, and that since the class is remedial, that means the instructor has to teach. A couple of the students explained what they meant by "teach." One student stated that her previous instructor did a problem on the board and then students did several like it. Another student adamantly declared that she needed "rules" for how to do every problem. After the uproar subsided, the instructor guided them through the test, assuring them that the student in question is doing the problem—thinking aloud and continually pointing out the rules or the significant places to which to pay attention.

Problem 1 *Compute:*

(*a*) $36 + (-20) + 50 - (-17) - 10 =$

(*b*) $2 - (-4 - 10)$

(*c*) $-18 - (-6 + 2)$

(*d*) $2 - (-13) + (-7) - 20$

(*e*) $8 - 5 \times 2 + 9 =$

(*f*) $6 \times 7(-1) - 3 \times 8(-2)$

(*m*) $7(-4)(8) - 9 \times 6(-2)$

(*n*) $15 - 2(-5) - (20 - 4) \div 8$

Each problem was solved/thought out aloud by the student selected by the instructor, and she/he read the problem, and when a symbol was stated, such as parenthesis, the student was asked for the meaning of the symbol (posing a problem). Once the whole problem was read aloud with meaning, the student had to determine the order in which to proceed and why (solving a problem), and then the student actually did the computation in question. It is important to recognize, here, that whether or not a question is a posed problem depends on the state of knowledge of the student. For a student who does not know the meaning of a symbol, the act of asking the question "what does this symbol mean?" is posing a relevant problem. For a student who understands the role of that symbol but has difficulty interpreting this particular case, the question about the symbol is directed toward clarifying that understanding, and hence would not be a posed problem.

At the end of the class, attention was brought back to the work done, how it constituted reading comprehension, paying attention to the structure of the problem and then paying attention to the meaning of individual symbols and thinking of structure and meaning together. There was clarity, satisfaction, and a turnaround in problem solving after this session.

What did this session do in the classroom? First, it debunked the myth that one has to memorize something in order to solve every problem. Second, it took away the authority of the teacher as the knowledgeable one (which the class was reluctant to give up), and finally when each person carefully read and translated/made sense of the problem in terms of symbols and structure, students saw the process of posing and solving working in unison with one of their own classmates carrying out all of the thinking. Hence, for example, when the student who was doing the problem, read "parenthesis," she was questioned about the meaning of "parenthesis," and what role it had to play in the problem (posing problems). The mathematical language with its various hidden symbols, many symbols with one meaning, or one symbol with many meanings are all sources of confusion for students. Situations such as the one narrated here provide for self-reflection, and clarification of the language and of the meaning of the language of mathematics. This approach required many posed questions along the way for clarification. Note, how affect, cognition, and metacognition—all three—enter the dialogic thinking that instructor and students went through together.

Problem-Posing Illustration 2

In this example, the class was Elementary Algebra. Students had trouble determining which rule of exponents was to be applied to the given problem. There was a tendency to use anything arbitrarily without justification. The class problems were followed by a quiz, in which students had much difficulty in determining which rule was applicable for the problem under consideration. Again, it was a matter of not being able to slow down the thinking sufficiently to observe the structure of the problem and the similarity of the structure with one or more rules. Students were asked to work on the following assignment:

Rules of Exponents

1. $a^n \times a^m = a^{n+m}$
2. $\dfrac{a^n}{a^m} = a^{n-m}$
3. $(a^n)^m = a^{nm}$
4. $a^0 = 1$
5. $a^{-n} = 1/a^n$

Make up your own problems using combinations below of the rules of exponents:

- Rules 1 and 2
- Rules 1 and 3
- Rules 1, 2, and 3
- Rules 1 and 4
- Rules 2 and 4
- Rules 1, 2, and 5
- Rules 1 and 5
- Rules 1, 2, 3, 4, and 5

Solve each of the problems you created.

In the work that students submitted, they created problems that had only one term that required the use of say Rule 1 (e.g., x^7y^8) and another term that required the use of Rule 2 (e.g., $\frac{y^5}{y^3}$) but there were no problems that had one term requiring the use of both rules (e.g., $\frac{y^5 \times y^7}{y^{10}}$). This gave the instructor in question a point from which to develop problem solving through deeper problem posing, i.e., through dialogic think-aloud face-to-face sessions, students were asked to observe the structure of the given problem and state the similarity to all rules, where similarity was observed (this led to examples of posed questions which, in turn, led the teacher–researcher to make more complex exercises). This increased students' repertoires in problem solving as evidenced in the quiz and test, described in Problem-Posing Illustration 3.

Problem-Posing Illustration 3

In this illustration, we provide the triptych used in Statistics classes (also used in Arithmetic and Algebra, but not included here), developed through Koestler's work on the development of creativity. A triptych in Koestler's usage is a collection of rows as shown in Figure 17.6, where the columns indicate humor, discovery, and art. In order to get to the discovery of the central concept, the learner can work their way into probing the concept through some word that is known and even funny. Students are provided with the triptych shown in Figure 17.6, with two rows completed. These completed rows were discussed in class as to whether they make sense. Students clarified their understandings in the discussion. It was then expected that students would complete all rows of the triptych and then write a couple of sentences of explanation of the connections between the three words. When all students had submitted their triptychs, the class triptychs were placed on an electronic platform, Blackboard, and students viewed and reflected on each other's work. Students then created a new triptych for the end of the semester and included a few sentences explaining the connections of the concept and its illustration across the row of the triptych.

These classes needed greater scaffolding with the triptych and here the elements of the triptych were introduced "Just-in-Time" as the topic under consideration was being covered in the class. Hence, for example, the triptych Powers ← → decimal representation ← → polynomial was discussed during the session on polynomials. Figure 17.7 shows the general strategy that was used to facilitate discovery and understanding from the teacher–researcher's perspective:

Trailblazer ←→ Outlier ←→ Original/ity
←→ Sampling ←→
←→ Probability ←→
←→ Confidence interval ←→
←→ Law of large numbers ←→
Lurking variable ←→ Correlation ←→ Causation

Figure 17.6. The statistics triptych.

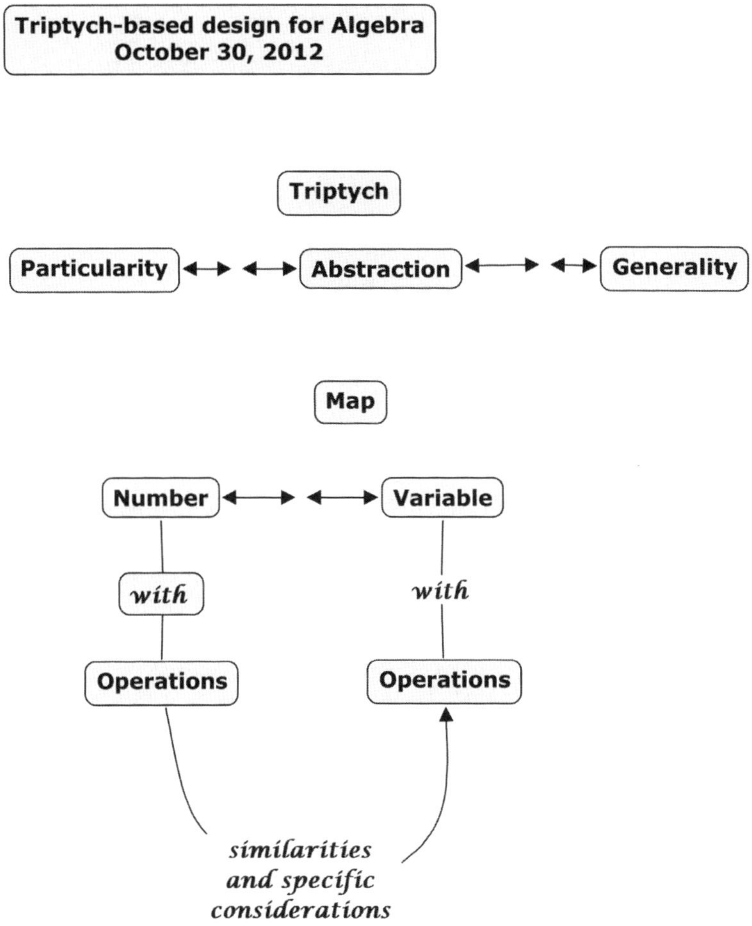

Figure 17.7. Algebra triptychs.

Problem posing was a constant in the discovery-oriented enquiry-based learning environment. Operations on integers and in particular, adding and subtracting with visualizing of the number line, formed the basis for ongoing questioning and posing of problems between students and teacher–researcher.

Algebra as the field of making sense of structure simultaneously with making sense of number provides opportunities for problem posing along the Particularity ←→ Abstraction ←→ Generality of the Arithmetic–Algebra spectrum. In Algebra classes, it was harder to introduce scaffolding, and problem posing occurred solely on the side of the teaching-research team as they explored ways to include triptychs in the Learning Environment mix. In the process, the triptych rows evolved into "simpler" usable forms.

Results and Discussion

The results discussed in this section were obtained after three teaching-research cycles. Consistent with this model for teaching, the results will be incorporated into the next TR cycle based on the described ideas and practice. We have discussed how our cyclical involvement in TR/NYC Model of teaching-research aims to solve the problem of our classrooms—students' understanding and mastery of mathematics led us to pose to ourselves a general question: *What are the necessary components of student success in mathematics*? Our answer to this problem was investigated in the teaching experiment *Jumpstart to Reform* which directed our attention to student creativity as the motivating factor for their advancement in learning. Quantitative analysis of the data is provided in Appendix 1. In turn, our facilitation of student creativity was scaffolded by a series of posed problems/questions designed either by the teacher or students of the classroom. (Doyle et al., in press) described the quantitative results of the teaching experiment *Problem Solving in Remedial Mathematics—Jumpstarting the Reform* supported by C^{3}IRG 7 awarded to the team in 2010. These results confirmed the impact of the approach for the improvement of student problem-solving capacity. These authors pointed out that the art of posing series of problems scaffolding student understanding depends strongly on the teacher's judgment concerning the appropriate amount of cognitive challenge.

Solving these problems in practice leads again to the posing of a general question, which, in agreement with the principles of TR/NYCity leads beyond the confines of our classroom: What is a learning trajectory (LT) of, for example, fractions in my classes? We illustrate a learning trajectory for fractions that developed over the period 2006–2012, with some movement at times, none at others, and a lot more when students are active learners. Problem posing has been an active element in that process within the student–teacher mutual understanding. A four-step approach was taken: (a) The meaning of fractions was established and revisited; (b) How should fractions be visualized? A fractions grid was developed as a visual tool (Czarnocha, 2008); (c) Proportional reasoning: Picture in various versions—seeing the interconnectedness of fractions in different representations: decimal, percent, pie chart; and (d) The meaning of fractions was revisited. Over time, the learning trajectory shown in Figure 17.8 was developed and can be summarized through the following six points:

1. Equivalent fractions visualized—operation: scaling—visualize with FG and then scaling
2. Increasing, decreasing order arrangement—prime factorization—common denominator—fraction grid and then reasoning; common denominators are meaningful before any other standard operations
3. Addition and subtraction
4. Multiplication
5. Division
6. Transition to language—what is half of 16?…

This learning trajectory will be refined through subsequent cycles of the course. Developing the LT for fractions is an illustration of how problem-posing works in

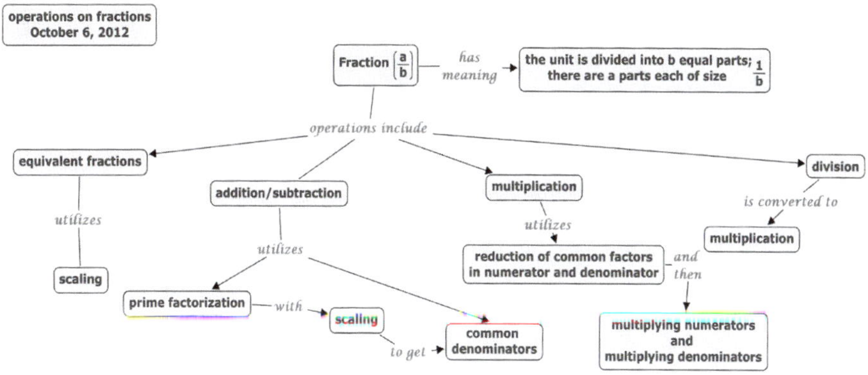

Figure 17.8, Learning trajectory for fractions.

the context of a satisfactory handshake on the part of learners. Problems utilizing exponents is an example of active problem posing leading to its successful integration by learners. The mastery of the language of mathematics through self-directed attention to reading comprehension is an example of how the repertoire needed for problem posing and solving needs to be consistently built up.

The development of several learning trajectories one of which is shown here demonstrate the usability of the methodology and developed materials for a much larger audience of students who fall in the category of self-proclaimed "no good at math," "don't like math," etc. The process of the development of learning trajectories proceeds through the elimination of learning difficulties in the collaborating classrooms.

Repeated problem-posing/problem-solving dynamics increases learners' repertoires for recognizing their own moments of understanding and the emerging patterns of understanding. Writing as the medium utilized for learning to write and writing to learn makes the understanding lasting, concrete, and reusable by learners (Luria & Yudovich, 1968).

The overarching result was that a discovery-based approach to the learning of basic mathematics, coupled with due attention to the cultivation of positive affect, was found to sustain development of learning "how to learn." The learning environment so created was thus a creative learning environment in that it was capable of stimulating creative moments of understanding and extending these to patterns of understanding that could transform learners' habits of doing/learning mathematics to an enquiry-oriented approach that fostered enjoyment and consequently boosted performance. Students' didactic contract/handshake toward their own learning markedly improved once they found mathematics to be enjoyable; their success in tests boosted their confidence; and their desire to achieve. Any fears which students had when the class started, and the accompanying resistance to learning, became nonexistent for the majority of the students. Two students who continued to hold some resistance were in a minority and slowly began to take greater interest. The emphasis on classroom creativity adopted in the teaching experiment outlined a possible pathway across the Achievement Gap.

Conclusion

Mathematics as the creative expression of the human mind is intrinsically questioning/wondering why and how, and through reflection/contemplation, gaining insight through careful justification of the answers to the questions posed. Problem posing and problem solving are thus the core elements of "doing mathematics." In contemporary contexts of teaching and learning of mathematics, this core of mathematics is hidden from sight, and a syllabus, learning objectives, learning outcomes, etc. are more prominent, making mathematics seem like a set of objectives and sometimes even called skills to be mastered by the student who is then considered proficient or competent in those skills. The high failure rate in mathematics starting as early as third grade (funded by MSP-Promyse, 2007), a dislike of mathematics reflected not just among students, but societally, and the low number of students seeking advanced degrees in mathematics are reflective of mathematics not being appreciated for what it is—the quest of the human mind toward knowing, and wanting to know why and how.

In the particular context of community colleges of the Bronx of the City University of New York, and analogously the large percentage of high school students who need remedial/developmental mathematics courses in college, problem posing has to be directly connected and on a regular basis with the classroom curriculum. The objective is urgent: closing the Achievement Gap. The problem as it exists is that an absence of proficiency in mathematics (i.e., scores on placement tests) could well prevent students from college education. The question is how to change this trend?

Knott (2010), in her paper *Problem posing from the foundations of mathematics*, stated:

> Recent developments in mathematics education research have shown that creating active classrooms, posing and solving cognitively challenging problems, promoting reflection, metacognition and facilitating broad ranging discussions, enhances students' understanding of mathematics at all levels. The associated discourse is enabled not only by the teacher's expertise in the content area, but also by what the teacher says, what kind of questions the teacher asks, and what kind of responses and participation the teacher expects and negotiates with the students. Teacher expectations are reflected in the social and socio-mathematical norms established in the classroom (p. 413).

Thus, for classroom environments to be effective, careful integration of simultaneous attention to cognition, affect and self-regulatory learning practices is needed (see also Barbatis et al., 2012). Vygotsky (1978) described the zone of proximal development (ZPD) as "the distance between the actual development level as determined by independent problem solving and the level of potential development as determined through problem solving under adult guidance or collaboration of more capable peers" (p. 86). The ZPD has to be "characterized from both cognitive and affective perspectives. From the cognitive perspective we say that material should not be too difficult or easy. From the affective perspective we say that the learner should avoid the extremes of being bored and being confused and frustrated" (Murray & Arroyo, 2002, p. 370).

Teaching and learning in a teaching-research environment is necessarily collaborative, as our work has demonstrated. This environment has to take account of the numerous difficulties faced in the classroom. The collaboration creates an open community environment in the classroom, which is beneficial to the problem-posing requirement. Mathematics as enquiry, as enjoyment, and as development of a thinking technology does not remain a collection of terms or unfamiliar notions to learners. In the span of one semester, college readiness has to be achieved so that the regular credit-bearing mathematics courses can be completed satisfactorily. Enquiry facilitating discovery becomes the modus operandi, possible now because of the creative learning environment. It is this environment that can provide learners with the keys to success in the learning and understanding of mathematics.

A problem-posing style of education in general whether it follows Freire's (2000) style of "reading the world," or in the style of Montessori (Montessori & Costelloe, 1972), in the design of the learning environment, all find use and applicability in Remedial Mathematics classrooms. Further, the discovery method, or Moore method (Mahavier, 1999) was applied successfully in calculus classes, and now finds a route into stimulating learners to enjoy and perform well in mathematics in remedial classes, thus paving the way toward closing the Achievement Gap and creating readiness for higher level mathematics classes.

Acknowledgement The work was partially supported by the Collaborative Community College Incentive Research Grant (C^3IRG 7) "Problem Solving in Remedial Mathematics: A Jumpstart to Reform." 2010/2011.

References

Akay, H., & Boz, N. (2010). The effect of problem posing oriented analyses-II course on the attitudes toward mathematics and mathematics self-efficacy of elementary prospective mathematics teachers. *Australian Journal of Teacher Education, 35*(1), 57–75.

Araujo, C. R., Andrade, F., Hazin, I., Falcao, J. T., do Nascimento, J. C., & Lessa, M. M. L. (2003). Affective aspects on mathematics conceptualization: From dichotomies to integrated approach. In N. A. Pateman, B. J. Dougherty, & J. T. Zilliox (Eds.), *Proceedings of the 27th International Group for the Psychology of Mathematics Education, held jointly with the 25th PME-NA* (Vol. 2, pp. 269–276). Honolulu, HI: PME.

Barbatis, P., Leyva, A. Prabhu, V., Sidibe, A., & Watson, J. (2012). Engagement in creative thinking. *Mathematics Teaching Research Journal Online*, *5*(2). www.hostos.cuny.edu/mtrj

Barbatis, P., Prabhu, V., & Watson, J. (2012). *A creative learning environment for making sense of number*. Retrieved from http://catalysis.eprints.iitm.ac.in/2293/1/special_issue.pdf

Brousseau, G., & Balacheff, N. (1997). *Theory of didactical situations in mathematics: Didactique des mathématiques, 1970–1990*. Dordrecht, The Netherlands: Kluwer.

Brown, S. I., & Walter, M. I. (1983). *The art of problem posing*. Philadelphia, PA: Franklin Institute Press.

Bruner, J. S. (1978). *Toward a theory of instruction*. Cambridge, MA: Belknap.

Czarnocha, B. (Ed.). (2008). *Handbook of mathematics teaching-research: Teaching-experiment-a tool for teacher-researchers*. Rzeszow, Poland: University of Rzeszow, Poland.

Czarnocha, B., & Prabhu, V. (2006). Teaching Research NYCity Model. *Dydaktyka Matematyki, 29*, 251–272.

Czarnocha, B., Prabhu, V., Baker, W., & Dias, O. (2010). *C3IRG 7 Problem solving in remedial arithmetic: Jumpstart to reform*. New York, NY: City University of New York.
Daro P., Mosher, F., & Corcoran, T. (2011, January). *Learning trajectories in Mathematics—A Foundation for Standards, Curriculum, Assessment and Instruction*. CPRE Research Report RR-#68. Consortium for Policy Research in Education, Teachers College, Columbia University.
DeBellis, V. A., & Goldin, G. A. (1997). The Affective Domain in Mathematical Problem Solving. In: Pekhonen, E. (Ed.) *Proceedings of the PME 21*, 2, 209–216.
Doyle K., Dias, O., Kennis, J., Czarnocha, B., Baker, W., & Prabhu, V. (in press). The rational number sub-constructs as a foundation for problem solving. *Adult Learning Mathematics*.
Duncker, K. (1945). On problem solving. *Psychological Monographs, 58*(5), 1–113.
Freire, P. (2000). *Pedagogy of the oppressed*. New York, NY: Continuum.
Furinghetti, F., & Morselli, F. (2004). Between affect and cognition: Proving at university level. In M. J. Hoines & A. B. Fuglestad (Eds.), *Proceedings of 28th Conference of the International Group for the Psychology of Mathematics Education* (Vol. 4, pp. 369–376). Bergen, Norway: International Group for the Psychology of Mathematics Education.
Goldin, G. A. (2002). Affect, meta-affect, and mathematical belief structures. In G. C. Leder, E. Pehkonen, & G. Törner (Eds.), *Beliefs: A hidden variable in mathematics education?* (pp. 59–72). Dordrecht, The Netherlands: Kluwer Academic.
Gomez-Chacon, I. M. (2000). Affective influences in the knowledge of mathematics. *Educational Studies in Mathematics, 43*, 149–161.
Knott, L. (2010). Problem posing from the foundations of mathematics. *The Montana Mathematics Enthusiast, 7*(2–3), 413–432.
Koestler, A. (1964). *The act of creation*. New York, NY: Macmillan.
Luria, A. R., & Yudovich, F. I. (1968). *Speech and the development of mental processes in the child*. London, United Kingdom: Staples Press.
Mahavier, W. (1999). What is the Moore method? *Primus, 9*, 254–339.
Montessori, M., & Costelloe, M. J. (1972). *The secret of childhood*. New York, NY: Ballantine Books.
Murray, T., & Arroyo, I. (2002). Toward measuring and maintaining the zone of proximal development in adaptive instructional systems. In S. Cerri, G. Guarderes, & F. Paraguacu (Eds.), *Intelligent tutoring systems: Lecture notes in computer science, Proceedings of the 6th International Conference ITS* (Vol. 2363, pp. 749–758). Biarritz, France: Intelligent Tutoring Systems.
Silver, E. A., Mamona-Downs, J., Leung, S. S., & Kenny, P. A. (1996). Posing mathematical problems in a complex task environment: An exploratory study. *Journal of Research in Mathematics Education, 27*(3), 293–309.
Singer, M. F., Pelcher, I., & Voica, C. (2011) Problem posing and modification as a criterion of mathematical creativity in M. Pytlak, T. Rowland, & E. Swoboda, (Eds.), *Proceedings of the Seventh Congress of the European Society for Research in Mathematics Education*, CERME 7, 9–13 February, Rzeszow, Poland: CERME.
Vygotsky, L. S. (1978). *Mind in society: The development of higher psychological processes*. Cambridge, MA: Harvard University Press.

Chapter 18
Problem Posing for and Through Investigations in a Dynamic Geometry Environment

Roza Leikin

Abstract This chapter analyzes three different types of problem posing associated with geometry investigations in school mathematics, namely (a) problem posing through proving; (b) problem posing for investigation; and (c) problem posing through investigation. Mathematical investigations and problem posing which are central for activities of professional mathematicians, when integrated in school mathematics, allow teachers and students to experience meaningful mathematical activities, including the discovery of new mathematical facts when posing mathematical problems. A dynamic geometry environment (DGE) plays a special role in mathematical problem posing. I describe different types of problem posing associated with geometry investigations by using examples from a course with prospective mathematics teachers. Starting from one simple problem I invite the readers to track one particular mathematical activity in which participants arrive at least at 25 new problems through investigation in a DGE and through proving.

Contents

R. Leikin (✉)
Faculty of Education, University of Haifa, Haifa, Israel
e-mail: rozal@edu.haifa.ac.il

F.M. Singer et al. (eds.), *Mathematical Problem Posing*,
Research in Mathematics Education, DOI 10.1007/978-1-4614-6258-3_18

Background

Mathematical Inquiry in the Mathematics Classroom

Inquiry and investigations are basic characteristics of the development of mathematics, science, and technology. According to Wells (1999), inquiry is a way of teaching and learning which integrates wonderment and puzzlement and arouses interest and motivation in learners. Investigation activities are associated with seeking knowledge, information, or truth through questioning.

Mathematical investigations are central to the activity of any research mathematician. In the past two decades mathematical investigations have become an integral part of mathematics teaching and learning in school (Da Ponte, 2007; Leikin, 2004, 2012; Silver, 1994; Yerushalmy, Chazan, & Gordon, 1990). Investigation tasks in mathematics classrooms are usually challenging, cognitively demanding, and enable highly motivated work by students (e.g., Yerushalmy et al., 1990). Borba and Villarreal (2005) stressed that "the experimental approach gains more power with the use of technology" (p. 75) by providing learners with the opportunity to propose and test conjectures using multiple examples, obtain quick feedback, use multiple representations, and become involved in the modeling process.

Both problem-posing and investigation problems in a broad range of types of mathematical tasks are called "open problems" (Pehkonen, 1995). This chapter focuses on problem posing associated with investigations in geometry. Yerushalmy et al. (1990) suggested to consider investigations in geometry as activities that include experimenting to arrive at a conjecture, conjecturing, testing the conjecture, and proving or refuting it. The conjectures raised by the students and teachers become new proof problems.

Investigations in geometry are naturally associated with the use of dynamic geometry environments (DGEs) (Mariotti, 2002; Schwartz, Yerushalmy, & Wilson, 1993; Yerushalmy et al., 1990). Numerous studies have explored the role of DGEs in the instructional process, specifically in concept acquisition, geometric constructions, proofs, and measurements (e.g., Chazan & Yerushalmy, 1998; Hölzl, 1996; Jones, 2000; Mariotti, 2002; Yerushalmy & Chazan, 1993). In this chapter, these problems will be referred to as *problems posed through investigation*.

Teachers Devolve Mathematical Investigations to the Classroom

Teachers' roles in integration of investigation tasks in teaching and learning cannot be overestimated. Teachers' knowledge, skills, and beliefs determine whether and how they implement mathematical investigations in their classes. To make systematic use of mathematical investigations in school, several potential pitfalls have to be overcome.

First, the majority of teachers of mathematics in school nowadays do not have personal experience in learning mathematics through mathematical investigations, while many teachers have limited experience in the use of dynamic software for mathematical investigations. Geometry investigations using DGEs require teachers to rethink teaching: they have to deal with unfamiliar or even new mathematical practices, and "take a more prominent role in designing learning activities for their students" (Healy & Lagrange, 2010, p. 288). When they are challenged by new (for them) teaching approaches, the teachers are often unenthusiastic and reluctant to adopt these practices and express preferences for the teaching methods used by their own teachers before them (e.g., Lampert & Ball, 1998; Leikin, 2008).

Second, implementation of investigation problems requires devolving investigation problems to the class (e.g., Da Ponte, 2007; Yerushalmy et al., 1990). Yet, often teachers cannot even find investigation problems in regular instructional materials. Thus, integration of mathematical investigations in the classroom means that teachers have to create investigation problems for their students.

Third, usually investigations in geometry are supported by DGEs that frequently lead to technological difficulties with the environment, or with classroom equipment, as well as other issues (Healy & Lagrange, 2010). Additionally, navigation of a lesson that engages students in investigation activities requires the teacher to possess diverse didactical skills, technological knowledge, and profound mathematical knowledge since these activities lead to unpredicted mathematical conjectures that sometimes require complex proving.

Da Ponte and Henriques (2013) and Ellerton (2013) stressed the importance of the integration of problem posing and investigation activities in teacher education programs. They demonstrated the effectiveness of these activities in the development of teachers' conceptions about the importance of problem-posing and investigation activities in school mathematics and the development of teachers' knowledge. When teachers themselves are involved in investigation activities, their thinking processes are stimulated so that they experience mathematical processes themselves (Da Ponte & Henriques, 2013). Teachers have to be educated for the generation of investigation tasks, for the classroom use of mathematical investigations, and for fluent management of mathematical lessons.

This chapter describes the integration of these activities in a geometry course for prospective secondary school mathematics teachers.

The Context

This chapter presents reflective insight from a long-term study conducted using design research methodology. As a design experiment it was a formative research study to examine and refine educational design (Collins, Joseph, & Bielaczyc, 2004). The setting was directed towards promoting learning, producing useful knowledge as well as modeling learning and teaching advancement (Cobb, Confrey, diSessa, Lehrer, & Schauble, 2003). That is, the present study had both a pragmatic and a theoretical orientation. The design experiment was performed in the context of a geometry course (within the teacher certificate program) aimed at the advancement of problem-solving and problem-posing expertise, by employing Multiple Proof Tasks (Leikin, 2008) and Mathematical Investigations (Leikin, accepted). The data in this chapter were collected from 22 prospective mathematics teachers aged 22–40, all of whom had a B.Sc. degree in mathematics prior to their participation in the program.

In this context an *Investigation Task* was defined as a complex task that includes:

1. Solving a proof problem in several ways;
2. Transforming the proof problem into an investigation problem;
3. Investigating the geometry object (from the proof problem) in a DGE for additional properties (experimenting and conjecturing); and
4. Proving or refuting conjectures.

The collected data included students' written work and protocols of group discussions and group interviews. In this chapter, similar to the exploration of investigation activities in calculus performed by Da Ponte and Henriques (2013), I provide theoretical analysis of problem-posing types associated with geometry investigations.

Investigation tasks of this type lead to three types of problem posing:

1. Problem posing through proving;
2. Problem posing for investigation; and
3. Problem posing through investigation, including problem posing through construction.

Types of Problem Posing Associated with Geometry Investigations: Definitions

Problem Posing Through Proving

Proving is an integral part of investigation activities in geometry. Through proving, one can also realize new and unforeseen properties of a given object that are proven at one of the proof stages. Then, proving each of such properties encompasses a new geometry problem. Problem posing through proving, if taken as a problem-posing strategy, is similar to the "chaining" strategy described by Hoehn (1993).

De Villiers (2012) analyzed the "looking back" discovery function of proof using specific advanced geometry examples. He noticed that it is "possible to design learning activities for younger students in the junior secondary school that allow acquainting students with the idea that a deductive argument can provide additional insight, and some form of novel discovery" (p. 1133). I provide such an example later in this chapter.

Problem Posing for Investigation

Several studies consider problem transformation (also called reformulation) as an instance of problem-posing activity (Stoyanova, 1998, with reference to Duncker, 1945; Leikin & Grossman, 2013; Mamona-Downs, 1993; Silver, 1994). Transformation of a proof problem into an investigation problem is considered herein as *problem posing for investigation.*

Problem posing related to problem transformation is explored by researchers focusing on systematic transformations of a given problem involving variations in goals and givens. The "what if not?" scheme is the most well-known problem-posing strategy (Brown & Walter, 1993, 2005). The "what if not?" strategy, which is based on changes in givens, leads to making room for conjecturing and producing new insights about problem outcomes. Leikin and Grossman (2013) pointed out an additional type of problem posing which they called the "what if yes?" strategy, which is based on the addition of properties to the given object (e.g., considering a special case of a square for a given parallelogram).

Leikin and Grossman (2013) classified problem transformations either as static or dynamic—with respect to the dynamic behavior of geometric figures in DGEs—as follows: *Dynamic changes* are those that can be obtained by dragging within a DGE, while *static changes* are those that cannot be obtained by dragging. Dragging (and thus dynamic change) does not change any of the critical properties of the figure constructed in the DGE (see distinction between figure and drawing by Laborde, 1992). For example, by dragging a rectangle, it can be transformed into a square ("what if yes?" strategy) but cannot be transformed into a parallelogram ("what if not?" strategy), which is not a rectangle. Static changes in a DGE usually require additional construction without changing the given figure, or constructing a new figure.

Problem transformations can also be obtained by the "goal manipulation" strategy (Silver, Mamona-Downs, Leung, & Kenny, 1996), in which the givens remain unchanged and only the goal is changed, or by the "symmetry" strategy (Hoehn, 1993; Silver et al., 1996) that leads to the creation of a problem in which the givens and the goals have been interchanged.

Leikin and Grossman (2013) found that *investigation* problems posed by teachers can be of *discovery* and *verification* types, depending on the degree of their openness. *Verification problems* do not require conjecturing but do ask for checking a proposition that needed to be proved. On the contrary, *discovery problems* are open problems that require conjecturing, analyzing conjectures, and proving. The problems posed by the teachers presented in this chapter are analyzed in terms of their openness.

Problem Posing Through Investigation in a DGE

Problem posing through investigation is usually associated with dragging and constructions in a DGE. Dragging is a critical feature of DGEs, which makes investigation possible. The two main functions of dragging are *testing* and *searching* (Hölzl, 2001):

- Testing verifies that a figure constructed in the process of experimentation satisfies all the conditions given in the task.
- Searching is aimed at finding new properties of a given figure and recognizing unforeseen regularities, relationships, and invariants.

In this context the distinction within DGEs between *drawing* and *figure* that was introduced by Parzysz (1988) and further developed by Laborde (1992) is especially important. Drawings and figures are visual images of geometric objects. Figures (rigorous constructions) are images of geometric objects constructed in such a way that all the necessary properties of the object are present. For example, if users drag any corner of a figure representing a square, the figure changes its size but remains a square. In this sense, a "figure does not refer to one object but to an infinity of objects" (Laborde, 1992, p. 128), which continuously preserve all critical properties under dragging. By contrast, drawings resemble the indented geometric object, with all its properties, but in a DGE they do not pass the drag test. In this way a corrected soft construction in a DGE is a drawing. Soft constructions have only part of the properties of a given object, and naturally—when corrected—do not pass the drag test. For example, when a drawing of a square is dragged it loses some of its properties and becomes some type of quadrilateral, i.e., a rectangle.

Based on the distinction between figures and drawings in a DGE, I suggested differentiation between two types of dragging: *figure dragging* and *correction dragging* (Leikin, 2012) that facilitate posing problems through two corresponding types of investigations in DGEs—a *figure investigation* and a *correction investigation*. Table 18.1 (based on Leikin, 2012) summarizes the differences between the two types of investigations.

Table 18.1
Distinctions Between Figure and Correction Investigations

	Investigation type	
Features	Figure investigation	Correction investigation
Dragging	Investigation dragging of the figure which is continuous and arbitrary	Correction dragging of the drawing (to achieve given conditions) which is discrete and purposeful
PP strategy	Searching for properties which are immune to dragging	"What if yes?" strategy Searching for properties that repeatedly occur in the corrected objects
Measurement	Exact	Approximate

Note here that figure investigations in DGEs cannot be performed using "what if not?" or "what if yes" schemes. "What if not?" is impossible since robust construction presumes that no properties of the figure can be "reduced" (Brown & Walter, 1993, 2005). "What if yes?" is impossible since adding properties to the constructed figure can only be done by means of soft constructions. "What if yes?" problem-posing strategies can be performed by means of correction investigations. Investigations in DGEs can also be performed based on static changes performed on the figure accompanied by subsequent dragging. Namely, investigations in a DGE can include performing auxiliary constructions. These constructions themselves can lead to unpredicted results. In this sense problem posing through investigation includes problem posing by construction.

In the next section I exemplify these findings through a reflective account of one particular mathematical activity when the participants arrived at least 25 new problems through investigation within a DGE and through proving. Most of the posed problems remain without proof, and the readers are invited to prove the problems, further perform geometry investigations and pose new problems related to the given mathematical object.

Tracking Geometry Investigation Through the Lens of Problem Posing

Problem Posing Through Proving

Prospective secondary school mathematics teachers (PMTs) were asked to produce at least two different proofs to Problem 1 (see Figure 18.1). As a rule, this part of the task was performed as homework with the subsequent classroom discussion focused on presentation of the solutions, analysis of similarities, and differences between the proofs and views on the elegance of the proofs and their level of difficulty. PMTs—as a group—produced two different solutions (Figure 18.1). As described below, one of these solutions appeared to be a source for a new problem.

In the discussion that took place during the lesson, PMTs regarded Proof 1.1 (Figure 18.1) as being easier than Proof 1.2 for two reasons: (a) In Proof 1.1 the auxiliary construction is performed "within the given figure" whereas in Proof 1.2 auxiliary construction is "outside the given figure"; and (b) Proof 1.1 is based on the problem givens and properties of the midline in the triangle and Thales theorem, whereas Proof 1.2 is based on the similarity of triangles.

At the same time, PMTs shared the opinion that "Proof 1.2 is *more interesting* since it shows additional properties of the given figure." They argued that Proof 1.2 leads to posing a new problem (Problem 2 shown in Figure 18.2). A statement in Problem 2 follows from Proof 1.2 that includes two facts: $CD \parallel GF$ and $DC = GF$.

Problem 1

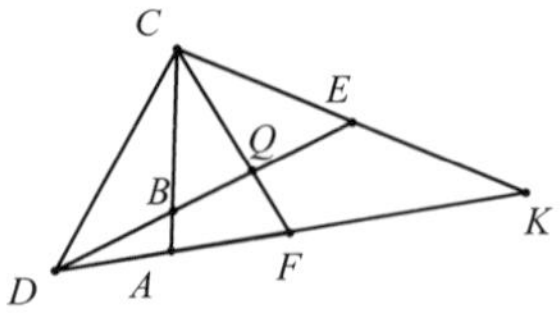

Given:
Δ DCK,
DE median in Δ DCK
CF median in Δ DCK
CA median in Δ DCF

Prove that
$\frac{DE}{DB} = \frac{5}{2}$
in at least 2 different ways

Proof 1.1

Auxilary construction $ET \| CA$

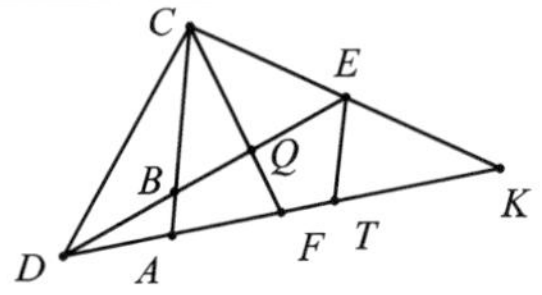

$DK = 20x \Rightarrow DF = 10x,\ DA = 5x \Rightarrow AK = 15x$

$CE = EK\ \&\ ET \| CA \Rightarrow ET$ midline in ΔACK

$\Rightarrow AT = \frac{1}{2} AK = 7.5x \Rightarrow DT = 12.5x$

$\frac{DE}{DB} = \frac{DT}{DA};\ \frac{DE}{DB} = \frac{12.5x}{5x} \Rightarrow \frac{DE}{DB} = \frac{5}{2}$

Proof 1.2

Auxilary construction EG through F (G on CA)

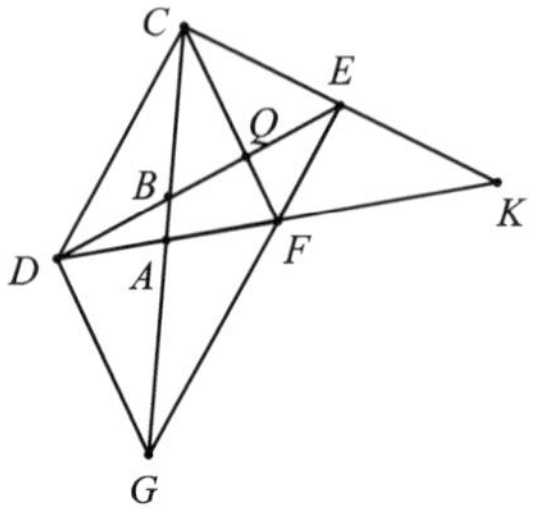

EF – midline $\Rightarrow DC = 2EF,\ DC \| EF$

$DA = AF, DC \| GF \Rightarrow \Delta DCA \cong \Delta FGA \Rightarrow DC = GF$

$EF = x, DC = GF = 2x \Rightarrow GE = 3x$

$\Delta DCB \sim \Delta EGB \Rightarrow DB = 2y, BE = 3y \Rightarrow DE = 5y$

$\frac{DE}{DB} = \frac{5}{2}$

Figure 18.1. Two proofs for Problem 1.

Problem 2

Given: ΔDCK,

DE median in ΔDCK

BF median in ΔDCK

CA median in ΔDCF

EG passes through F, G on CA

Prove:

$DCFG$ - parallelogram

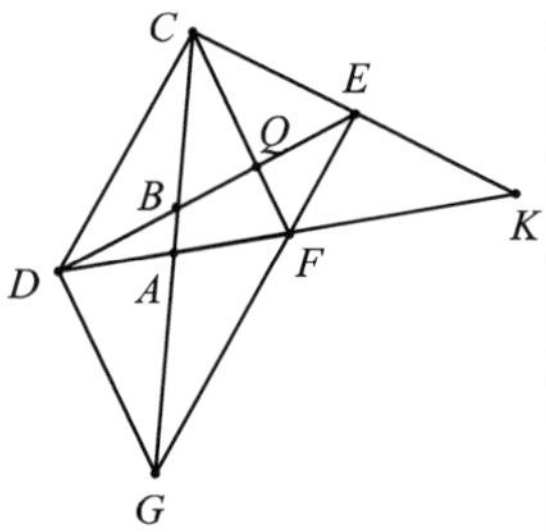

Figure 18.2. Problem posed through Proof 1.2.

Problem Posing for Investigation

At the second stage of coping with Problem 1, participants were required to transform the proof problem into an investigation problem. Figure 18.3 demonstrates two of the investigation problems (3A and 3B) created by PMTs. Problem 3A exemplifies a *verification problem* since it does not require conjecturing but only checking a proposition that had to be proved. Problem 3B illustrates a *discovery problem*, as it is formulated as an open problem that requires conjecturing, analyzing conjectures, and proving (see additional examples of discovery problems in Figure 18.5).

Problem 3

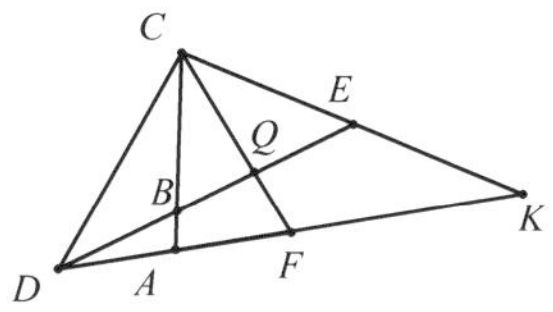

Given:

ΔDCK,

DE median in ΔDCK

BF median in ΔDCK

CA median in ΔDCF

3A. Verification problem:
Is it true that $\frac{DE}{DB} = \frac{5}{2}$?

3B. Discovery problem:
Find different relations between the elements in the given figure.

Figure 18.3. Transforming Problem 1 into new investigation-oriented problems.

Problem 3B allowed participants to search for all possible relationships between elements in the given figure and other figures that can be achieved by auxiliary constructions from the given figure. The investigation and the constructions were performed in different DGEs (e.g., Geo-Gebra, Geometry Sketchpad or Geometry Investigator) according to the PMTs' preferences. The PMTs were allowed to perform investigations with robust as well as soft construction. Investigations were mostly directed at searching for those relationships and properties of a robust construction which are immune to dragging in DGE.

Problem Posing Through Investigation

Overall PMTs discovered more than 20 properties related to the geometrical object from Problem 1. Figures 18.4, 18.5, and 18.6 depict examples from the collective problem-posing space related to the properties discovered by PMTs. Figure 18.4 demonstrates properties discovered with auxiliary constructions "inside" the given geometry object. In contrast, Figure 18.5 depicts properties which are based on the auxiliary constructions "outside" the given geometry object. Thus, properties in Figure 18.5 are considered as requiring more advanced thinking. Discovery of properties presented in both Figures 18.4 and 18.5 was based on the *figure investigation* that included carrying out auxiliary constructions, measurements, and search for the invariants (properties which are immune to dragging).

Problem 4: Given: ΔDCK, DE median in ΔDCK, BF median in ΔDCK, CA median in ΔDCF

Auxilary construction "within the given figure" EF, $ET \parallel CA$

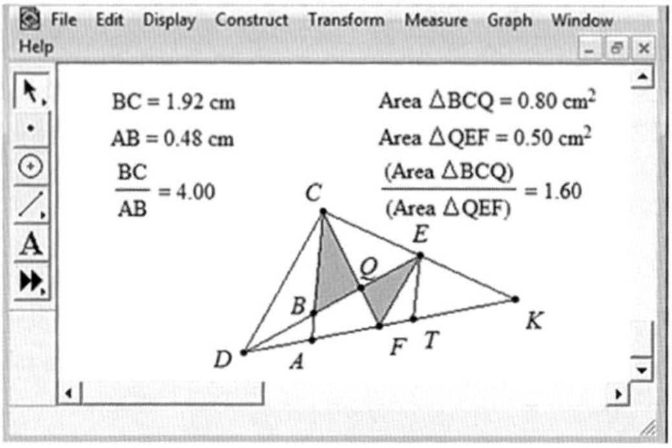

Prove that

4a. $\frac{BC}{AB} = 4$ 4c. $\frac{A(BCQ)}{A(EQF)} = \frac{8}{5}$ 4e. $\frac{AT}{DK} = \frac{3}{8}$ 4g. $\frac{DA}{DT} = \frac{2}{5}$

4b. $\frac{CQ}{QF} = 2$ 4d. $\frac{A(DCQ)}{A(EQF)} = 4$ 4f. $\frac{QE}{BQ} = \frac{5}{4}$ 4h. $\frac{A(DCK)}{A(ETK)} = \frac{16}{3}$

The new posed problems ask to prove discovered properties.

4*b*, 4*d* are trivial discoveries:
Q is intersection of medians

Figure 18.4. Posing a problem through investigation: Looking within the figure.

The whole group discussion focused on the *newness of the discovered properties* and the connections between the properties. Some of the discovered properties were evaluated as *trivial* ones since PMTs ought to know these properties without investigation. Properties 4*b, d, g* are trivial for different reasons: $\frac{CQ}{QF} = 2$ since Q is the point of intersection of medians in triangle *DCK*. For the same reason $\frac{A(DCQ)}{A(EQF)} = 4$ is associated with similarity of the triangles with a coefficient of similarity equal to 2. Property $\frac{DA}{DT} = \frac{2}{5}$ follows immediately from the property proven in Problem 1. Note that at advanced stages of the course, trivial discoveries were given a negative evaluation as an indicator of a lack of basic geometry knowledge and an absence of PMTs' critical reasoning.

Properties 4*a, c, e, f, h* are *nontrivial* since they do not constitute geometric theorems from the geometry course, and they do require proving in several stages. The PMTs were asked to prove properties that were nontrivial. I invite the readers also to perform these proofs.

Problem 5:

Given: ΔDCK, DE median in ΔDCK, BF median in ΔDCK, CA median in ΔDCF

Auxilary construction A:

$CN = DC$ (N on continuation of DC); NK, NF, L, M, R

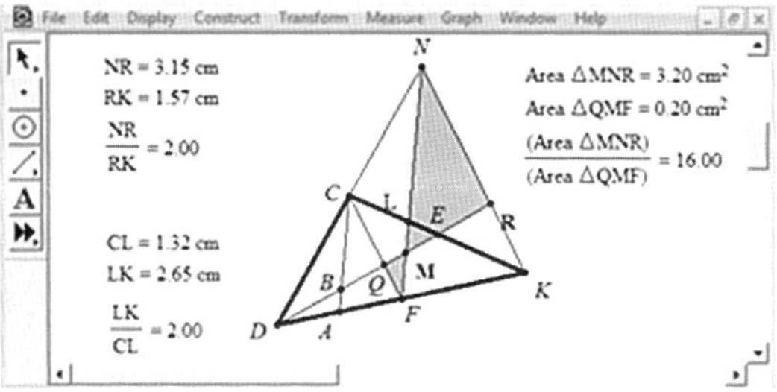

5a. $\frac{NR}{RK} = 2$

5c. $\frac{A(MNR)}{A(QMF)} = 16$

5e. $\frac{A(MNR)}{A(CFK)} = \frac{32}{30}$

Prove that
(The new posed problems ask to prove discovered properties.)

5b. $\frac{A(MNR)}{A(BCQ)} = 4$

5d. $\frac{A(MLE)}{A(QMF)} = 1$

5f. $\frac{A(MNR)}{A(CLF)} = \frac{16}{5}$

Auxilary construction B:

EM through T, $M = ET \cap CQ$

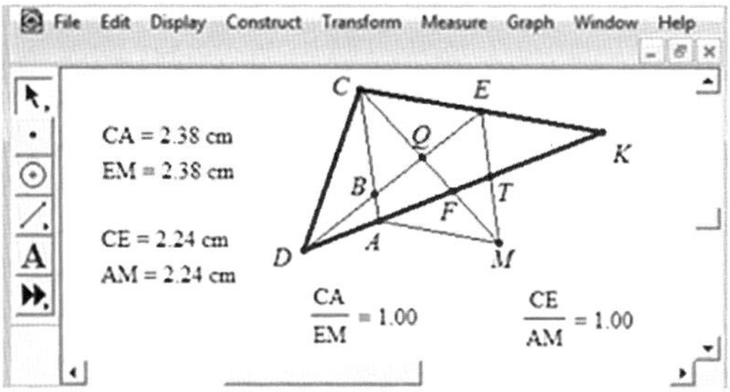

5h. $CAME$ is a parallelogram

5i. $\frac{A(DCK)}{A(ACEM)} = \frac{4}{3}$

Figure 18.5. Posing a problem through investigation: Looking beyond the figure.

As noted above, discovery of additional nontrivial properties is associated with auxiliary constructions "outside the triangle" (Figure 18.5). The participants agreed that most of these properties were surprising and that *surprise is one of the special characteristics of a nontrivial discovery*. The PMTs found property 5*h*: *CAME* is a parallelogram, to be the most surprising. They were asked to prove all the discovered nontrivial properties (see Appendix A for proof that *CAME* is a parallelogram). Note here that problem 5*h* can be considered as *posed through construction* since property "*CAME* is a parallelogram" was discovered accidentally when line *ET* was drawn (Auxilliary construction B in Figure 18.5).

Problem 6:
Under which conditions fulfilled by the given triangle is the parallelogram ***CAME***

(i) a rhombus?
(ii) a square?

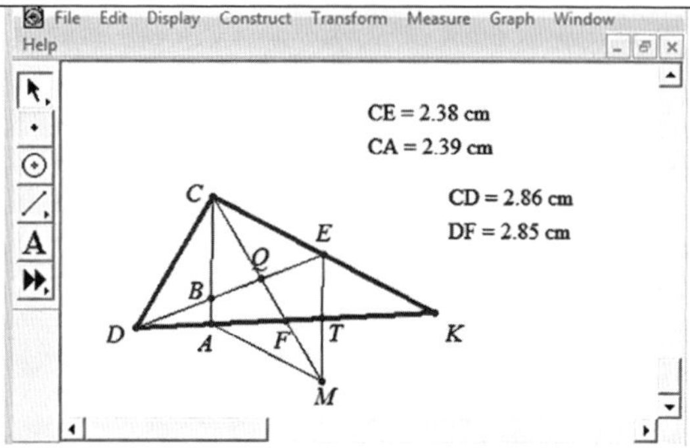

Dragging triangle ***DCK*** to the condition ***CA* = *CE***

Figure 18.6. Transforming Problem 5*h* into a discovery problem.

Overall, about 20 nontrivial properties were discovered by the participants; thus, in this way about 20 new *problems were posed through investigations*. The richness of the collective spaces of the nontrivial discovered properties and thus of the problem-posing space was almost shocking for the participants. While they doubted that each participant alone can discover such a rich collection of properties however, they arrived at the conclusion that "collaborative work is essential in order to discover many properties" and that the collective space of discovered properties serves as a source for the development of their problem-posing and problem-solving expertise.

Back to Problem Posing for Investigation

The discovery that *CAME* is a parallelogram (property 5*h* in Figure 18.5) led one of the PMTs to pose a new investigation problem (Problem 6 in Figure 18.6).

This investigation problem differs significantly from Problem 3B (Figure 18.3) posed for investigation previously. While both problems are open and belong to the category of discovery problems, Problem 3B is unfocused and allows solvers to search for all possible invariants. In contrast, Problem 6 directs solvers to discover special conditions of the given figure that are sufficient for the nearly constructed parallelogram to be a rhombus (or a square). Problem 6 is posed based on Problem 5*h* by the combination of two problem-transformation strategies: symmetry changes (when goals and givens are interchanged) and the "what if yes?" strategy (Leikin & Grossman, 2013). Last but not least important, Problem 6 requires *correction investigation*.

Back to Problem Posing Through Investigation

In contrast to figure investigation performed in a DGE for Problem 3B that was directed at searching for robust constructions (Figures. 18.4 and 18.5), the investigation related to Problem 6 was performed by correction strategy, in which triangle *DCK* was dragged to obtain the drawing of a rhombus (a square) from the parallelogram *ACEM*. In this way, by dragging the triangle to a state in which in parallelogram *ACEM* sides *CA* and *CE* are equal (*ACEM* becomes a rhombus), the participants conjectured that $CD=DE$; in other words $DK=2CD$ (Figure 18.6) is based on the repeated observation of the properties in "corrected drawing." This strategy did not allow for "exact" measuring but did allow for raising the conjecture based on the repeating properties in the corrected situations.

Investigation related to Problem 6 was also performed (with the instructor's guidance) with robust constructions by searching for properties that are immune to dragging. One of the robust constructions started out with the construction of a rhombus/a square and the consequent construction of the triangle *DCK* so that segments *CF* and *CA* will be medians in triangle *DCK* and triangle *DCK* respectively (see the diagrams for Problems 7*A* and 7*B* in Figure 18.7). In this way participants posed Problem 7a: "If rhombus *CAME* is given and triangle *DCK* is constructed so that *DK* intersects *EM* at the midpoint *T* on *EM*, *F* (intersection of *DK* and *CM*) is a midpoint on *DK*, *A* is a midpoint on *DF*, then $DK=2DC$." When the rhombus is a square (Problem 7b) then angle *CDA* is 36.87°.

Problems 7*a* and 7*b* are nontrivial ones with complex proofs (see Appendix B). These problems and the investigations (Figure 18.7) are associated with necessary

Problem 7: Given: ΔDCK, DE median in ΔDCK, BF median in ΔDCK, CA median in ΔDCF

7A. $CAME$ is a rhombus

7B. $CAME$ is a square

T is the midpoint on EM, $F = AT \cap CM$

DK on AT: D: $DA = AF$, K: $KF = FA$

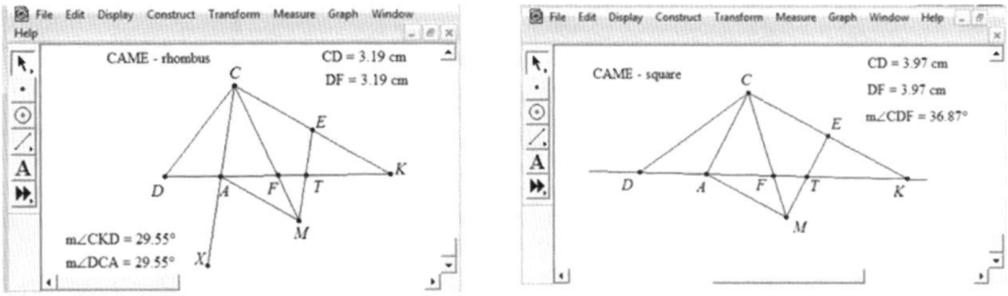

Prove that

7*a*. $DC = DF$ ($\Leftrightarrow DK = 2DC$)

7*b*. $DK = 2DC$ and $\angle CDK = 36.87°$

Figure 18.7. Problem posing through investigation: Focusing on new givens and goals.

Problem 8: Given:

ΔDCK, $DK = 2DC = 4DA$, $CE = EK$, $ACEM$ – parallelogram, $J = DK \cap CM$

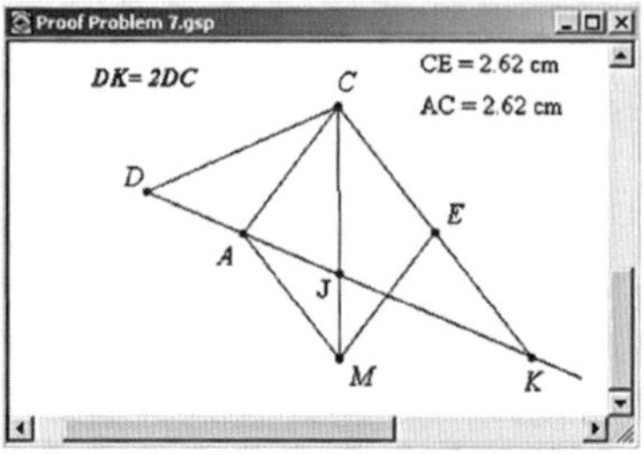

8*a* What can be said about quadrilateral $ACEM$?

8*b* Prove: $DJ = JK$

Figure 18.8. Problem 8*a* is an inverse problem to Problem 7*a*.

conditions that the triangle should satisfy for *CAME* to be either a rhombus or a square. As an alternative, PMTs suggested investigating Problem 8, which was an inverse problem to Problem 7*a*. In this case the construction started with a triangle *DCK* in which *DK*=2*DC* and resulted with a verification that *ACEM* is a rhombus (Figure 18.8). Interestingly, the PMTs found this problem better connected to Problem 1 "since the triangle in this problem is given and the proof focuses on the properties of the quadrilateral."

Concluding Comments

In this chapter I have demonstrated the power of investigations in DGEs as an effective problem-posing tool. Problem posing in mathematics is one of the central mathematical tasks directed at the development of mathematical knowledge and creativity. Not less importantly, problem posing is an important pedagogical skill that enhances teachers' proficiency and makes teaching more flexible. This chapter has presented three types of problem-posing acts associated with geometry investigations: (a) problem posing through proving; (b) problem posing for investigation; and (c) problem posing through investigation. These three types of problem posing are mutually dependent and interrelated (see Figure 18.9).

The PMTs who participated in the activity described in this chapter were encouraged to perform geometry investigations of this type during a 56-hour course. Throughout the course their competencies developed gradually, and by the end of the course PMTs were able to design activities of this kind for their peers (see Appendix C "PMTs' posed problems" in support of this finding). The participants expressed their willingness to "teach their students in a similar way," though

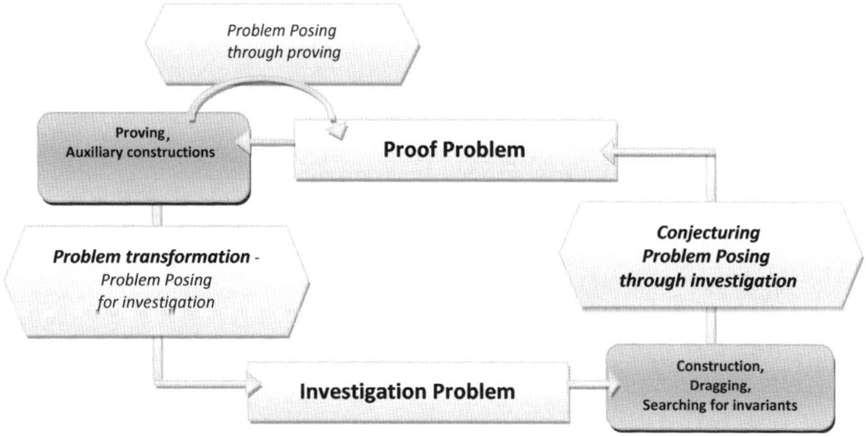

Figure 18.9. Problem-posing types associated with investigations in DGE.

(not surprisingly) they were skeptical whether, under the pressure of meeting the demands of the school mathematics curriculum, these activities could be implemented systematically in a regular mathematics classroom. The contrast between PMTs' enjoyment from coping with investigation problems, problem posing for and through investigation and their uncertainty with regard to the usefulness of similar activities in the classroom setting is rooted in the stable nature of teachers' beliefs (Cooney, Shealy, & Arvold, 1998) and the "conviction loop": "To implement new pedagogical approaches, teachers must be convinced of the suitability of those approaches in their work with students and, at the same time, to be convinced of the suitability of those approaches they have to implement them in school" (Leikin, 2008, p. 80). I suggest that, in order to break the conviction loop, PMTs should be assigned to implement geometry investigations with individual students or with classes during their school practicum.

In my view, the majority of proof problems from school textbooks, when opened for investigations and formulated as discovery problems, lead to doing mathematics rich in surprises, discoveries, and proofs. At the same time, finding sufficiently rich examples to support the emergence of a variety of ways of problem posing is critical for effective work with PMTs and school students. Therefore, teacher educators and mathematics teachers should execute a critical choice of the tasks for their learners.

The PMTs were astonished by the number of new problems formulated during the session described in this chapter. This type of activities led them to the conclusion that "through investigations in a DGE, a teacher can solve multiple problems related to one particular geometric object and prepare more interesting lessons for his/her students." Students and teachers involved in the real doing of mathematics find that they enjoy mathematical discovery at the level which is appropriate to their own abilities.

Appendix A

Proof for Problem *5h* (Figure 18.5)

(1) $ET \parallel CA$ thus triangles *CBQ* and *MEQ* are similar;

(2) $\frac{QE}{BQ} = \frac{5}{4}$ (*2b*, Figure 18.4);

(3) From (1) and (2) $\frac{EM}{BC} = \frac{5}{4}$;

(4) $\frac{AC}{BC} = \frac{5}{4}$ (*1b*, Figure 18.4) thus $CA = EM$.

(5) Hence $EM \parallel CA$ and $EM = CA$; that is *CEMA* is a parallelogram.

Appendix B

Proof for Problems 7*a*, 7*b* (Figure 18.7)

Construction outline:

CAME is a rhombus, *T*-midpoint on *EM*, $F = AT \cap CM$
DK on $AT: D: DA = AF,\ K: KF = FA$

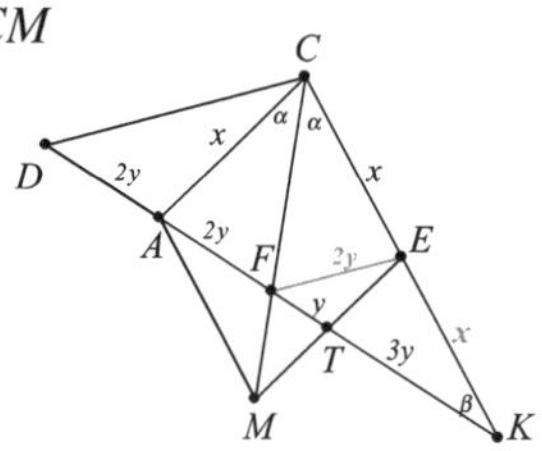

Prove:
7a. $DC = DF\left(\Leftrightarrow DK = 2DC\right)$
7b. *CAME* is a square when $\angle CDK = 36.9°$
7c. K' on *CE*: $K'E = EC \Leftrightarrow K'$ coincides with *K*

Proof

1. According to the construction: $AC = CE = EM = AM = x$,
   $ET = TM = \frac{1}{2}x$, $FA = AD = 2y$, $FK = DF = 4y \Rightarrow DK = 8y$;
2. *CAME* is a rhombus $\Rightarrow \Delta CAF \cong \Delta CEF \Rightarrow FE = 2y$;
3. *MF*-bisects angle AMT, $\mathrm{AM} = 2\mathrm{MT} \Rightarrow \mathrm{AF} = \mathrm{FT}$; $\mathrm{FT} = \mathrm{y}$; $\mathrm{TK} = 3\mathrm{y}$
4. $\Delta TEK \cong \Delta TMA$; AT = TK, ME $\parallel$ CA $\Rightarrow$ *TE* midline on $\Delta ACK \Rightarrow EK = CE$ (18.7c)
5. *TE* midline on $\Delta ACK \Rightarrow CD = 2EF \Rightarrow DC = 4y$ (18.7a)
   $\Rightarrow DC = AF\left(\Leftrightarrow DK = 2DC\right)$
6. $\Delta FEK \sim \Delta DCK \sim \Delta DAC \Rightarrow \angle DCA = \angle CKD$
7. If *CAME* is a square $\alpha = 45°$; $\tan\beta = \frac{1}{2} \Leftrightarrow \beta = 26.57° \Leftrightarrow \angle CDK$ (18.7b)
   $= 71.57° \angle CDK = 36.9°$

Appendix C

Problems posed for and through investigations by a PMT who participated in the study

Rasha's Problem

Initial problem: *Midline in a triangle*

The segment connecting the midpoints of two sides of a triangle is parallel to the third side and is half as long. (Given: $\Delta ABC,\ AE = EB,\ AP = PC$; Prove: $EP \parallel BC,\ EP = \frac{1}{2}BC$)

Posed problems:

Given:

$\Delta ABC,\ AE = EB,\ AP = PC$

PD is a continuation of $EP,\ ED = 2EP,\ F = EC \cap BP,\ G = BD \cap AC,$
$S = EC \cap BD,\ O = FG \cap SP$

Prove:

$\frac{ED}{FG} = 3;\ \frac{BA}{SP} = 4,\ FO = OG,\ OP = 2OS$

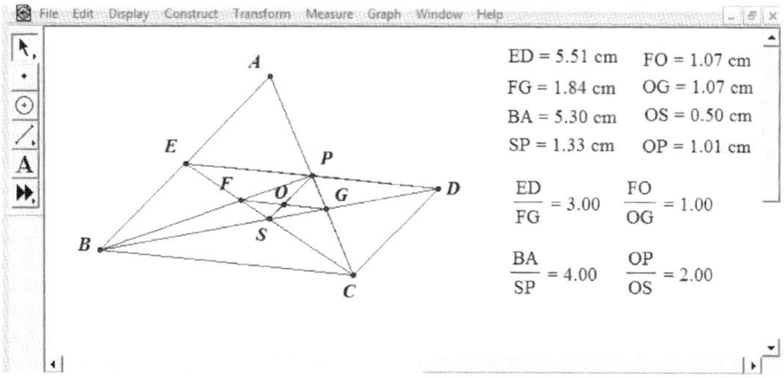

References

Borba, M. C., & Villarreal, M. E. (2005). *Humans-with-media and the reorganization of mathematical thinking: Information and communication technologies, modeling, visualization, and experimentation*. New York, NY: Springer.

Brown, S., & Walter, M. (1993). *Problem posing: Reflections and applications*. Hillsdale, NJ: Lawrence Erlbaum.

Brown, S., & Walter, M. (2005). *The art of problem posing* (3rd ed.). New York, NY: Routledge.

Chazan, D., & Yerushalmy, M. (1998). Charting a course for secondary geometry. In R. Lehrer & D. Chazan (Eds.), *Designing learning environments for developing understanding of geometry and space* (pp. 67–90). Hillsdale, NJ: Lawrence Erlbaum.

Cobb, P., Confrey, J., diSessa, A., Lehrer, R., & Schauble, L. (2003). Design experiments in educational research. *Educational Researcher, 32*(1), 9–13.

Collins, A., Joseph, D., & Bielaczyc, K. (2004). Design research: Theoretical and methodological issues. *The Journal of the Learning Sciences, 13*(1), 15–42.

Cooney, T. J., Shealy, B. E., & Arvold, B. (1998). Conceptualizing belief structures of preservice secondary mathematics teachers. *Journal for Research in Mathematics Education, 29*, 306–333.

Da Ponte, J. P. (2007). Investigations and explorations in the mathematics classroom. *ZDM: The International Journal on Mathematics Education, 39*, 419–430.

Da Ponte, J. P., & Henriques, A. C. (2013). Problem posing based on investigation activities by university students. *Educational Studies in Mathematics, 83*, 145–156.

De Villiers, M. (2012). An illustration of the explanatory and discovery functions of proof. In *Proceedings of the 12th International Congress on Mathematical Education. Regular Lectures* (pp. 1122–1137). Seoul, Korea: COEX.

Duncker, K. (1945). On problem solving. *Psychological Monographs, 58*(5), 270.

Ellerton, N. F. (2013). Engaging pre-service middle-school teacher-education students in mathematics problem posing: Development of an active learning framework. *Educational Studies in Mathematics, 83*, 87–101.

Healy, L., & Lagrange, J.-B. (2010). Introduction to section 3. In C. Hoyles & J.-B. Lagrange (Eds.), *Mathematics education and technology: Rethinking the terrain. The 17th ICMI Study* (pp. 287–292). New York: Springer.

Hoehn, L. (1993). Problem posing in geometry. In S. Brown & M. Walter (Eds.), *Problem posing: Reflections and applications* (pp. 281–288). Hillsdale, NJ: Lawrence Erlbaum.

Hölzl, R. (1996). How does "dragging" affect the learning of geometry? *International Journal of Computers for Mathematical Learning, 1*, 169–187.

Hölzl, R. (2001). Using dynamic geometry software to add contrast to geometric situation—A case study. *International Journal of Computers for Mathematical Learning, 6*(1), 63–86.

Jones, K. (2000). Providing a foundation for deductive reasoning: students' interpretations when using dynamic geometry software and their evolving mathematical explanations. *Educational Studies in Mathematics, 44*, 55–85.

Laborde, C. (1992). Solving problems in computer based geometry environments: The influence of the features of the software. *ZDM: The International Journal on Mathematics Education, 92*(4), 128–135.

Lampert, M., & Ball, D. (1998). *Teaching, multimedia, and mathematics: Investigations of real practice. The Practitioner Inquiry Series*. New York, NY: Teachers College Press.

Leikin, R. (2004). Towards high quality geometrical tasks: Reformulation of a proof problem. In M. J. Hoines & A. B. Fuglestad (Eds.), *Proceedings of the 28th International Conference for the Psychology of Mathematics Education* (Vol. 3, pp. 209–216). Bergen, Norway: International Group for the Psychology in Mathematics Education.

Leikin, R. (2008). Teams of prospective mathematics teachers: Multiple problems and multiple solutions. In T. Wood (Series Ed.) & K. Krainer (Vol. Ed.), *International handbook of mathematics teacher education: Participants in mathematics teacher education: Individuals, teams, communities, and networks* (Vol. 3, pp. 63–88). Rotterdam, The Netherlands: Sense.

Leikin, R. (2012). What is given in the problem? Looking through the lens of constructions and dragging in DGE. *Mediterranean Journal for Research in Mathematics Education, 11*(1–2), 103–116.

Leikin, R., & Grossman, D. (2013). Teachers modify geometry problems: From proof to investigation. *Educational Studies in Mathematics, 82*(3), 515–531.

Mamona-Downs, J. (1993). On analyzing problem posing. In I. Hirabayashi, N. Nohda, K. Shigematsu, & F. L. Lin (Eds.), *Proceedings of the 17th International Conference for the Psychology of Mathematics Education* (Vol. 3, pp. 41–47). Tsukuba, Japan: International Group for the Psychology in Mathematics Education.

Mariotti, M. A. (2002). The influence of technological advances on students' mathematics learning. In L. D. English (Ed.), *Handbook of international research in mathematics education* (pp. 695–723). Hillsdale, NJ: Erlbaum.

Parzysz, B. (1988). Knowing vs seeing: Problems of the plane representation of space geometry figures. *Educational Studies in Mathematics, 19*(1), 79–92.

Pehkonen, E. (1995). Using open-ended problem in mathematics. *ZDM: The International Journal on Mathematics Education, 27*(2), 67–71.

Schwartz, J. L., Yerushalmy, M., & Wilson, B. (Eds.). (1993). *The geometric supposer: What is it a case of?* Hillsdale, NJ: Lawrence Erlbaum.

Silver, E. A. (1994). On mathematical problem posing. *For the Learning of Mathematics, 14*, 19–28.

Silver, E. A., Mamona-Downs, J., Leung, S. S., & Kenny, P. A. (1996). Posing mathematical problems in a complex environment: An exploratory study. *Journal for Research in Mathematics Education, 27*, 293–309.

Stoyanova, E. (1998). Problem posing in mathematics classrooms. In A. McIntosh & N. F. Ellerton (Eds.), *Research in mathematics education: A contemporary perspective* (pp. 164–185). Perth, Australia, Australia: MASTEC.

Wells, G. (1999). *Dialogic inquiry: Towards a sociocultural practice and theory of education.* Cambridge, England: Cambridge University Press.

Yerushalmy, M., & Chazan, D. (1993). Overcoming visual obstacles with the aid of the Supposer. In J. L. Schwartz, M. Yerushalmy, & B. Wilson (Eds.), *The geometric supposer: What is it a case of?* (pp. 25–56). Hillsdale, NJ: Erlbaum.

Yerushalmy, M., Chazan, D., & Gordon, M. (1990). Mathematical problem posing: Implications for facilitating student inquiry in classrooms. *Instructional Science, 19*, 219–245.

Chapter 19
Problem-Posing Activities in a Dynamic Geometry Environment: When and How

Ilana Lavy

Abstract In this chapter, results obtained from previous studies on the issue of problem posing in a dynamic software environment using the "what if not?" strategy are presented. These results include outcomes received from prospective teachers' engagement in problem-posing activities both in plane and solid geometry, and outcomes received by the engagement of the researcher in the problem-posing activity. The above-presented results are followed by discussion and a list of implications for instruction. Problem-posing activities should follow activities of problem solving through which the content knowledge of the learnt topic is built. Students should experience problem-posing activities starting at elementary school. In these activities they should be provided with opportunities to develop cognitive processes needed for problem posing such as filtering, comprehending, translating, and editing. When students are exposed to geometrical objects, they should be provided with the option to make sense of the objects via dynamic geometry software.

Contents

I. Lavy (✉)
Department Head of Management Information Systems,
The Academic College of Yezreel Valley, Yezreel Valley, Israel
e-mail: ilanal@yvc.ac.il

F.M. Singer et al. (eds.), *Mathematical Problem Posing*,
Research in Mathematics Education, DOI 10.1007/978-1-4614-6258-3_19

Introduction

Many researchers have discussed the importance of incorporating problem-posing activities into mathematics lessons and have emphasized the benefits students might gain from such activities (Cunningham, 2004; English, 1997). However, little attention has been paid to the question of the stage of development at which such activities should be incorporated into mathematics lessons. Relying on Mestre (2002) who asserted that problem posing is intellectually more demanding than problem solving and on my experience in research concerning problem posing both in plane and solid geometry (Lavy & Bershadsky, 2003; Lavy & Shriki, 2010; Shriki & Lavy, 2012), I believe that problem-posing activities are more efficient after students have gained some experience in problem solving. Engagement in problem-posing activities challenges both the meta- and the actual knowledge students have about the learning materials (Lavy & Shriki, 2010). Hence, one should gain first the required knowledge about the learning materials. To be able to think "out of the box" and "produce" meaningful new problems, one has to develop problem-posing skills. Meta-knowledge is essential for the process of problem posing since in this process one has to be able to judge whether the new created problem is mathematically valid. Engagement in problem posing without having the sufficient meta- and actual knowledge of the examined topic may result in poor outcomes (Cemalettin, Tuğrul, Tuğba, & Kıymet, 2011).

Problem posing can be done in an arbitrary or in a structured manner. One of the structured ways to pose new problems is the "what if not" (WIN) strategy (Brown & Walter, 1993). This strategy is based on the idea that each of the attributes of a given problem (the base problem) can be negated and replaced by alternative one—an action that can yield in a new problem situation. The above process can be perceived as a technical one, but in order to yield a valid new problem, students have to think carefully about the alternative suggestion by recalling the attributes of the given, and by considering the relationship between the original problem and the "new" posed problem. Through such considerations, a student's understanding of the problem-posing process may be deepened.

Engagement in problem-posing activities in dynamic geometry environments becomes richer and more useful when technology is involved. The software frees students from the technical work involving computing and graphing, enabling them to invest more efforts in the inquiry process (Ranasinghe & Leisher, 2009).

In order to be able to create meaningful and useful problem-posing activities for their students, prospective and in-service teachers need to develop their own self-confidence regarding their ability to handle such activities successfully. Developing this self-confidence can be achieved by appropriate training in which prospective and in-service teachers can themselves experience various problem-posing activities as students. Studies which discuss problem posing activities for prospective mathematics teachers in dynamic geometry environments include, for example, those by Lavy and Bershadsky (2003), Lavy and Shriki (2010), and Shriki and Lavy (2012); my own experiences on problem posing have been presented in Lavy and Shriki (2009).

Theoretical Background

In this section a brief literature survey is presented in the following related areas: the role of problem posing in students' mathematics education; the role of problem posing in mathematics teacher's education; problem posing activities in a dynamic computerized environment and the "what if not?" strategy.

The Role of Problem Posing in Students' Mathematics Education

In many cases, during their study of mathematics at school, students experience mainly problem solving. Researchers in mathematics education have emphasized the importance of integrating activities of problem posing and have suggested the incorporation of such activities in school mathematics (Brown & Walter, 1983; Ellerton, 1986; Goldenberg, 1993; Leung & Silver, 1997; Mason, 2000; NCTM, 2000; Silver, 1994; Silver, Mamona-Downs, Leung, & Kenney, 1996). The importance of an ability to pose significant problems was recognized by Einstein and Infeld (1938), who wrote:

> The formulation of a problem is often more essential than its solution, which may be merely a matter of mathematical or experimental skills. To raise new questions, new possibilities, to regard old questions from a new angle, require creative imagination and marks real advance in science. (Quoted in Ellerton and Clarkson, 1996, p. 1010).

Engagement in problem-posing activities can result in students becoming enterprising, creative, and active learners. They have the opportunity to navigate the problems they pose to their domains of interest according to their cognitive abilities (Mason, 2000) and improve their reasoning and reflection skills (Cunningham, 2004).

The importance of incorporating problem-posing activities in mathematics lessons is also supported by the National Council of Teachers of Mathematics (NCTM) in the United States of America (NCTM, 2000) who recommended that students should make and investigate mathematical conjectures and learn how to generalize and extend problems by posing follow-up questions.

Problem posing can also promote a spirit of curiosity and create diverse and flexible thinking (English, 1997). Studies have shown that problem posing might reduce common mathematics fears and anxieties (Brown & Walter, 1993; English, 1997; Moses, Bjork, & Goldenberg, 1990; Silver, 1994). The inclusion of problem-posing activities might improve students' attitudes toward mathematics, reduce erroneous views about the nature of mathematics, and help to encourage students to be more responsible for their own learning. Problem posing can also help to broaden students' perception of mathematics and enrich and consolidate their knowledge of basic concepts (Brown & Walter, 1993; English, 1997; Silver et al., 1996).

Engagement in problem posing may help students to "reason by analogy" when presented with similar questions (English, 1997) and may help them reduce their dependency on their teachers and textbooks and provide them with a sense of responsibility for their own education. By providing students with the opportunity to formulate new problems, the sense of ownership that they need in order to construct their own knowledge is fostered (Cunningham, 2004). This ownership of the problems can result in a high level of engagement and curiosity, as well as enthusiasm towards the process of learning mathematics.

In the process of problem posing, students might end with new problem situations whose mathematical validity has to be checked. For that matter, students need to rethink the mathematical relationships between the concepts involved and, as a result, they might develop and deepen their mathematical and meta-mathematical knowledge. Examining possible links between problem posing and mathematical competence, Mestre (2002) asserted that problem posing can be used to study the transfer of concepts across contexts and to identify students' knowledge, reasoning, and conceptual development.

Researchers have emphasized the inverse process in which the development of problem-solving skills can be helpful in developing problem-posing skills (Brown & Walter, 1993; English, 1997; Skinner, 1990). Research conducted by Philippou, Charalambous, and Christou (2001) have revealed that their study participants realized the importance of developing problem-posing competencies. The participants considered problem posing as harder than problem solving and valued problem posing as the ultimate goal of mathematics learning. However, there are few didactical tools and activities for developing students' skills in problem posing (Yevdokimov, 2005).

Silver (1994) classified problem posing according to whether it takes place before, during, or after problem solving. He argued that problem posing could take place prior to problem solving when problems are being generated as a reaction to a given stimulus such as a picture, a diagram, or a story; during the process of problem solving when students are asked to change the goals and conditions of a problem, or after solving a problem when experiences from the problem-solving context are applied to new situations. Four main cognitive processes are involved in the process of problem posing: filtering (e.g., posing a problem that its answer is 325 sticks); translating (e.g., write a problem based on a given diagram); comprehending (e.g., write an appropriate problem for: $(150-70)+14=x$); and editing (e.g., write an appropriate problem based on a given picture) (Pittalis, Christou, Mousoulides, & Pitta-Pantazi, 2004).

The Role of Problem Posing in Mathematics Education

Teachers have an important role in the incorporation of problem-posing activities into the mathematics lessons (Gonzales, 1996). Nevertheless, although problem posing is recognized as an important teaching method, many students are not provided with the opportunity to engage in problem-posing activities while studying mathematics (Silver et al., 1996). In many cases, teachers tend to emphasize skills, rules, and procedures, which become the essence of learning, instead of focusing on instruments for developing understanding and reasoning (Ernest, 1991). Consequently, mathematics teachers fail to take advantage of the opportunity both to support their students in developing problem-solving skills and to help them build/acquire the required confidence in managing unfamiliar mathematics situations. Teachers rarely use problem posing because they find it difficult to implement in classrooms and because they themselves do not possess the required confidence and skills (Tichá & Hošpesová, 2013; Leung & Silver, 1997). Contreras and Martinez-Cruz (1999) also found that prospective teachers' problem-posing abilities are often underdeveloped, and they should be encouraged to develop their own problem-posing skills (Leung & Silver, 1997; Silver et al., 1996; Southwell, 1998). These skills will enable them to create tasks that include opportunities for their students to be engaged in problem posing (Gonzales, 1996). Southwell (1998) found that posing problems based on given problems could be a useful strategy for developing the problem-solving ability of preservice mathematics teachers. Integrating problem-posing activities in their mathematics lessons enabled preservice teachers to become better acquainted with their own students' mathematical knowledge and understanding. In order to address some of the concerns noted in the literature, it is important that problem-posing activities are included in teacher education programs for prospective teachers.

Problem posing can be integrated in various settings. Crespo and Sinclair (2008) found that engaging prospective teachers in exploratory mathematical activities improved both the range and quality of problems they posed. These authors claimed that such engagement in exploration work enabled the prospective teachers to pose problems that were both interesting and challenging even to them.

Problem-Posing Activities in a Dynamic Computerized Environment

Problem-posing activities become richer and more profound when technology is involved, since the technical work involving computing and graphing is executed by the software more rapidly and efficiently (Ranasinghe & Leisher, 2009). One of the distinctive features of dynamic geometry software (DGS) is the facility to construct geometrical objects and specify relationships between them. Within the DGS, geometrical objects created on the screen can be manipulated, moved, and reshaped interactively (Christou, Mousoulides, Pittalis, & Pitta-Pantazi, 2005). Hence, when working in interactive computerized environments, students can do mathematics differently (Aviram, 2001) and in ways that they could not do with paper and pencil. Their interaction with dynamic geometry software enables students to focus on courses of inquiry without investing time and effort on calculating and drawing, which one cannot avoid while working with paper and pencil. The computerized environment includes tools that mediate students' actions and bridges between the students and the mathematical world (Artigue, 2002). Moreover, dynamic geometry software enables students to represent situations visually and therefore to identify patterns (McKenzie, 2009). Problem posing using computerized environments provides teachers with research-like skills in the development of instructional materials for school mathematics (Abramovich & Cho, 2006).

Problem posing using dynamic geometry software involves unique interactions between the software's interface and the students' actions and understandings. The students are provided with the opportunity to utilize visual reasoning in mathematics, helping them through the dragging facilities, and can help them to generalize problems and relationships, or to examine the validity of a new problem situation (Sinclair, 2004). The exploration techniques—tools, definitions, and visual representations associated with dynamic geometry—contribute to the construction of rich learning environments (Laborde, 1998). Two systems are involved in this interaction between the students and the software: the first system involves the students attempting to pose a problem, and the second system involves the environment, which provides opportunities for students to act and react (Brousseau, 1997).

The "What If Not?" Strategy

Posing new problems can be based on free, semi-structured, and structured situations (Stoyanova, 1998). A *free* problem-posing situation refers to the case in which the student has a free hand in formulating new problems. A *semi-structured* problem-posing activity relates to an open situation in which the student is asked first to explore its structure and complete it, and then to pose new problems. A *structured*

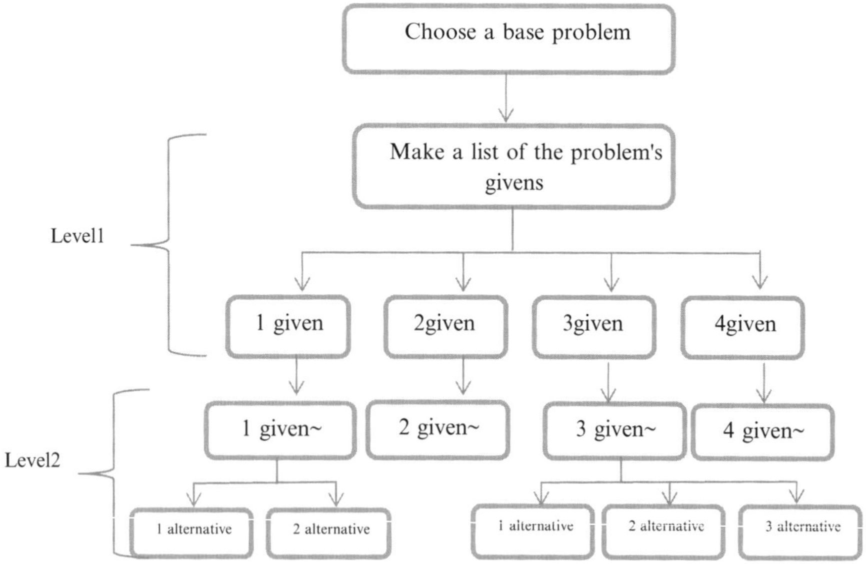

Figure 19.1. Schematic description of the WIN strategy.

problem-posing situation refers to the case in which the learner is asked to suggest new problems that rely on a given base problem.

The "what if not?" (WIN) strategy (Brown & Walter, 1983, 1993) is an example of a structured tool for problem posing. According to WIN strategy, each component of the problem data and the problem question is examined and manipulated through the process of negating one of the base problem's givens.

In fact, the strategy consisted of two levels: level 1 and level 2 (Figure 19.1). In level 1, one has to list all of the problem's givens including the question of the problem, and in level 2, one has to negate each of the listed givens by asking "what if not given *k*?" Then she has to make a list of alternatives to the negated given. Part of the offered alternatives results in a new problem situation. Implementing the WIN strategy enables teachers and students to move away from a rigid teaching format that makes students believe that there is only one "right way" to refer to a given problem. Using this problem-posing strategy provides students with the opportunity to discuss a wide range of ideas and to consider the meaning of the problem rather than merely focusing on finding its solution (Brown & Walter, 1993).

Results

In this section, a brief summary of the results obtained from previous studies is presented. These results refer to problem posing done by prospective teachers and to problem posing done by the researcher. The purpose of this comparison is to emphasize that the main difference in the process of problem posing between ones who had not previously experienced problem posing (in this case, the prospective teachers) and those who did (in this case, the researcher) stemmed from a lack of self-confidence in their own mathematical ability. Since the activities in which the prospective teachers were engaged involved mathematical subjects that they were proficient in, it can be assumed that the source of their difficulties stemmed from a lack of confidence in their ability to perform such tasks. Therefore, to enable students to build self-confidence in their ability to perform such tasks, there should be a frequent engagement with problem-posing activities.

Prospective Teachers' Engagement in Problem Posing

I was involved in several studies in which prospective teachers (PTs) had to pose problems using the WIN strategy (Lavy & Bershadsky, 2003; Lavy & Shriki, 2010; Shriki & Lavy, 2012). In Lavy and Bershadsky (2003) the PTs had to pose problems in solid geometry, while in Lavy and Shriki (2010) they had to pose problems in plane geometry. In Lavy and Bershadsky (2003) the following results were obtained: the majority of the PTs changed one of the givens of the base problem and only a few of them changed the question of the base problem. In the case where the given of the base problem was numerical, most of the PTs suggested another numerical value which was very close to the original one. In the case where the given was a geometrical shape such as a right triangle, most of the PTs suggested to replace it with another shape from the same group of shapes (e.g., from right to isosceles triangle). Although some of the PTs suggested replacing one of the givens of the base problem by a generalization of it, for example, instead of a height of 10 cm they suggested *h* cm, none of them chose to explore this new problem situation. These findings are in line with those of Tichá and Hošpesová (2013) who found that many preservice and inservice teachers tended to regard problem posing as a very unusual activity. Some of them encountered difficulties in coping with such activities, feeling that it was beyond their capabilities.

For the PTs to be able to suggest the above-mentioned alternatives, they applied the cognitive processes of editing, filtering, comprehending, and translating quantitative information. Data obtained from PTs in the Lavy and Bershadsky (2003) study for Problem 1 (Figure 19.2) and Problem 2 (Figure 19.3) are summarized in

What is the distance from the center of the base of a regular triangular pyramid to the pyramid's lateral face given that the pyramid's height is 10 cm and the dihedral angle (the angle between the two faces) is 67^0.

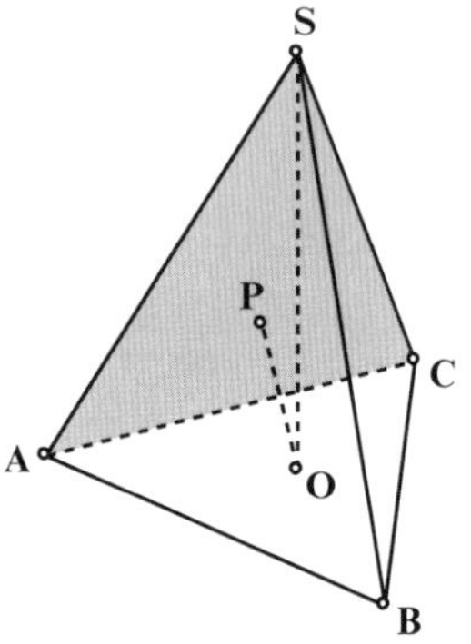

Figure 19.2. Problem 1.

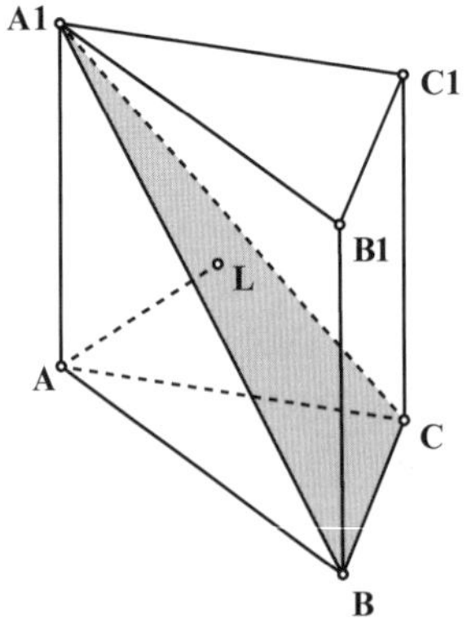

In a regular right prism $ABCA_1B_1C_1$ with isosceles triangular base (AB=BC), AB=5 cm and $\angle ABC=40^0$. The prism height is 8.2 cm. What is the distance between A and the plane A_1BC?

Figure 19.3. Problem 2.

Table 19.1
Distribution of Posed Problems and the Cognitive Processes Involved

Main category				Problem 1 (18 prospective teachers)	Problem 2 (10 prospective teachers)	Cognitive processes
Changing one aspect of the problem's data	Changing of the numerical value of data	Another specific value		6	12	F
		A range of values		4	2	FC
		Negation		2	–	F
		Generalization	Implicit	1	–	FC
			Formal	0	4	FC
	Changing of the data kind	Another specific data kind		15	26	FCE
		Negation		4	1	
		Generalization	Implicit	3	9	FCE
			Formal	1	3	FCE
	Eliminating of one of the problem's data			–	5	F
Changing of the problem question	Another specific question			6	3	FTCE
	Inverting of the given problem into a proof problem			1	–	FTCE
Total				43	65	
Average number of posed problems per one PT				2.4	6.5	

Table 19.1. The PTs' suggestions for changing the problems were categorized according to the four cognitive processes (Table 19.1).

The first five columns in Table 19.1 have been taken from page 377 in Lavy and Bershadsky (2003). A sixth column was added on the right which refers to the cognitive process(es) applied by the PTs while posing new problems. The letters appearing in the sixth column are abbreviations: "E" stands for editing; "F" for filtering; "C" for comprehending; and "T" for translating.

Before discussing the cognitive processes applied by the PTs while posing problems using the WIN strategy, it should be mentioned that there were two sessions of posing problems. In the first session, 18 PTs had to pose new problems while Problem 1 served as a base problem. In this case the PTs were asked not to solve the base problem and only to pose as many problems as they could, based on the given problem. In the second session, ten PTs were asked to solve Problem 2 first, and only then were they asked to pose as many problems as they could, based on the given problem.

Interpretation of Table 19.1 reveals that the average number of posed problems per PT increased from 2.4 to 6.5 (from Problem 1 to Problem 2). This increase may be attributed to the different situations involved in obtaining the two sets of posed problems. The fact that the PTs had to solve the base problem first (in the case of Problem 2) appears to have had a significant impact on the number of problems they were able to pose. While the PTs attempting to solve the base problem, they had to recall the relevant attributes of the geometrical shapes involved, and they had to examine the interrelations between the givens of the base problem. As a result, they could develop some understanding about the various possible modifications that might be applied to the base problem in order to yield new problems.

In what follows I refer to the cognitive processes applied by the PTs in the process of posing problems (sixth column). All of the following examples relate to the first session (Problem 1). Using the WIN strategy, the PTs changed either one of the base problem's givens or the base problem's question. In the case where the base problem included numerical givens, the PTs changed it to another value which was close to the original one. For example: "Change the pyramid height from 10 into 12 cm." In this case it can be said that the cognitive process applied is filtering since they had to choose certain values that would fit the question of the problem which remained the same. However, by changing a numerical given to a range of values, for example: "Change the angle between the lateral faces from 67° into an angle between 67° and 90°," in addition to filtering, the PTs had to think of possible values that could be suggested to replace the given one and yet end with a mathematically valid problem. The dragging facility provided by the dynamic geometry software also enabled the PTs to verify whether their suggestions yield mathematically valid problems or not.

When PTs changed the data type for one of the base problem givens—for example: "Change from a triangular base pyramid to a square base pyramid" (Problem 1), the PTs applied filtering, comprehending, and editing. In this case the PTs had to draw a new sketch of the problem which was completely different from the sketch of the base problem, while at the same time, they did not change the problem's question. To suggest such a given, the PTs had to comprehend the interrelationships between the problem's givens and decide whether such a suggestion could yield a mathematically valid new problem.

All four cognitive processes (filtering, comprehending, editing, and translating) were involved when PTs changed the base problem's question as in this example: "Find the pyramid base area" or: "Prove that sin $\alpha/2 = 5/8$ while the relation between a lateral edge of the regular triangle pyramid to the base edge is 5/9." By leaving the givens of the base problem untouched, they had to filter the possible questions that could be asked to yield a mathematically valid problem. Moreover, to be able to pose a reasonable new question for the given situation, PTs had to demonstrate comprehension of the geometrical shapes involved and their attributes, and they also had to understand the interrelationships between the problem's givens. In changing the question of the base problem, the PTs applied the cognitive process of translating in which they had to write a problem based on a given situation which is composed of certain geometrical shapes (e.g., triangular pyramid) and givens (e.g., pyramid height of 10 cm). Also a process of editing was applied since in this process one has to write an appropriate problem based on a given sketch, and since the givens of the base problem were not changed, the sketch of the base problem remained the same.

Similar results were also reported by Lavy and Shriki (2010). After posing problems using the WIN strategy, the PTs had to choose one of the new posed problems and provide its solution. Most of them chose a problem with a trivial change.

Although the students were familiar with the examined topics, they chose not to challenge themselves with intriguing situations. By making minor change in one of the base problem's givens, the PTs avoided the need to examine the correctness and validity of the new posed problem. This phenomenon can be attributed to their insufficient experience with problem-posing activities. These results are in line with those reported by Cemalettin et al. (2011) who found that prospective teachers' success in problem posing was low. Effective engagement in problem posing necessitates a profound examination of the definitions of the mathematical objects and their interrelationships. To avoid such an engagement, the PTs chose to suggest alternatives which minimized the need to probe the attributes of, and interrelationships between, the mathematical objects involved. Mason (2000) asserted that providing students with the opportunity to pose problems enabled them to navigate the problems they posed to their domains of interest according to their cognitive abilities. However, the results obtained in the above studies revealed that the PTs did not necessarily focus on what they found to be interesting, nor did they always utilize their cognitive abilities in full. Observations of the PTs' initial stages of inquiry (which they soon discarded) suggest that they had the opportunity to develop their mathematical knowledge far beyond what actually occurred. The fact that they overemphasized the need to provide solutions to the new posed problems prevented them from exploring less common shapes and unfamiliar situations.

The Researcher's Engagement in Problem Posing

Before a colleague and I decided to engage our PTs in problem-posing activities, each of us decided to experience this process first. We chose the following to be the base problem for our investigation: *The three medians of a triangle divide it into 6 triangles possessing the same area*. By using the WIN strategy and dynamic geometry software, we experienced a fascinating process and ended with some interesting new insights. Starting from the base problem, we negated the number of divisions of the triangle sides by raising the question: What if each of the triangle sides will be divided into three instead of two segments?

The division of each of the triangle sides into three equal segments created a new posed problem including four triangles and three quadrangles inside the given triangle (Figure 19.4).

Based on measurements taken by means of dynamic geometry software, the following conjectures with respect to the areas and segments were raised:

$$\frac{BK}{KI}=\frac{AL}{LG}=\frac{CJ}{JE}=6;\quad \frac{S_1}{S_2}=3$$

$$KJ=JB;\quad LK=KA;\quad JL=LC;\quad S_2=S_4=S_6;\quad S_3=S_5=S_7$$

In order to investigate and ultimately prove the above conjectures, segments *KD*, *LH*, and *JF* (Figure 19.4) were added to generate triangles *BDK*, *AHL*, and *JCF* and

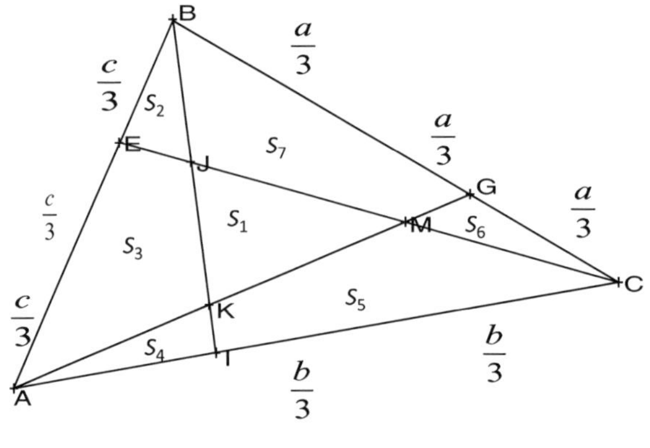

Figure 19.4. Schematic description of the case: $n=3$.

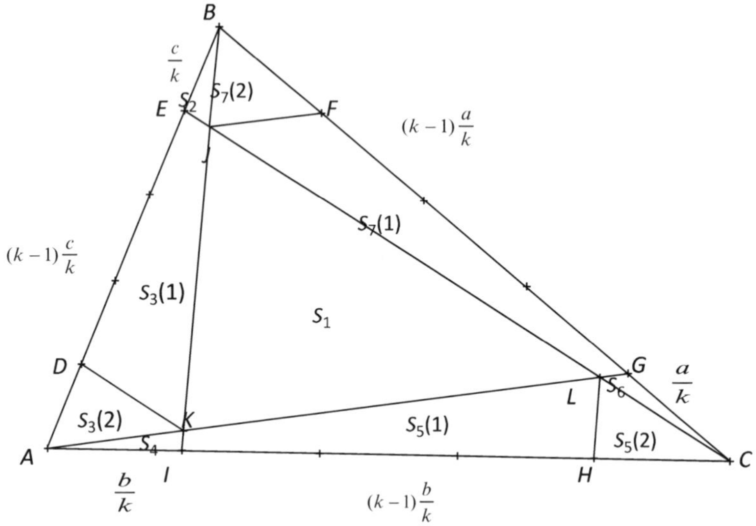

Figure 19.5. Schematic description of the case of $n=k$.

to our surprise we found that: $EJ \parallel DK$; $IK \parallel HL$; $GL \parallel FJ$. Only by using the principles of affine geometry were we able to succeed in proving the parallelism of these segments. Then we examined the general case in which each of the triangle sides is divided into k equal segments (Figure 19.5) and generated the following attributes: $S_2=S_4=S_6$; $S_3=S_5=S_7$; $\frac{S_1}{S_2}=k(k-2)^2$; $\frac{BK}{KI}=k(k-1)$; $\frac{JK}{BJ}=k-2$

Finally we examined the case in which each side of the triangle is divided into a different number of equal segments (k-ians) (see Figure 19.6).

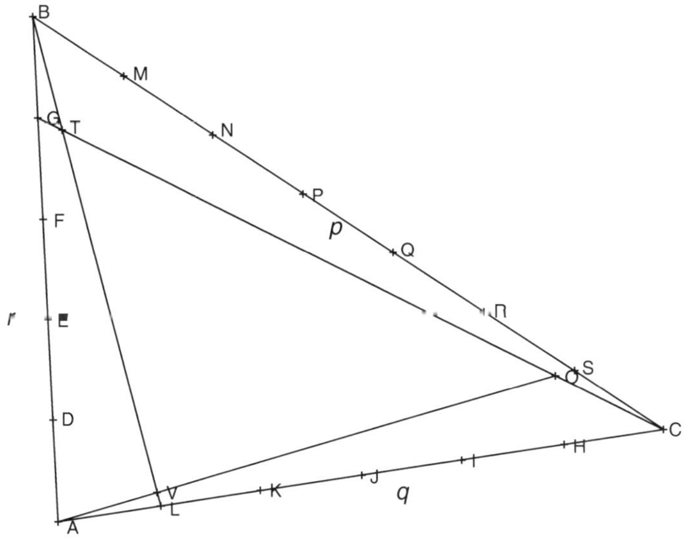

Figure 19.6. Schematic description of the *k*-ians.

In this case we found the following attributes:

$$\frac{BY}{YL}=(p-1)\cdot q;\ \frac{AU}{US}=(r-1)\cdot p;\ \frac{CT}{TG}=(q-1)\cdot r$$

The above-described process demonstrates a sequence of modifications in data from the given base problem that yielded new posed problems with new surprising regularities. Reflection on the above process reveals the potential of applying a sequence of simple modifications to one of the base problem's givens—in this case, yielding supersizing regularities. The question we asked ourselves was: Why do our PTs seem to avoid acting on given problem situations in a similar way?

Discussion and Implications for Instruction

The incorporation of problem-posing activities into the mathematics curriculum is highly recommended by the educational community (NCTM, 2000). Most students, however, are not provided with the opportunity of experiencing problem posing while studying mathematics (Silver et al., 1996; Wilson & Berne, 1999). Hence, when PTs enter teacher education programs, many of them are not yet acquainted with problem-posing activities and when they are exposed to such activities, they refer to them as unusual ones (Tichá & Hošpesová, 2013). PTs should first experience innovative teaching approaches such as problem-posing activities as learners during teacher education programs before they are able to incorporate them effectively in their teaching (Abramovich & Cho, 2006; Crespo & Sinclair, 2008). This is especially true when they encounter unfamiliar approaches which stems from the fact that they were not exposed to them as high school students (Crespo & Sinclair, 2008). If teacher educators wish PTs to implement problem posing in their future classes, they should provide them with opportunities to gain experience in it within various settings. While engaging in various activities associated with problem posing, PTs might become aware of the encompassed cognitive processes involved, discuss and reflect on them, and as a result improve their instruction skills.

My research interest in the issue of problem posing has focused on PTs' engagement in problem-posing activities in geometry. When exposing the PTs to problem-posing activities I noticed that this was often their first exposure to problem posing in mathematics in general and in geometry in particular. Informal conversations with high-school mathematics teachers have confirmed that learners of mathematics in high school rarely engaged in such activities (Lavy & Shriki, 2010). As a result, most university-level students who study to become teachers of mathematics are not familiar with problem-posing activities. My recommendations below relate to PTs who have not had the opportunity to acquire previous experience in such activities. However, in order to create a situation in which PTs will feel most comfortable in acquiring problem-posing knowledge and skills, appropriate activities should be employed earlier—when they are still school students. Considering the results on problem posing by PTs reported in this chapter, I believe that, in order to help students develop problem-posing skills, students should be engaged in problem-posing activities on a regular basis—starting in elementary school.

Problem-posing activities should be planned in a way that they will provide PSTs with the opportunity to apply the cognitive processes of filtering, editing, comprehending, and translating (Pittalis et al., 2004), which are important for the development of problem-posing skills. Teachers should choose various problems relating to the current content topic and initiate problem-posing activities in which the above cognitive processes could be developed. Posing new problems can be based on free, semi-structured, and structured situations (Stoyanova, 1998). Based on my experience, I believe that high school students should be engaged in problem-posing activities in geometry basing on structured situations. Many students find geometry challenging and encounter difficulties when attempting to solve geometrical problems (Gal & Linchevski, 2010; Lin, 2005). Moreover, since many preservice teachers tend to refer to problem posing as a very unusual and complex activity (Tichá & Hošpesová, 2013), I believe that working in a structured situation can make the process of problem posing easier for the PTs. For this reason, I found the WIN strategy (Brown & Walter; 1993) to be useful. Problem posing using the WIN strategy encompasses the four cognitive processes (Pittalis et al., 2004) as was demonstrated in the results section. Changing one of the givens of the base problem or the problem's question can result in a process of filtering, translating, comprehending, or editing. The use of a structured approach to problem posing should provide a gentler transition from problem-solving activities in which students have to cope with valid and solvable problems to problem-posing activities in which new problem situations can be neither mathematically valid nor solvable.

Class discussions on activities involving problem posing are essential for the development of PTs' problem-posing skills (Lavy & Shriki, 2007). Problem-posing activities should be followed by class discussions in which the posed problems can be discussed and guided by the class teacher. In such discussions PTs could reflect on the process they had gone through and ask questions such as: "Does the suggested alternative result in a mathematically valid problem situation?" or "What can be the consequences of changing one of the givens to another geometrical shape?" or "Does the new problem situation include missing/redundant data to solve the new problem?" will be discussed. One of the advantages of class discussions following problem-posing activities is the PTs' exposure to classmates' ideas that they themselves had not thought about. Christou et al. (2005) noted that the discussions which followed a problem-posing activity helped the students to reconsider their generalizations. Before the class discussion the students seemed to over-generalize their solutions, based on particular cases, and they failed to extend the problem to all possible situations. Only after the discussion were the students able to generalize correctly.

One of the important skills PTs have to develop in order to be effectively engaged in problem-posing activities is reflection. Among the means by which reflection skills can be developed are class discussions (McDuffie & Slavit, 2003). Class discussions, in which the participants exchange ideas regarding the attributes and interrelationships of the mathematical objects under examination with other members in class, may stimulate the development of their reflection skills. Each decision students make in the process of problem posing necessitates reflective thoughts regarding the meanings and consequences of such a decision. Cunningham (2004) found that engagement in problem-posing activities improves students' refection skills. Relying on my own experience I believe that a certain degree of reflection skills are needed a priori for engagement in effective problem posing. These skills are essential for probing the attributes and interrelationships of the mathematical object under examination and the possible consequences of replacing any one of these by another.

Frequent engagement in problem-posing activities can contribute to the development of the PTs' self-confidence in their mathematical abilities. This confidence is required especially in cases PTs have to cope with complex situations which may draw upon advanced mathematical topics they have not yet mastered (e.g., affine geometry) in order to investigate a regularity they have discovered. PTs' self-confidence in their mathematical abilities can also help them to develop their ability to think "outside the box" and be free of some traditional constraints. Self-confidence in one's mathematical abilities also applies to meta-knowledge of mathematics. In order to be able to think of possible alternatives to a negated given, one has to be able to probe into the given's attributes and possible interrelationships, as well as understand the possible consequences of suggesting other data with different attributes and different interrelationships. Moreover, PTs' engagement in problem-posing activities can also help them build their self-confidence in their ability to handle problem-posing activities and to manage follow-up class discussions effectively.

Structured and guided activities of problem posing have an important role in shaping PTs' inquiry habits. They need to develop systematic inquiry habits progressing by small steps. Moving forward through a sequence of small changes can help PTs observe whether a mathematical regularity can be unfolded.

From the beginning of students' exposure to geometrical objects, they should be introduced to dynamic geometry environments in which they can create new objects and move and reshape them interactively. The process of problem posing in geometry can be facilitated when using DGS which frees the PTs from technical work and enables them to focus on the inquiry process. The DGS enables the PTs to experiment, observe the stability or instability of phenomena, and state and verify conjectures easily and rapidly (Marrades & Gutiérrez, 2000). The visual aspect provided by the software is also crucially important. By freeing the PTs from technical work which is time-consuming, they are able to invest more efforts into examining interrelationships between the problem's givens, and think of potentially interesting changes. The dragging facilities of the software and the fact that the geometric objects can be easily manipulated and reshaped interactively (Sinclair, 2004) enable the PTs to view on the computer screen a kind of a proof.

Engagement in problem posing necessitates the organization of the PTs' existing knowledge in such a way that they will be able to draw on this knowledge—in not just a technical manner. Problem posing should be implemented in ways that students will be able to make sense of the activity via the cognitive tools already at their disposal. Problem-posing activities should be presented to students in ways that allow them experience a content-related sense of purpose, and that bring them to see the point of extending their existing conceptual knowledge and experiences in fruitful directions.

References

Abramovich, S., & Cho, E. (2006). Technology as a medium for elementary pre-teachers' problem posing experience in mathematics. *Journal of Computers in Mathematics and Science Teaching, 25*(4), 309–323.

Artigue, M. (2002). Learning mathematics in a CAS environment: The genesis of a reflection about instrumentation and the dialectics between technical and conceptual work. *International Journal of Computers for Mathematical Learning, 7*(3), 245–274.

Aviram, A. (2001). From "computers in the classroom" to mindful radical adaptation by education systems to the emerging cyber culture. *Journal of Educational Change, 1*, 331–352.

Brousseau, G. (1997). *Theory of didactical situations in mathematics. Didactique des Mathematiques 1970–1990*. Dordrecht, The Netherlands: Kluwer.

Brown, S. I., & Walter, M. I. (1983). *The art of problem posing*. Hillsdale, NJ: Erlbaum.

Brown, S. I., & Walter, M. I. (1993). Problem posing in mathematics education. In S. I. Brown & M. I. Walter (Eds.), *Problem posing: Reflection and applications* (pp. 16–27). Hillsdale, NJ: Erlbaum.

Cemalettin, I., Tuğrul, K., Tuğba, Y., & Kıymet, Z. (2011). Prospective teachers' skills in problem posing with regard to different problem posing models. *Procedia—Social and Behavioral Sciences, 15*, 485–489.

Christou, C., Mousoulides, N., Pittalis, M., & Pitta-Pantazi, D. (2005). Problem solving and problem posing in a dynamic geometry environment. *The Montana Mathematics Enthusiast, 2*(2), 125–143.

Contreras, J., & Martinez-Cruz, A. A. (1999). Examining what prospective secondary teachers bring to teacher education: A preliminary analysis of their initial problem-posing abilities within geometric tasks. In F. Hitt & M. Santos (Eds.), *Proceedings of PME-NA XXI* (Vol. 2, pp. 413–420).

Crespo, S., & Sinclair, N. (2008). What makes a problem mathematically interesting? Inviting prospective teachers to pose better problems. *Journal of Mathematics Teacher Education, 11*, 395–415.

Cunningham, R. F. (2004). Problem posing: An opportunity for increasing student responsibility. *Mathematics and Computer Education, 38*(1), 83–89.

Einstein, A., & Infeld, L. (1938). *The evolution of physics from early concepts to relativity and quanta*. New York, NY: Simon & Schuster.

Ellerton, N. F. (1986). Children's made-up mathematics problems—A New perspective on talented mathematicians. *Educational Studies in Mathematics, 17*, 261–271.

Ellerton, N. F., & Clarkson, P. C. (1996). Language factors in mathematics teaching and learning. In A. J. Bishop, K. Clements, C. Keitel, J. Kilpatrick, & C. Laborde (Eds.), *International handbook of mathematics education* (pp. 987–1033). Dordrecht, The Netherlands: Kluwer.

English, L. D. (1997). Promoting a problem-posing classroom. *Teaching Children Mathematics, 4*, 172–179.

Ernest, P. (1991). *The philosophy of mathematics education*. London, United Kingdom: Palmer Press.

Gal, H., & Linchevski, L. (2010). To see or not to see: Analyzing difficulties in geometry from the perspective of visual perception. *Educational Studies in Mathematics, 74*, 163–183.

Goldenberg, E. P. (1993). On building curriculum materials that foster problem posing. In S. Brown & M. Walter (Eds.), *Problem posing: Reflections and applications* (pp. 31–38). Hillsdale, NJ: Erlbaum.

Gonzales, N. A. (1996). Problem formulation: Insights from student generated questions. *School Science and Mathematics, 96*(3), 152–157.

Laborde, C. (1998). Visual phenomena in the teaching/learning of geometry in a computer-based environment. In C. Mammana & V. Villani (Eds.), *Perspectives on the teaching of geometry for the 21st century* (pp. 113–122). Dordrecht, The Netherlands: Kluwer.

Lavy, I., & Bershadsky, I. (2003). Problem posing via "What if not?" strategy in solid geometry—A case study. *The Journal of Mathematical Behavior, 22*(4), 369–387.

Lavy, I., & Shriki, A. (2007). Problem posing as a means for developing mathematical knowledge of prospective teachers. In W. Jeong-Ho, P. Kyo-Sik, L. Hee-Chan, & S. Dong-Yeop (Eds.), *Proceedings of the 31st International Conference on the Psychology of Mathematics Education* (Vol. 3, pp. 129–136). Seoul, South Korea: PME.

Lavy, I., & Shriki, A. (2009). Small change—Big difference. *The Montana Enthusiast, 6*(3), 395–410.

Lavy, I., & Shriki, A. (2010). Engagement in problem posing activity in a dynamic geometry setting and the developing of mathematical knowledge of prospective teachers. *The Journal of Mathematical Behavior, 29*(1), 11–24.

Leung, S. S., & Silver, E. A. (1997). The role of task format, mathematics knowledge and creative thinking on the arithmetic problem posing of preservice elementary school teachers. *Mathematics Education Research Journal, 9*(1), 5–24.

Lin. F. L. (2005). Modeling students' learning on mathematical proof and refutation. In H. Chick & J. Vincent (Eds.), *Proceedings of the 29th Conference for the Psychology of Mathematics Education* (Vol. 1, pp. 3–18), Melbourne, Australia: PME.

Marrades, R., & Gutiérrez, Á. (2000). Proofs produced by secondary school students learning geometry in a dynamic computer environment. *Educational Studies in Mathematics, 44*(1–2), 87–125.

Mason, J. (2000). Asking mathematical questions mathematically. *International Journal of Mathematical Education in Science and Technology, 31*(1), 97–111.
McDuffie, A. R., & Slavit, D. (2003). Utilizing online discussion to support reflection and challenge beliefs in elementary mathematics methods classrooms. *Contemporary Issues in Technology and Teacher Education, 2*(4), 446–466.
McKenzie, J. (2009). *Beyond cut-and-paste: Engaging students in making good new ideas.* Bellingham, WA: FNO Press.
Mestre, P. J. (2002). Probing adults' conceptual understanding and transfer of learning via problem posing. *Applied Developmental Psychology, 23*, 9–50.
Moses, B. E., Bjork, E., & Goldenberg, P. E. (1990). Beyond problem solving: Problem posing. In T. J. Cooney & C. R. Hirsch (Eds.), *Teaching and learning mathematics in the 1990's* (pp. 82–91). Reston, VA: NCTM.
National Council of Teachers of Mathematics. (2000). *Principles and standards for school mathematics*. Reston, VA: Author.
Philippou, G. N., Charalambous, C., & Christou, C. (2001). Efficacy in problem posing and teaching problem posing. In M. Van den Heuvel-Panhuizen (Ed.), *Proceedings of the 25th Conference of the International Group for the Psychology of Mathematics Education* (Vol. 4, pp. 41–48). Utrecht, The Netherlands: PME.
Pittalis, M., Christou, C., Mousoulides, N., & Pitta-Pantazi, D. (2004). A structural model for problem posing. In M. J. Høines & A. B. Fuglestad (Eds.), *Proceedings of the 28th Conference of the International Group for the Psychology of Mathematics Education* (Vol. 4, pp. 49–56). Bergen, Norway: PME.
Ranasinghe, A. I., & Leisher, D. (2009). The benefit of integrating technology into the classroom. *International Mathematical Forum, 4*(40). Retrieved December 2010, from http://www.m--hikari.com/imf-password2009/37-40-2009/ranasingheIMF37-2009.pdf.
Shriki, A., & Lavy, I. (2012). Problem posing and the development of mathematical insights. *International Journal of Learning, 18*(5), 61–70.
Silver, E. A. (1994). On mathematical problem posing. *For the Learning of Mathematics, 14*(1), 19–28.
Silver, E. A., Mamona-Downs, J., Leung, S. S., & Kenney, P. A. (1996). Posing mathematical problems: An exploratory study. *Journal for Research in Mathematics Education, 27*(3), 293–309.
Sinclair, M. (2004). Working with accurate representations: The case of preconstructed dynamic geometry sketches. *Journal of Computers in Mathematics and Science Teaching, 23*(2), 191–208.
Skinner, P. (1990). *What's your problem?* Melbourne, Australia: Nelson.
Southwell, B. (1998). Problem solving through problem posing: The experiences of two teacher education students. In C. Kanes, M. Goos, & E. Warren (Eds.), *Teaching mathematics in new times* (pp. 524–531). Brisbane, Australia: MERGA.
Stoyanova, E. (1998). Problem posing in mathematics classrooms. In A. McIntosh & N. F. Ellerton (Eds.), *Research in mathematics education: A contemporary perspective* (pp. 164–185). Perth, Australia: Edith Cowan University.
Tichá, M., & Hošpesová, A. (2013). Developing teachers' subject didactic competence through problem posing. *Educational Studies in Mathematics, 83*(1), 133–143.
Wilson, S. M., & Berne, J. (1999). Teacher learning and the acquisition of professional knowledge: An examination of research on contemporary professional development. In A. Iran-Nejad & P. D. Pearson (Eds.), *Review of research in education* (Vol. 24). Washington, DC: AERA.
Yevdokimov, O. (2005). On development of students' abilities in problem posing: A case of plane geometry. In *Proceedings of the 4th Mediterranean Conference on Mathematics Education* (pp. 255–267). Palermo, Italy: Cyprus Mathematical Society.

Chapter 20
Developing the Problem-Posing Abilities of Prospective Elementary and Middle School Teachers

Todd A. Grundmeier

Abstract This chapter describes the results of an exploratory study incorporating problem posing in a mathematics content course for prospective elementary and middle school teachers. Problem posing was incorporated as problem generation (posing problems from a set of given information) and problem reformulation (posing problems related to a given problem). The content coverage of the course included problem solving, data analysis and probability, discrete mathematics, and algebraic thinking. Exposure to problem posing had two effects on those who posed the problems. First they began using more sophisticated problem reformulation techniques as the course progressed. Second, with regard to problem generation, participants developed efficient ways of posing problems when time constraints were imposed, and they developed greater aptitude for posing multi-step problems. The development of participants' problem-posing abilities will be described in detail, and qualitative data will be presented to highlight participants' views of the relationship between problem posing and school mathematics.

Contents

T.A. Grundmeier (✉)
Department of Mathematics, California Polytechnic State University, San Luis Obispo, CA, USA
e-mail: tgrundme@calpoly.edu

F.M. Singer et al. (eds.), *Mathematical Problem Posing*,
Research in Mathematics Education, DOI 10.1007/978-1-4614-6258-3_20

If we change the question in the title to 'Where do good mathematics problems come from?', the answer ought to be readily apparent to any competent high school graduate. Mathematics problems obviously come from mathematics teachers and textbooks, so good mathematics problems must come from good mathematics teachers and good mathematics textbooks. The idea that students themselves can be the source of good mathematics problems has probably not occurred to many students or to many of their teachers. (Kilpatrick, 1987, p. 123)

Introduction

Kilpatrick (1987) suggested that instruction rich in formulating problems that requires students to become problem posers is essential throughout mathematics education. The landscape of mathematics education has encountered much change since Kilpatrick (1987) wrote these words and many educators and authors have considered mathematical problem posing as a skill. However, it can still be argued that students are not "required" to become problem posers.

Through the early 2000s both mathematics educators and professional organizations continued to advocate for the inclusion of problem posing in mathematics classrooms and curricula (Kilpatrick, Swafford, & Findell, 2001; NCTM 1991, 2000; Silver, 1994). Literature within the mathematics education community also focused on the importance of problem posing and research has demonstrated the problem posing capabilities of K-6 students (English, 1997; Silver, 1997; Silver & Cai, 1996; Winograd, 1997). Problem-posing research in the late 1990s likely led to the following suggestion for the incorporation of problem posing in mathematics classrooms and curricula by The National Council of Teachers of Mathematics (NCTM), in *Principles and Standards for School Mathematics* (2000):

> Posing problems comes naturally to young children …Teachers and parents can foster this inclination by helping students make mathematical problems from their worlds … In such supportive environments, students develop confidence in their abilities and a willingness to engage in and explore problems, and they will be more likely to pose problems and persist with challenging problems. (p. 53)

Although much problem-posing research occurred during the 1980s and 1990s, problem posing became prominent again in the mathematics education research literature in the early 2000s (see, e.g., Barlow & Cates, 2006; Stickles, 2010–2011; Whiten, 2004a, 2004b). In order to facilitate the suggestions made by Kilpatrick (1987) and others, to incorporate problem posing at all levels of mathematics instruction, prospective teachers need problem-posing experiences as part of their preservice education (Leung & Silver, 1997).

Gonzales (1994, 1998) examined the incorporation of problem posing in instruction for prospective teachers. Gonzales (1994) suggested engaging prospective elementary and middle-school teachers in problem posing by posing related problems and posing story problems. Gonzales (1994) found that prospective teachers could be guided through a transition from problem solver to problem poser and based on this transition called for the increased use of problem posing with this audience. Gonzales (1998) described a "blueprint" to help teachers and teacher educators include problem posing in their classrooms. The "blueprint" started with posing related problems and after exposure to problem reformulation asked students to generate problems. The project described here incorporated some aspects of this blueprint, as well as subsequent work of Gonzales (1994, 1998), but extended her ideas by formally exploring the outcomes of incorporating problem posing into students' mathematical experiences. This extension of Gonzales' work addresses the need to develop the problem-posing abilities of prospective elementary and middle-school teachers by engaging them in problem posing throughout a mathematics content course designed for prospective teachers. The goal of this work, as suggested by Leung and Silver (1997), was to carry out a careful evaluation of empirical problem posing and to describe changes in the characteristics of participants' posed problems as they gained problem-posing experience.

Methodology

The instructional treatment for this study was the incorporation of problem posing into the expectations of a mathematics content course for prospective teachers ($n = 19$). The main components of the methodology are the working definitions used by the author, the course setting, the participants, the instructional treatment, and data collection. These components will be discussed below.

Working Definitions

In this study problem posing took two forms: (a) the generation of new problems; and (b) the reformulation of given problems (Silver, 1994). It is important to define statement, problem, problem reformulation, and problem generation to give a sense of how the ideas were utilized for the purpose of this research. Definitions are summarized in Table 20.1.

Table 20.1
Working Definitions

Term	Working definition
Statement	A statement will refer to the outcomes of student problem-posing tasks. Statements are all text that is produced as a response to a problem-posing task and is not necessarily a mathematics problem or question
Problem	A mathematical statement for which a valid solution exists
Problem reformulation	The process of posing a problem related to a problem that is or was the focus of problem solving
Problem generation	The process of posing a problem based on a set of given information. Generated problems may include additional information to the original set but must be related to the original set of information

Course Setting

The course was the second in a required sequence of mathematics content courses for prospective elementary and middle-school teachers. The content coverage of the course included problem solving, data analysis and probability, discrete mathematics, and algebraic thinking. The course instructor routinely modeled a mathematics classroom environment that was student-centered, included group work and discussion, and was a safe environment for participants to ask questions and pose conjectures. The class met twice a week for 1 hour and 50 minutes and a typical class would consist of a brief lecture followed by a group activity that asked participants to explore the mathematics content from the lecture in depth. Each class generally concluded with a discussion of the content covered and the goals of the instructional situation. The daily class activities could be described as inquiry-oriented and focused on participant problem solving. Inquiry-oriented in this context refers to the definition offered by Silver (1997) and was "… characterized as one in which some of the responsibility for problem formulation and solution is shared between teacher and students" (p. 77). The inquiry-oriented nature of the class was important because engaging in such problem-solving activities can help students develop more mathematical creativity (Silver, 1997).

Participants

Students enrolled in the course were the participants in this study. Past research has shown that preservice teacher-education students have the ability to pose mathematics problems (Gonzales, 1994). Also, if problem posing is going to become predominant in mathematics classrooms and curricula as suggested by NCTM (2000) and Kilpatrick, Swafford & Findell (2001), it is the author's belief that prospective teachers should not only have experience in posing mathematics problems but also the opportunity to reflect on the role of problem posing in school mathematics. Twenty students were enrolled in the semester-long course and 19 of those students agreed to serve as participants in this study. Participants included 17 undergraduates who were mathematics education or family-studies majors and 2 graduate students working towards their master's degree in education. All participants were prospective teachers and intended to seek certification to teach in the elementary or middle school.

Instructional Treatment and Data Collection

The classroom instructor and the author met before the semester and agreed on incorporating problem posing in the course in the form of a pre- and post-assessment, problem reformulation, problem generation, and journal writing. During the semester the instructor and author met weekly. The goal of these meetings was to examine the course content and plan for the coming week. These meetings led to the development of problem-posing tasks for the instructional treatment and agreement between the instructor and author about how these tasks would be incorporated into the course expectations. All tasks related to problem posing were incorporated into the expectations of the course as part of collected homework assignments, except for the pre- and post-assessments. The only class time directly dedicated to problem-posing tasks was for the pre- and post-assessments, each of which took 25 minutes.

Problem reformulation occurred as an extension of Polya's (1957) four-step problem-solving heuristic. After solving problems using the four step heuristic on the first problem set assigned as homework, participants were asked to use a five-step problem-solving heuristic adding the fifth step—"pose a related problem"—on the remainder of the homework problem sets. Participants were asked to apply this heuristic to a subset of each problem set and in all cases were able to choose the problems to which they applied the heuristic. Problem reformulation occurred on 7 problem sets during the semester and related to a total of 22 problems that were assigned to students to solve.

Problem generation occurred on the pre- and post-assessments, a journal entry, and two problem sets during the semester. The sets of given information provided participants with the context of possible mathematics problems but did not include any questions. The first problem-generation task was presented in a prompted journal entry that was completed as homework and included reflection on the problem-posing process. The final two problem-generation tasks were part of assigned homework problem sets.

The goal of the problem-reformulation and problem-generation activities was to provide participants with opportunities to pose mathematics problems. Therefore, the instructor checked participants' problem-posing work for completeness, but did not grade the assignments or count the assignments in the determination of their course grade. It was also a goal of the project to explore participants' views of the relationship between problem posing and school mathematics. The catalysts for this exploration were journal prompts assigned as homework. The remainder of this chapter will focus on outcomes related to participants' problem posing, while participants' views of the relationship between problem posing and school mathematics will be presented to add context to the problem-posing results.

Problem Generation Coding

An adaptation of Leung and Silver's (1997) scheme was chosen to code problems because they determined that "...the multi-stage analytic scheme ... functioned in a reasonable way..." (p. 18). Based on this, a statement determined to be mathematical from a problem-generation activity was coded along three dimensions: plausibility, sufficiency of information, and complexity. An implausible problem is one that contains an invalid assumption and hence is not plausible to solve even with more information. Similar to Leung (1993), implausible problems were not coded further, since the author of this paper was interested in problems that contained a plausible solution. If a posed problem was plausible, the author then determined whether there was sufficient information to solve the problem. Problems with extraneous information were coded as having sufficient information since they were solvable. If a problem was both plausible and contained sufficient information, it was then determined if multiple steps were necessary for solution. Arithmetic steps were not the determining characteristic of a multi-step problem because, as suggested by Silver and Cai (1996), counting steps is easy but could cause simple arithmetic problems to be coded as fairly complex. To solve a multi-step problem, the problem solver must be required to perform at least two mathematical tasks. Problems posed from problem-generation activities were assigned a score as shown in Table 20.2.

Table 20.2
Criteria for Scores from Problem-Generation Coding

Score	Criteria
0 pts	Non-mathematical statement or mathematical statement but not a plausible problem
1 pt	Plausible problem without sufficient information
2 pts	Single-step plausible problem with sufficient information
3 pts	Multi-step plausible problem with sufficient information

Problem Reformulation Coding

Classification of problems began with four posing techniques (switch the given and wanted, change the context, change the given, add information) that describe the relationship between the posed and original problem. Techniques were added until all problems could be described as being posed using at least one technique. It is also important to note that a single problem reformulation could have employed

Table 20.3
Problem Reformulation Techniques

Category	Description
Switch the given and the wanted	A problem in the same context as the original problem with the given and wanted information switched
Change the context	A problem with the same structure but context changed
Change the given	Same problem context and structure but the given information is changed
Change the wanted	Same problem context and structure but what the question asks for is changed
Extension	An extension of the given problem
Add information	Same problem context and structure with added information
Re-word	Same problem with different wording

two or more techniques. For instance, a participant could change the given and change the wanted of the same problem to produce a new related problem. Table 20.3 describes the techniques that exhausted the coding of all problems posed as problem-reformulation.

Inter-Rater Reliability

To check the validity of the coding, two additional raters volunteered to code problems from problem-posing tasks. Raters used the developed schemes to code 90 problems from problem generation tasks and 75 problems from problem reformulation tasks. With regard to problem-generation coding, the author and raters agreed on the plausibility of 96.7% of the problems. Of the 87 problems the raters and author agreed were plausible, all agreed that 87.4% contained sufficient information. Of the 76 problems agreed upon as containing sufficient information, the author and raters agreed that 80.3% required a multi-step solution. All problems not agreed upon were discussed and consensus was reached between the author and raters.

With regard to problem-reformulation tasks, the author asked the raters to code problems based on the seven techniques developed during the initial coding and to report if they believed other techniques were used. Neither rater suggested another technique. Of the 75 problems the author and raters agreed on the coding of 74.7% of the problems. The main discrepancies in coding occurred when the raters coded problems into multiple categories and often considered changing the given as an extension of a problem. The 19 problems coded differently were discussed and coding was agreed upon.

Similar to Leung and Silver (1997), there were high levels of inter-rater agreement on the coding schemes for both problem-reformulation and problem-generation. Based on this the author continued coding all posed problems using the schemes described for problem-generation and problem-reformulation.

Problem-Posing Results

Pre- and Post-Assessment

The same problem generation assessment was administered on the first and then final day of the semester. Participants had 25 minutes in class to pose as many problems as they could. The measure (see Figure 20.1) consisted of a set of information with numeric content and a set of information without numeric content.

The assessment was coded using the described problem generation scheme and each participant received a score for numeric and non-numeric posing based on the scores described in Table 20.2. Aggregate data from the assessments are summarized in Table 20.4.

Directions: Consider the possible combinations of pieces of information given below and pose as many mathematical problems as you can think of.

Numeric Set of Information: You have decided to purchase a computer for college. The new top-of-the-line laptop costs $2500. You have two options for purchasing the computer, you can use your credit card, which has an annual interest rate of 13.99%, or you can finance it through the university computer store for 48 months at $70 a month. You have saved $500, but you need to be able to pay for your books next semester.

Non-Numeric Set of Information: The university has decided to build a parking garage for the use of students and staff. The university has a maximum amount of land that they can use and also a minimum number of faculty/staff spots and a minimum number of student spots needed at certain hours of the day. The university has done research that shows that a fixed number of faculty/staff and a fixed number of students arrive at 8am and 12 noon. The university is also restricted by a fixed budget for paving and general construction.

Figure 20.1. Pre- and post-assessment of problem posing.

Table 20.4
Aggregate Pre- and Post-assessment Problem-Posing Data

	Pre-assessment	Post-assessment
Statements	101	133
Plausible	96 (95%)	122 (92%)
Sufficient information	55 (54%)	87 (65%)
Multi-step solution	16 (16%)	37 (28%)
Numeric average	5.21	8.72
Non-numeric average	3.47	4.89

Results on the pre- and post-assessment were compared using a Tukey–Kramer multiple comparisons matched-pairs test at the alpha equals 0.05 level. The data from the participant who did not complete the post-assessment were not included in the analysis. The statistical analysis showed that the difference in the means of Numeric pre and Numeric post ($q=0.97$), as well as Numeric post and Non-numeric post ($q=1.41$), is statistically significant.

With regard to numeric problem-posing ability, the participants' average increased from 5.21 on the pre-assessment to 8.72 on the post-assessment. For non-numeric posing the participants' average changed from 3.47 to 4.89. It is important to consider if these changes are because participants were writing more situations or were generating more plausible problems with sufficient information that required multi-step solutions. The data in Table 20.4 suggest that participants' efficiency in posing problems increased, as they posed 122 plausible problems on the post-assessment compared to 96 on the pre-assessment. It is also clear that participants posed more problems with sufficient information requiring multi-step solutions on the post-assessment. The results of the pre- and post-assessments suggest that, after this course, which included exposure to problem posing, participants became more efficient at posing problems when problems were generated under a time constraint, and they posed more multi-step problems with sufficient information for solution. The remaining results related to problem generation will highlight that the characteristics of participants' problem generation were consistent with the tasks collected during the semester.

Problem Generation

Other than the pre- and post-assessments, participants engaged in problem generation during the fifth, seventh, and tenth weeks of the semester. The first problem generation task was assigned as part of a journal entry and asked students to pose three to five problems. The set of given information and typical problems follow in Table 20.5.

Table 20.5
Typical Week 5 Problem Generation

Given information	Mrs. Smith's and Mr. Jones' fifth-grade classes took the same mathematics test last week. You have been given all the graded exams and the answer key
Not plausible (0 pts)	Do you feel by the overall grades, that it would be fair to scale the grades or should students get the grade they earned?
Plausible without sufficient information (1 pt)	There are 15 students in Mrs. Smith's class and 12 students in Mr. Jones' class. The median of all the tests from both classes is an 82. How many students scored above the median? How many students scored below the median?
Plausible with sufficient information (2 pts)	Mr. Jones' class has an average of 80 and there are 18 students who have taken the exam, but Suzy was absent that day. If she takes the test and gets a 99 what is the new average?
Plausible, sufficient information and multi-step (3 pts)	Does the mean, median, or mode best reflect the class test scores in Mrs. Smith's class [test data was included]? Explain why you feel as you do?

Sixty-seven percent of the problems generated on this task were multi-step problems and only 19% were not plausible or did not contain sufficient information for solution. Participants also added information to 29% of the problems, most likely due to the non-numeric nature of the set of given information.

The week 7 problem generation task was assigned as part of the homework problem set and asked participants to pose at least three problems. The set of given information and typical problems follow in Table 20.6.

Table 20.6
Typical Week 7 Problem Generation

Given information	You arrive at your friend's home and they are sitting at a table with $20, a deck of cards, and red, white, and blue die
Plausible without sufficient information (1 pt)	If there are 4 red chips, 8 white chips, and no blue chips in a pile on the table and the betting is $8 so far what do the red and white chips stand for?
Plausible with sufficient information (2 pts)	If the cards Ace, 2, 3, … up through J, Q, K are each given a value 1–13, in order, what is the probability that a card picked at random will have a value greater than 10?
Plausible, sufficient information and multi-step (3 pts)	Your friend offers to give you $10 if you get a sum of 9, 10, 11, or 12 when all 3 die are rolled. You have four chances. If you do not roll any of these sums in your four chances you owe him $10. Are you going to accept this challenge? Why or why not?

Participants continued the trend of posing multi-step problems on this task, as 56% of the generated problems required a multi-step solution. All problems posed were plausible, only 12% did not contain sufficient information or information was not added to any of the posed problems.

The week 10 problem-generation task was assigned as part of the homework problem set and participants were asked to pose at least two problems. Table 20.7 includes the set of given information and typical posed problems from the task.

Table 20.7
Typical Week 10 Problem Generation

Given information	A roulette wheel has 18 red numbers, 18 black numbers, and 2 green numbers. A person bets on either an individual number or a color. A one dollar bet on an individual number pays $35, on black or red pays $1, and on green pays $12
Not plausible (0 pts)	A roulette wheel has 18 red numbers, 18 black numbers, and 2 green numbers. If Annie puts $5 on one red number and $5 on two black numbers what is the probability that she will win $10 in 2 spins?
Plausible without sufficient information (1 pt)	How many bets would you have to make to win $80?
Plausible with sufficient information (2 pts)	If you bet on black 23 times in a row and win 12 times. Do you have more or less money than when you started?
Plausible, sufficient information and multi-step (3 pts)	Would you bet on an individual number, black, red, or green? Explain your decision using probability

Similar to the first two problem generation tasks, 61% of the posed problems required a multi-step solution and only 13% were not plausible or did not contain sufficient information. As with the second task, information was not added to any of the posed problems. Table 20.8 summarizes the results from the three problem-generation tasks.

Table 20.8
Aggregate Problem Generation Results from the Semester

	Statements	Plausible	Sufficient information	Multi-step solution	Added information
Week 5	42	39 (93%)	34 (81%)	28 (67%)	12 (29%)
Week 7	48	48 (100%)	42 (88%)	27 (56%)	0
Week 10	23	21 (91%)	20 (87%)	14 (61%)	0

Participants were not under a time constraint as they had at least 5 days to complete each of the tasks. The data in Table 20.8 suggest that the characteristics of participants' problem generation during the semester were consistent. These participants were able to pose plausible, multi-step problems from sets of information regardless of whether they contained numeric information. This consistency may have been a pre-cursor to participants' apparent aptitude for posing multi-step problems on the post-assessment.

Problem Reformulation

Participants engaged in problem reformulation on seven homework problem sets during the semester. The participants utilized two distinct types of problem reformulation techniques. "Surface" techniques consisted of adding information, changing the given, changing the wanted, and re-wording. Surface reformulation techniques did not require the problem poser to change the structure of the problem; they required only a change of the surface features of the problem (e.g., numbers, what is asked for). "Structure" techniques included switching the given and wanted, changing the context, and extending the original problem. Structure reformulation techniques required more creativity and a deeper understanding of mathematical content on the part of the problem poser, as they required changing the structure of the problem. The utilization of these two types of problem reformulation techniques will be discussed in this section. Table 20.9 provides an overview of participant problem reformulation.

The mathematical content focus of the problem sets in weeks 4 and 5 was problem solving and data analysis. Reformulation on these problem sets was dominated by changing the given and changing the wanted. Structure reformulation techniques were 22% of the techniques utilized in week 4 and increased to 35% of the techniques utilized in week 5. On both problem sets switching the given and the wanted was the most popular structure technique. The increase in use of structure techniques

Table 20.9
Aggregate Problem Reformulation Data

	Week 4	Week 5	Week 6	Week 7	Week 10	Week 11	Week 15
Total posed problems	49	56	42	40	37	35	32
Total techniques	59	62	45	45	42	37	35
Switch given and wanted	7	11	6	3	3	0	2
Change context	2	5	3	1	4	3	5
Extend original	4	6	0	2	5	13	5
Structure techniques	13 (22%)	22 (35%)	9 (20%)	6 (13%)	12 (29%)	16 (43%)	12 (34%)
Change given	27	26	25	17	16	13	17
Add information	3	3	3	5	2	3	3
Change wanted	15	10	8	15	11	5	3
Re-word original	1	1	0	2	1	0	0
Surface techniques	46 (78%)	40 (65%)	36 (80%)	39 (87%)	30 (71%)	21 (57%)	23 (66%)

Table 20.10
Problem Reformulation from Weeks 4 and 5

Original problem Week 3	A special rubber ball is dropped from the top of a wall that is 16 feet high. Each time the ball hits the ground it bounces back only half as high as the distance it fell. The ball is caught when it bounces back to a high point of 1 foot. How many times does the ball hit the ground?
Switch the given and the wanted	If a special rubber ball is dropped from a wall with an unknown height and bounces four times and is caught at the height of its fourth bounce at 2 feet. If we know that every time the ball bounces it only bounces back half the distance as the distance it fell. How high is the wall the ball dropped off of originally?
Change the given	A special rubber ball is dropped from the top of a wall that is 768 feet tall. Each time the ball hits the ground it bounces back only one-fourth as high as the distance it fell. The ball is caught when it bounces back to a high point of 3 feet. How many times does the ball hit the ground?

in week 5 may be attributed to it being the second problem set related to the mathematical content of problem solving and data analysis. Table 20.10 includes typical examples of problem reformulation on these problem sets.

Data representation and analysis was the mathematical content focus of the problem set in week 6. Participants were still relying heavily on changing the given information and structure techniques were 20% of the techniques utilized. This similarity to reformulation in week 4 may be attributed to this being the only problem set

Table 20.11
Problem Reformulation from Week 6

Original problem Week 5	The average of 7 numbers is 49. If 1 is added to the first number, 2 is added to the second number, 3 is added to the third number, 4 is added to the fourth number, and so on up to the seventh number, what is the new average?
Switch the given and the wanted	The average of 7 numbers is 49. Each of the data points were increased by the same amount. The new average is 53, what value was each data point increased by to raise the mean?
Change the given	The average of 11 numbers is 121. If 1 is added to the first number, 2 to the second number, and so on up to the eleventh number, what is the new average?

Table 20.12
Problem Reformulation from Weeks 7 and 10

Original problem Week 7	In a random drawing of one ticket from a set numbered 1–1,000, you have tickets 8,775–8,785. What is your probability of winning?
Switch the given and the wanted	You have a probability of 3/20 of winning and received the following numbers from a drawing 122–136. What was the total number of tickets distributed for the event?
Change the given and change the wanted	If Beth has 19 tickets for a drawing with 100 total tickets and Veronica has 4 tickets for a drawing with 20 tickets, who has a better probability of winning?
Original problem Week 10	Six people enter a tennis tournament. Each player played each other person one time. How many games were played?
Extension	Three different tournaments, one with four people, one with five people, one with six people. Each player played the other person one time. How many games were played in each tournament? Is there a pattern? Can you find a rule?

on which participants were asked to reformulate problems-related data representation. As with the week 4 and the week 5 problem sets participants favored the structure technique of switching the given and the wanted. Typical examples of posed problems on the week 6 problem set are presented in Table 20.11.

The mathematical content focus of the problem sets during weeks 7 and 10 was counting, chance, and probability. As with previous problem sets, participants relied heavily on the reformulation techniques of changing the given and changing the wanted. During week 7 participants continued to favor switching the given and wanted as a structure reformulation technique, but this gave way to favoring extension during problem reformulation in week 10. Structure techniques were only 13% of the techniques utilized in week 7, but were 28.5% of the techniques utilized in week 10. This increase continued the trend from weeks 4 and 5 of an increased use of structure techniques on the second problem set related to specific course content. Week 7 was the only occurrence of less that 20% structure techniques, and this could be attributed to the difficulty of the material related to probability. Table 20.12 includes typical examples of reformulated problems from weeks 7 to 10.

The mathematical content focus of the problem set in week 11 was discrete mathematics and in week 15 was algebraic thinking. Participants relied on the

Table 20.13
Problem Reformulation from Weeks 11 and 15

Original problem Week 11	Consider networks with 0, 1, 2, 3, and 4 odd vertices. Make a conjecture about the number of odd vertices that are possible in a network. Explain your thinking
Change the given	Consider networks with 0, 1, 2, 3, and 4 even vertices. Make a conjecture about the number of even vertices and the traverse ability of the network. Explain
Extension	Knowing that you can create a network with an even number of odd vertices, is it possible for these types of networks to be traversable?
Original problem Week 15	A whole brick is balanced with ¾ of a pound and ¾ of a brick. What is the weight of the whole brick?
Change the context	If a bottle and a glass balance with a pitcher, a bottle balances with a glass and a plate, and two pitchers balance with three plates, can you figure out how many glasses will balance with a bottle?

reformulation techniques of changing the given and extension in week 11 and structure problem reformulation techniques were 43% of the techniques utilized. Changing the given was the most utilized reformulation technique during week 15 and structure techniques were utilized 34% of the time. Both problem sets were the only problem sets related to the specific mathematical content and there is an increase in the use of structure techniques from previous problem sets. Participant's problem-posing experience on the first five problem sets may have prepared them to use structure techniques when considering new mathematical content. Typical examples of problem reformulation during weeks 11 and 15 can be found in Table 20.13.

In summary, although surface techniques dominated reformulation throughout the semester, changes are evident in participants' problem reformulation. Participants' problem reformulation in weeks 11 and 15 suggest that, as they gained problem-posing experience, they relied more on structure techniques when problems sets were related to course content for the first time. Participants' choice of structure techniques also became more diverse—switching the given and wanted dominated structure reformulation early in the semester, but this gave way to the use of both extension and changing the context later in the semester. These changes in use of structure reformulation techniques suggest that participants developed problem-posing creativity and the ability to generate a more diverse set of problems.

The Relationship Between Problem Posing and School Mathematics

Data related to participants' beliefs about the relationship between problem posing and school mathematics was collected on the pre and post-assessment of beliefs and five journal entries. This data will highlight participants' articulated beliefs that problem posing is a beneficial task for their future students and that they will utilize problem posing in their future classrooms. On the pre-assessment of beliefs

Table 20.14
Participants Pre-assessment Views of Problem Posing

Category	Participants' responses
Problem solving	Help students better understand word problems; students will understand designing problems; create problems that relate to them; develop a better understanding of problem solving; helps students think beyond problem solving
Understanding	Consider information on multiple levels; better understanding of material; help teachers assess student understanding; helps students recognize pertinent information
Feelings	Alleviate student fear of word problems; develop ownership of mathematics; freedom and creativity with numbers and relationships
Negatives	Students may be confused or frustrated at first; may pose unsolvable or non-mathematical questions; questions may take lessons off track; students may take easy way out and ask simple questions; not practicing math directly

instrument, participants were asked to consider problems posed by an elementary-age student and respond to the question, "Do you believe that problem posing from sets of given information is a worthwhile task for elementary school students?" Participants had the prior experience of the pre-assessment of problem posing before completing this task. Participants' descriptions of the possible benefits of problem posing can be organized around three themes, the relationship to problem solving, aiding student understanding, and influencing student feelings about mathematics. Responses from the task are included in Table 20.14, which also includes potential negatives of problem posing suggested by participants.

At the beginning of the semester, participants seemed to believe that although there were some potential drawbacks to problem posing in school mathematics, problem posing had the potential to help students with their problem-solving ability, help students develop understanding, and affect students' creativity and ownership of mathematics. The remainder of this chapter will examine how participants were better able to articulate their beliefs as they gained experience posing problems. This will be highlighted by participants' abilities to discuss possibilities for the utilization of problem posing in school mathematics.

During week 5 of the semester participants responded to the following journal prompt as part of their assigned homework.

> Imagine that you are teaching and someone comes in to observe your classroom and a mathematics lesson that you are teaching. Write a description of your classroom and the lesson from the eyes of the observer. What would they see you doing during the lesson, what would they see the students doing, what would they notice about your classroom?

In response to this prompt only two participants suggested utilizing problem posing in their future classrooms. In the description of her lesson one participant stated that she would have students write word problems for division facts that she had on the chalkboard. Another participant stated that she would give students a journal prompt that asked them to think of a division problem, solve it, and then write in their own words how they would explain the problem to a third grader.

Participants' next journal entry was collected a week later; the prompt asked them to respond to the following:

> Please write a brief reflection on how you think class is going so far this semester, what aspects have you found the most helpful, least helpful and why?, how is the workload?, what aspects would you change?, what additional topics would you like to see covered?

Responses for this journal activity suggested some further reflection on problem posing and its relationship to school mathematics had taken place. Four participants commented that their problem-reformulation and problem-generation experiences have caused them to think beyond the activities and start to relate problem posing to their future classrooms. Other responses related to problem posing included comments that problem posing seemed to be an effective teaching tool and that students should want to pose and solve their own problems in and out of the classroom.

During week 10 participants responded to a journal prompt that specifically asked them to consider problem posing:

> As you are posing related problems or posing problems from a given set of information who is your intended audience? Why? Does the audience change depending on the problem? Would you consider yourself better at posing problems as reformulations or posing problems from sets of given information? Why?

Responses showed evidence that, when prompted, participants were capable of reflecting on the relationship between problem posing and school mathematics. Eleven of the 16 participants who responded stated that they were posing problems for their future students and indicated what they believed an appropriate grade-level range for the problems they created was from second to eighth grade. Ten participants also said that the grade level for which they posed problems was dependent on the original problem or the original set of given information. Participant reflection is highlighted through the following quotes: "When I'm actually teaching, I will need to pose appropriate problems for all children in my class to best facilitate their growth in mathematics," and "What I try to keep in mind most as I am problem posing is whether or not most students at a particular grade level will be able to find a solution with meaning and understanding."

Participants engaged in problem posing for another month before the journal entry collected during week 14 asked them to consider if they would utilize problem posing in their future classrooms through the following prompt. "Do you think you will utilize problem posing in your future classroom? If so, in what ways? Please try to be as specific as possible."

All participants articulated a role for problem posing as a future classroom resource and suggested that they saw potential for student and teacher problem posing. Participants suggested many possibilities to promote student problem posing in their future classrooms including as a whole class, as problem reformulation, as an introduction to new material, on homework, as an extra credit assignment, as a device to give fast students something to do, and by using a "problem-posing box." The most common suggestion was whole-class problem reformulation followed by assigning problem generation tasks when students were more comfortable with problem reformulation. Suggestions also included students posing problems related to a new topic and having the class research answers to these problems to gain introductory knowledge about the topic. Finally, one participant suggested that students could pose problems for homework or during class activities and collect them in a "problem posing box." When time permitted in class, students could choose a problem from the "problem posing box" and attempt to solve it.

Participants' reflection also included possible outcomes and benefits of student problem posing. Participants suggested that problem posing could promote student thinking and could allow for deeper understanding of content. One participant stated "By the problem posing process, students begin to identify key terms and concepts that define a topic, and by structuring problems around these topics, they begin to make connections, which enhances the learning process." Participants also supported their ideas from the pre-assessment of beliefs that problem posing would allow for student control and autonomy and can give students a sense of ownership over a problem. Two statements from participants illustrate these ideas: "I think that when students inquire about topics they are taking learning into their own hands, and that is one of the best things that problem posing can bring to a classroom," and "The questioning can help students determine their level of knowledge and helps students to develop metacognition."

As a tool for teachers, participants suggested using problem posing for assessment, to take advantage of "teachable moments," to accommodate all learning styles more effectively in their classroom, and to help develop activities, problems, tests, and quizzes. One participant described how and why a teacher would utilize problem posing when she wrote, "A teacher must be able to predict what students will find easy and difficult to do, and know her students well enough to be able to pose problems that will be thought provoking and meaningful to them."

In these journal entries participants described similar benefits of problem posing to those identified on the pre-assessment of beliefs, but extended these ideas by articulating specific ways to incorporate problem posing in their classrooms and reasons why problem posing may influence their teaching, student understanding, and student feelings about mathematics. This implies that further problem-posing experience may influence participants' abilities to reflect on and articulate potential roles of problem posing in school mathematics.

Participants' final prompted journal entry was collected in week 15 and participants responded to the following prompt:

> Please write a reflection on your experiences in this course this semester. The following questions might help to guide your reflection: (1) What have I learned about myself as a learner of mathematics? (2) What have I learned about myself as a prospective teacher of mathematics? (3) How has my conception of mathematics or teaching changed? (4) What questions do I still have?

A few participants' quotes stand out to highlight the ideas about problem posing already mentioned in this chapter. Even when they were not specifically prompted to do so, participants still reflected on their problem-posing experiences.

- With problem posing, I as the architect developed the concepts that should be incorporated into the problems and determined the age groups to be assessed, and as the carpenter I wrote the problems, determining what style would suit the students needs best, much like a carpenter must do when building a piece of furniture, or a house.
- I also learned how beneficial it is to having children pose problems, something I didn't like before this class. It is extremely important to give the students a sense of ownership over a problem and a better understanding of the problem.
- Uses in the classroom and importance of problem posing are the biggest thing that I have learned.
- I can also have students pose their own problems to be solved by their classmates. This allows more freedom and power for the students in owning their learning.

In summary, as they gained problem-posing experience, participants articulated detailed beliefs about the relationship between problem posing and school mathematics. It should be noted that the description of these beliefs occurred in both journal entries that specifically prompted for information about problem posing and those that did not. Similar to the results of Akay and Boz (2009), participants saw engaging in problem posing as a beneficial task for their future students and viewed problem posing as a tool that they would utilize in their future teaching. Journal entries collected as homework suggested that participant reflection on problem posing and teaching and learning occurred throughout the semester. This reflection allowed participants to provide detailed descriptions of their new beliefs about problem posing as they developed their own problem-posing skills.

Discussion

The development of participants' abilities and creativity as problem posers was highlighted through quantitative data related to the characteristics of their problem posing. One student summarized class changes with respect to problem posing with clarity in the final journal entry of the semester in the following way:

> However the greatest thing that I will take from this class is my newly discovered talent of problem posing. I remember back to the first class this semester when we were asked to do some problem posing for Dr. G's research project. I was stumped by this task. Posing a problem from the given information was like another language to me. As the problem sets were assigned throughout the semester, I truly dreaded problem posing. But about half way through the semester, it was like a light turned on in my head and I was suddenly able to create problems without all that difficulty. This allowed me to focus on posing valid challenging problems. It was great to have the same packet handed out once again the last day of class for Dr. G's research project, and being asked to pose as many problems as I could. This was such a valuable task for me because I could literally see my growth as a problem

poser first hand! I sat there and posed problems for minutes without even taking a breather! It was a great feeling to have actually seen how much I grew in this one area of math throughout the course of the semester.

Problem posing was not explicitly discussed in class but as participants' gained problem posing experience they became more efficient problem posers and became more creative at problem reformulation. The changes in participants-posed problems support Leung and Silver's (1997) hypothesis "… that further experience [beyond a single assessment] with problem posing would lead to better, and more sophisticated performance" (p. 17).

Participants' problem generation was consistent when they were not posing problems under a time constraint. On the three problem generation tasks collected, 61% of participants-posed problems required a multi-step solution. As evidenced by the post-assessment of problem posing, participants in this study became more efficient problem posers and were able to pose more complex problems under a time constraint. These results for problem generation support Leung and Silver's (1997) conclusion that prospective elementary school teachers are capable of posing appropriate mathematical problems and are also capable of posing complex mathematics problems.

At the beginning of the semester these prospective teachers relied on surface problem-reformulation techniques the first time they reformulated problems related to specific course content. This reliance gave way to an increased use of structure-reformulation techniques later in the semester. The participants continued to develop their creativity in posing problems as problem reformulation even though the quality of their problems did not influence their grade. Therefore, it seems possible to develop the problem reformulation abilities of prospective teachers through problem-posing experiences that do not have to include explicit instruction in problem posing. Therefore, engaging prospective teachers in problem posing as was done in this study has the potential to help develop both problem generation and reformulation abilities.

Consistent with the work of Leung and Silver (1997) and Stickles (2010–2011), the research reported in this chapter has shown that prospective elementary and middle-school teachers were able to pose more problems and more complex problems on problem generation tasks when the set of information included numeric content. This is evident by the statistically significant difference in their posing from pre-assessment to post-assessment. While there was a positive change in participants' non-numeric problem posing, this change was not statistically significant and after gaining experience posing mathematics problems participants still favored posing problems when numeric content was included. Developing participants' ability to pose problems on tasks that do not include numeric content should be a focus of future problem-posing research.

Participants developed beliefs about the potential benefits of problem posing in school mathematics and developed problem-posing abilities to support the incorporation of problem posing in their future classrooms. As suggested in journal entries, these prospective teachers share Silver's (1997) view that problem posing can be included in mathematics instruction to "… develop in students a more creative disposition towards mathematics" (p. 76). Although problem-posing tasks during the instructional treatment focused on problem reformulation and problem generation, participants' reflection on the role of problem posing in school mathematics went beyond these two techniques and included potential benefits for student learning. The benefits these preservice teachers articulated were consistent with research and writing in mathematics education (Silver, 1994). This understanding of the benefits of problem posing may help these preservice teachers develop practices that mirror their beliefs and incorporate problem posing in their future classrooms.

Conclusion and Implications

Based on this and past work, it is reasonable to assume that prospective teachers have some capability for posing mathematics problems. This research extends previous studies by showing that problem-posing ability and creativity can be developed further by engaging prospective teachers in two forms of problem posing as part of the expectations of a mathematics content course. This incorporation of problem posing does not reduce time in class to discuss mathematics content and does not require class time devoted explicitly to teaching problem posing. Therefore, teacher educators should consider incorporating problem posing in mathematics content courses for prospective elementary and middle-school teachers as problem generation and problem reformulation. Preparing problem-generation tasks and requiring students to pose problems as problem reformulation does not add significant time to the instructor's development of course materials.

The incorporation of problem posing as described in this chapter has the potential to be in the vanguard for the incorporation of problem posing at all levels of mathematics education. Problem-posing experiences may help prepare teacher-education students who are poised to engage their students in problem posing. It may also serve to educate mathematics teacher-educator colleagues about mathematical problem posing. Further research is needed to extend this work. First, due to the study design we cannot directly attribute participants' changes to the problem-posing experience in the class. A study that collects similar data without incorporating problem-posing in the context of the class may be able to shed light on whether the problem-posing experience is key to create change. Second, the prospective teachers in this study were poised to incorporate problem posing in their future classrooms, but did they? Longitudinal studies that incorporate problem posing in classes for prospective teachers are needed. Participants would then be followed into the classroom to determine if and how, as teachers, they implement problem-posing practices.

References

Akay, H., & Boz, N. (2009). Prospective teachers' views about problem-posing activities. *Procedia—Social and Behavioral Sciences, 1*, 1192–1198.

Barlow, A. T., & Cates, J. M. (2006). The impact of problem posing on elementary teachers' beliefs about mathematics and mathematics teaching. *School Science and Mathematics, 106*(2), 64–73.

English, L. D. (1997). Promoting a problem-posing classroom. *Teaching Children Mathematics, 4*(3), 172–179.

Gonzales, N. A. (1994). Problem posing: A neglected component in mathematics courses for prospective elementary and middle school teachers. *School Science and Mathematics, 94*(2), 78–84.

Gonzales, N. A. (1998). A blueprint for problem posing. *School Science and Mathematics, 98*(8), 448–456.

Kilpatrick, J. (1987). Problem formulating: Where do good problems come from? In A. Schoenfeld (Ed.), *Cognitive science and mathematics education* (pp. 123–147). Hillsdale, NJ: Lawrence Erlbaum.

Kilpatrick, J., Swafford, J., & Findell, B. (Eds.). (2001). *Adding it up: Helping children learn mathematics*. Washington, DC: National Academy Press.

Leung, S. (1993). *The relation between mathematical knowledge and creative thinking to the mathematical problem posing of prospective elementary school teachers on tasks differing in numerical information content* (Unpublished doctoral dissertation). University of Pittsburgh, Pittsburgh, PA.

Leung, S., & Silver, E. A. (1997). The role of task format, mathematics knowledge, and creative thinking on the arithmetic problem posing of prospective elementary school teachers. *Mathematics Education Research Journal, 9*(1), 5–24.

National Council of Teachers of Mathematics. (1991). *Professional standards for teaching mathematics*. Reston, VA: Author.

National Council of Teachers of Mathematics. (2000). *Principles and standards for school mathematics*. Reston, VA: Author.

Polya, G. (1957). *How to solve it: A new aspect of mathematical method*. Princeton, NJ: Princeton University Press.

Silver, E. A. (1994). On mathematical problem posing. *For the Learning of Mathematics, 14*(1), 19–28.

Silver, E. A. (1997). Fostering creativity through instruction rich in mathematical problem solving and problem posing. *International Reviews on Mathematical Education, 29*(3), 75–80.

Silver, E. A., & Cai, J. (1996). An analysis of arithmetic problem posing by middle school students. *Journal for Research in Mathematics Education, 27*(5), 521–539.

Stickles, P. R. (2010–2011). An analysis of secondary and middle school teachers mathematical problem posing. *Investigations in Mathematics Learning, 3*(2), 1–34.

Whiten, D. J. (2004a). Building a mathematical community through problem posing. In R. N. Rubenstein & G. W. Bright (Eds.), *Perspectives on the teaching of mathematics: Sixty-sixth yearbook* (pp. 129–140). Reston, VA: National Council of Teachers of Mathematics.

Whiten, D. J. (2004b). Exploring the strategy of problem posing. In G. W. Bright & R. N. Rubenstein (Eds.), *Professional development guidebook for perspectives on the teaching and learning of mathematics: Companion to the sixty-sixth yearbook* (pp. 1–4). Reston, VA: National Council of Teachers of Mathematics.

Winograd, K. (1997). Ways of sharing student-authored story problems. *Teaching Children Mathematics, 4*(1), 40–47.

Chapter 21
Problem Posing in Primary School Teacher Training

Alena Hošpesová and Marie Tichá

Abstract The chapter reports results of a survey whose aim was to contribute to research in the area of problem posing in teacher training. The core of the research project was empirical survey with qualitative design. Preservice and in-service teachers were posing problems in the environment of fractions and reflected on this activity in writing. Analysis of the posed problems and participants' reflections were to answer the following questions: (a) What shortcomings can be identified in the posed problems? (b) How are the posed problems perceived by preservice and in-service teachers? (c) What relations are there between quality of the posed problems and perception of this activity by their authors?

Contents

A. Hošpesová (✉)
University of South Bohemia, České Budějovice, Czech Republic
e-mail: hospes@pf.jcu.cz

M. Tichá
Institute of Mathematics of the Academy of Sciences of the Czech Republic,
University of South Bohemia, Prague, Czech Republic

F.M. Singer et al. (eds.), *Mathematical Problem Posing*,
Research in Mathematics Education, DOI 10.1007/978-1-4614-6258-3_21

Introduction

Development of professional competences of primary school teachers has been one of our focal points for a long time (e.g., Tichá & Hošpesová, 2010). We understand teachers' professional competences as a set of specific knowledge needed for teaching, as an amalgam of skills that teachers should master in order to be able to perform and continuously improve their teaching. We consider that the key to teachers' professional competence is subject didactic competence combining content knowledge, its didactical treatment and application of this knowledge in practice. Teachers, however, often perceive this competence in the narrow sense of simply having knowledge of methodological approaches to teaching a given content, but they are unaware of the need of its didactical analysis (see, e.g., Klafki, 1967). Teachers must not to be satisfied by looking for answers to the question *how*? They must also consider other questions, like *what*? and *why*? Like other researchers, we stress not only theoretical knowledge, but also the capability to act adequately in the development and delivery of mathematics lessons. We are convinced that only with insight and deep knowledge will a teacher be able to assess what can be improved, and how this can be achieved.

Primary school mathematics is the period when foundations for concept, imagery and future understanding of mathematics, and for positive attitudes to the discipline are formed. Teaching mathematics at the primary school level can be understood as a system of propaedeutics[1] of mathematical concepts and solving methods. From this perspective, special demands are made on a teacher's subject didactic competence. Our experience as teacher educators shows that teachers are sometimes unaware of deficits in their competences. This results in inadequate self-efficacy (Bandura, 1997; Gavora, 2010) and in a tendency to overestimate one's own proficiency. This phenomenon is of special importance for in-service teacher training. We came across the following two extremes:

- Teachers who are convinced that they have mastered a sufficient repertoire of methodological approaches which they have tried and tested in their teaching practice; they do not think that any change (let alone improvement) is necessary; and they do not realize that their subject didactic knowledge should be deeper.
- Teachers who are well aware of their weaknesses, who see their knowledge of certain topics are not at a sufficiently high level, and who want to do something about it. There are two categories of teachers who fall under this extreme:
  - Some of them expect that there are ready-made universal recipes they can easily and effortlessly learn by drill, and that will naturally lead to improvement; they fail to see that a change in grasping a problem with

[1] We understand propaedeutics as an introduction to knowledge of preparatory instruction, similarly to Webster's definition "pertaining to or of the nature of preliminary instruction; introductory to some art of science" (Webster's Encyclopedic Unabridged Dictionary of the English Language. New York 1996: Gramercy Books, p. 1152).

understanding is not very likely in these cases; they do not realize that there are no simple recipes leading to understanding.

- Others are well aware of the fact they will have to work hard to gain true insight.

It is often difficult for a teacher educator to discuss any deficiencies with in-service teachers. These teachers may take it very personally and are closed to any new stimuli. Preservice teachers, in general, seem to be more open.

In our practice as teacher educators, we have explored various ways of guiding primary school teachers in their quest to handle mathematical content with deep insight. At first, our focus was on joint reflections of selected teaching episodes (Tichá & Hošpesová, 2006). Later, we turned our attention to problem posing and started exploring the role of joint reflections in the process of assessment of problems posed to meet given criteria (e.g., Tichá and Hošpesová, 2010, 2013; Toluk-Ucar, 2008).

In this chapter, we will describe the rationale, conduct, and outcomes of a study, in which problem posing was used as a teaching method with a group of preservice and a group of in-service teachers. Much attention was paid to posing problems in the environment of fractions. This content area was selected because we are convinced of the importance of the concept of fractions in the curriculum in general, and of the relation between the whole and its parts, in particular. It is vital that teachers realize that they must not only master procedures of calculations with factions, but that they also need to see the importance of concept building and understanding the concept of fractions, and grow aware of the differences between fractions and natural numbers (see, e.g., Toluk-Ucar, 2008). At the end of the course, the participating teachers were asked to reflect on problem posing and on its benefits for teacher training.

Rationale of the Study: Problem Posing in Teacher Training

It is common practice for mathematics to be taught predominantly through problem solving. We understand problem solving as an activity directed towards a certain specific target and that this activity is continuously corrected by this target. The problem-solving process can be perceived as a "dialogue" between the solver and the problem. While solving a problem, solvers ask questions like "How shall I begin?" or How can I carry on at this point? (i.e., they ask themselves questions before starting the solving process, while solving it and having solved it). This can be seen as one of the early forms of problem posing.

We first became involved in the issue of problem posing when doing research on the structure of the process of grasping real-life situations (Koman & Tichá, 1998). By "grasping a real-life situation" we meant the process of thinking, involving especially identification of key objects and phenomena, relationships between them, and identification of problems growing/emerging from the situation, as well as of growing awareness of questions that may be asked—in other words, the process of understanding aspects of a situation that are needed for posing a problem. These stages are followed by solving the posed problem, answering questions, articulating the results, and interpreting and analyzing the results.

Our earlier research established a narrow link between grasping a situation and problem posing. Problem posing is closely interwoven with grasping a situation. If we recall Polya's (2004) characterization of the four heuristic steps in problem solving—getting an insight into the problem, designing a solution plan, implementing the plan, and verifying and critically assessing the solution. Comparing these steps with the above described stages of grasping situations, we can clearly see that some components are characteristic for both activities (i.e., both for grasping a situation and for problem solving).

The interrelatedness of the processes of problem solving and of problem posing has been the focus of a number of studies. A wide range of aspects of these processes have been scrutinized, e.g., their structure, stages, and interaction.

Like a number of mathematics educators (see, e.g., Cai & Brooke, 2006; English, 1997; Freudenthal, 1983; Kilpatrick, 1987; Pittalis, Christou, Mousolides, & Pitta-Pantazi, 2004; Ponte & Henriques, 2013; Silver & Cai, 1996; Singer, Ellerton, Cai, & Leung, 2011), we understand problem posing in teacher training to be a method leading to enhancement of preservice teachers' subject didactic competence. The complex nature of this competence implies that problem posing can have several functions:

- It is an educational tool because a teacher must often pose problems that are, for example, related to a specific situation in the class (Silver & Cai, 2005).
- It is also a diagnostic tool which helps teachers to uncover deficits and obstacles in students' knowledge (English, 1997; Harel, Koichu, & Manaster 2006; Silver & Cai, 1996; Tichá & Hošpesová, 2013).
- On the basis of our experience, we have recently begun stressing that problem posing can be a significant motivational element leading to deeper inquiry into mathematical content areas, resulting in deeper study and effort to improve one's knowledge base for teaching, a deeper understanding of concepts, and in boosting one's repertoire of interpretation.

Goals of the Research Project, Its Conception, and Methodology

The goal of the project reported in this chapter was to explore and describe the role of problem posing in teacher training. Special attention was paid to gaining insight into how preservice and in-service teachers perceive and interpret problem posing because it is these subjective interpretations that play a decisive role in the dynamics of the process of their improvement.

The core of the project was a qualitatively designed empirical survey. Data were collected from an analysis of problems posed by preservice and in-service teachers and from their written reflections on this activity.

Research Questions

The following three research questions were asked at the outset of the project:

1. Which deficits in grasping the concept of fraction hinder teachers' subject didactic competence (and therefore their teaching) can be identified in problems posed by preservice and in-service teachers? Are these shortcomings the same or different in problems posed by preservice and in-service teachers?
2. What are preservice and in-service teachers' beliefs about the importance of problem posing in their own training?
3. What is the relationship between a respondent's view of the role, importance, and benefit of problem posing and quality of problems he/she poses?

When trying to answer the first of these questions, we tried to describe objectively which phenomena were inherent in the process of posing problems involving fractions by preservice and in-service teachers. Other questions were directed to tackling the participants'—preservice and in-service teachers'—perspectives. Data used for objective scrutiny were the posed problems. Participants' subjective perspectives were recorded in the form of their written reflections. Collection of the two different types of data enabled triangulation of the collected data. These data were also the basis for answering Research Question 3.

Participants and Data Collection

The study was carried out with two groups of respondents:

- The first group consisted of 32 preservice teachers, students who posed problems in a compulsory course of Didactics of Mathematics in the second half of their undergraduate studies[2]. This group will be referred to as *students*.
- The other group included 24 participants in the course Didactics of Mathematics, who enrolled to this course to deepen their knowledge. All of these participants were in-service teachers with several years of work experience as primary school teachers. They chose this lifelong learning course for professional development voluntarily. This group will be referred to as *teachers*.

We worked with both respondent groups for approximately 3 months using similar methods.

The course was not originally conceived as a teaching experiment. Participants in both groups studied arithmetical content of primary school mathematics. They solved and posed problems on topics related to the mathematics content taught at

[2] Primary school teachers in the Czech Republic must study 4- or 5-year-long undergraduate courses designed especially for primary school teachers. Their undergraduate teacher education program includes courses in all subjects taught at primary school level.

the primary school level and discussed a variety of questions related to teaching mathematics. In one of the last seminars, the participants' task was *to pose several problems which include fractions ½ and ¾* (similar to that reported in Tichá and Hošpesová (2013) but with a different group of respondents). Deliberately, other constraints were not imposed on the problem-posing task given to the participants, as this was investigated in our earlier studies (e.g., Hošpesová & Tichá, 2010; Tichá & Hošpesová, 2006, 2010). We assumed that this open situation would enable the students and teachers to create such situations deliberately in which fractions were used in different contexts (in the sense of Behr, Lesh, Post, & Silver, 1983).

Transcription of Posed Problems and Approach to Data Analysis

We started by designing a table with three columns in which we matched the posed problems and the written reflections to their authors. Thus were formed triplets [author-problem-reflection]. These were then analyzed. As the design was qualitative, no coding system had been developed in advance and analysis was conducted using open coding. Thus:

- In each of the posed problems, we looked for characteristics showing to which concrete situations the author had linked fractions, how he/she had interpreted them in different situations, and what sub-constructs of fractions had been incorporated into the problems.
- Texts of reflections were classified according to the topics addressed; we tried to find suitable codes for the meaning they connoted (for details about open coding, see Švaříček and Šeďová, 2007); we examined what perspective students and teachers selected when posing word problems, and what opinions they expressed about the process.

We scrutinized word problems from two perspectives: (a) we tried to treat posed problems from an external perspective, indicating their strengths and weaknesses from our perspective; and (b) we tried to determine the respondents' (problem authors') perspective.

Having first analyzed the data individually (each of the authors of this chapter on her own), we then met to discuss our findings, to link the individually created codes, and then to code both types of data again. The subsequent categorization of codes identified substantial topics that were relevant to our study; this was done, as stated above, without any explicit preconceptions or clear ideas. This form of analysis allowed the emergence, for example, of the category "refusal to pose problems," whose occurrence we had not been anticipated and had not been included in our research questions.

Discussion

Shortcomings in the Posed Problems

The analysis of the posed problems showed substantial deficits in respondents' knowledge, and revealed their misconceptions, misunderstandings, and shortcomings. This confirmed the conclusions reached in our previous surveys (Hošpesová & Tichá, 2010; Tichá & Hošpesová, 2013). We are convinced that the reasons for this are to be found in how students and teachers themselves were introduced to the concept of fraction; in other words, in their own evolution of the concept of fraction from its very beginnings. In addition, we believe that any deficits can also be linked to earlier demands there had been on their knowledge of fractions. This assumption is confirmed by the fact that the same or similar shortcomings were characteristic for both groups of respondents. For that matter, the same misconceptions could be seen in pupils and students from different age groups, as our former research has shown (Tichá, 2003). These shortcomings and misconceptions are very pervasive; they can be seen in all of the sets of problems posed by our respondents.

The posed problems indicate that teachers do not realize that:

- It is not sufficient to master arithmetical operations; they lack conceptions, imagery that would enable them to grasp the concept of fraction, solve and pose problems, even application problems (see also Prediger, 2006; Toluk-Ucar, 2008).
- A problem must be carefully formulated and its authors must be very accurate when wording it; Cai & Cifarelli (2005) refer to "ill-structured problems" (p. 47), and cite Kilpatrick (1987), who claims that these problems "lack a clear formulation or a specific procedure that will guarantee a solution, and criteria for determining when a solution has been achieved" (p. 134).
- Problem posing requires knowledge of the curriculum—what knowledge is prerequisite and what the aim of teaching is.

Similarly to our previous research (e.g., Tichá & Hošpesová, 2013), we noted that most of the posed problems were of markedly monotonous nature of situational context (cakes, marbles, etc.), of properties of the environment (either discrete, or continuous, but rarely both), and of interpretation (fraction as operator, quantity (measure), magnitude of physical quantity (in these cases the problems tended to be well formulated, but it was not clear whether their authors were aware of any differences between them, e.g., problems F1 and F2 posed by teacher Filo later in this text), quantity of physical value, and part/whole construct, etc.)

The base was often given ambiguously. Both students and teachers did not realize the need to consider the whole and the part-whole relationship. This can be observed, for example, in the students' and teachers' failure to realize that they must consider the role of the whole (e.g., problems posed by Cecily).

Interference of work with fractions with knowledge of calculations with natural numbers (in which fractions-operators were handled as natural numbers) was another source of mistakes. The core, key deficits spring from the fact that fractions were perceived as quantitative data, as natural numbers and they were handled this way in arithmetical operations. In this context, Streefland (1991) used the term *N*-distractor which warned of the possible interference by knowledge of work with natural numbers, stemming from immersion into the world of natural numbers.

> Dad decorated ½ of the guest-room. Granddad decorated ¾ of the living room. Who decorated more and how much more? (Cecily)

The majority of students and teachers posed problems of an additive nature. Multiplicative problems were very rare. These often showed that their authors did not realize that "whereas multiplication always makes bigger for natural numbers (apart from 0 and 1), this cannot be applied to fractions" (Prediger, 2006, p. 377). Toluk-Ucar (2008) also reported similar findings.

> Honza had ¾ of some dessert. Jana had ½ times less than Honza. How much did they have together? (Vlasta)

When looking for an answer to the first research question, we compared problems posed by students with problems posed by teachers. Problems posed by teachers differed in two aspects: (a) in-service teachers usually asked for specification of pupils of what grade the problems were posed for, and (b) their repertoire of problems was richer. In line with requests that are usually made for sets of problems with fractions (Lamon, 2006) teachers posed more varied *n*-tuples of problems in which different sub-constructs of fractions, various environments (discrete, continuous), various representations, problems related to evidence, construction, etc. were used. Teachers would have been influenced by their experience with problems found in good textbooks and from collections of problems. They may also have been influenced by problems that their pupils had been assigned in various competitions and tests.

For illustration, we can present problems posed by teacher Filo, although her pentad of problems also included the above mentioned misconceptions and confusing formulations, e.g., in problem F1 (what is the whole?) and F5 (are the fractions in the function of an operator or do they give the number of passengers?):

- F1. Children ate cakes. One of them ate ½, the other ¾. How many quarters did they eat?
- F2. In one vessel, there is ½ L of liquid, in another one ¾ L of liquid. How much liquid is in both vessels?
- F3. The sides of a rectangle are ½ cm and ¾ cm. Calculate its area.
- F4. ¾ of a field was seeded with corn but ½ did not germinate. How many quarters germinated?
- F5. There were 20 passengers on a plane. ¾ of the passengers left the plane during the stopover, ½ boarded the plane. How many passengers continued the journey?

What are Preservice and In-service Teachers' Opinions of Problem Posing in Their Own Training?

The majority of comments about the inclusion of activities related to problem posing into preservice and in-service teacher training were positive. In the student group, about two thirds of the participants talked about this question. Mostly, they appreciated inclusion of problem posing into their undergraduate training. There were only two sceptical opinions. In the in-service teacher group, almost everybody answered this question and all answers were positive.

However, we must ask whether the respondents' answers were not mere proclamations. We suspect the respondents may have felt that it was "desirable" to say that problem posing was useful and beneficial as the seminars focused on the issue. Comparison of problems posed and opinions declared by the participants (see below) seems to confirm this suspicion.

When analyzing written reflections on the posed problems, we found that respondents tended to express their opinions on two topics: on subjective feelings when posing problems (codes 1–4) and on the impact of problem posing on a teacher's subject didactic competence (codes 5 and 6). Open coding resulted in the following codes:

1. Problem posing is important
2. Problem posing is surprisingly difficult
3. The teacher finds it easier to work with problems he/she has posed (posing the problem makes it easier to solve)
4. It is not a teacher's task to pose problems
5. Problems posed by a teacher are more appealing for the children and more up-to-date
6. Problems posed by a teacher help children's comprehension

These codes are discussed in more detail in the following section.

Problem posing is important. Students' and teachers' comments showed they were either well aware of the importance and benefit of activities associated with problem posing or at least hint at being aware. Reflections where there were no reasons or justification of this opinion made us ask whether students and teachers had really become convinced about the importance of problem posing in the course, or whether they were just repeating what we had discussed in the course.

> I think it is very important for a teacher to develop this as it enables him/her to understand the structure of already existing problems problem and to be able to carry them out. (Student Soňa)

None of the participants, however, tried to formulate what the prerequisites for successful problem posing are, what knowledge, skills and experiences, etc. a person posing problems for their pupils' needs. The comments, the posed problems, and the following joint discussion showed that these issues were not considered either by the students or by the teachers.

Problem posing is surprisingly difficult. Some of the participants admitted to having had difficulties when they posed the problems. Some of them stressed that the task had been very demanding. Some of their comments showed that the task of posing problems made participants reflect on the adequacy of their knowledge base of mathematics for teaching.

> I think it is very important because when posing problems one often grasps it or starts to understand it but also grows aware of one's deficits. At this moment I feel a bit down as I find it very difficult. The more I think about it and try to come up with something, the more lost I get in it and look for complexities and things that I normally find simple, comprehensible, are now confusing and I have a lot of doubts. (Student Beruška)

> I tried to come up with something but it didn't work too well. (Teacher Filo)

> The "test" today clearly showed how important this is. I've never come across posing problems with fractions and had no idea how difficult it could be. (Student Tereza)

The teacher finds it easier to work with problems he/she has posed (problem posing helps problem solving). This area of comments only confirmed our idea of the closeness of the relationship between problem posing and problem solving.

It is not a teacher's task to pose problems. Some participants expressed their anxiety and refused to do the activity of posing problems entirely. They stated that they did not like problem posing because it was time-consuming; they expected textbooks, i.e., authority, to furnish them with problems. For example, one student (Gábina) spoke of her fear that she would not be able to pose problems:

> Word problems are undoubtedly important for children but no teacher will want to spend their time posing problems when they can find millions of them in textbooks or on the Internet. I will rather be advised. Or I will just modify some existing problems. (Gábina)

Problems posed by a teacher are more appealing for the children and more up-to-date. The students' comments often included the assertion that the competence of and skills in problem posing are an important part of teacher's knowledge. Students often associated problem posing with arousal of pupils' interest, and stressed creativity and inquiry-based mathematics education (see Dagmar's words below). However, we cannot again rule out the possibility that these are not the true beliefs of the participants, but are echoing what they understood to be part of the course.

> Mathematics is an important discipline. Its basis is being able to carry out basic operations such as +, −, *, /. When we go shopping, when we want to solve riddles and puzzles. The teacher should plan his/her lessons playfully and the lessons should be entertaining. Nobody is interested in boring lessons. That is why it is important for the teacher to pay attention to lesson planning, development of skills and competence. (Student Martina)

> I think it is crucial that the teacher understand and be able to pose problems. Teaching could then be more creative and enjoyable. For example, problem posing with pupils and so on. (Student Linn)

> I think it is good to get engaged in problem posing. Problems can correspond to children's hobbies and then they find their solution more attractive because they are more personal. For example, Anička—gymnastics, Pepíček——soldiers, … The sky is the limit. (Student Cecily)

Problems posed by a teacher help children's comprehension. The most common comments in the teachers' group (e.g., teacher Hanka) were related to search for ways of supporting pupils' comprehension. The comments seemed to reflect everyday practical problems teachers that teachers face.

> I based the problem on the math for primary school level. Clarity is crucial. I tried to formulate clear assignment, which the children are able to solve. (Hanka)

What Was the Relationship Between a Respondent's Opinion on Problem Posing and "Quality" of Problems He/She Poses?

We identified several phenomena in our analysis.

Discrepancy: Quality problems vs. opinion of the respondent. Discrepancies between a given participant's beliefs on importance and role of problem posing and the problems he/she actually posed were noteworthy, especially in the student group. We came across proclaimed appreciation of the importance of problem posing accompanied by posed problems of very low quality. We also came across expressions of fear and anxiety of problem posing accompanied by *n*-tuples of problems we evaluated as being of high quality.

For example, above we noted the negative attitude to problem posing expressed by the student Gábina. Paradoxically, this student was very proficient in posing "variegated" problems (Problems G1–G6), in which various interpretations of fractions (quantity, operator, etc.) in the sense presented in Behr et al. (1983). This student also used different contexts. The solutions to her problems required the use of a number of different operations (comparison, addition, multiplication). In spite of the range and form of the problems she posed, some problems included ambiguities and misleading formulations of the definition of a whole (it seems in some cases she did not think of the whole, and posed questions like "How much …?" instead of "What proportion?" or "Which part?"

G1. How much does an iron rod melded of two rods measure? One rod is ½ m long, the other ¾ m long.

G2. There is ½ in the swimming pool. We add ¾ more in the swimming pool. Will the swimming pool overflow?

G3. Mum and Andulka ate together ½ of a cake; Dad and Petřík ate ¾ of the cake. Who ate most?

G4. Mum and Andulka ate together ½ of a cake; Dad and Petřík ate ¾ of the cake. How much cake is there left?

G5. Pepíček ate ¾ out of ½ of a cake. How much cake was there left?

However, it was more common for students to reflect on the high demands of posing problems (inquiry-based approach, creativity, interest, etc. (see, for example, the following statements by Dagmar and Hortenzia) and then posed problems similar to those common in textbooks (it seems as if these students wanted "to be safe").

It is crucially important to develop teachers' abilities and competence because the more interesting, creative and inquiry based the problem is, the more the children enjoy its solution. (Dagmar)

I think it is very important to develop the ability to pose problems attractive for children. It is not enough just to use textbooks. (Hortenzia)

D1. Mark in the circles such a part that the first represents ½ of the size and the second ¾ of the size.

D2. When you add ½ of 1 m and ¾ of 1 m, how many meters do you get?

D3. What part of a cake is there left if you subtract ½ from ¾?

H1. There was one half of a cake on the table. Petr came and ate ¾ of it. What part of the cake was left on the table?

H2. Mum bought ½ kg of apples. Dad brought ¾ kg of apples. How many apples did they have in total?

H3. Granny baked sponge cake; Tomáš ate ¾ of the sponge cake; Anička later ½ of what was left. What part of the sponge cake was left in the end?

Awareness of the deficit in knowledge. We can say that the participants were unaware of their insufficient knowledge of the content and of their misconceptions. For example, having posed the set of problems (J1–J5), the teacher Jana wrote: "I was guided by my experience from work with children. Most important for me are illustrativeness and understanding. I was focusing on grasping mathematical operations with understanding. The goal was to empathize with the children."

J1. Jane put one half of sweets into one bag and also put there 3/4 from another bag. How many did she have in total?

J2. Compare the following fractions: ½ and ¾.

J3. A dressmaker cut one half from a ¾ long strip of fabrics. How much was she left with?

J4. Two children had their own halves of cake. Can you calculate how much they had together when they got ¾ more?

J5. Calculate the area of an estate whose measurements are ½ (side a) and ¾ km (side b).

Comprehensibility, reality. Even problems posed by the teachers included some inaccurate, ambiguous formulations, despite the fact the teachers often stressed the importance of the comprehensibility of problems. Other teachers' proclaimed demand was that problems be "real" but the posed problems did not meet this criterion. For example, Svatava wrote: "I was guided by my experience, most important: to understand it, to try and use real ideas, the goal comprehensibility," but the problems she posed included the following:

S1. I need ½ m of fabrics for a jacket. I need ¾ m for trousers. How much fabrics will I have to buy? How much will I pay if 1 m costs 200 CZK?

S2. Petr jumped ½, Pavel ¾ m. Who jumped more? By how much?

Discussion of the Role of Problem Posing in Teacher Training

The goal of the study presented in this chapter was to assess the benefit of problem posing in preservice and in-service teacher training and to compare the posed problems with their authors' beliefs. We tried to show that analysis of problems posed in the environment of fractions can lead to identification of misconceptions and major deficits in preservice and in-service teachers' mathematical content knowledge. In the classroom, such misconceptions and deficits could result in the teacher's single-tracked, simplified or even erroneous interpretation of the subject matter, or in his/her failure to react adequately to pupils' valuable contributions.

Problem posing on its own is by no means a sufficient tool for remedy. It must be supplemented by some reaction from others. That is why we used joint reflection, discussed in detail in our preceding studies (Tichá & Hošpesová 2006), as one of the possible ways leading to discovery of deficiencies (especially those of a conceptual nature). This is also called for by other authors. For example, Selter (1997), following Bromme (1994), stated that:

> offering teacher students the necessary background knowledge surely is a precondition for their professionality as teachers. However, teachers actually become professionals while they are teaching and reflecting on their teaching … teacher education … should first and foremost assist prospective teachers in developing their autonomy. This implies to support them in increasing their degree of awareness—about mathematics, about children's mathematical learning, about the quality of teaching material and so forth. (p. 57)

The survey used in our study concluded with a joint reflection. In fact, the participants in both groups were asked for some evaluation of the problems they had posed. Only then, when they were asked to reflect on their own posed problems, were some of the participants willing to admit deficits in their knowledge of mathematics (specifically the concept of fractions), lack of creativity and insufficient knowledge of what their pupils might be interested in.

This survey also confirmed that the teacher educator leading the seminar should have input into the joint reflection, as his/her questions could guide the direction of the discussion. It is possible to proceed in more ways:

- The educator could offer examples of posed (*n*-tuples) problems, point out flaws, mistakes, misconceptions (e.g., Hošpesová & Tichá, 2010). This approach would stress especially the use of problem posing as an educational tool in teacher training.
- The educator could ask the participants to choose the problems they want to discuss. They should always be able to justify their choice (an interesting problem, open problem, ambiguously formulated problem, etc.). Then problem posing would be used as a diagnostic and re-educational tool.

The question how to persuade in-service and preservice teachers that problem posing should become an integral part of their teaching needs to be answered yet. We came across the belief (especially in case of in-service teachers) that inclusion of "problem posing" distracts from "appropriate, genuine mathematics" oriented on "mastering of craftsmanship—carrying out calculations." Some of the teachers are afraid that their training is not sufficient as to enable them inclusion of these activities in their teaching. Others object that it would require intellectually and time-demanding planning. Many are hindered by the fact they would not know how to evaluate problem posing.

We found out that posing problems in groups ceases to be stimulating. It seems much more convenient for students to work individually or in pairs. However, it is crucial they have a chance to present their problems and discuss them from different points of view (the choice of mathematical topic, continuity, difficulty, symmetry).

The following questions, asked by a number of researchers, still need to be answered:

- What knowledge (mathematical and general) is a prerequisite to successful problem posing?
- How can we assess the benefit of "problem posing" for their authors and the "change" in professional competences of these authors?
- What help or guidance can we offer to teachers who decide to include "problem posing" in their teaching?
- How can teachers and students be persuaded about the potential and benefit of "problem posing" for mathematics education and for the development of mathematical literacy?

Acknowledgement Elaboration of the chapter was supported by RVO 67985840.

References

Bandura, A. (1997). *Self-efficacy. The exercise of control*. New York, NY: W. H. Freeman.

Behr, M., Lesh, R., Post, T., & Silver, E. A. (1983). Rational number concepts. In R. Lesh & M. Landau (Eds.), *Acquisition of mathematics concepts and processes* (pp. 91–125). New York, NY: Academic.

Bromme, R. (1994). Beyond subject matter: A psychological topology of teachers' professional knowledge. In R. Biehler et al. (Eds.), *Didactics of mathematics as a scientific discipline* (pp. 73–88). Dordrecht, The Netherlands: Kluwer.

Cai, J., & Brooke, M. (2006). New perspectives of looking back in mathematical problem solving. *Mathematics Teaching, 196*, 42–45.

Cai, J., & Cifarelli, V. (2005). Exploring mathematical exploration: How do two college students formulate and solve their own mathematical problems? *FOCUS on Learning Problems in Mathematics, 27*(3), 43–72.

English, L. D. (1997). The development of fifth-grade children's problem-posing abilities. *Educational Studies in Mathematics, 34*, 183–217.

Freudenthal, H. (1983). *Didactical phenomenology of mathematical structures*. Dodrecht, The Netherlands: Reidel.

Gavora, P. (2010). Slovak preservice teacher self-efficacy: theoretical and research considerations. *The New Educational Review, 21*(2), 17–30.

Harel, G., Koichu, B., & Manaster, A. (2006). Algebra teachers' ways of thinking characterizing the mental act of problem posing. In J. Novotná, H. Moraová, M. Krátká, & N. Stehlíková (Eds.), *Proceedings of the 30th Conference of the International Group for the Psychology of Mathematics Education* (Vol. 2, pp. 241–248). Prague, Czech Republic: Charles University.

Hošpesová, A., & Tichá, M. (2010). Reflexion der aufgabenbildung als weg zu erhöhung der lehrerprofesionalität. In C. Böttinger, K. Bräuning, M. Nührenbörger, R. Schwarzkopf, & E. Söbbeke (Eds.), *Mathematik im Denken der Kinder; Anregung zur mathematik-didaktischen Reflexion* (pp. 122–126). Seelze, Germany: Klett/Kalmeyer.

Kilpatrick, J. (1987). Problem formulating: Where do good problems come from? In A. H. Schoenfeld (Ed.), *Cognitive science and mathematics education* (pp. 123–147). Hillsdale, NJ: Erlbaum.

Klafki, W. (1967). *Studie k teorii vzdělání a didaktice (Studien zur Bildungstheorie und Didaktik)*. Praha, Czech Republic: SPN.

Koman, M., & Tichá, M. (1998). On travelling together and sharing expenses. *Teaching Mathematics and its Applications, 17*(3), 117–122.

Lamon, S. (2006). *Teaching fractions and ratios for understanding* (2nd ed.). Mahwah, NJ: Erlbaum.

Pittalis, M., Christou, C., Mousolides, N., & Pitta-Pantazi, D. (2004). A structural model for problem posing. In M. J. Hoines & A. J. Bishop (Eds.), *Proceedings of 28th Conference of the International Group for the Psychology of Mathematics Education* (Vol. 4, pp. 49–56). Bergen, Norway: PME.

Polya, G. (2004). *How to solve it: A new aspect of mathematical method (with a new foreword by John H. Conway)*. Princeton, NJ: Princeton University Press.

Ponte, J. P., & Henriques, A. (2013). Problem posing based on investigation activities by university students. *Educational Studies in Mathematics, 83*(1), 145–156.

Prediger, S. (2006). Continuities and discontinuities for fractions: a proposal for analysing in different levels. In J. Novotná et al. (Eds.), *Proceedings of the 30th Conference of the International Group for the Psychology of Mathematics Education* (Vol. 4, pp. 377–384). Praha, Czech Republic: PME.

Selter, C. (1997). Instructional design for teacher education. In M. Beishuizen, K. P. E. Gravenmeijeer, & E. C. D. M. van Lieshout (Eds.), *The role of contexts and models in the development of mathematical strategies and procedures* (pp. 55–77). Utrecht, The Netherlands: Freudenthal Institute.

Silver, E. A., & Cai, J. (1996). An analysis of arithmetic problem posing by middle school students. *Journal for Research in Mathematics Education, 27*, 521–539.

Silver, E. A., & Cai, J. (2005). Assessing students' mathematical problem posing. *Teaching Children Mathematics, 12*(3), 129–135.

Singer, F. M., Ellerton, N. F., Cai, J., & Leung, E. (2011). Problem posing in mathematics learning and teaching: A research agenda. In B. Ubuz (Ed.), *Proceedings of the 35th Conference of the International Group for the Psychology of Mathematics Education* (Vol. 1, pp. 137–166). Ankara, Turkey: PME.

Streefland, L. (1991). *Fractions in realistic mathematics education: A paradigm of developmental research*. Dordrecht, The Netherlands: Kluwer.

Švaříček, R., & Šeďová, K. (2007). Kvalitativní výzkum v pedagogických vědách. (Qualitative research in education.) Praha: Portál.

Tichá, M., & Hošpesová, A. (2006). Qualified pedagogical reflection as a way to improve mathematics education. *Journal for Mathematics Teachers Education, 9*, 129–156. (Special Issue: Inter-relating theory and practice in mathematics teacher education).

Tichá, M., & Hošpesová, A. (2010). Problem posing and development of pedagogical content knowledge in preservice teacher training. In V. Durand-Guerrier, S. Soury-Lavergne, & F. Arzarello (Eds.), *Proceedings of the Sixth Congress of the European Society for Research in Mathematics Education* (pp. 1941–1950). Lyon, France: Institut National de Recherche Pédagogique.

Tichá, M., & Hošpesová, A. (2013). Developing teachers' subject didactic competence through problem posing. *Educational Studies in Mathematics, 83*(1), 133–143.

Tichá, M. (2003). Following the path of discovering fractions. In J. Novotná (Ed.), *Proceedings of SEMT'03* (pp. 17–27). Praha, Czech Republic: United Kingdom PedF.

Toluk-Ucar, Z. (2008). Developing preservice teachers understanding of fractions through problem posing. *Teaching and Teacher Education, 25*, 166–175.

Webster's Encyclopedic Unabridged Dictionary of the English Language (1996). New York, NY: Gramercy Books.

Chapter 22
What Do High School Teachers Mean by Saying "I Pose My Own Problems"?

Michal Klinshtern, Boris Koichu, and Avi Berman

Abstract The aim of this chapter was to identify mathematics teachers' conceptions of the notion of "problem posing." The data were collected from a web-based survey, from about 150 high school mathematics teachers, followed by eight semi-structured interviews. An unexpected finding shows that more than 50% of the teachers see themselves as problem posers for their teaching. This finding is not in line with the literature, which gives the impression that not many mathematics teachers are active problem posers. In addition, we identified four types of teachers' conceptions for "problem posing." We found that the teachers tended to explain what problem posing meant to them in ways that would embrace their own practices. Our findings imply that most of the mathematics teachers are result-oriented—as opposed to being process-oriented—when they talk about problem posing. Moreover, many teachers who pose problems doubt the ability of their students to do so and consider problem-posing tasks inappropriate for their classrooms.

Contents

M. Klinshtern (✉) • B. Koichu • A. Berman
Department of Education in Science and Technology,
Technion—Israel Institute of Technology, Haifa, Israel
e-mail: mklinst@tx.technion.ac.il; bkoichu@technion.ac.il; berman@technion.ac.il

F.M. Singer et al. (eds.), *Mathematical Problem Posing*,
Research in Mathematics Education, DOI 10.1007/978-1-4614-6258-3_22

Introduction

Mathematical problem posing is widely recognized as one of the central activities in mathematics and as a useful tool in the teaching and learning mathematics (e.g., National Council of Teachers of Mathematics [NCTM], 2000). However, a glimpse at the professional literature reveals that the notion of "problem posing" is used with a variety of not-always-compatible meanings and is applied to a variety of not-always-comparable teaching/learning situations. In addition, existing conceptualizations of problem posing, as diverse as they are, reflect the researchers' and mathematics educators' points of view on what counts (or not counts) as a worthwhile result of problem posing. The question of what problem posing means for mathematics teachers is still unexplored. The study presented in this chapter aims at partially closing this gap by exploring what the notion of "problem posing" and the associated notion of "my own problem" mean for in-service mathematics teachers.[1]

Theoretical Background

Problem posing as a teaching/learning tool has been extensively studied with students (e.g., Brown & Walters, 1983; English, 1997a, 1997b, 2003; Lowrie, 2004; Mestre, 2002; Silver, 1994) and with preservice and in-service teachers (e.g., Crespo, 2003; Koichu, Harel, & Manaster, 2013; Lavy & Bershadsky, 2003; Silver, Mamona-Downs, Leung, & Kenney, 1996). In this section we review how problem posing has been conceptualized in the aforementioned studies, with particular attention to studies in which teachers act as problem posers.

[1] This study is part of a Ph.D. dissertation, in progress, by the first-named author under the supervision of the two other authors. A brief version of this paper was accepted as a research report at PME-37.

Conceptualization of Problem Posing

Kilpatrick (1987) conceptualized problem posing as reformulating an existing problem in order to make it your own. This conceptualization is deliberately poser-centered and depends on one's decisions about whether an existing problem is modified enough to be perceived by the poser as his or her "own." From this perspective, one may decide that the problem is his or her own after making only a cosmetic change, whereas another person may feel that even the changes that look essential to the readers or solvers of the modified problem are not enough in order to claim that a "new" problem has been born.

Stoyanova and Ellerton (1996) considered situations in which students pose problems and defined problem posing as "the process by which, on the basis of mathematical experience, students construct personal interpretations of concrete situations and formulate them as meaningful mathematical problems" (p. 518). The subjective nature of this definition—one should decide in which meaning the problem is meaningful and for whom—is apparent (see Koichu & Kontorovich, 2013, for an elaborated discussion about this issue).

Silver (1994) referred to problem posing either as generating new problems and questions for exploring a given situation or reformulating a given problem during the process of solving it. This conceptualization leaves room to inquire in which sense the processes involved in generating new problems and reformulating the given ones could be seen as instantiations of the same process tagged "problem posing." The same question can be asked in relation to the highly inclusive definition of problem posing by Crespo (2003), who referred to *selecting* worthwhile problems and *designing* challenging tasks for teaching as particular cases of problem posing.

Teachers as Problem Posers

Crespo's conceptualization brings us to the question of whether there is room (and need) for using "self-made problems" in teaching, given that rich collections of expert-made problems are readily available, for instance, in mathematics textbooks. Extensive research on problem posing by mathematics teachers has not provided an unequivocal answer to this seemingly simple question. Indeed, in the majority of studies, mathematics teachers pose problems "on request" in laboratory conditions (e.g., Koichu et al., 2013; Silver et al., 1996) or pose problems in the framework of professional developmental workshops aimed at enhancing their problem-posing skills (e.g., Crespo, 2003; Lavy & Shriki, 2007). Moreover, probably the most frequently reported finding on problem posing by mathematics teachers is that not many teachers have skills to pose worthwhile problems (e.g., Singer & Voica, 2013).

A promising finding on mathematics teachers' willingness to modify textbook problems and create their own problems was reported in Nicol and Crespo (2006).

These scholars found that two participants in their study attempted to extend the mathematical content of the chosen textbook problems in order to make them more complex. When the teachers were asked to prepare collections of problems for teaching in fourth grade based on the available textbooks, they preferred to create some of their own. Based on these results, Nicol and Crespo distinguished between three ways of using textbooks by the teachers: "adhering" (i.e., do not see self as a resource), "elaborating" (i.e., seeing self as a resource), and "creating" (i.e., seeing self as a knowledgeable resource for designing problems). The latter way of using the textbooks "brought forth opportunities to consider connections within and beyond mathematical topics" (p. 347). Note that this study was conducted with only four preservice elementary school teachers.

Research Questions

To our knowledge, evidence about whether and how in-service high school mathematics teachers pose problems for real use in their classrooms does not yet exist. Accordingly, and in light of the reviewed literature, our study pursues the following interrelated research questions:

1. To what extent do high school mathematics teachers see themselves as posers of problems for their teaching? For what purposes do they pose problems?
2. How do the teachers perceive the notions "problem posing" and "my own problem"?

Methodology

The data were collected from a web-based survey, which was filled in by 151 mathematics teachers. In addition, a semi-structured interview was carried out with eight of the survey participants. The collected data also included two classroom observations and problems that some of the teachers sent us by email. Details of the survey and participants will be provided in the next section.

Survey and Participants

The SurveyMonkey tool[2] was used in order to administer an online survey. The survey consisted of an introduction, six background questions, and four questions about teaching practices (see Appendix). The goal of the background questions was to collect data on the participants' teaching experience and academic education. The

[2] See http://www.surveymonkey.com.

goal of the questions on teaching practices was to collect data on how the teachers select mathematical problems for their teaching. The central question of the survey (Question 4) was: "*To what extent do you use the following resources for selecting mathematical problems for your teaching*?" The teachers were offered nine resources, one of which was "Pose my own problems." The other resources were: "Textbooks," "Other books," "Internet resources," "My prior academic study," "Professional development workshops (PDW)," "Fellow teachers," "Problems posed by my students," and "Other sources." For each resource, the participants were asked to choose one of the following five options: "Almost never," "Rarely," "Sometimes," "Often," and "Almost Always." The central question of the survey was formulated in this way (i.e., problem posing was put in line with the other possible sources of problems for teaching) in order to avoid a situation in which the teachers would overestimate the role of problem posing in their practice by trying to guess the "correct" answer to the question.

Methodological advice presented in "Response Rates and Surveying Techniques" (SurveyMonkey, 2009) and in Cook, Heath, and Thompson (2000) was followed when validating and administering the survey. First, the formulations included in the survey were validated by five experts in mathematics education. Then, the survey was trialed with 20 high school mathematics teachers. Next, individually named e-mails with the invitation to respond to the survey were sent to about 500 secondary school mathematics teachers whose names appeared in the departmental database of mathematics teachers. These letters contained a brief outline of the study and a request to fill in the survey. Some of the respondents informed us that they sent the survey to their colleagues. One hundred and fifty one teachers responded to the survey during 2011–2012 school year. That is, we achieved a response rate of about 30%, which is compatible with the result of Cook et al. (2000), who found in their meta-analysis of response rates that the mean response rate for electronic surveys is about 34%.

From the responses to the background questions of the survey, we know that more than 80% of the respondents teach in high school (grades 10–12); 76% teach the advanced versions of the Israeli mathematics curriculum; and 82% of the teachers had teaching experience of ten or more years. Thus, the research sample represents well a cluster of experienced in-service mathematics high school teachers in Israel.

Interviews

Eight participants representing the groups of teachers who indicated that they pose their own problems "Rarely," "Sometimes," "Often," and "Almost always" (two teachers per group) took part in the individual in-depth, semi-structured interviews. These teachers were chosen because they showed interest in the study, that is, they provided us with their contact information (see Appendix, Question 10), positively answered Question 11 of the survey and agreed to continue their participation in the study. Seven of the teachers had taught for more than 20 years and one teacher for more than 5 years. Thus, the interviewees represent well the participants of the survey in terms of their experience.

Three interviews were face-to-face, and the others were carried out by phone in order to get wide geographical access (Opdenakker, 2006). According to Opdenakker (2006), despite the absence of some social cues in phone interviews (e.g., body language) there are still enough social cues that can be used as valuable information (e.g., words, voice, and intonation). The interviews lasted between 20 and 60 minutes and were recorded, transcribed, and inductively analyzed (in the meaning specified, for instance, in Thomas, 2006).

During the interviews, teachers were asked to describe how they planned their lessons and selected mathematics problems for teaching. They were also asked to provide examples of their "own problems" and explain what "problem posing" means for them. Only two teachers gave an example of their own problems during the interviews. Two other teachers sent their problems by email after the interview. All other interviewees invited the interviewer to visit their classes.

Observations, a Lecture, and Teachers' Examples

As a result of the teachers' invitations to visit their classes, observations in two different classes of one of the teachers were carried out. This teacher was chosen because she indicated in the questionnaire that she poses her own problems "Often." In addition, this teacher gave a lecture about her problem-posing practices in a course for preservice mathematics teachers. The lecture was videotaped and served as a complementary data source.

Findings

The findings have been organized in accordance with the research questions. First, we report the extent to which the teachers saw themselves as posers of problems for their teaching and the purposes for which they posed their own problems. We then devote a section to the question of how the teachers perceived the notions "problem posing" and "my own problem." In the final section, we present an interesting result concerning the teachers' opinions about problem posing as a learning activity for their students.

Mathematics Teachers as Problem Posers

The percentages of using the different resources for choosing problems for teaching (see Appendix, Question 4) are presented in Table 22.1.

Table 22.1
Percentage of Responses to Problems Resources

	Problem resource	Almost never (%)	Rarely (%)	Sometimes (%)	Often (%)	Almost always (%)	No. of responses (100%)
1	Textbooks	1.3	0	2.6	23.8	72.2	151
2	Other books	10.1	9.4	26.8	39.6	14.1	149
3	Internet resources	14.4	17.8	37.0	19.9	11.0	146
4	Teacher PDW	18.1	24.2	38.9	11.4	7.4	149
5	Fellow teachers	12.2	18.9	36.5	25.7	6.8	148
6	My academic study	30.6	26.4	27.1	10.4	5.6	144
7	Pose my own problems	19.9	22.6	29.5	21.2	6.8	146
8	Problem posed by my students	51.7	27.9	16.3	3.4	0.7	147
9	Other	56.3	6.3	14.6	14.6	8.3	48

In view of past studies about teachers' usage of textbooks (e.g., Ball & Cohen, 1996; Ball & Feiman-Nemser, 1988; Ben-Peretz, 1990), it is not surprising that curriculum-based textbooks are the most frequently mentioned by the participants as a source of mathematical problems (see Table 22.1 and Figure 22.1). However, in light of the literature about teachers' difficulties with problem posing (e.g., Koichu et al., 2013; Silver et al., 1996; Singer & Voica, 2013), we were surprised to find that about 57% of the teachers indicated that they pose their own problems at least "sometimes," and 7% (11 of 146) "Almost always." The graph presented in Figure 22.1 summarizes the frequencies of using each resource at least "sometimes."

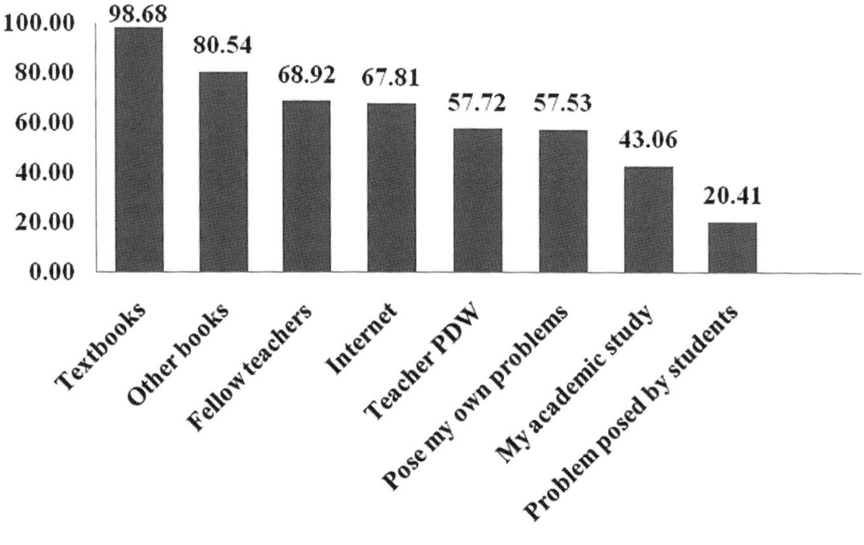

Figure 22.1 Frequencies of teachers' use of different problem resources.

The frequencies presented in Figure 22.1 can imply that the teachers tended to use the most available resources and turn to less readily available resources or to posing their own problems when they could not easily find problems that fitted their teaching needs. Indeed, the most popular problem resources appeared to be "Textbooks" and "Other books," which were most readily available. One may comment that the internet is the most available resource, given that in Israel all the teachers, as a rule, are proficient users of internet resources. However, it should be taken into account that using the problems found on internet resources requires preparing and printing work sheets, and it may not be the most parsimonious strategy. This may explain why the internet was only the fourth most popular resource, after "Textbooks," "Other books," and "Fellow teachers."

The survey included a request to explain when and why the teachers posed their own problems (see Appendix, Question 6). One hundred and nine out of 146 teachers (about 75%) responded to this request. Two main groups of reasons for problem posing were found: "In order to adapt a problem to my students' needs" (hereafter, *adaptations*) and "to interest myself" (hereafter, *self-interest*). The categories and subcategories of the reasons are presented in Figure 22.2. The distribution of the subcategories for the *adaptations* category is presented in Figure 22.3. In addition, three categories presented in Figure 22.3 unpack the category "others" from Figure 22.2. The most popular reason for posing problems, as reflected in Figure 22.3, is "for tests." A typical explanation for posing problems for tests was: "… because you don't want to be in a situation in which they [the students] saw the problem before …".[3]

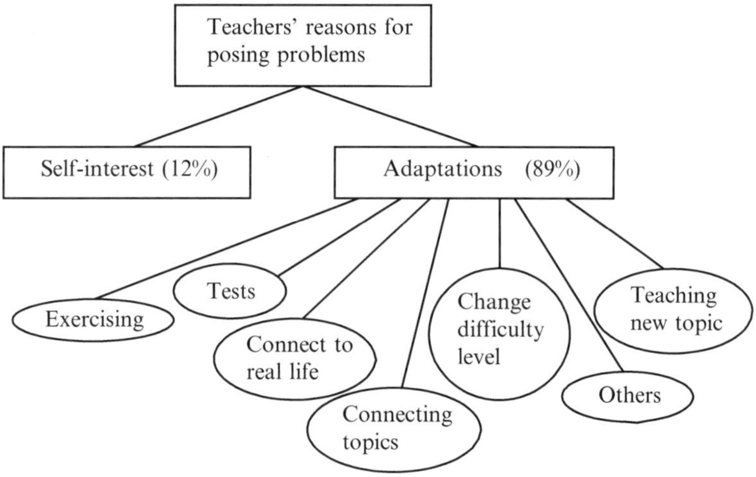

Figure 22.2 Teachers' reasons for posing problems.

Another finding that can be derived from Figure 22.3 is that about 27% of the teachers felt that they had to change the difficulty level of textbook problems in order to adapt them to their students' level. Some teachers claimed that textbook

[3] The quotations have been translated from Hebrew by the authors.

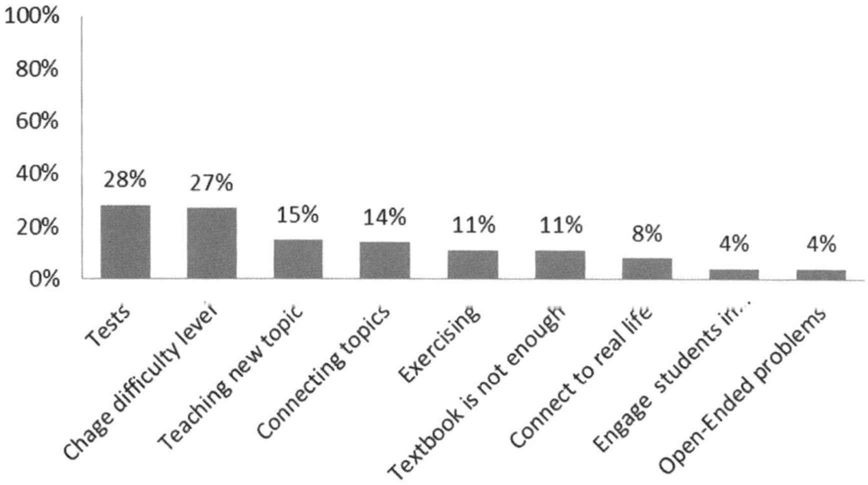

Figure 22.3 Teachers' reasons for posing problems in the *adaptations* category.

problems were too easy (e.g., "[I pose problems] when I feel that the level in the textbook problems was too low"), whereas others said that they needed to simplify textbook problems (e.g., "When the problem is not structured enough and there is a need to simplify it").

Rarely mentioned reasons, indicated by only one or two teachers (and thus not in Figure 22.3), were: "for reflecting the class work," "for integrating the education for values," "for inquiry by students," and "for didactical games."

Perceptions of the Notion of Problem-Posing

The teachers' perceptions were inductively distilled from the eight interviewees' explanations of what "posing a mathematical problem" meant to them, from examples of problems that they supplied as their own, and from their responses to the interview questions. Two main categories were defined; each category includes two subcategories. The names of the categories were given by us, the authors of the paper, based on the teachers' examples and explanations (Table 22.2).

Routine problem posing. Generally speaking, this category refers to textbook-like problems posed by the teachers for a test or in order to adapt a textbook problem to the students' needs.

Cosmetic change. For example, Teacher B posed for a test the following problem: "Sketch a graph of the function $y = \frac{(x-2)(x+3)}{(x+4)(x-5)}$ without formally exploring it."

Table 22.2
Categories of the Meaning of the Notion "Problem Posing" for the Teachers

Category		Description
Routine problem posing	Cosmetic changes	Changing an existing problem by replacing some of its parameters or its story without changing the idea of the solution
	In-the-moment problems	Unplanned using a known mathematical problem or spontaneously generated questions and examples during the lesson
Innovative problem posing	Combining ideas	Creating a new problem that, as a rule, requires for its solving the use of ideas or techniques that have been previously studied and used separately
	Using the same problem in different contexts	Using the same, probably known, problem when teaching different topics in order to encourage the students to solve it by using different tools, or as a starting point for explaining a new concept

During the lessons preceding the test, the students of Teacher B discussed problems of the same formulation, but with other functions. The above function was "invented" by Teacher B.

In the moment problems. All of the interviewees indicated that, during lessons, they deal with students' misunderstandings and obstacles by means of generating or recalling mathematical questions, problems or examples on the spot, without planning to do so prior to the lesson. About 5% of the teachers, who explained in the survey their reasons for posing their own problems, mentioned that it was done during the lessons, or in the moment. Interestingly, the interviewees could not provide examples related to this category during the interview. Some of them explained that this was because the questions looked too uninteresting out of the context of the lessons.

An example, however, was found, in a lesson in which Teacher N introduced the notion of functions to students. During this lesson, Teacher N realized that the concept of a linear function was not clear to her low-level ninth-grade students and decided to tell a story (see Figure 22.4).

Suppose you did a math test and I decided to give a 4 point bonus to all students' final grades. What will be your grade now if you had 82? 95? 74? …
Can you build a rule describing your new grade after the bonus?

Figure 22.4 An example of a problem posed during the lesson.

After the lesson Teacher N was asked if she planned this problem. She said: "I *invented* it during the lesson."

Seven of the eight interviewees, who had teaching experience of more than 20 years, indicated that their knowledge and experience enabled them to pose *in the moment problems*. Two teachers referred to this practice as problem posing from the beginning, and the others were unsure whether this practice could be called such. Two teachers started by saying that they did not pose many problems, but then they decided, probably due to the context of the interview, that recalling or formulating examples, routine problems, and questions during the lessons could be considered as problem posing (e.g., "Mostly, I don't bring examples from a book. I invent them ... Then the students say: it is not from the book, it's B's [problem]. So it is like I pose it, if it can be considered as problem posing. I have never thought of this ... Naturally, I pose examples"). It looked like the teachers wanted to figure out how their perceptions could go along with the scholars' definitions, and if they could consider themselves as problem posers.

Innovative problem posing. This category fits more readily the conceptualizations of problem posing as described in the literature. The main reasons for problem posing underlying the problems in the "*Combining ideas*" and "*Using the same problem in different contexts*" categories was the need to connect different mathematical topics.

The teachers' need for a "feeling of innovation" (Kontorovich & Koichu, 2012) manifested itself in the 12% of the survey participants who wrote that they pose problems for their self-interest. Indeed, among the teachers' explanations of the reasons for which they posed their own problems, we found the following examples: "It [posing problems] gives me an opportunity to use my creativity," or, "The main reason [for posing problems] is to do something non-routine for me," or "I enjoy it [problem posing], otherwise I would be bored."

Combining ideas. About 14% of the teachers who responded to Question 6 of the survey (see Appendix) mentioned "Combining ideas for connecting topics" as a reason for posing their own problems. The teachers indicated that they need problems combining several ideas for tests, exams, and lessons aimed at summarizing particular topics. For example, Teacher R and her colleagues combined the previously taught ideas related to arithmetic and geometric sequences in the problem for the test presented in Figure 22.5.

Given an arithmetic sequence containing $2n+1$ elements. The first element of this sequence is equal to k, and the sequence difference is equal to d. From the given sequence, a new sequence was built as follows: The even elements were doubled, and the odd elements were increased by 4.

a) Using k and d, write the five first elements of the new sequence.
b) Prove that the odd elements of the new sequence form an arithmetic sequence.
c) Prove that the sum of the new sequence is $3n^2d + 3kn + nd + k + 4n + 4$
d) The second element in the new sequence is 7 times bigger than the first element in the original sequence. The elements in places 1, 6, and 97 in the new sequence form a geometric sequence. Find k and d if it is given that the elements in the sequence are integers.

Figure 22.5 An example of problem invention.

Another example was provided by Teacher M, who posed a problem for teaching a new subject: "One year I posed a problem about a magician that pulls scarves from a hat with or without returning. We defined when the magic succeeded and then you can ask many questions. In my opinion, it is better [than the problems] from the book because in the textbooks there are headlines and they [the students] don't see connections." This assertion is reminiscent of the comment by Ball and Feiman-Nemser (1988), who noted that in textbooks "problem solving is often trivialized and math portrayed as a collection of algorithms to be followed" (p. 402).

The need for posing problems with the potential to combine ideas and connect topics also manifested in the survey. Nine out of 24 teachers who selected "Other resource" for mathematical problems indicated that they used problems from past years' exams. The exams in Israel frequently include problems connecting ideas and topics.

It is interesting that two teachers mentioned, in their interviews, the difficulties students faced with problems that connected different ideas and topics. In the words of Teacher M, "There was a year that I posed problems on many occasions. I stopped doing so when I realized that the students did not enjoy them. Now, I do only small changes in existing problems and sometimes add challenging items [to the existing problems]." Teacher A described in the interview a situation in which "one student said [after the test] that it [the test] was too difficult, and when we [the students] solve textbook problems, they are different, they are simpler." These two comments suggest that the teachers' sensitivity to the students' needs may either encourage or discourage them from posing innovative problems.

Using the same problem in different contexts. The first example of a problem from this category is taken from the interview with Teacher M. She used a problem about finding the area of a triangle given the coordinates of its vertices when teaching plane geometry, vectors, and complex numbers. She mentioned the repeated use of the same problem when asked to provide examples of the problems she posed. An additional example came from Teacher N. She told us, as an example of a posed problem, that at the beginning of the topic "Functions" she reminds the students of one of the problems on sequences, and just changes the notation from an to $f(n)$.

Miscellaneous: The Teachers' Views of Their Students as Problem Posers

An interesting result is related to the extent to which the teachers used their own problems in class, as compared to the extent of using problems posed by their students. The result was derived from juxtaposing the responses to Questions 4 and 5 of the survey and from the interviews. In the graphs presented in Figure 22.6 we can see that there is a strong connection between the frequency of posing problems and the frequency of using them in class (based on Questions 4 and 5 of the survey). The x-axis is the frequency of posing problems, according to Question 4 ("To what

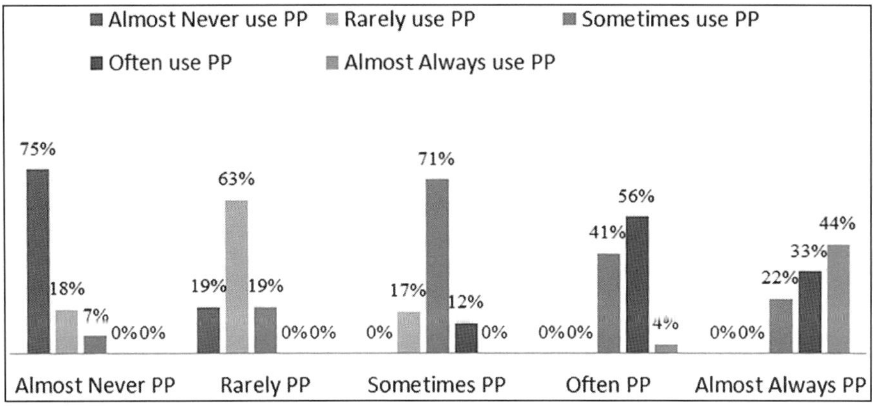

Figure 22.6 Using teachers' own problems for teaching.

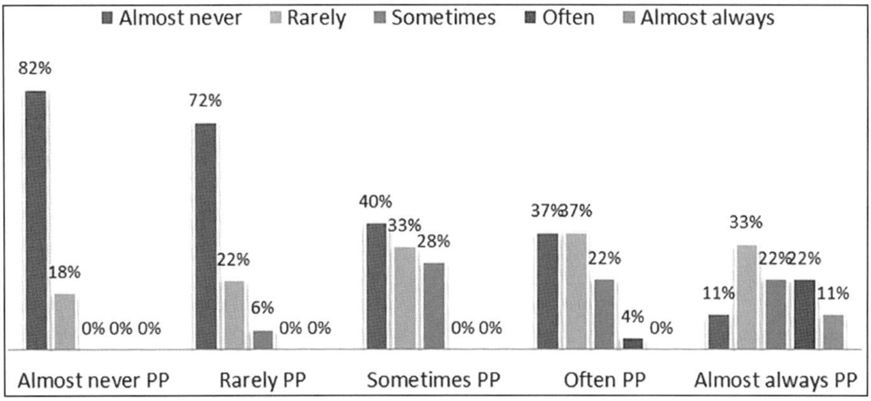

Figure 22.7 Using problems composed by students.

extent do you use your own problems as a resource for problems?"), and the bars indicate the frequency of using the posed problems in class.

It is reasonable to suggest that the teachers who posed problems tended to use them in their classes, and indeed Figure 22.6 shows that about 75% of the teachers who "sometimes" posed their own problems, "sometimes" use them. To us, it was reasonable to expect that the teachers who posed their own problems would encourage their students to pose problems. However, our results show that the teachers seldom used problems posed by their students. Figure 22.7 shows that only 28% of the teachers who posed problems "sometimes" tended to use their students' problems.

The interviews revealed that some teachers thought that their students were incapable of composing "good" problems. Teacher A said, for example: "I don't think that the students are capable of doing so [posing problems], they are still not mature enough so they don't have patience to sit and think about problems. So I don't want to waste time for this [problem posing activities]." With respect to the teacher's perception of problem posing, this assertion implies that, although teachers may appreciate the results of problem posing, this does not mean that they appreciate the learning potential inherent in the problem-posing process which is strongly emphasized in the professional literature.

Discussion and Conclusions

Both past and recent studies (e.g., Koichu et al., 2013; Silver et al., 1996; Singer & Voica, 2013) give the impression that not many mathematics teachers are active problem posers. In light of this, it is quite surprising that more than half of the participants in our survey indicated that they posed problems at least "Sometimes" and about 28% at least "Often." This supports Nicol and Crespo's (2006) finding that teachers can choose problems for their teaching either by elaborating or by creating (cf. Leikin & Grossman, 2013, for an example of how the teacher can creatively modify problems from geometry textbooks).

Our study also resulted in categorization of what problem posing means for the teachers. Two main categories were defined—"*routine problem posing*," and "*innovative problem posing*." The first category ("*routine problem posing*") consists of "*cosmetic changes*" to existing problems and "*in-the-moment problems*." The emergence of "*cosmetic changes*" and "*in-the-moment questions*" categories in our data supports Silver et al.'s (1996) observation that teachers tend to pose textbook-like problems and can also do so spontaneously during the lessons. In addition, the "*cosmetic changes*" category of problem posing is reminiscent of what Singer and Voica (2013) called "not interesting, being just scholastic" problems, and what Crespo (2003) called "non-problematic" or "avoiding pupils' errors" problems. However, our data suggest that problems created by "*cosmetic changes*" do not bear a negative connotation from the teachers' perspective. This is because this type of problem is instrumental in everyday teaching, especially for preparing tests and exams. This is in line with the comment by Prestage and Perks (2007), who claimed that teachers have to be able to make "in-the-moment shifts in a task in relation to learners' needs" (p. 382). Interestingly, the teachers in our study regarded this type of problem posing as belonging to their craft knowledge (in the words of one of the interviewees, "it is something that you do naturally") and connect it to their teaching experience.

The second category ("*innovative problem posing*") consists of "*combining ideas*" and "*using problems in different contexts*." These teachers' perceptions of problem posing are in line with conceptualizations of problem posing by Kilpatrick (1987) and Silver (1994) (see section "Introduction"), as well as with the findings of Nicol and Crespo (2006), who mentioned that teachers can pose problems in order to connect different topics.

The presented findings imply that most of the mathematics teachers are result-oriented—as opposed to being process-oriented—when they talk about problem posing. That is, they give high value to problem posing when it leads to creating worthwhile problems for real use, and less value when it is an activity with the potential to develop their or their students' mathematical skills (cf. Silver, 1997, for the analysis of how problem posing can foster mathematical creativity). This suggestion is indirectly supported by the fact that most of the participants in our study indicated that they rarely used problems posed by their students in teaching, and as a rule, did not ask their students to pose problems.

The identified conceptions suggest that there can be some discrepancy between how mathematics teachers treat the role of problem posing in their practice and how problem posing is treated in the research literature, especially in studies where the teachers pose problems "on request." This merits further research attention, and probably, revisiting some of the ways by which the quality of the problems posed by teachers is evaluated in the frameworks of various studies on problem posing conducted in laboratory conditions. Specifically, we believe that the role of relevance of the posed problems to the teachers' needs should be given attention and merit in future studies. With regard to teacher education it seems that understanding what experienced teachers mean by posing problems will be useful in training preservice teachers to use their own problems in their teaching practice.

In summary, we offer a general conceptualization of what problem posing seems to mean for the participants in our study: problem posing means an accomplishment that consists of *constructing a problem that satisfies the following three conditions*: (a) *it somehow differs from the problems that appear in the resources available to the teacher*; (b) *it has not been approached by the students*; *and* (c) *it can be used in order to fulfill teaching needs that otherwise could be difficult to fulfill*. Of course more comprehensive research is needed in order to understand whether this definition can be applied widely.

Appendix

The Survey

Selecting Problems to Be Used in Mathematics Teaching

This questionnaire is part of a research done in the Department of Education in Technology and Science at the Technion. The research aims at understanding how mathematics teachers select problems for their teaching. None of the questions has a "right" or "wrong" answer. It is very important that you will answer all the questions of this brief questionnaire.

We would like to thank you for the time you dedicated to answer this questionnaire.

1. The last three years I teach grades:

7	8	9	10	11	12

2. Usually I teach class levels

Strong	Medium	Weak

3. Describe all your special teaching project if any ____________________

4. To what extent do you use the following resources for selecting mathematical problems for your teaching?

	Almost always	Often	Sometimes	Rarely	Almost never
Textbooks					
Other books					
Internet resources					
Professional development workshops					
Fellow teachers					
My prior academic study					
Pose my own problems					
Problem posed by students					
Others					

Point out any other resources that you use ____________________________

5. To what extent do the following situations occur in your teaching?

	Almost always	Often	Sometimes	Rarely	Almost never
Use my own problems					
Activate students in problem posing					
Promote class discussion					
Encourage group work					

6. For what purposes do you pose your own problems?

___

7. Seniority in mathematics teaching

1–2 years	3–5 years	6–10 years	More than 10 years

8. To what extent the following situations are in your responsibility:

	Little	Much
Planning the school year		
Planning the lessons		
Execute my planning		
Select the mathematical problems to be used		

9. Education

	Math	Math Ed.	Science Ed.	Computer science	Engineering	Biology	Physics	Others
B.A./ B.Sc.								
M.A./ M.Sc.								
Ph.D.								
Other								

10. Personal details (optional)

Name: ___

Email: ___

Phone: ___

11. I am interested in receiving updates about the results of the study

Yes	No

References

Ball, D. L., & Cohen, D. K. (1996). Reform by the book: What is—or might be—The role of curriculum materials in teacher learning and instructional reform? *Educational Researcher, 25*(9), 6–8, 14.

Ball, D. L., & Feiman-Nemser, S. (1988). Using textbooks and teachers' guides: A dilemma for beginning teachers and teacher educators. *Curriculum Inquiry, 18*(4), 401–423.

Ben-Peretz, M. (1990). *The teacher-curriculum encounter: Freeing teachers from the tyranny of texts*. Albany, NY: SUNY.

Brown, S., & Walters, M. (1983). *The art of problem posing*. Hillsdale, NJ: Lawrence Erlbaum.

Cook, C., Heath, F., & Thompson, R. L. (2000). A meta-analysis of response rates in web- or internet-based surveys. *Educational and Psychological Measurement, 60*(6), 821–836.

Crespo, S. (2003). Learning to pose mathematical problems: Exploring changes in preservice teachers' practices. *Educational Studies in Mathematics, 52*(3), 243–270.

English, L. D. (1997a). The development of fifth-grade children's problem posing abilities. *Educational Studies in Mathematics, 34*(3), 183–217.

English, L. D. (1997b). Promoting a problem posing classroom. *Teaching Children Mathematics, 4*(3), 172–179.

English, L. D. (2003). Engaging students in problem posing in an inquiry-oriented mathematics classroom. In F. Lester & R. Charles (Eds.), *Teaching mathematics through problem solving* (pp. 187–198). Reston, VA: National Council of Teachers of Mathematics.

Kilpatrick, J. (1987). Problem formulating: Where do good problems come from? In A. H. Schoenfeld (Ed.), *Cognitive science and mathematics education* (pp. 123–147). Hillsdale, NJ: Lawrence Erlbaum.

Koichu, B., Harel, G., & Manaster, A. (2013). Ways of thinking associated with mathematics teachers' problem posing in the context of division of fractions. *Instructional Science, 41*(4), 681–698. Retrieved May 31, 2013, from http://link.springer.com/article/10.1007/s11251-012-9254-1#page-1.

Koichu, B., & Kontorovich, I. (2013). Dissecting success stories on mathematical problem posing: A case of the Billiard Task. *Educational Studies in Mathematics, 83*(1), 71–86.

Kontorovich, I., & Koichu, B. (2012). Feeling of innovation in expert problem posing. *Nordic Studies in Mathematics Education, 17*(3–4), 199–212.

Lavy, I., & Bershadsky, I. (2003). Problem posing via "What if not?" strategy in solid geometry—A case study. *The Journal of Mathematical Behavior, 22*(4), 369–387.

Lavy, I., & Shriki, A. (2007). Problem posing as a means for developing mathematical knowledge of prospective teachers. In J. H. Woo, H. C. Lew, K. S. Park, & D. Y. Seo (Eds.), *Proceedings of the 31st Conference of the International Group for the Psychology of Mathematics Education* (Vol. 3, pp. 129–136). Seoul, Korea: PME.

Leikin, R., & Grossman, D. (2013). Teachers modify geometry problems: From proof to investigation. *Educational Studies in Mathematics, 82*(3), 515–531.

Lowrie, T. (2004). Designing a framework for problem posing: Young children generating open-ended tasks. *Contemporary Issues in Early Childhood, 3*(3), 354–364.

Mestre, P. J. (2002). Probing adults' conceptual understanding and transfer of learning via problem posing. *Journal of Applied Developmental Psychology, 23*(1), 9–50.

National Council of Teachers of Mathematics. (2000). *Principles and standards for teaching mathematics*. Reston, VA: Author.

Nicol, C. C., & Crespo, S. M. (2006). Learning to teach with mathematics textbooks: How preservice teachers interpret and use curriculum materials. *Educational Studies in Mathematics, 62*(3), 331–355.

Opdenakker, R. (2006). Advantages and disadvantages of four interview techniques in qualitative research. *Forum Qualitative Sozialforschung/Forum: Qualitative Social Research*, *7*(4), Art. 11.

Prestage, S., & Perks, P. (2007). Developing teacher knowledge using a tool for creating tasks for the classroom. *Journal of Mathematics Teacher Education, 10*(4), 381–390.

Silver, E. A. (1994). On mathematical problem posing. *For the Learning of Mathematics, 14*(1), 19–28.
Silver, E. A. (1997). Fostering creativity through instruction rich in mathematical problem solving and problem posing. *ZDM, 29*(3), 75–80.
Silver, E. A., Mamona-Downs, J., Leung, S. S., & Kenney, P. A. (1996). Posing mathematical problems: An exploratory study. *Journal for Research in Mathematics Education, 27*(3), 293–309.
Singer, F. M., & Voica, C. (2013). A problem-solving conceptual framework and its implications in designing problem-posing tasks. *Educational Studies in Mathematics, 83*(1), 9–26.
Stoyanova, E., & Ellerton, N. F. (1996). A framework for research into students' problem posing in school mathematics. In P. Clarkson (Ed.), *Technology in mathematics education* (pp. 518–525). Melbourne, Australia: Mathematics Education Research Group of Australasia.
SurveyMonkey (2009). *Response rates & surveying techniques—Tips to enhance survey respondent participation*. Retrieved May 2, 2010, from http://s3.amazonaws.com/SurveyMonkeyFiles/Response_Rates.pdf
Thomas, D. R. (2006). A general inductive approach for analyzing qualitative evaluation data. *American Journal of Evaluation, 27*(2), 237–246.

Chapter 23
A Review on Problem Posing in Teacher Education

Helena P. Osana and Ildiko Pelczer

Abstract Over the last two decades, researchers have shown increased interest in problem posing in mathematics professional development. In the context of teaching mathematics, problem posing can entail asking questions during classroom interactions to assess student understanding, modifying existing problems to adjust the difficulty level of a task, and creating problems to meet instructional objectives. In this chapter, we review the research conducted between 1990 and 2012 on problem posing in mathematics methods courses in elementary teacher education. Despite the range of foci, goals, and theoretical perspectives in the literature, we describe ways in which problem posing has been investigated in the preservice teacher population. Despite the paucity of empirical studies, we were able to group these studies into three distinct categories: (a) problem posing as a skill integral to the practice of teaching mathematics; (b) problem posing as an activity separate from teaching; and (c) problem posing as a tool to assess an outcome variable (for researchers) or as a tool for teaching or assessing the development of preservice teachers' knowledge or beliefs. Implications for mathematics teacher educators that stem from the review of the literature are discussed.

Contents

H.P. Osana (✉)
Department of Education, Concordia University, Montréal, QC, Canada
e-mail: osana@education.concordia.ca

I. Pelczer
Department of Mathematics and Statistics, Concordia University, Montréal, QC, Canada
e-mail: ildiko.pelczer@gmail.com

F.M. Singer et al. (eds.), *Mathematical Problem Posing*,
Research in Mathematics Education, DOI 10.1007/978-1-4614-6258-3_23

A Review on Problem Posing in Teacher Education

Contemporary research on problem posing can be traced back to the 1980s and has steadily been gaining interest since that time. Results from cognitive psychology and a reorientation in mathematics education were at the root of this emerging interest, as evidenced by researchers' emphases on identifying the processes underlying mathematical thinking (see Schoenfeld, 1992, for a review on the topic). A growing movement in mathematics education that placed problem solving at the center of school mathematics further contributed to researchers' focus on problem posing, particularly its role in teaching and learning. In this context, problem solving was simultaneously viewed as a means for teaching mathematical reasoning and as a learning objective in and of itself (see, for example, the first *Standards* document published by the National Council of Teachers of Mathematics [NCTM], 1989). Scholars' understanding of the nature and effects of mathematical problem solving over the last several decades, with its roots in cognitive science (e.g., Newell & Simon, 1972), gave rise to the notion that problem posing is a process that is embedded within (and difficult to separate from) problem solving. Indeed, Kilpatrick (1987) claimed that problem posing is an important constituent of mathematical thinking, but some years before, Brown and Walter (1983) had already argued for the central position of problem posing in learning and thinking about mathematics. Similarly, interest in the nature and role of creativity—and its link to other elements of mathematical thinking—further contributed to the study of problem posing. From this perspective, problem posing was seen by many as a way to assess and enhance creativity (Silver, 1997). The history of problem posing, its varied uses in research and teaching, and its inherent cognitive and creative components together attest to its complex nature.

Early research on problem posing was centered on children's thinking and reasoning. In particular, scholars studied the cognitive processes used by children during problem posing (e.g., English, 1997), the types of problems posed (e.g., Gonzales, 1996), and comparisons of the behaviors and attitudes of students from different cultural settings (e.g., Cai & Hwang, 2002). In this line of research, and in parallel with the study of children's problem solving, problem posing was also seen as a way to assess children's mathematical understanding. At a somewhat slower rate, an interest in the problem-posing abilities of teachers was emerging, particularly in mathematics. A series of studies published by Silver and his collaborators in the 1990s looked at relationships between teachers' knowledge, task format, and creativity (e.g., Leung & Silver, 1997). These studies focused on investigating the commonalities and distinguishing features of mathematics teacher knowledge and task-related conditions for posing problems. Although these findings have been informative for the study of teacher knowledge from a theoretical perspective, the data were not, at least in these cases, directly applied to improve professional development initiatives.

Shulman's (1986) well-known article introduced the construct of *pedagogical content knowledge* (*PCK*) as being a critical type of knowledge, along with *subject-matter* and *curricular knowledge*, needed by teachers for effective practice. Shulman's work opened up a new line of research and brought forth efforts by many mathematics educators to specify and clarify the PCK construct further. For example, the construct of *mathematical knowledge for teaching*, and its accompanying conceptual framework, was introduced by Ball, Thames, and Phelps (2008); it describes the nature of knowledge needed for mathematics teachers in their practice. Their framework is useful for our purposes because it provide an analytical framework for our review of problem posing in the context of preservice teacher preparation in mathematics.

The Ball et al. (2008) framework has several components, including common content knowledge (CCK), specialized content knowledge (SCK), knowledge of content and students (KCS), and knowledge of content and teaching (KCT). CCK is mathematical knowledge that is not unique to teaching, but is also invoked by professionals in other fields (hence the term, "common"). CCK is required by teachers in a variety of tasks, such as solving the problems that they assign to their students and identifying when textbooks provide inaccurate descriptions of mathematical content. In other words, as stated by Ball et al. (2008), "teachers need to know that material they teach" (p. 399). SCK represents a body of mathematical knowledge beyond what is taught to students and is different from the mathematical knowledge used by research mathematicians or applied in professions where mathematics is used as a means to solve practical problems. It is mathematical knowledge that is specifically needed in teaching, or in the words of Ball and Bass (2001), knowledge that is "pedagogically useful." It is needed in carrying out tasks such as, "recognizing what is involved in a particular representation, finding an example to make a mathematical point, or modifying tasks to be either easier or harder, asking well-chosen follow-up questions" (Ball et al., 2008, p. 400).

KCS centers on student thinking in mathematics: teachers must know the common pitfalls in student thinking about specific topics and problems, the types of thinking afforded by specific tasks, and what will be challenging or trivial for students at any given point in a unit. KCS also entails being able to interpret the ways in which students describe their thinking about mathematical ideas, which, given the frequent incompleteness and complexity of students' explanations, is necessarily based on knowledge of student thinking more generally. Finally, KCT involves making key decisions about instruction, and how specific aspects of instruction will affect student learning. For example, KCT involves knowledge about content sequencing, which means knowing what examples and activities to begin with, and which ones to use to explore the content in more depth.

Problem posing has been viewed as a pedagogical tool for teachers who aim to enhance their students' learning of mathematics (English, 1998; Pirie, 2002), and as a mechanism used by teachers to engage in productive mathematical conversations with their students (Boaler & Brodie, 2004; Franke et al., 2009). With preservice teachers, it has been found to enhance pedagogical content knowledge (Ticha & Hošpesová, 2009) and to have positive influences on their beliefs about and attitudes toward mathematics (Bragg & Nicol, 2008). What is meant by the term *problem*, however, differs in this literature; a problem can be considered a formal word problem written and presented to students to solve (e.g., Crespo, 2003). In this case, problem posing refers to the act of designing such a problem either during planning or in the middle of a lesson. A problem could also be a question that is verbally stated for a specific purpose, such as delving more deeply into the reasoning used by a child or as a way to extend a student's application of a mathematical concept. In this latter conception of the term *problem*, all questions and follow-up questions a teacher asks to test or further the mathematical thinking of her students can be considered problems (e.g., Ball & Forzani, 2009).

Consider the following scenario as an illustration of how problem posing lies at the heart of teacher questioning (the mathematics in this example is offered by Ball & Forzani, 2009). In Mr. Clay's fifth-grade classroom, a student claims that $.2 \times .3 = .6$. It is relatively easy to see what the child did to arrive at her answer (used the same procedure for multiplying decimals as for adding them), but Mr. Clay needs to get at the mathematical reasoning behind the strategy. Given this objective, what would be the best follow-up problem? Giving $.5 \times .2 = \square$ next would likely result in the student answering .10 (the correct answer) for very incorrect reasons, after which the teacher may think that $.2 \times .3 = .6$ was just a calculation error. Instead, giving $.1 \times .5 = \square$ would likely result in the student producing .5 as the answer, confirming that her difficulty lies in understanding that multiplying tenths by tenths produces hundredths. By posing the right follow-up question, Mr. Clay can thereby open key learning opportunities.

Preservice teachers have difficulty engaging in problem posing because it is a skill with which they are not familiar, both as students and as educators (Crespo & Sinclair, 2008). The problems they pose are not cognitively or structurally complex (Stein, Smith, Henningsen, & Silver, 2000; Vacc, 1993) and they often do not align with targeted mathematical concepts (Osana & Royea, 2011). Recent studies on teacher questioning have revealed similar challenges. Franke et al. (2009), for example, found that, after the initial "how did you figure that out?" question, teachers had difficulty asking follow-up questions to delve more deeply into student thinking. Osana et al. (2012) found that, although inservice teachers are quite adept at asking students how they solved problems, they were largely unable to ask specific probing questions that would challenge students' misconceptions about the equal sign. It is clear, therefore, that there is a pressing need to improve teachers' ability to ask purposeful questions and pose useful problems in the context of teaching mathematics.

The newly acknowledged role of problem posing in teaching, coupled with converging findings on prospective teacher's difficulties in problem posing, has prompted teacher educators to include problem posing skills in the elementary mathematics methods curriculum (we shall restrict our discussion here to the elementary level). Although this area of inquiry is still in its infancy, our review of the literature suggests that a synthesis of the existing research on how problem posing has been incorporated into the professional development of prospective teachers would be beneficial for researchers and mathematics teacher educators. Therefore, the purpose of the present review is to analyze and synthesize reported research on problem posing in elementary teacher education. From a theoretical perspective, we aim to interpret the existing research through Ball et al.'s (2008) conceptual framework of mathematical knowledge for teaching. Our interpretations promise to shed light on teacher educators' views of problem posing in mathematics and how problem posing allows them to operationalize mathematics knowledge for teaching in their methods courses. On a practical level, we aim to identify key questions at the heart of using problem posing in teacher education and to identify those practices that have promise for the development of this skill in the preservice teacher population. More generally, therefore, our review can generate useful guidelines for reflection on essential objectives in mathematics teacher education and ways to achieve them.

Below, we describe our selection criteria for the reviewed articles and the ways in which we categorized the selected articles. We also provide a synthesis of the articles using a coding rubric that emerged throughout the reviewing and selection process, and we interpret the selected articles using the framework of mathematical knowledge for teaching presented by Ball et al. (2008). We close by discussing implications of the review for teacher preparation in elementary mathematics.

Method

Selection Criteria

The articles were selected through major indexing databases, available at a large university in Canada. The SCOPUS, Web of Science, Web of Knowledge, Springer, ERIC, JSTOR, OpenDOAR from Sherpa (scopus.com, springer.com, jstor.org, http://www.eric.ed.gov/) repositories were searched using the following keywords: *problem posing*, *preservice/prospective/future teachers*, and *mathematics*. We only included studies published between (1990), which marked a shift in research interest toward teachers' problem posing, to present (2012). From this group of articles, we then selected only those that presented empirical research on problem posing: We excluded those that did not report at least one data set related to the phenomenon, but no specific methodology was excluded.

Finally, we further reduced the sample to include only those articles with elementary preservice teachers as participants; in particular, we were only interested in studies situated in "elementary" teacher education—that is, the professional development of future teachers as generalists who would teach all subjects (e.g., mathematics, language arts, science, social studies) starting in Grade 1 through, in North America, Grade 6. The main characteristic of these teacher education programs is that future teachers receive training for a series of "school subjects." Generally, the majority, if not all, of their mathematics-related courses focus on general guidelines related to the teaching and learning of mathematics (i.e., "methods" courses) and focus on such overarching principles as problem solving and communication in the classroom. Fewer courses are specifically targeted to the content itself, such as the conceptual basis of computational algorithms or the underlying mathematical structure of word problems.

In contrast, secondary teachers receive subject-specific training, which entails enrolling in a number of content courses, such as mathematics and physics. Preservice teachers of elementary mathematics, who do not receive such training, are left with the mathematics they learned when they were students themselves, and the extent to which they engage in additional experiences with mathematics is highly dependent on the teacher education programs they attend. Because the training of preservice teachers differs considerably between elementary and secondary teacher education programs, at least with respect to the emphasis on content, we targeted articles only at the elementary level. Those articles in which it was impossible to determine the type or level of teacher education program from which the participants were selected (e.g., "generalist" vs. "specialist" teacher education) were excluded.

Coding and Analysis

Using the selection criteria, we were left with 8 articles for the review presented in this chapter. Our coding of these articles was guided by our specific focus on teacher educators' practices related to problem posing: in particular, the ways in which they attempted to either foster problem posing among their students or the ways in which they used problem posing as an approach to their instructional practice. The specific codes, based on a careful reading of the articles, were generated using a grounded theory technique (Strauss & Corbin, 1998) in which we engaged in successive rounds of coding, each time followed by discussions that served to resolve any discrepancies and to refine the rubric by generating subcategories for the main codes. The coding process resulted in three major categories of the ways in which problem posing has been incorporated by elementary mathematics teacher educators in their practice. Specifically, these three categories are: (a) fostering problem posing as a skill integral to teaching practice; (b) problem posing as an activity separated from teaching, but conducted by preservice teachers; and (c) problem posing as a tool for researchers (e.g., as the basis for the design of outcome measures) or as a tool for teacher educators to change or enhance preservice teachers' knowledge. The rubric is listed in Table 23.1.

In the first category, problem posing as a skill integral to teaching practice (code: TP), we placed articles in which one of the objectives for the preservice teachers was to generate problems for actual or hypothetical students. In these studies, problem posing was seen as an integral part of mathematics teaching in the sense that the preservice teachers were to use certain criteria to make the problems cognitively appropriate, mathematically suitable, and motivating for their students. The articles we placed in the second category, problem posing as a separate activity (code: SEP), were those in which the process of problem posing itself was the object of study. While the data were collected from preservice teachers, problem posing was not conducted with the mathematical learning of their future students in mind; the focus of the data collected and their analyses was on cognitive aspects of problem posing

Table 23.1
Coding Rubric

Code name	Code	Description
Problem posing as a skill integral to teaching practice	TP	Generating strategically and pedagogically targeted problems to uncover student thinking or mobilize specific mathematical concepts during teaching (e.g., follow-up problems)
Problem posing as an activity separated from teaching	SEP	Cognitive processes involved in problem posing and the factors that are tied to them (e.g., beliefs, epistemology, context)
Problem posing as a tool		
Problem posing as a tool Research tool	T-R	Problem posing used as a way to assess a variable of interest (e.g., conceptual knowledge)
Problem posing as a tool Tool for teacher educators	T-TE	Problem posing used to assess the development of preservice teachers' beliefs, perceptions, knowledge; problem posing used to foster and enhance preservice teachers' pedagogical knowledge

as a form of problem solving, as well as on factors that play a role in posing problems, such as beliefs and context. Finally, the third category was created for studies in which problem posing was used either as a tool for researchers (e.g., as a way to assess a specific variable of interest in preservice teachers, such as conceptual knowledge) or as a tool for teacher educators to assess the development of the preservice teachers in their mathematics methods classes (e.g., learning, changes in beliefs or epistemology) or to foster growth in their pedagogical knowledge. Table 23.2 lists the eight articles included in the present review, including the number of participants in each study, the codes we assigned to the articles according to the rubric in Table 23.1, and the main findings for each.

Analysis and Synthesis of Problem Posing Literature

Problem Posing as Integral to Teaching Practice

In an important article, Ball and Forzani (2009) addressed the role of teachers' follow-up questions during classroom interactions for the purposes of clarifying students' understanding of specific concepts or skills. Problem posing, or the ability to "pose strategically targeted questions," as the authors put it, is central to teaching by offering a way to access and understand students' thinking. This kind of problem posing, along with the ability to "choose tasks, examples, models or analogies, and materials" (p. 501), is particular to the teaching practice. It is used by teachers to mobilize important concepts, test hypotheses about student thinking, and assist students to move through their learning challenges. Moreover, problem posing is considered as a "high leverage practice"—that is, one of the "practices [that is] most likely to equip beginners with capabilities for the fundamental elements of professional work and

Table 23.2
Summary of Study Characteristics

Study	*N*	Code	Main findings
Bragg and Nicol (2008)	33	T-TE	Preservice teachers' views on mathematics teaching and learning changed as a consequence of an open-ended problem-posing task. Curriculum specifications appeared to be both inhibiting and facilitating factors.
Chapman (2012)	40	SEP	Preservice teachers approached problem posing from a variety of perspectives, which in turn influenced the nature of the problems they posed. As a result of engaging in problem posing, preservice teachers became aware of the limitations of their mathematical knowledge for teaching.
Crespo (2003)	20	TP	Preservice teachers' strategies for posing problems changed from posing simple, single-step problems to proposing open-ended, cognitively complex problems. The presence of a "real" audience and discussions with their peers about their problems were possible factors.
Crespo and Sinclair (2008)	22	TP	The quality of problems posed by preservice teachers improved as a consequence of an activity in which a given mathematical situation was explored prior to the posing task. Whole-class discussions on the ways to evaluate problems appeared related to problem posing.
Nicol and Bragg (2009)	33	SEP	Preservice teachers posed open-ended problems, but had difficulties in identifying the intended learning objective targeted by the problem. The specific context (basing problems on digital photographs) was useful in raising awareness of the use of problem posing in mathematics teaching.
Osana and Royea (2011)	8	T-R	The authors assessed improvement of conceptual knowledge in preservice teachers after instruction. A problem-posing task was used as a transfer task of conceptual knowledge. Although there was improvement on one measure of conceptual knowledge, preservice teachers did not improve in their problem-posing abilities.
Ticha and Hošpesová (2009)	24	T-TE	Problem posing was found to be a useful tool for assessing preservice teachers' level of understanding and misconceptions. Personal beliefs about problem posing seem to influence preservice teachers' predisposition when engaging in problem-posing tasks.
Toluk-Uçar (2009)	50	T-TE	Problem posing is a useful tool for assessing preservice teachers' knowledge of specific mathematical concepts (fractions). In addition, it likely had a positive impact on teachers' views about knowing and doing mathematics.

that [is] unlikely to be learned on one's own through experience … [a] teaching [practice] in which the proficient enactment by a teacher is likely to lead to a comparatively large advances in student learning" (p. 460). As such, Ball and Forzani see problem posing as integral to teachers' interactions with students, which, in good teaching, necessitates considering their students' prior knowledge and interests. By viewing problem posing in this way, future teachers must learn, among other things, to take into account feedback received from the student.

Two articles were placed in this category. In the first of these articles, Crespo (2003) examined the processes used by preservice teachers as they posed problems for elementary students in a letter-exchange project. Over the course of a semester-long methods course, Crespo required her students to engage in a pen-pal activity with fourth-grade students at a local elementary school. The letter-writing activity, in the author's view, allowed future teachers to focus on three aspects of mathematics teaching: creating and presenting tasks to students, analyzing pupils' work, and reacting to their ideas outside the context of a busy classroom, thereby affording time for reflection and revision. By excluding the need for classroom management, the prospective teachers were able to concentrate all their efforts on these three aspects of teaching.

Crespo (2003) required her students to pose problems that were modifications of existing ones and to create their own problems "from scratch." She observed that her students used three approaches when they began this activity: (a) making existing problems easier to solve (by simplifying existing problems they had found, for example); (b) posing problems that were structurally similar to familiar ones (such as "typical" textbook word problems); or (c) posing problems "blindly" without reflecting on the mathematics at the heart of the problem or children's thinking about associated concepts. After 11 weeks of letter writing, however, the author reported a significant shift in their strategies: The preservice teachers were more likely to present problems that were less "traditional" to their letter-writing student partners, pose problems that would challenge the children's mathematical thinking, and pose problems designed to gain insight into their thinking. Crespo attributed this change to the authentic nature of the letter-writing activity—the preservice teachers had a "real" audience who received and tackled their problems, and they were confronted with actual responses from children.

Crespo's research illustrates one mathematics teacher educator's view of problem posing: as an activity undergone by teachers as they planned activities in advance of the lessons they conducted in the classroom. As such, she views problem posing as a component that is integral to a teacher's practice. Furthermore, Crespo saw great potential for the letter-writing activity to enhance the student teachers' KCS; the problem posing activity supported their awareness of how problem posing could be tailored specifically to children's needs and interests and increased the preservice teachers' sensitivity to the types of problems that are likely to elicit specific responses. Further, Crespo engaged the preservice teachers in discussions about the characteristics of non-traditional mathematical problems that emerged during the semester. This allowed her students to gain a deeper understanding of what mathematics children can learn in relation to the problems they are given to solve, again emphasizing the critical role of KCS in the preparation of mathematics teachers.

A recurring theme in the research on teachers' problem posing is the challenges they experience posing correct problems (from a mathematical point of view) as well as those that are pedagogically suitable. In another study aimed at enhancing preservice teachers' problem posing, Crespo and Sinclair (2008) set out to investigate the potential effects of holding discussions with preservice teachers on what constitutes good, interesting, appropriate, and even "beautiful" problems. The authors' hypothesis that mathematical exploration is effective in fostering appropriate problem-posing processes lies in the work of Hawkins (2000) and Dewey (1933), who both claimed that perceiving a certain situation as "problematic" (Hiebert et al., 1997)—one that presents a dilemma—is a necessary condition for proposing alternative solutions. Crespo and Sinclair (2008) suggested that considering exploration as a distinct activity from problem posing leaves the poser unable to recognize potential sources for problems, thereby preventing the generation of suitable and interesting problems. Ultimately, the authors proposed that viewing mathematical exploration and problem posing as separate processes can explain the preponderance of ill-formulated or uninteresting problems in the preservice teacher population.

In this context, Crespo and Sinclair (2008) engaged preservice teachers in explorations by encouraging them to make judgments, using mathematical and pedagogical criteria, about a number of mathematical problems during class discussions. Then, the authors compared the problems posed by preservice teachers before and after the exploration intervention and concluded that the quality and range of the problems they posed increased. Their problems were more cognitively demanding relative to the problems posed by those who did not receive the exploration intervention (a comparison group); their problems would ostensibly require considerable effort and mathematical reasoning to solve. These results provided some evidence that the teachers' KCS was enhanced considerably: They became more sensitive to characteristics of the problems themselves and how the problems' features and structure could be modified to make them either more or less challenging for children.

The authors designed a second intervention that was based on classroom discussions about esthetic criteria used by mathematicians in judging problems. Drawing on Dewey's (1934) interpretation, Sinclair (2004) stated that the term "esthetic" can be understood as related to a "sensibility in combining information and imagination" (p. 262). Crespo and Sinclair argued that esthetic features of problems, such as novelty, surprise, and "fruitfulness," are important for teachers to consider, even though their objectives are in many ways different than those of mathematicians. Sinclair (2004) argued that teachers can generate rich contexts for mathematical exploration by taking such esthetic criteria into account in their tasks and interactions with students. Crespo and Sinclair (2008) noted the tension that emerged when the preservice teachers grappled with both the pedagogical and mathematical potential of the problems. Seemingly, the prospective teachers generally tended to the pedagogical values of problems and often ignored their mathematical qualities.

By encouraging preservice teachers to engage in open-ended explorations of previously posed problems, Crespo and Sinclair (2008) underscored the importance of teachers reflecting on problems before incorporating them into their mathematics lessons. Thus, their study illustrates that part of problem posing involves the ability to evaluate existing problems on a variety of mathematical and pedagogical dimensions. As such, along the same lines as Crespo (2003), the study places problem posing squarely in the domain of KCS because it can assist teachers to attend to the features of the problems themselves and their affordances for learning. Finally, the conclusions drawn by Crespo and Sinclair underscored the distinct nature, but yet important connection between content knowledge (CCK and SCK) and KCS, an element of pedagogical knowledge. As such, we argue that Crespo and Sinclair view problem posing as a catalyst for the mobilization of both content knowledge and pedagogical knowledge, thereby further promoting the power of problem posing in mathematics methods courses.

The Sense Preservice Teachers Make of Problem Posing

In this category, we placed articles in which problem posing itself is the focus of the research. Although the studies were conducted with preservice teachers, the researchers' emphasis was on the individual cognitive processes of the preservice teachers as they engaged in problem posing activity as well as the sense they made of such activity. In these articles, participants are not tasked with fine-tuning problems to delve at specific aspects of their students' thinking or to further elaborate on a mathematical concept; rather, the researchers' purpose is to reveal how future teachers make sense of a problem posing task, what they find difficult, and how they cope with these difficulties. In terms of mathematical knowledge for teaching, the focus is on the design of instruction, and more specifically, the challenges encountered by preservice teachers in their attempts to engage in various aspects central to the practice of creating mathematics problems for students.

In the first of the two articles we placed in this category, Nicol and Bragg (2009) argued that the types of problems future teachers pose, without any previous training on this skill, act as expressions of the sense they make not only of the task, but also of the beliefs and knowledge they bring to their teacher education programs. The idea that a task, whether attended to or created, is a manifestation of the beliefs and perceptions teachers hold for images about mathematics is reflected in Schoenfeld (1992), who argued that the mathematical tasks given to students can have an impact on the conceptions and beliefs they hold about the discipline.

Nicol and Bragg (2009) presented 33 preservice teachers from a Canadian university with photographs of real life scenes and asked them to pose open-ended problems that incorporated the photograph and its contents. In addition, once they had posed problems, the students were required to connect them to a specific learning objective (using the definition of "learning objective" presented by Martin and Booth, 1997). Using a three-point scale, Nicol and Bragg independently rated the problems according to how closely they actually connected to the learning objectives identified by the preservice teachers. The authors also investigated three specific aspects of the preservice teachers' problem posing: the types of problems posed, the factors that influenced them in the process, and what they found challenging about the task. The analyses of the data revealed, in part, that students experienced difficulty posing open-ended problems, partly also because of their general unfamiliarity with such problems. A more difficult aspect proved to be the correct identification of the intended learning objective—the authors' own ratings demonstrated that only 26% of the problems posed were strongly connected to the preservice teachers' stated learning objective.

A second result pertinent to this discussion was the way in which the preservice teachers incorporated the photographs in the problems they posed. The authors used the term *interactive* for problems in which the photograph was essential for answering them and *illustrative* for the other cases. Although designing interactive problems proved to be challenging for the preservice teachers (59% of the posed problems were of this type), the experience generating interactive problems had an impact on their perceptions of the mathematical potential of their everyday environments, a point to which we return later in the chapter. A second analysis entailed an examination of the strategies that the future teachers employed to generate open-ended problems. The authors classified the strategies as follows: (a) removing information from closed questions (i.e., those problems with only one correct answer); (b) using major curricular areas (e.g., geometry, measurement) as starting points; (c) starting from more specific curricular topics (e.g., patterns) and taking a photo they thought was suitable; (d) imagining being a young child and posing a problem he or she might ask; and (e) focusing on the wording and other linguistic aspects of the problem.

Although this study, like the two previously reviewed, illustrates the authors' view of problem posing as an activity conducted by teachers as they plan their mathematics lessons, Nicol and Bragg (2009) placed the practice of teaching itself in the background. Rather than focusing on problem posing as a teaching task, the authors emphasized the processes undergone by the prospective teachers as they posed problems and described the obstacles encountered during the activity. The authors' analyses of the challenges encountered by the preservice teachers highlight the connection between problem posing and KCT. More specifically, their research points to the ways in which problem posing requires a teacher to consider the types of problems that would be appropriate for specific topics and learning objectives.

Chapman (2012) investigated the sense preservice teachers make of problem posing in the absence of any instruction on it. As part of a larger study, she presented 40 preservice teachers with a variety of tasks, modeled after those found in the problem-posing literature, that entailed posing problems and reformulating given problems. By giving a range of problem posing tasks to the same groups of participants, Chapman was able to make more valid comparisons of problem-posing behavior by task type. The tasks were given to the preservice teachers one at a time, in alternating order of problem posing and problem reformulating.

By analyzing the participants' work, Chapman identified five viewpoints on problem posing held by the preservice teachers. The *paradigmatic* perspective characterizes problem posing as the creation of problems "with universal interpretation, a particular solution and an independent existence from the problem solver" (p. 140). These problems resemble "traditional" word problems and reflect the preservice teachers' own prior experiences as elementary students. The *objectivist* perspective characterizes problem posing as the creation of a problem in a backward fashion, starting with a fact (e.g., a multiplication fact such as "$3\times4=12$") and then writing a word problem around it. From the *phenomenological* perspective, the creation of a problem emerges from a given situation through the construction of meaning; this involves the individual's point of view and the production of "personalized interpretations and solutions." The *humanistic* perspective is similar to the phenomenological one, but the context in which the problem is situated reflects the individual's experience and interests, including hobbies and personal preferences. Finally, the *utilitarian* perspective characterizes problem posing as a mechanism for getting at students' mathematical thinking, such as problems that target specific representations or that require students to articulate their knowledge.

Chapman focused her discussion on three particular tasks: (a) one in which the students were required to "create a word problem that you think is open ended" (p. 138); (b) another in which the students were asked to pose problems that embody different meanings of multiplication; and (c) a final one that required the creation of a problem based on a given diagram. Similar to the conclusions of other researchers, Chapman highlighted the difficulties experienced by preservice teachers in creating open-ended problems (see Nicol & Bragg, 2009) and creating problems that reflected a different meaning of multiplication (see e.g., Toluk-Uçar, 2009). The specific areas of difficulties she observed revealed weaknesses in both CCK and SCK: a reliance on closed problems, singular interpretations of mathematical concepts, and a lack of awareness of the importance of mathematical structures and representations in problem posing.

Once again taking the perspective that problem posing is an activity conducted in advance of teaching, Chapman nevertheless concluded that the five perspectives on problem posing that emerged from her data could be useful for mathematics teacher educators because they make explicit the affordances and inhibitions in the development of preservice teachers' mathematical thinking. Her focus on disciplinary thinking is a significant departure from the emphasis on the pedagogical aspects of a teacher's knowledge in the studies reviewed earlier in this chapter. Chapman's findings, therefore, allowed her to view problem posing as an activity that is dependent on, and reflective of, the content components of mathematical knowledge for teaching, such as CCK and SCK.

Problem Posing as a Research or Pedagogical Tool

In the articles in this category, problem posing is used either as a research tool or as a pedagogical tool for the teacher educator, whose objective is to foster change in preservice teachers' cognition or affect. When problem posing was used as a research tool, the focus was not on problem posing per se, but rather on a different construct that the researchers believed was correlated with problem posing. For instance, Osana and Royea (2011) used problem posing to measure preservice teachers' conceptual understanding of fractions, which was the focus of their study. When problem posing is used as a pedagogical tool, it is used as a means by which teacher educators either assess preservice teachers' knowledge and beliefs or as a context in which teachers' conceptual understanding of attitudes about mathematics can be fostered. In all the studies in this category, problem posing was not itself the object of investigation, but its role was examined insofar as it could support instructional practice in mathematics methods courses.

Problem posing as a research tool. Osana and Royea (2011) reported the results of a one-on-one intervention with eight preservice teachers on the topic of conceptual and procedural knowledge of fractions. Using a single group pretest–posttest design, the authors gave five individual tutoring sessions to each participant and measured changes in fractions knowledge after the tutoring. During their first session, each participant completed a paper-and-pencil assessment that included items measuring fractions knowledge as well as four problem-posing items, which required the participants to write word problems that corresponded to a given mathematical equation (e.g., $5 \times 1/3$). Given that problem posing was not part of the fractions intervention, the authors used problem posing as the basis for a transfer task to measure conceptual knowledge. During the last session, Osana and Royea administered the same paper and pencil test, which included isomorphically similar problem-posing items.

The authors found significant improvement on the conceptual knowledge scale, but not on the procedural knowledge scale, nor on the problem-posing task. Error analyses highlighted the preservice teachers' difficulty in posing problems for number sentences involving division, and in particular, those in which the divisor is a fraction. The second most frequent error was in cases in which the number sentence involved subtraction: The preservice teachers most often did not attend to the unit when considering fractional parts.

Osana and Royea (2011) viewed problem posing as having a utilitarian function (Chapman, 2012)—that is, although the authors examined in detail the problems posed by the preservice teachers and carefully catalogued their errors, the authors' primary objective was to enhance their fractions knowledge. Problem posing was used as a way to assess whether the preservice teachers were able to apply their conceptual understandings to create problems that accurately reflected specific mathematical operations with fractions. It is clear, therefore, that Osana and Royea (2011) focused their intervention on enriching the CCK and SCK of preservice teachers. In so doing, the authors positioned problem posing as a tool to evaluate their development in these areas, but not as a part of mathematical knowledge for teaching itself. Nevertheless, Osana and Royea's findings prompted them to recommend including explicit instruction on problem posing in mathematics methods courses, which could constitute "a context for highlighting the connection between mathematical concepts and procedures used to solve … problems" (p. 350). The effect of direct instruction on the development of SCK and other aspects of teacher knowledge, however, remains open for further discussion.

Problem posing as a pedagogical tool in mathematics methods courses. Bragg and Nicol (2008), in the same study described earlier, investigated to what extent problem posing could challenge the preservice teachers' views on mathematics as a discipline. At the completion of their methods course that included the problem-pictures assignment, the authors administered a 15-item online questionnaire. Four of the items were designed to measure the participants' perceptions of the assignment. The preservice teachers reported that they found the task challenging, and they indicated that, over the course of the semester, they found it easier to start with a mathematical concept and then find photographs to illustrate it, which was not intended by the original task. Regarding their view of mathematics, the preservice teachers reported that, after engaging in the problem-posing task, they were better able to see the mathematics in their everyday surroundings. In addition, designing open-ended problems gave a quarter of the participants a feeling of "empowerment," which they described as gaining ownership of their created problems and an accompanying sense of confidence in their ability to use them in their future teaching.

In this study, Bragg and Nicol (2008) viewed problem posing as a pedagogical tool for the mathematics teacher educator—that is, they predicted that engaging in problem posing would influence the preservice teachers' attitudes and perceptions about mathematics and their ideas about what is involved in teaching it. Moreoever, the challenges and experiences reported by the preservice teachers bring the authors' view of mathematical knowledge for teaching to the fore. This insight into the broader impact of problem posing on preservice teachers' perceptions demonstrates how Bragg and Nicol viewed teacher preparation as more than "acquiring" specific types of knowledge. Their research on problem posing demonstrates that mathematical knowledge for teaching should also incorporate more affective constructs, such as teachers' personal connections with the mathematics in the world and their confidence in helping children learn mathematics.

In a similar fashion, Ticha and Hošpesová's (2009) main focus was on using problem posing as a means for fostering preservice teachers' pedagogical knowledge in a mathematics methods course. At one point during the course, the authors asked 24 preservice teachers to pose three word problems corresponding to a given symbolic expression involving fraction multiplication (e.g., $\frac{1}{4} \times 2/3$). The problems were collected and stored in a database, and the students then produced ratings of the suitability and the correctness of each other's problems, which were also stored in the database and available for their peers to consult. The authors then selected three problems posed by one preservice teacher and brought them to the class for discussion. The discussions entailed analyses of the problems as well as ways to view them as diagnostic tools in the elementary mathematics classroom. Ticha and Hošpesová observed that once the preservice teachers posed their problems, they did not verify whether the problems reflected the target mathematical expressions, again pointing to preservice teachers' difficulty in invoking KCT, possibly as a result of incomplete mathematical knowledge. The authors were, however, able to correct the problems once any discrepancies were brought to their attention during the class discussions by the authors or their peers.

Ticha and Hošpesová viewed problem posing in a similar way to that demonstrated by Bragg and Nicol (2008)—as a means to foster in preservice teachers specific elements of what they considered to be central to mathematics teaching. In this case, the authors illustrated how problem posing could assist the teacher educator to support prospective teachers in the development of KCT. In particular, through explicit reflection and communication, problem posing can reveal gaps in the teachers'

KCT, which can then be addressed by the teacher educator in a methods course. This said, the use of problem posing as a tool for teacher educators was not conceptualized by Ticha and Hošpesová as part of mathematical knowledge for teaching.

In her mathematics methods course, Toluk-Uçar's (2009) aim was to enhance her students' understanding of fractions, and she hypothesized that this could be achieved by engaging her students in instruction centered on problem posing. The participants were 95 preservice teachers enrolled in a year-long elementary methods course in Turkey. The participants were split into two non-random groups, with 50 preservice teachers in the problem-posing group and 45 in a comparison group. During the first half of the methods course, all participants were given a 2-hour lecture on problem posing that included a description of different types of problem posing as well as general problem posing strategies. After the lecture, participants completed a pretest on fractions that comprised ten items designed to measure their conceptual understanding of fractions concepts and one question requiring the preservice teachers to write about how confident they felt about their knowledge of fractions. At the end of the course in which problem-posing activities were implemented, the same instrument was administered again to both groups. Toluk-Uçar also collected the participants' weekly mathematics journals to gain insight into the development of their fractions knowledge and views of mathematics.

The fractions intervention lasted for 6 weeks and consisted of asking participants to pose problems during class, either individually or in small groups, that would invoke certain fractions concepts, such as equivalence or comparison of fractions and fractions expressions (e.g., $\frac{3}{4}-\frac{1}{2}$). Toluk-Uçar then selected specific problems posed by her students and engaged them in whole-class discussions about these problems. The discussions focused on the appropriateness of the problems to the given situations, the solvability of the problems, and their appropriateness for students at specific developmental levels. The discussions also included possible ways to modify problems that were inappropriate. Throughout the intervention, participants were encouraged to use different representations for fractions and their operations. The comparison group followed the instructional approach that had been traditionally used in the course previously, which entailed developing lesson plans for the same fraction topics as those covered in the problem-posing group.

The results revealed a significant difference in students' conceptual understanding of fractions after the problem-posing intervention, especially with respect to multiplication and division. Furthermore, Toluk-Uçar found that on the pretest, most participants from both groups reported that fraction multiplication and division are characterized by their corresponding algorithms and do not have any connection to real life situations. On the posttest, in contrast, the students in the problem-posing group were better able than those in the comparison group to pose word problems corresponding to those operations using real life contexts. On this measure, the performance of the comparison group did not change.

Along the same lines as the other studies included in this category, the view of problem posing taken by Toluk-Uçar was as a tool to enhance the content knowledge of preservice teachers. In this case, the Toluk-Uçar objective was to use problem posing as a vehicle for the enhancement of their CCK and SCK in the area of fractions. The author's approach illustrates how she chose to mobilize the content component of Ball et al.'s (2008) framework for teacher knowledge. Furthermore, Toluk-Uçar's study is another example of how problem posing can reflect the ways in which teacher educators can target mathematical knowledge for teaching in their methods courses, and at the same time, conceptualize problem posing itself as an activity separate from teaching.

Discussion

It is clear from the present review of the literature that problem posing is a complex endeavor that can, and has been, studied from many angles. One of our observations is that the term "problem posing" has been interpreted in different ways in the literature. This is partly because of the disparate research objectives of the authors, but also because the act of posing a problem (or asking a question or reasoning through an argument) manifests itself in a large number of daily and professional situations. Building on previous definitions of the term, we propose a working definition in the context of teacher education: the act of formulating a new task or situation, or modifying an existing one, with a specific mathematical learning objective and a targeted pedagogical purpose in mind.

The research suggests that efforts to enhance preservice teachers' problem posing must take into account a variety of factors that appear to be related to its development. These factors can be grouped as follows: (a) a focus on the teachers' content and curricular knowledge; (b) the extent to which teachers are required to use a variety of strategies for posing problems; (c) the degree to which teachers are asked to reflect on criteria for the evaluation of problems, which could include mathematical and pedagogical criteria; and (d) a focus on the development of their metacognition, which entails, in part, reflections on personal beliefs and attitudes related to mathematics. The almost ubiquitous reference to pedagogical knowledge (or other conceptualizations of teacher knowledge in line with Ball et al.'s (2008) model), as well as the considerable attention paid to metacognition in preservice teachers, is what makes problem posing in mathematics different from that observed in other contexts. Considering the unique aspects of problem posing in mathematics teaching, the complexity of introducing it into professional development quickly becomes overwhelming, especially with the limited time constraints and multiple goals within any given mathematics methods course (Sierpinska & Osana, 2012).

Most of the authors of the articles reviewed here explicitly stated the need to include problem posing as an objective of teacher education, which entails helping preservice teachers "to build on, reconstruct, and extend their sense-making of it" (Chapman, 2012, p. 144). This conclusion is in line with the general view that professional development is most effective when preservice teachers are actively engaged in the very practices they are expected to carry out when they enter the workforce (Wilson & Berne, 1999). The question remains, however: How can a mathematics teacher educator come to grips with the teaching of problem posing, a construct that is admittedly not well understood, in practical terms? We look for answers by highlighting common themes and key points from the studies reviewed in this chapter and use these to form implications for professional development.

Implications for Professional Development

Often, preservice teachers know that, as practitioners, they will need to select and implement mathematical tasks from textbooks and other sources for their students. Sometimes, they will "use them as they are," but many preservice teachers are unaware that they will often need to assess the instructional value of the tasks and modify them according to specific learning objectives or student needs. It becomes incumbent on teacher educators, therefore, to inform preservice teachers about these and related responsibilities and help them develop the requisite strategies to meet them. Ball and Forzani (2009) introduced the term "high-leverage practice" that "include[s] tasks and activities that are essential for skillful beginning teachers to understand, take responsibility for, and be prepared to carry out in order to enact their core instructional responsibilities" (p. 504). Problem posing, in its most general form of involving the reformulation of existing problems, but also in the generation of new ones, is one of those practices.

A major theme in the literature reviewed here is the impact of the preservice teachers' prior experiences with mathematics on the perceptions, beliefs, attitudes, and skills they bring to problem posing. As a result, we propose that metacognition is an important factor in the development of their skills in this area. For example, as Ticha and Hošpesová (2009) observed, some preservice teachers reject the idea of posing a problem or modifying an existing one. This can often be a consequence of the mathematics instruction they had themselves received when in school—that is, they had been exposed primarily to ready-made problems and rarely, if ever, to teachers who came up with problems themselves. Nicol and Bragg (2009) also identified teachers' unfamiliarity with open-ended problems as contributing to their challenges with problem posing. Such difficulties stem from exposure to problems with only one "right answer" and tasks that require a known procedure for solution (i.e., closed problems), as opposed to problems that are open-ended and that require exploration and inquiry.

Because it has been well established in previous research that the experiences of most preservice teachers are in many ways discrepant with the types of thinking expected from them in their teacher education programs, Ball (1988) mentioned the need for preservice teachers to "unlearn to teach mathematics." Along the same lines, Toluk-Uçar (2009) concluded from her research that, "methods courses can be used as a setting to challenge and revise pre-service teachers' mathematical knowledge and beliefs" (p. 174). To deveop a better understanding of how this might be achieved, we turn to Wilson and Berne (1999), who argued for engaging preservice teachers in the very activities that they will need in their future classrooms. Indeed, researchers have attempted to achieve this objective by engaging preservice teachers in reflections of their own problem posing (through journals, portfolios, and written assessments) and by simulating real audiences (through letter writing, for example). In turn, such activities appear to have honed their problem-posing skills, and the mathematical knowledge and beliefs of preservice teachers changed substantially, even in a relatively short period of time.

At the same time, the initial understandings preservice teachers bring into their methods courses shape their learning. As Chapman (2012) suggested, therefore, it is important for teacher educators to identify how preservice teachers make sense of problem posing and to build on that understanding to develop their students' abilities more fully. She further observed, however, that "…their sense-making of posing 'word problems' often excluded intentional or conscious consideration of mathematical structure or context of the problems or the relationship to the problem situation" (p. 144) and pointed to the importance of mathematical content knowledge in problem posing. Indeed, content knowledge has been shown to be a factor in the quality and quantity of posed problems (e.g., Leung, 1994) and has for some time been viewed as critical in the process. Participants with weak content knowledge, for example, were more likely to pose unsolvable problems or simple ones, while those with stronger content knowledge were able to pose collections of related problems (which required an understanding of the structural relationships between problems) and problems that were structurally more complex (Leung, 1994). Currently, the conception of preservice teachers' problem posing is more targeted: Preservice teachers need to learn how to pose problems that attend to a given learning objective or a certain interpretation of an operation spontaneously during classroom interactions with students. This skill relies heavily on content knowledge (Ball et al., 2008) because it requires teachers to see the mathematical potential of their environment and to build instruction around it.

The literature also points to the fact that content knowledge, although apparently necessary, is not sufficient for teachers' problem posing. As Chapman (2012) argued, "Problem posers have to appropriately combine problem contexts with key concepts and structures in solutions along with constraints and requirements in the task. Thus, both contextual settings and *structural features of problems* are recognized as crucial" (our emphasis, p. 137). Otherwise said, attending to the structure of problems appears to be a necessary element in teacher education, which implies, more broadly, that a focus on appropriate strategies for reasoning about problem posing is necessary. Osana and Royea (2011) made a similar argument in supporting preservice teachers' ability to see the connections between intuitive solutions to fractions problems and formal algorithmic representations of the same solutions. Their data revealed that, although students improved in this respect, they were still unable to transfer their learning to a problem-posing situation. The implication of this finding for teacher educators is that it has been found helpful to engage preservice teachers in class discussions that involve reflecting on the underlying structure of problems, comparing different problems vis-à-vis their structure, and connecting problem structure to targeted learning objectives. Identifying and reflecting on the deep structure of problems can lead preservice teachers to develop schemata for a variety of problem types which, in turn, can be activated under favorable instructional conditions.

Tasks that require preservice teachers to reformulate problems and to assess their modifications on a range of dimensions (solvability, accessibility, solution methods, correctness, contextual features, potential errors, and learning objectives) can result in a greater awareness of the critical aspects of problem posing. Because problem reformulation can occur on a variety of levels (e.g., the same mathematical expression can be worded in different ways), preservice teachers must also learn to consider the relative advantages and disadvantages of each (re)formulation.

A related task is one that requires the modification of problems for specific pedagogical purposes, such as mobilizing a concept for instruction or testing hypotheses about children's mathematical knowledge. These added constraints would focus reformulation on changing specific elements of a problem in response to students' needs (e.g., Crespo, 2003). Elsewhere, such elements have been termed "didactic variables" (Brousseau, 1997). The following excerpt from Brousseau (1997) is helpful in describing the process of problem adaptation:

> A field of problems can be generated from a situation by changing some variables which, in turn, are changing the characteristics of solution strategies (cost, validity, complexity, etc.). … Only changes that affect the hierarchy of strategies should be considered as relevant variables and among the relevant variables, those that can be modified by a teacher are particularly interesting: these are the didactical variables. (p. 208)

Thus, a didactic variable has its values assigned by the teacher, who, by modifying its values, can have an impact on her students' learning. Clearly, problem posing can support preservice teachers in their efforts to identify those variables and modify their values to achieve specific objectives during their teaching (Ball & Forzani, 2009).

Additionally, connecting mathematics to real life has played a major role in the development of problem posing in professional development (Verschaffel, Greer, & De Corte, 2000) because it can function as a criterion for the evaluation of the quality of a problem. As Nicol and Bragg (2009) suggested, a way to foster such connections is by encouraging preservice teachers to "see" the mathematics around them. Within this perspective, teacher educators should design their professional development activities so that preservice teachers are inspired by the environment, art, or science, for example. This could lead naturally to explorations of open-ended situations, known to be notoriously difficult for preservice teachers (Chapman, 2012; Nicol & Bragg, 2009). From activities that are situated in real-life contexts, a variety of different types of mathematical explorations can emerge, such as assessing different solutions according to a variety of criteria and considering situations from different perspectives. Indeed, as Crespo and Sinclair (2008) found, mathematical exploration can be highly beneficial for preservice teachers when it comes to developing criteria for judging the quality of a problem. Open-ended tasks, especially if preceded by exploration, are more conducive to problem posing than tasks that are constrained by specific criteria (such as writing a problem for a number sentence, for example). We argue that mathematical explorations of real-life situations, when combined with both open-ended inquiry and specific pedagogical constraints, could be highly productive for the development of preservice teachers' problem-posing abilities.

In the process of learning how to assess the problems they pose, preservice teachers need to attend to the pedagogical and mathematical fruitfulness of a given problem (Crespo & Sinclair, 2008), and doing so provides opportunities for them to modify problems according to specific criteria or teaching goals. As Crespo and Sinclair argued, however, preservice teachers would also benefit from occasions to appreciate the "mathematical beauty" in a problem. Given that most preservice teachers have not previously been exposed to mathematical esthetics, discussions in methods courses could be effective in this regard. Such explorations could be linked to examinations of non-routine problems, problems with multiple solutions and the relative suitability of the solutions, and extensions of problems to other topics or domains.

Because preservice teachers must acquire a body of professional knowledge specific to their future practice (e.g., Ball et al., 2008; Shulman, 1986), they need to develop an ability to pose and adapt problems with specific pedagogical purposes in mind, such as probing the thinking of their students or extending a specific mathematical concept. For these purposes, preservice teachers need curricular and content knowledge, another factor we identify as critical to the development of their problem posing. Content knowledge supports their understanding of the conceptual underpinnings of the algorithms in the school curriculum (and subsequent explanations of these concepts; e.g., Ball et al., 2008), and as the current review has shown, assists them to produce a larger number of creative and interesting problems for their students. Without understanding the conceptual rationales for topics and procedures in the school curriculum, preservice teachers' problem posing is seriously hindered by the knowledge they bring to their methods courses—knowledge which is usually fragmented and procedural (e.g., Livy & Vale, 2011; Newton, 2008; Simon, 1993; Tirosh & Graeber 1990; Zazkis & Campball, 1996). As such, they rely primarily on reproducing problems they have already seen—a finding that is not uncommon in the problem-posing literature.

We considered it important to highlight in the literature reviewed here the emphasis teacher educators placed on metacognition with their students when engaging in problem posing. From the research designs used in most of the problem-posing literature, however, it is difficult to determine the relative contributions of metacognitive activity and the act of problem posing itself. It may be that by engaging in metacognitive activities such as keeping portfolios, reflecting in written journals, and responding to open-ended questions about their thinking, the preservice teachers learned as much, or perhaps more, about how to pose problems than simply practicing the skill itself. It is clear that more research is needed on the factors that contribute to the development of preservice teachers' problem posing. Whether the objective is problem posing or not, however, we assert that reflective activity is paramount for preservice teachers' professional development, both as teachers-in-training as well as throughout their careers; indeed, much of the literature in teacher professional development would support this assertion (e.g., Van Zoest & Stockero, 2008). Teachers need to remain open to their students' inquiry and allow themselves to capitalize on fruitful comments and questions that arise during classroom interactions. From our analysis of the literature, we see problem posing as an effective vehicle for such growth and a springboard for the further development of curiosity and a continual willingness to learn.

References

Ball, D. L. (1988). Unlearning to teach mathematics. *For the Learning of Mathematics, 8*(1), 40–48.

Ball, D. L., & Bass, H. (2001). Making mathematics reasonable in school. In G. Martin (Ed.), *Research compendium for the principles and standards for school mathematics*. Reston, VA: National Council of Teachers of Mathematics.

Ball, D. L., & Forzani, F. (2009). The work of teaching and the challenge for teacher education. *Journal of Teacher Education, 60*(5), 497–511.

Ball, D., Thames, M., & Phelps, G. (2008). Content knowledge for teaching: What makes it special. *Journal of Teacher Education, 59*(5), 389–407.

Boaler, J., & Brodie, K. (2004). The importance, nature and impact of teacher questions. In D. E. McDougall, & J. A. Ross (Eds.), *Proceedings of the Twenty-Sixth Annual Meeting of the North American Chapter of the International Group for Psychology of Mathematics Education* (Vol. 2, pp. 773–782). Toronto, ON: PMENA.

Bragg, L., & Nicol, C. (2008). Designing open-ended problems to challenge preservice teachers' views on mathematics and pedagogy. In O. Figueras, J. L. Cortina, S. Alatorre, T. Rojano, & A. Sepulveda (Eds.), *Proceedings of the Joint Meeting of the 32nd Conference of the International Group for the Psychology of Mathematics Education and 20th Conference of the North-American Chapter* (Vol. 2, pp. 256–270). Morelia, Mexico: PME.

Brousseau, G. (1997). *Theory of didactical situations in mathematics*. Dortrecht, The Netherlands: Kluwer.

Brown, S. I., & Walter, M. I. (1983). *The art of problem posing*. Hillsdale, NJ: Erlbaum.

Cai, J., & Hwang, S. (2002). Generalized and generative thinking in US and Chinese students' mathematical problem solving and problem posing. *The Journal of Mathematical Behavior, 21*, 401–421.

Chapman, O. (2012). Prospective elementary school teachers' ways of making sense of mathematical problem posing. *PNA, 6*(4), 135–146.

Crespo, S. (2003). Learning to pose mathematical problems: Exploring changes in preservice teachers' practices. *Educational Studies in Mathematics, 52*(3), 243–270.

Crespo, S., & Sinclair, N. (2008). What makes a problem mathematically interesting? Inviting prospective teachers to pose better problems. *Journal of Mathematics Teacher Education, 11*(5), 395–415.

Dewey, J. (1933). *How we think*. Boston, MA: D.C. Heath and Company.

Dewey, J. (1934). *Art as experience*. New York, NY: Perigree.

English, L. (1997). Development of seventh-grade students' problem posing. In E. Pehkonen (Ed.), *21st Conference of the International Group for the Psychology of Mathematics Education* (Vol. 2, pp. 241–248). Lahti, Finland: PME.

English, L. D. (1998). Children's problem posing within formal and informal contexts. *Journal for Research in Mathematics Education, 29*(1), 83–106.

Franke, M. L., Webb, N. M., Chan, A. G., Ing, M., Freund, D., & Battey, D. (2009). Teacher questioning to elicit students' mathematical thinking in elementary school classrooms. *Journal of Teacher Education, 60*(4), 380–392.

Gonzales, N. A. (1996). Problem formulation: Insights from student generated questions. *School Science and Mathematics, 96*(3), 152–157.

Hawkins, D. (2000). *The roots of literacy*. Boulder, CO: University Press of Colorado.

Hiebert, J., Carpenter, T. R., Fennema, E., Fuson, K. C., Wearne, D., Murray, H., … Human, P. (1997). *Making sense: Teaching and learning mathematics with understanding*. Portsmouth, NH: Heinemann.

Kilpatrick, J. (1987). Problem formulating: Where do good problems come from? In A. H. Schoenfeld (Ed.), *Cognitive science and mathematics education* (pp. 123–147). Hillsdale, NJ: Lawrence Erlbaum.

Leung, S. S. (1994). On analyzing problem posing processes: A study of prospective elementary teachers differing in mathematics knowledge. In J. P. da Ponte, & J. F. Matos (Eds.), *Proceedings of the 18th International Conference of the International Group for the Psychology of Mathematics Education* (pp. 168–175). Lisbon, Portugal: PME.

Leung, S. S., & Silver, E. A. (1997). The role of task format, mathematics knowledge, and creative thinking on the arithmetic problem posing of prospective elementary school teachers. *Mathematics Education Research Journal, 9*(1), 5–24.

Livy, S., & Vale, C. (2011). First-year pre-service teachers' mathematical content knowledge: Methods of solution for a ratio question. *Mathematics Teacher Education and Development, 13*(2), 22–43.

Martin, F., & Booth, S. (1997). *Learning and awareness*. Mahwah, NJ: Erlbaum.

National Council of Teachers of Mathematics. (1989). *Principles and standards for school mathematics*. Reston, VA: Author.

Newell, A., & Simon, H. A. (1972). *Human problem solving*. Englewood Cliffs, NJ: Prentice-Hall.

Newton, K. J. (2008). An extensive analysis of preservice elementary teachers' knowledge of fractions. *American Educational Research Journal, 45*, 1080–1110.

Nicol, C., & Bragg, L. A. (2009) Designing problems: What kinds of open-ended problems do preservice teachers pose? In M. Tzekaki, M. Kaldrimidou, & H. Sakonidis (Eds.), *Proceedings of the 33rd Annual Meeting of the International Group for the Psychology of Mathematics Education* (Vol. 4, pp. 225–232). Thessaloniki, Greece: PME.

Osana, H. P., Cooperman, A., Adrien, E., Rayner, V., Bisanz, J., Watchorn, R., & Sherman LeVos, J. (2012, April). *Examining teacher knowledge and classroom practices during inquiry teaching on the equal sign*. Paper presented at the American Educational Research Association, Vancouver, Canada.

Osana, H., & Royea, D. (2011). Obstacles and challenges in preservice teachers' explorations with fractions: A view from a small-scale intervention study. *The Journal of Mathematical Behavior, 30*(4), 333–352.

Pirie, S. (2002). Problem posing: What can it tell us about students' mathematical understanding? (ERIC Accession No. ED 471 760).

Schoenfeld, A. H. (1992). Learning to think mathematically: Problem solving, metacognition, and sense-making in mathematics. In D. A. Grouws (Ed.), *Handbook for research on mathematics teaching and learning* (pp. 334–370). New York, NY: Macmillan.

Shulman, L. S. (1986). Those who understand: Knowledge growth in teaching. *Educational Researcher, 15*(2), 4–31.

Sierpinska, A., & Osana, H. (2012). Analysis of tasks in pre-service elementary teacher education courses. *Research in Mathematics Education, 14*(2), 109–135.

Silver, E. (1997). Fostering creativity through instruction rich in mathematical problem posing and problem solving. *Zentralblatt für Didaktik der Mathematik, 29*(3), 75–80.

Simon, M. A. (1993). Prospective elementary teachers' knowledge of division. *Journal for Research in Mathematics Education, 24*(3), 233–254.

Sinclair, N. (2004). The roles of the aesthetic in mathematical inquiry. *Mathematical Thinking and Learning, 6*(3), 261–284.

Stein, M. K., Smith, M. S., Henningsen, M. A., & Silver, E. A. (2000). *Implementing standards-based mathematics instruction: A casebook for professional development*. New York, NY: Teachers College Press.

Strauss, A., & Corbin, J. (1998). *Basics of qualitative research: Techniques and procedures for developing grounded theory* (2nd ed.). Thousand Oaks, CA: Sage.

Ticha, M., & Hošpesová, A. (2009). Problem posing and development of pedagogical content knowledge in pre-service teacher training. In V. Durand-Guerrier, S. Soury-Lavergne, F. Arzarello, & F. Arzarello (Eds.), *Proceedings of the Sixth Congress of the European Society for Research in Mathematics Education* (pp. 1941–1950). Lyon, France: Institut National de Recherche Pédagogique.

Tirosh, D., & Graeber, A. O. (1990). Evoking cognitive conflict to explore preservice teachers' thinking about division. *Journal for Research in Mathematics Education, 21*, 98–108.

Toluk-Ucar, Z. (2009). Developing pre-service teachers understanding of fractions through problem posing. *Teaching and Teacher Education, 25*(1), 166–175.

Vacc, N. (1993). Questioning in the mathematics classroom. *Arithmetic Teacher, 41*(2), 88–91.

Van Zoest, L., & Stockero, S. (2008). Synergistic scaffolds as a means to support preservice teacher learning. *Teaching and Teacher Education, 24*(8), 2038–2048.

Verschaffel, L., Greer, B., & De Corte, E. (2000). *Making sense of word problems*. Lisse, The Netherlands: Swets & Zeitlinger.

Wilson, S., & Berne, J. (1999). Teacher learning and acquisition of professional knowledge: An examination of contemporary professional development. In A. Iran-Nejad & D. Pearson (Eds.), *Review of research in education* (pp. 173–209). Washington, DC: American Educational Research Association.

Zazkis, R., & Campbell, S. (1996). Divisibility and multiplicative structure of natural numbers: Preservice teachers' understanding. *Journal for Research in Mathematics Education, 27*, 540–563.

Chapter 24
A Collection of Problem-Posing Experiences for Prospective Mathematics Teachers that Make a Difference

Sandra Crespo

Abstract Without significant work on problem posing during teacher preparation, prospective teachers will enter the profession with limited vision and strategies for mathematics teaching. Based on previous and ongoing research on problem posing, the author proposes three essential strands for a problem posing framework that strives to teach prospective teachers to: (a) mindfully pose problems to students; (b) engage in problem posing with their students; and (c) pose personally and socially relevant mathematics problems. These strands engage prospective teachers with enduring questions for teachers of mathematics: What makes a mathematics problem educational? Who poses mathematics problems in the classroom? and Why do people spend time posing and solving mathematics problems? These three strands, individually and combined, can empower prospective teachers as problem posers and as teachers of mathematics who will pose rich and engaging problems to and with their future students.

Contents

S. Crespo (✉)
Michigan State University, East Lansing, MI, USA
e-mail: crespo@msu.edu

F.M. Singer et al. (eds.), *Mathematical Problem Posing*,
Research in Mathematics Education, DOI 10.1007/978-1-4614-6258-3_24

Introduction

The quality of mathematics teaching and learning is partially dependent on the mathematical tasks teachers use in their classrooms. Yet in many classrooms the majority of problems and questions posed to students focus on memorization and procedural understanding rather than on mathematical reasoning and inquiry. Even when teachers have access to high-quality textbooks and curricula, they often transform potentially rich, worthwhile problems in ways that lower their cognitive demand (Henningsen & Stein, 1997). If teachers and prospective teachers are to provide rich and deep learning experiences to their students, it is important that they develop some principled ways of deciding on the relative worth of problems—what makes some problems better than others. More importantly, they also need experiences generating such problems themselves.

In this chapter I synthesize and reflect on past research projects (Crespo, 2003a, 2003b; Crespo & Sinclair, 2008; Nicol & Crespo, 2006; Phillips & Crespo, 1996) to describe and theorize how prospective teachers can grow their problem-posing practices throughout their teacher preparation courses and experiences. Here I offer a cross-project reflection on the impact of three different types of problem-posing experiences that were offered to prospective elementary teachers across different institutional settings, albeit all at the time when they were taking mathematics methods courses in their respective programs. These three problem-posing experiences included: (a) Classifying and adapting textbook problems (Nicol & Crespo, 2006); (b) Interactive problem posing *with* students (Phillips & Crespo, 1996; Crespo, 2003a, 2003b); and (c) Generating personally and socially relevant problems (Crespo & Sinclair, 2008; Jacobsen & Mistele 2009). Although these are diverse types of problem-posing activities, they share the common goals of broadening the kinds of problems prospective teachers will consider using in their mathematics classrooms and to encourage them to see themselves as problem posers *to* and *alongside* their students. Next, I offer a rationale and brief history for why and how problem posing must claim its stake in the school mathematics curriculum as well as in mathematics teacher preparation programs.

Why Problem-Posing Education?

The central role of problems and problem solving to the discipline of mathematics is easily agreed upon among diverse mathematics communities. Problem posing is perhaps considered one of the highest forms of mathematical knowing and a sure path to gain status in the world of mathematics. The generation of new mathematical problems is a much less defined practice within the discipline of mathematics, perhaps considered much more as a creative act or an artistic endeavor, than a systematic practice. However, rather than seeing problem solving and problem posing as two distinct or parallel tracks of mathematical activity, or perhaps in opposition to one another, most mathematicians and mathematics educators would argue that the two are inextricably intertwined. The process of solving problems naturally gives rise to new problems. In other words, in the process of solving, problem reformulations are necessary. By the same token problem posing also entails the work of problem solving, as problems do not simply arise out of thin air or without some significant mathematical explorations. Many authors have vividly described the process of knowledge generation that occurs when mathematicians work at solving challenging and worthy mathematical problems, such as Fermat's famous conjecture (Singh, 1998).

Sadly a parallel track of problem solving and of problem posing still prevails in the context of school mathematics. In the 1980s, mathematics reform efforts called attention to and promoted problem solving as central to learning mathematics based on the earlier work of George Pólya (1945). Stephen Brown and Marion Walter (1983) drew attention to the role of problem posing in the discipline of mathematics and in learning mathematics but it would be a decade later (in the mid-1990s) that researchers in mathematics education turned their attention to students' problem posing and the relation between their solving and posing abilities. By the year 2000 researchers began to focus on teacher problem posing and how it opens and closes students' learning opportunities. However, problem posing is yet to gain the same status that problem solving has in school mathematics curricula.

Making problem solving the goal and focus of school mathematics has opened up opportunities for classrooms to become learning environments that, in turn, provide students with opportunities to learn mathematics in ways that lead to more and deeper understanding of key concepts and ideas. A problem-solving focus also allows students to become savvy users of mathematics and to have access to classically taught content and proven methods that are widely known and practiced in the discipline of mathematics. However, the pedagogy of problem solving positions the textbook and the teacher as the sole authority on what counts as a worthy mathematical problem. It suggests an intellectual stratification and division of labor between those who pose problems and those who solve them.

In contrast, the pedagogy of problem posing is associated with a pedagogy of empowerment and transformation (Aguirre, 2009; Freire, 1970; Gutstein & Peterson 2005). Problem-posing education considers students as co-constructors and as active participants in the design of their educational experiences. In a problem-posing classroom, students generate problems and questions as much as the teacher or textbook; they are encouraged to raise questions that are personally and socially meaningful which contrast to the often ahistorical and decontextualized types of mathematical problems written in generic mathematics textbooks. Mathematics teaching via problem posing provides students with opportunities to become producers of knowledge, not only recipients of already known content. It encourages students to use mathematics to raise and to answer questions that are deeply personal and socially relevant.

Why Problem Posing in Teacher Education

If we agree that problem-posing education is important for all of the reasons suggested in the previous section, then problem posing also needs to play a central role in teacher preparation. Furthermore, if it is true that the quality of teaching and learning depends on the quality of mathematics problems, then prospective teachers need firsthand experience with problem posing. But these posing experiences require more than sending prospective teachers to design mathematics problems impromptu or without some significant work in their teacher preparation classes, such as developing criteria and strategies for identifying and judging the quality of mathematical problems.

Consider the following example of teacher problem posing featured in a professional journal (Williams & Copley, 1994). The authors used the task below to illustrate how teachers could help students explore division with decimals conceptually rather than with a focus only on calculations (Thompson et al., 1994). The following questions were suggested to guide the students' mathematical investigations: *What is the same*? *What is different*? *What patterns can you identify*? *What do you think is causing these patterns to occur*? These are open-ended and reasoning types of teacher questions (Vacc, 1993) that suggest the mathematical goal of the task is to invite students to notice and search for patterns and then attempt to explain the reasons behind those patterns using what they already know about division with whole numbers.

Given $2{,}726 \div 58 = 47$

Predict: $272.6 \div 58 =$ ______

$27.26 \div 58 =$ ______

$2.726 \div 58 =$ ______

$0.2726 \div 58 =$ ______

In contrast, the questions prospective teachers in my courses have generated to use alongside this task are quite different. Although explicitly prompted to consider how to use this task as an opportunity to engage students in "doing mathematics" (Stein & Smith, 2011), inexperienced prospective teachers have a hard time conceptualizing mathematics instruction that shares intellectual authority with students. Consider one prospective teacher's (Susan) anticipated teaching using this task.

> The first thing I would do is ask the student "is 272.6 bigger or smaller than 2726?" Then, ask "by how much?" Then, I would ask them what did we do to make 2726 10x smaller? (Divide by power of 10). So, if 272.6 is divided by power of 10, is the answer to 272.6÷58, still 47. If say 'yes,' then try it. If say no, ask why? How much smaller does it have to be? (10x smaller). Therefore, divide by what? (power of 10). Then, ask about 27.26÷58. Is it bigger/smaller than 2726? By how much? (100)\divide by 2726 by what, to get 27.26 (power of 100)\what happens to 47? Stay same? Smaller? Try it. Soon, the kids may see a pattern arising and realize that making the number smaller (dividing by power 10, then 100, then 1000, etc.) requires making the answer smaller, as well (47, 4.7, 0.47, etc.).

This hypothetical teaching plan speaks volumes about Susan's emerging problem-posing practice. Notice the very narrow and prescriptive nature of the questions

Susan intends to ask to her students to lead them towards the mathematical punch line of the task without allowing them to do significant mathematical thinking and reasoning. Susan's problem-posing approach is fairly typical of prospective teachers who have not had significant experiences as problem posers themselves. They tend to reduce mathematics problems into small and digestible bite sizes to the point where the mathematical meat of the problem is lost in translation.

It is important to note that this is not solely an issue for those studying elementary school mathematics teaching. Secondary school prospective teachers also make similar kinds of uninformed choices when they pose mathematical tasks to their students. Consider the example from a seventh grade mathematics lesson a prospective teacher designed with a focus on rational numbers and the topic of terminating and repeating decimals. The driving mathematical question of the lesson was: "How do we know when a fraction will convert to a terminating or a repeating decimal?" The following questions were going to be used as a written assessment to evaluate the extent to which students understood the mathematical point of the lesson: (a) *What are you noticing about the fractions whose decimals terminate*? (b) *What are you noticing about the denominators of the fractions with decimals that terminate*? (c) *How can mathematicians tell accurately when fractions will convert to decimals that terminate*? Sadly, these very interesting and important mathematical questions were overshadowed by a poorly posed mathematics problem that did not align with the clearly stated goals and assessments of the lesson. An excerpt of the problem read as follows:

> It's almost Thanksgiving and Susie's family is getting together for dinner. Each person is bringing their favorite Thanksgiving dish to pass. However, a few days before Thanksgiving, your cousin Will tells the family he has a last-minute business trip he must attend and he and his family will not be coming to the dinner. This cuts the amount of people eating dinner in half. Therefore everyone decides to make half of their original recipe. Susie's family needs your help figuring out how much they need of certain ingredients.
>
> (a) Grandpa Joe's famous pumpkin pie recipe calls for 1/3 tablespoon of nutmeg. How much does he need if only half of the recipe is made?
>
> (b) Let's examine the fraction 1\3 from above and your answer for part a. Use long division to convert 1/3 and your answer for part a into decimals. Do these decimals terminate?

Seeking to make mathematics relevant to students' lives, the prospective teacher working on this lesson designed a mathematical problem that was set in the context of the American Thanksgiving dinner, since the lesson would be taught a few days before this holiday, and the scenario was that those bringing food to the Thanksgiving dinner had to scale up or down their recipe based on the number of family members coming to the dinner. Although this is indeed a very interesting context, it is more suited for a lesson on multiplication of fractions or scaling up or down fractions than a lesson about converting fractions to decimals. This context was used as the hook to entice students to engage with the task, but ultimately students were to leave that context behind and work on a series of number problems that asked them to pull out the fractional parts of the recipe and convert them to decimals and determine—using long division without a calculator—whether the resulting decimal terminated or not. These computational types of questions made no sense within this particular problem context.

This is another typical example of what happens when prospective teachers who themselves have not experienced problem posing as school students are in a position to create problems for their future students. Although they espouse well-meaning and well-intentioned beliefs and visions of mathematics teaching and see themselves as teachers who pose interesting and worthwhile mathematics problems, they are not well positioned to translate their vision into classroom reality. Their ill-formulated mathematics problems tend to result from a lack of close attention to the competing and sometimes problematic relation between a problem's context and its mathematical goal. Without significant experiences with problem posing in and for mathematics teaching, prospective teachers will continue to design what might be called "spoon feeding" instruction whereby the mathematical core of the lesson has been cut up into so many bite-size pieces that students end up answering piece-meal questions rather than engaging in genuine mathematical problem solving.

In my previous problem-posing research projects including those synthesized for this chapter, I have found that, without spending significant time doing problem posing themselves, prospective teachers generate the prototypical types of simplistic school mathematics problems that have been documented in the research literature as focused more on memorization and procedures than on substantive mathematical inquiry. I characterize these inexperienced problem-posing approaches as disempowered and disempowering because they unknowingly and unintentionally reproduce the "banking model" (Freire, 1970) or transmission function of school mathematics instruction whereby the teacher or textbook reduces students to the role of a technician taught to apply techniques and algorithmic procedures to solve problems that are given to them. Characteristics of such disempowered and disempowering approaches to posing mathematics problems are included in Table 24.1.

Table 24.1
Problem Posing of Prospective Teachers Without Significant Posing Experience

Disempowered/disempowering problem posing	Features
Posing closed problems	Problems are quick-translation story problems or computational exercises. Generated questions, if any, test for speed and accuracy
Posing simplified problems	Adaptations narrow mathematical scope of original version of problem and the work of solvers. Generated questions take the form of hints and lead solvers to a solution strategy or answer
Posing problems blindly	Mathematical complexity of the posed problem is underestimated. Problems are posed without solving beforehand or deeply understanding the mathematics. Generated questions suggest unawareness of the mathematical potential and scope of problem

Although many prospective elementary teachers report traumatic experiences as mathematics students that had turned them away from pursuing mathematics-related careers and wish to do things differently with their students, without some significant experiences that build new relationships with mathematics they will continue to reproduce the kind of mathematics instruction they grew to dislike. In Crespo (2003a, 2003b) I illustrate this point with Mitch, a prospective teacher who struggled to move away from what he remembered about his elementary school days when working with students in an interactive letter-writing project offered as part of his mathematics methods course. The very first mathematics problem Mitch assigned to his fourth-grade student was based on his memory of his elementary school days. He set a multiplication exercise that included: $1\times1=$___, $2\times2=$___, … $12\times12=$___. "A good test," Mitch said in his journal, "to see where she is in math right now, since I remember Grade 3 and Grade 4 composed partially of doing and re-doing times tables." Yet in his Mathematics Autobiography, Mitch spoke with disdain about the dreaded times table.

> Believe it or not I have no recollections whatsoever from elementary school regarding this subject except for Grade four's hated times tables. 'Two times two equals four, three times three equals nine, four times four equals sixteen,' and on it went, ad nauseam. Why did we always stop at 'twelve times twelve' anyways?

Without some significant intervention in these patterns of narrow and simplistic problem posing, prospective teachers will continue the trend of traditional mathematics instruction that perpetuates limited conceptions of the teaching-and-learning of mathematics. Given this lack of prior experience with problem posing during their formative years studying mathematics, it is important for teacher preparation programs to provide explicit experiences with problem posing.

In the following section I describe examples of the kinds of problem-posing experiences I have constructed for prospective teachers and found to make a difference in the ways in which they come to understand and conceptualize the role of problem posing in the teaching and learning of mathematics. I argue that, either individually or combined, these problem-posing experiences empower prospective teachers as problem posers themselves and are potentially empowering to their future students. Table 24.2 summarizes features that characterize empowered and

Table 24.2
Problem Posing of Prospective Teachers with Significant Posing Experience

Empowered/empowering problem posing	Features
Posing open problems	Problems require solvers to explain their work and communicate their ideas. Added questions invite solvers to share, explain, and reflect on their thinking
Posing mathematically challenging problems	Problems introduce new ideas, push solvers' thinking, or challenge their understanding. Added questions and adaptations scaffold rather than lead the solver's thinking. Adaptations open the mathematical work and scope of the original version of problem rather than narrow it down
Posing mathematically interesting problems	Exploring and mathematizing situations to generate "interesting" problems. Using mathematics aesthetic criteria such as: surprise, novelty, simplicity, fruitfulness—to decide on the quality of generated problems
Posing socially relevant mathematics problems	Exploring and mathematizing real world situations to engage in understanding and addressing social issues with mathematics

empowering problem-posing practices and that I suggest are essential and possible for prospective teachers to experience and learn during teacher preparation.

Empowering Problem-Posing Experiences for Prospective Teachers

It is reasonable to assume that identifying and recognizing worthy mathematics problems is difficult in all circumstances, but especially so when prospective teachers have been fed a steady diet of closed and simplified mathematics problems throughout their school years—the kind that provides practice applying a taught method or procedure for solving problems. Sociologist Dan Lortie (1975) called these schooling experiences a teachers' "apprenticeship of observation" and warned of the powerful effects these schooling experiences have on how teachers eventually teach in their own classrooms. The following problem-posing experiences aim to disrupt prospective teachers' "apprenticeship of observation" as mathematics problem solvers in school classrooms and invite them to pay close attention and be more reflective about the kinds of problems they will, in turn, use with their future mathematics students.

By participating in problem-posing activities during teacher preparation, prospective teachers can overcome their lack of experience as problem posers. Each of the following problem-posing activities was designed to address prospective teachers' tendencies to pose problems blindly and without serious considerations about what, how, or why problems matter in the teaching and learning of mathematics. Each of these experiences aims to help broaden the types of problems prospective teachers will consider posing to students and to gain strategies and tools for identifying, classifying, and reformulating problems. These experiences challenge prospective teachers to do more than mindlessly reproduce the typical types of school mathematics problems or the kinds of problems they remember from their elementary school days. Broadly speaking, each of these problem-posing experiences can be thought of as preparing prospective teachers to broaden and refine their professional vision and practice of mathematics teaching by learning to: (a) pose problems *to* students mindfully, (b) pose problems *with* students, and (c) pose problems *for* personal and/or social relevance. These three types of problem-posing experiences will lead to problem-posing pedagogies that give students more empowering and affirming experiences with mathematics.

Learning to Pose Problems Mindfully *to* Students

One problem-posing experience I continue to use in my mathematics methods courses engages prospective teachers in selecting, analyzing, and adapting textbook math problems. This assignment is designed to help prospective teachers recognize differences in the context, content, and quality of mathematical tasks. With this project, preservice teachers make a collection of mathematics problems that include a range of tasks, not just a single type. For example, Laurie, a prospective teacher featured in Nicol and Crespo (2006), included mathematics problems that could be considered procedural learning tasks that had single answers but she also included rich learning tasks that could be used to, in Laurie's words, "have a discussion on 'thinking about our thinking' during problem solving." Figure 24.1 shows an example of such a problem in Laurie's collection.

Sarah and Carla want to buy 15 cans of fruit that cost $0.62 each. They aren't sure they have enough money, and neither of them has a calculator. This is how each figures out how much the fruit costs:

Sarah	Carla
10 x 62 = 620	Each can costs a little bit more than $0.60
Half as much is 310	We need 15 cans
620 + 310 = 930	15 x 60 = 900
The fruit costs $9.30	So the fruit costs a little more than $9.00

Figure 24.1. Sample problem in a prospective teacher's collection of math problems.

I have continued to use this type of problem-posing assignment with prospective teachers and now include more explicit frameworks and classification schemes, such as Smith and Stein's (1998) "cognitive demand of tasks" framework, to help guide prospective teachers' analysis and selection of tasks they might use with their students, but also to develop their strategies for making adaptations that sustain or increase the cognitive demand of their chosen tasks. Figure 24.2 shows an example of such a problem-posing assignment, along with a sample analysis and adaptation to a mathematics task by a prospective teacher, with discussion about how to transform low-level tasks into higher level ones.

Inviting prospective teachers to analyze, sort, and classify mathematical tasks published in mathematics textbooks, standardized district or State exams, or unit tests provides them with opportunities to learn that mathematical tasks vary not only in the content of their questions but also in their quality. Learning how to assess the quality of mathematics tasks before they are tried out with students is part of what every prospective teacher must learn in their teacher preparation courses. However, the work of analyzing mathematical tasks cannot stay at the classification and sorting of tasks level. Prospective teachers also need experiences tinkering with and making changes to low quality tasks and attempt to improve them. These two activities combined offer an empowering experience for prospective teachers. Within this strand of problem posing for teaching experience, prospective teachers can learn criteria for judging the instructional quality of mathematics problems, but also learn to use those criteria to identify high/low mathematics tasks and to reformulate problems in ways that increase their instructional quality.

Problem-posing Prompt

Pick a task from above that you classified as a **LOW** level task and change it so as to make the task more high level (either procedure with connections or doing mathematics). Write the revised task and explain (using what you learned from Smith and Stein's classification schemes) the differences you see between the original task and the revisions you made to it.

Sample Task Analysis and Adaptation by a Prospective Teacher:
Original Task: Using the edge of a triangle pattern block as the unit of measure, determine the perimeter of the following pattern-block trains...
Revised Task: Using the edge of all or some of the triangles given, make different patterns and determine the perimeter. Explain why some might have the same perimeters and some have different perimeters. What is the biggest perimeter you could make? What is the smallest?

Explanation: The original task is straight and to the point. It is just asking the students to count the edges and doesn't make connections to anything else. Yes, the student might understand the perimeter is how many segments are on the outside, but other than that a student is not able to fully understand the concept of a perimeter. The revised version allows the students to play around with them a bit and see what they can come up with on their own. They may find what could make a perimeter smaller and what the best way is to make it as large as possible. This gives students a chance to utilize what they already know and build upon it.

Figure 24.2. Problem-posing prompt and sample explanation.

Learning to Pose Problems with Students

Prospective teachers also need opportunities to try their selected and adapted mathematics problems in order to learn from and with students their perspective on what makes a good mathematics problem. A few years ago I offered one such opportunity by engaging prospective teachers in an interactive mathematics letter exchange with Eileen Phillip's fourth-grade students focused on the posing and solving of mathematics problems (see Phillips & Crespo, 1996; Crespo, 2003a, 2003b). Each letter included at least one mathematics problem for the fourth-grade student to work on, and the mathematics work the prospective teacher had done on the mathematics problem the fourth-grade student had sent to them. Through this interactive posing and solving experience prospective teachers had opportunities to practice selecting, adapting, and posing tasks to students and to read actual students' mathematical work and feedback on those problems. The prospective teachers were then in a better position to understand the instructional quality and potential of their posed tasks. In addition, this posing experience gave prospective teachers opportunities to learn from their students how to pose problems and to see them as problem posers too, since the fourth graders created or selected mathematics problems in their letters for the prospective teachers to solve as well.

Through this iterative process of exchanging mathematics problems over a period of 8 weeks, prospective teachers had many insights on their problems' instructional quality by reading their students' work. One prospective teacher, for example, asked a fourth grader to complete the following number sentence: $\frac{1}{2} + \_\_\_ = 1$. The fourth-grader's response was that the answer could be either ½, 2/4, 3/6, 4/8, 5/10, etc. … The prospective teacher who had expected a single-number answer was surprised to learn that such a seemingly closed question could be read so much more broadly and open-mindedly. Another example occurred when a prospective teacher's imprecise mathematical language led a fourth grader to solve the task in two different ways and explain in her solution that twice the size and double the size did not mean the same thing, and that depending on the intended meaning, the problem's answer would either be this one or this other one. Figure 24.3 shows an excerpt from one problem-posing exchange between a prospective teacher and a fourth-grade student.

Prospective teacher:
Imagine having a pizza that is twice as big as another. Should this larger pizza be twice the price? In other words, when the circumference of a pizza pan doubles, should the price double?

Fourth-grade student:
If the pizza's size doubles, then I would double the price because I would get the same amount of money as I would if I sold two small pizzas. When I read this [arrow pointing to the "In other words . . . " part of the question], I got confused because this is not the same as the one above. When the circumference doubles, the size is more than doubled. I looked at John's [another fourth grader] letter to help me along. Then I understood. If I were selling the pizza, I would do more than double the price because it is more than twice as much pizza.

Prospective teacher:
Sorry I wasn't very clear. I realized when I reread the question that it was pretty confusing.

Figure 24.3. Letter exchange between a prospective teacher and a fourth-grade student.

Additionally, prospective teachers were the recipients of mathematics problems generated by the fourth graders and this allowed them to see that young students welcomed and enjoyed playing the role of problem posers. The fourth graders' problems also provided prospective teachers with clues about the type and the quality of mathematics problems their students enjoyed writing. One fourth grader, for example, posed the following mathematics problem: "If you feed a baby 8 ounces of milk in one feeding how many ounces would it be in two feedings?" The prospective teacher who received this task replied with the answer of 16 ounces and included the mathematics work she had done to come up with that answer. In the next letter, the fourth grade student responded: *You were close for the answer of my question about the baby but the right answer is 15 oz. or less because the baby will either throw up or the milk will dribble out the sides.*

Over time prospective teachers generated a collection of mathematics problems and had student data about the quality and potential of those problems. They also increased their repertoire of posing moves (hints, questions, examples, contextualizations, etc.) that made their problems accessible and enticing enough for their students to share their mathematical work and thinking even when they had not completely solved the problem or were not sure if they had the right answer. Within this interactive problem-posing experience, prospective teachers realized they were underestimating their young students' abilities and they were often surprised by the work and explanations that the fourth graders provided and also by the challenging problems the youngsters in turn had selected or generated for them to solve. More importantly, prospective teachers experienced firsthand the power of sharing mathematical authority with their students, and that mathematical problems could be generated by them and their students rather than from a mathematics textbook.

Learning to Pose Personally and Socially Relevant Mathematics Problems

Because the previous two problem-posing activities set up prospective teachers to think about mathematics problems pedagogically (i.e., what kinds of mathematics problems are good for students, what is the instructional value or cognitive demand of tasks I might give to my students?), I have also explored ways of supporting prospective teachers' own personal explorations of mathematics problems. This exploration has included the question "What makes problems mathematically interesting?" as well as considered the connections between mathematics and the world outside of school, especially ways in which mathematics can be used to sway public opinion, to distort the truth, and to study social issues that have real consequences such as fairness and injustice around us.

Posing personally relevant and interesting mathematics problems. In more recent problem-posing projects and in collaboration with Nathalie Sinclair, I sought to support prospective teachers' explorations of and deliberations on the quality of mathematics problems by developing explicitly mathematical criteria, not only pedagogical ones for judging mathematics problems (see Crespo & Sinclair, 2008; Sinclair & Crespo, 2006). In that research project we tested two different problem posing structures—"explore first and pose second" vs. the opposite sequence of "pose first and explore second"—and examined the quantity and quality of mathematics problems generated by prospective teachers in each of the two structures. In the *explore first* condition, prospective teachers explored the given situation and from that experience generated potential mathematics problems. For example, "What happens if?" or "Can I make this other kind of shape or number with these examples?" The second condition invited prospective teachers to generate mathematics problems after just a few minutes of glancing over the given materials or mathematical situation.

Unsurprisingly, and consistent with the research literature that guided the design of the project, we found that prospective teachers generated a better collection of mathematics problems and questions when they had personally engaged in open explorations of given situations first to mathematize those situations themselves—for example, exploring a group of shapes or exploring a collection of number sequences—before they were asked to generate mathematics problems from those situations. When given a set of shapes, the *pose first* prospective teachers generated the prototypical types of low-level (name, identify, count) mathematics problems, such as: "What is the name of that shape?"or "How many of these shapes are triangles?" Students in the *explore first* condition generated more cognitively demanding and reasoning types of mathematics problems, such as: "Can you put two or more of these shapes to make a new shape?" or "How many shapes can you create that are symmetrical?"

More importantly, those in the *explore first* condition were more likely to offer a mathematically and personally relevant reason (for example, "not sure if it can be solved," or "this problem mushrooms into more questions") for why they thought their generated mathematics problem was interesting, in contrast to considering only their pedagogical interest (e.g., "students can find different solutions" or "will force students to explain their thinking"). To generate further support for the development of their personal and mathematical interest in generating problems, we introduced mathematical aesthetic criteria (such as novelty, surprise, fruitfulness) into our study by using a food metaphor of "nutritious" and "tasty" to describe the mathematical quality of problems. This metaphor/criteria provided more accessible language to discuss and describe what could be considered to be mathematically tasty problems.

In addition to helping prospective teachers generate a broader collection of mathematics problems and to develop personal connections with core mathematical practices and aesthetic criteria for judging the quality of mathematics problems—what makes mathematics worth doing—this type of problem-posing experience gives prospective teachers empowering opportunities to learn that human aesthetics apply to generating mathematically interesting problems. People are drawn to surprise, novelty, simplicity, generalizing, disproving, and fixing things in their daily lives and this translates to mathematics work as well. Considering mathematics and problem posing in particular to be accessible to everyone, not just to the so-called mathematically talented people, is important, especially for future teachers who need to see themselves as capable of generating and judging the mathematical quality of mathematics problems, and of seeing the same in their future students of mathematics. Furthermore, prospective teachers can also learn that "mathematically interesting" and "pedagogically sound" problems are not necessarily one and the same, and instead look for and generate mathematics problems that share both of these qualities as more powerful and empowering types of problems.

Exploring and posing socially relevant mathematics problems. More recently I have worked on a different kind of problem-posing project, one that seeks to make explicit the role mathematics can play in addressing larger out-of-school problems, such as issues with inequities and power struggles between private and public interests and between privileged and oppressed groups. This work grew from a collaboration with Laura Jacobsen's and Jean Mistele's (2009–2011) project "Mathematics Education in the Public Interest" which in turn was inspired by Gutstein and Peterson's "Rethinking Mathematics" (2005) proposal that mathematics can be taught in a way that helps students "more clearly understand their lives in relation to their surroundings and to see mathematics as a tool to help make the world more equal and just" and to prepare them to be "critical and active participants in a democracy" (p. 1). Inspired by these mathematics educators' work I have been exploring ways to engage prospective teachers in posing problems that have social relevance, in particular problems that are related to their young students' lives and communities, as well as to the schools where they teach.

Because most students in general often question the usefulness of mathematics when they experience it as disconnected from real world happenings—what is in the news and what is talked about outside the mathematics classroom—it would seem that this kind of alternative problem-posing project would be welcomed and embraced by most prospective teachers as the answer to the dreaded student question of "Why do I need to know this?" However, as Gutstein and Peterson (2005) also noted, the many years of working with supposedly neutral mathematics contexts makes the more explicitly socially relevant contexts seem too controversial and perhaps even less mathematical. Using simple mathematics story problems about addition with regrouping as examples, Gutstein and Peterson pointed out that the dominant forms of mathematics problems tend to appear as non-controversial contexts, such as "kids going to the store to buy candy that cost 43¢ each …" whereas less commonly used contexts will appear to be more controversial, such as: "Youth working on a sweat shop factory make 43¢ an hour on a 14 hour/day shift …" (p. 6).

Although the latter is readily recognized as a politically charged context, the former is considered to be neutral. Yet, as Gutstein and Peterson argue, no mathematics teaching is ever neutral; the subtext of the first problem is consumerism and unhealthy eating habits that are so prevalent in the U.S. and many other societies, but we fail to see the subtext as controversial. Similarly, Felton (2010) pointed out the prevalence of mathematics story problems comparing boys to girls and how the volume of such problems reflect and reinforce dominant views about gender as a binary category, rather than to challenge or question such a perspective. Upon encountering these ideas and new kinds of mathematics problems, prospective teachers in my courses range in the extent to which they consider anything other than the supposedly "neutral" contexts as suitable for elementary school mathematics story problems. This is illustrated in the following two journal entries:

> I understand that there is a push for transgender and homosexuality but to start involving those into math I think is taking it a little too far. I think parents and others would be very upset and I don't think that needs to be involved in math class. (Sample response 1).

> Math problems communicate what is considered "not normal" and "normal" in the world, as stated by Felton. Too often, they send messages about race, gender, consumerism, etc. that are often stereotypes. For example, in my current placement, my mentor teacher used a problem with our students that talked about a girl named Juanita who made tortillas for her brothers and sisters. I believe that this presents a stereotype about Hispanic children and their families; it is possible that a boy named John may have made tortillas for his brothers and sisters as well. (Sample response 2).

In spite of some initial discomfort with the idea that mathematics problems do more than give students opportunities to manipulate the quantitative information embedded in the text, prospective teachers are curious about this new perspective on mathematics problems. They continue to explore and then design mathematics problems that could engage their students in using mathematics to better understand and perhaps address larger social problems and issues in the world today—lowering crime rates, seeking a fairer tax system and distribution of wealth, addressing climate change, and so forth.

To support prospective teachers in designing mathematics problems that involve a significant social issue while also addressing significant and rigorous mathematics, I provide several in-class *mathematics in the public interest* experiences. For example, one of these experiences entailed comparing the relative size and geographical location of countries as represented in world maps. This problem highlights the use of mathematics to question land ownership and how land is measured in the global (but also local) contexts, and how those measurements are then represented in world maps commonly displayed in school classrooms.

To many of the prospective teachers in my classroom this was a new way of thinking about mathematics. They had not considered the role that mathematics plays in understanding how countries determine and negotiate their borders or that mathematics could be used to explore and question world maps and how well they represent the size and geography of many countries, in particular those often labelled as "third world" or "developing countries." In a follow-up in-class experience prospective teachers engaged with another *mathematics in the public interest* task also focusing on the topic of area measurement. This time we studied the problem of school overcrowding (class size and room size—what is a reasonable square footage per student?) and talked about how classroom space might have an impact on student learning. For this lesson I asked prospective teachers to bring the measurements of their field placement classrooms and the number of students in their classroom. This information was then compiled into a table to contrast with specified building codes and policies for class size and room size in schools.

In the conversations that happened during the first and the second experiences described above, it was clear that this was a new perspective about mathematics and mathematics teaching for all of the prospective teachers taking my class. When asked if they had ever discussed or considered that the study of mathematics could also include topics that are relevant to students' social world, everyone's answer was that they had not experienced this as students themselves or in their mathematics content courses while prospective teachers in our program. They were intrigued about the experience but found it really challenging to re-imagine a mathematics lesson that was so different from what they had experienced themselves as mathematics students, and from the mathematics lessons they were observing in their field classrooms. When the time came for prospective teachers to then create their own *mathematics in the public interest* problem, many still needed support. Consider the email exchange (see Figure 24.4) between "Emily" and myself over the question of what would count as an issue of public interest.

In spite of these challenges Emily and the other prospective teachers were able to carry out the course assignment of designing and teaching a *mathematics in the public interest* mathematics lesson in their field placement classroom. With support, Emily ended up revising her fifth-grade lesson, which continued the focus on the topic of adding and subtracting three-digit and larger whole numbers, but was

Prospective Teacher:
Sandra, I have a question about the mathematics in public interest lesson—would this one work? Students can practice adding and subtracting whole numbers. Have them use an almanac to find the populations of the three largest cities in their state. Then have them do the following:
- Round the population of each city to the greatest place-value.
- Estimate the sum of the cities' populations.
- Compare the estimated sum with the rounded total population of their state.

Ask students if the estimated sum of the cities' populations is more than half the rounded population of the state.
Thanks!
"Emily"

Teacher Educator:
Hi Emily –
What issue of public interest will you want them to explore with this information?
Sandra

Prospective Teacher:
That's what I was hoping you could help me with... Isn't exploring their city in terms of the other large cities in the state and all together those in comparison to the population of the whole state, an issue of public interest?
Emily

Figure 24.4. Email exchange between a prospective teacher and her teacher educator.

revised to be set within the context of exploring the problem of nutrition and poor health associated with fast foods. She planned to play a small video clip from the documentary "SuperSize Me" and engage students in the core of the mathematics lesson in which they had to add and subtract the number of calories in a chosen McDonald's meal and then analyse the nutritious value of these meals. Emily stated the following mathematical and social goals for her lesson: (a) Students will be able to add and subtract whole numbers using "counting up" method; (b) Students will understand when to use estimation; and (c) Students will understand that some McDonald's food might harm your health and has a lot more calories than many of the things we eat. She stated:

> I will tell them we are going to watch a little part of a movie called SuperSize Me and I will ask the children if they have seen it. If there are some students that have seen the movie I will allow them to explain what happened in the movie. If students have not seen the movie or heard of it, I will explain to them that it is a true story about a man that ate McDonalds and only McDonald's three meals a day for one month straight. Then I will show the portion of the movie where the doctors are informing the man of the bad effects the food is having on his body. We will then debrief the bad things to happen to his body. He became very unhealthy, and his body was weaker than it ever had been before. This was because of the type of food he had been eating. I will then point out that some fast food does not meet the normal calorie amount for our food we should be eating and it also can make us very sick if we eat extreme amounts of it. I will ask if there is anyone else whose meal was more than our class's. If there are any students that were I will have them share what they ate and have them bring up their paper to the projector so we can see their adding. I will also go the other way and ask if there are any students who think their meal had the least amount of calories in the class, and I will have them share. We will go over these students' papers as a class and focus on what they did great and what they might need help on.

Implications and Future Direction

All of the problem-posing experiences described and discussed here are potentially transformative to all involved in mathematics instruction because they each target more centrally the student solver, the teacher poser, or the mathematics curriculum. The student solver's experience with mathematics is very different from typical mathematics instruction when the teacher is mindful and clear about the instructional quality of the problems they choose to assign to their students. The teacher poser's experiences are also very different when they are willing to share with students the intellectual authority of generating or improving mathematics tasks together. The mathematics curriculum is also transformed when the teacher and students explore personally interesting and socially relevant mathematics problems.

Collectively, the problem-posing experiences in the studies shared in this chapter show that it is possible—within a short period of time—for prospective elementary teachers to make important gains in the range and quality of the mathematical problems and questions they generate and use in their mathematics teaching when they participate in purposeful and empowering problem-posing experiences. Each of the proposed problem-posing strands provides a different lens on the roles and responsibilities of the teacher as a poser of mathematics problems—attending to the instructional quality of mathematics tasks, seeing themselves and their students as capable of generating worthy mathematics problems, and exploring the mathematical interest and social relevance of a mathematics task. Together, these three different problem-posing strands make a beginning framework (see Table 24.3) that can be used as an initial blueprint for guiding other teacher educators as they design problem-posing experiences for prospective teachers. Table 24.3 summarizes the foci of prospective teachers' problem posing before and after they have had a significant and empowering problem-posing experience within the three proposed strands.

There is still much to explore about the potential impact these different kinds of problem-posing experiences can have in the eventual teaching of prospective teachers. A more elaborated and refined version of this framework is needed to help describe and track the prospective teachers' growth and progress in improving their problem-posing practices over time. The studies synthesized here suggest that providing prospective teachers with opportunities to explore and discuss the mathematical potential of problems and evaluating their mathematical as well as pedagogical interest can be a positive step towards preparing teachers who will design rich but also empowering mathematical opportunities for their students. Since each activity described focused on a different aspect of designing problems, a single type of problem-posing experience is not sufficient and therefore all three strands of experiences are important. However, questions about the sequencing and relative impact of these experiences remain. Furthermore, different types of problem-posing experiences will lead prospective teachers to create different kinds of relationships, expectations, and stances about themselves and their students' relationship with

Table 24.3
A Starting Framework of Problem Posing Experiences for Prospective Teachers

Strand	Driving question	Description	Expected outcome
Posing problems *to* students mindfully	What makes mathematics problems educational?	Engage in experiences that prompt for analysis of the instructional quality of mathematics tasks and to articulate reasons for choosing, adapting, and using tasks with students grounded in a mindful analysis of the task's educational quality	*From*: Posing low-quality mathematics tasks and blindly selecting or adapting tasks. *To*: Posing high-quality mathematics tasks with solid and well-argued instructional goals and reasons
Posing problems *with* students	Who poses problems in the mathematics classroom?	Engage in interactive problem-posing activities with students by exchanging problems over a sustained period of time and in ways that both teacher and student play the role of poser and of solver to each other's problems	*From*: Considering mathematics textbooks as the source of all mathematics problems *To*: Considering the teacher and the students as creators of mathematics problems themselves
Posing personally interesting mathematics problems	What makes problems mathematically interesting?	Explore in pairs or small groups the mathematical interest of situations that are not explicitly mathematical to generate potential mathematics problems. Nominate problems for public discussion and examination of their mathematical interest	*From*: From solely providing pedagogical reasons for selecting and posing tasks to students *To*: Using mathematical interest as criteria for judging the quality of mathematics problems
Posing socially relevant mathematics problems	Why spend time posing and solving mathematics problems?	Explore and discuss mathematics problems that are situated within contexts that raise social awareness and interest about the world we live in. Generate mathematics problems that are situated within appropriate contexts for school age students that help them to better understand with mathematics the issue raised in the mathematics task	*From*: Using context solely to generate interest or as an entry point into solving mathematics problems *To*: Using context to support mathematical understanding; and mathematics to better understand the context or situation in the task

mathematics, their mathematical authority, and the ways in which mathematics can be used to develop a better understanding of the world around them. All of these are worthy and important questions for the next generation of studies focused on improving the problem-posing practices of prospective teachers.

Acknowledgement This work was supported by the National Science Foundation under Grant No. 0546164. Any opinions, findings, and conclusions or recommendations expressed in this material are those of the author(s) and do not necessarily reflect the views of the National Science Foundation.

References

Aguirre, J. (2009). Privileging mathematics and equity in teacher education: Framework, counter-resistance strategies and reflections from a Latina mathematics educator. In B. Greer, S. Mukhopadhyay, S. Nelson-Barber, & A. Powell (Eds.), *Culturally responsive mathematics education* (pp. 295–319). New York, NY: Routledge.

Brown, S. I., & Walter, M. I. (1983). *The art of problem posing*. Hillsdale, NJ: Lawrence Erlbaum.

Crespo, S. (2003a). Learning to pose mathematical problems: Exploring changes in prospective teachers' practices. *Educational Studies in Mathematics, 52*, 243–270.

Crespo, S. (2003b). Using math pen-pals to promote mathematical communication. *Teaching Children Mathematics, 10*(1), 34–40.

Crespo, S., & Sinclair, N. (2008). What makes a problem mathematically interesting? Inviting prospective teachers to pose better problems. *Journal of Mathematics Teacher Education, 11*, 395–415.

Felton, M. D. (2010). Is math politically neutral? *Teaching Children Mathematics, 17*(2), 60–63.

Freire, P. (1970). *Pedagogy of the oppressed*. New York, NY: Continuum.

Gutstein, E., & Peterson, B. (2005). *Rethinking mathematics: Teaching social justice by the numbers*. Milwaukee, WI: Rethinking Schools.

Henningsen, M., & Stein, M. K. (1997). Mathematical tasks and student cognition: Classroom-based factors that support and inhibit high-level mathematical thinking and reasoning. *Journal for Research in Mathematics Education, 28*, 524–549.

Jacobsen, L., & Mistele, J. (2009). *Mathematics education in the public interest*. National Science Foundation (NSF)—Course, Curriculum, and Laboratory Improvement Grant Program.

Lortie, D. (1975). *Schoolteacher: A sociological study*. Chicago, IL: University of Chicago Press.

Nicol, C. C., & Crespo, S. M. (2006). Learning to teach with mathematics textbooks: How prospective teachers interpret and use curriculum materials. *Educational Studies in Mathematics, 62*(3), 331–355.

Phillips, E., & Crespo, S. (1996). Developing written communication in mathematics through math penpals. *For the Learning of Mathematics, 16*(1), 15–22.

Pólya, G. (1945). *How to solve it: A new aspect of mathematical method*. Princeton, NJ: Princeton University Press.

Sinclair, N., & Crespo, S. (2006). What makes a good problem: An aesthetic lens. In J. Novotná, H. Moraová, M. Krátká, & N. Stehlíková (Eds.), *Proceeding of the 2006 International Group for the Psychology of Mathematics Education* (Vol. 5, pp. 129–136). Prague, Czech Republic: PME.

Singh, S. (1998). *Fermat's enigma: The epic quest to solve the world's greatest mathematical problem*. New York, NY: Anchor Books.

Smith, M. S., & Stein, M. K. (1998). Selecting and creating mathematical tasks: From research to practice. *Mathematics Teaching in the Middle School, 3*(5), 344–350.

Stein, M., & Smith, M. P. (2011). *Five practices for orchestrating productive mathematics discussions*. Reston, VA: National Council of Teachers of Mathematics.

Thompson, A. G., Philipp, R. A., Thompson, P. W., & Boyd, B. A. (1994). Calculational and conceptual orientations in teaching mathematics. In A. Coxford (Ed.), *Professional development for teachers of mathematics* (1994 Yearbook of the National Council of Teachers of Mathematics, pp. 79–92). Reston, VA: National Council of Teachers of Mathematics.

Vacc, N. (1993). Questioning in the mathematics classroom. *Arithmetic Teacher, 41*(2), 88–91.

Williams, S. E., & Copley, J. V. (1994). Promoting classroom dialogue. *Mathematics Teaching in the Middle School, 1*(1), 72–75.

Chapter 25
Problem Posing as an Integral Component of the Mathematics Curriculum: A Study with Prospective and Practicing Middle-School Teachers

Nerida F. Ellerton

Abstract Details are presented of a study in which problem posing was an integral component of a mathematical modelling class for preservice and practicing middle-school teachers. One of the activities involved a project in which students individually planned and drafted mathematical modelling problems. Students then shared their draft problems with their peers before developing and presenting final versions of their problems to the class. Their personal reflections on the project formed an important part of the activity. Results are discussed in terms of an *Active Learning Framework*, and characteristics of a pedagogy for problem posing are proposed.

Contents

N.F. Ellerton (✉)
Department of Mathematics, Illinois State University, Normal, IL, USA
e-mail: nellert@ilstu.edu

F.M. Singer et al. (eds.), *Mathematical Problem Posing*,
Research in Mathematics Education, DOI 10.1007/978-1-4614-6258-3_25

Prelude

> ***The Candy Manufacturer's Problem***
> Ms C. is in charge of planning for her company which makes and sells candy. The company prides itself in making interesting containers for the candy it sells. Part of Ms C.'s responsibility is to ensure that the company makes the greatest possible profit on each "box" of candy sold.
>
> Ms C. is able to order fancy cardboard, suitable for making cylindrical containers for one type of candy sold by the company. The cardboard comes precut into rectangles, ready to form the cylinders which will be made by the candy company. Ms C. must decide on the dimensions of the cylinders, and how they will be formed from the rectangular sheets of cardboard—tall and slim or short and fat. The cylinders must be no longer than 11 inches and no shorter than 4 inches, and have a diameter of at least 1 inch. The large sheets of cardboard measure $22'' \times 17''$, and Ms C. wants to have the minimum number of cuts, with no wastage of cardboard.
>
> Your task is to help Ms C. decide on the dimensions of the precut cardboard she should order, and on the size of the cylinders to be made, so that the company can maximize its profits.
>
> Investigate, as fully as you can, the mathematics that underlies this contextualized problem. Develop a report for Ms C. that she could use to justify her decisions about the production of suitable cylindrical containers. You can assume that Ms C. has told you that she will be able to provide plastic tops and bottoms that are manufactured separately for the containers.

The *Candy Manufacturer's Problem* was the first problem that I presented to a mathematical modelling class that comprised preservice and practicing middle-school teachers. Some instructors may use a task such as this to give students experience with extended problem-solving tasks. Others may prefer a task which they consider to be defined in more specific detail. I designed the task as a way of integrating problem posing into the curriculum. The semester's work built on the environment established through students' collaborative work on tasks such as the *Candy Manufacturer's Problem*. In this chapter, I describe how students embraced the challenge of creating their own mathematics modelling problems, and discuss ways to integrate problem posing into the teaching and learning of mathematics.

Introduction

Solving mathematical problems in efficient and creative ways has long been recognized as an essential outcome of learning mathematics at all levels. Until recent years, however, it has usually been tacitly assumed that the posing of mathematics problems should be left to textbook writers, teachers, or researchers. Now, "problem posing" appears alongside "problem solving" in curriculum documents which recommend that teachers introduce problem-posing activities and opportunities for their students (see e.g., National Council of Teachers of Mathematics [NCTM], 2000; National Governors Association Center for Best Practices & Council of Chief State School Officers, 2010; Victorian Curriculum and Assessment Authority, 2008).

It is therefore not only important that preservice teachers understand what is involved in posing mathematical problems, but also that they themselves become competent and consistent problem posers. This chapter examines several ways in which problem posing became an integral component of the mathematics curriculum for some prospective and practicing middle-school teachers.

The study will be considered from four perspectives: curriculum planning, modes of communication, assessment, and cooperative group work. Curriculum planning needs to take account of the intended, textbook, implemented, assessed, and learned curricula, as proposed by Tarr, Grouws, Chávez, and Soria (2013), which extended the framework of intended, implemented, and attained curricula originally put forward by Westbury (1980). In particular, problem posing needs to be included as an integral component of the mathematics curriculum experienced by prospective and practicing teachers so that they, in turn, can make it an integral component of their students' mathematics curriculum. It follows that the experiences that these prospective and practicing teachers have in posing mathematical problems should provide a model for them to use in their own schools and classrooms.

The roles of students and teachers in the mathematics classroom can be considered in terms of receptive and expressive modes of communication. Del Campo and Clements (1987) noted that, in most classrooms, students use mainly receptive modes of communication, while their teachers adopt more expressive modes. However, when *students* pose problems, they use expressive modes of communication. In problem-solving situations, it is usually the teacher who puts scaffolding in place and judges when and how to remove it. By contrast, in problem-posing situations, students themselves decide on possible constructions (Ellerton, 1986), and build their own scaffolding to achieve their goals.

Inherent in the task of posing mathematics problems is the assessment of that task—whether or not the problem is complete, makes sense, is solvable, or is elegant and creative in its design (Clements & Ellerton, 2006; Stoyanova & Ellerton, 1996). The problems created not only provide both the student and the teacher with data that will assist in an assessment of that student's problem-posing performance, but also an assessment of the student's understanding of the mathematical structure of the problem.

In this study, most problem-posing tasks were carried out within cooperative small groups. Important factors, which were identified in the design of the problem-posing tasks, will be discussed. Examples of problems created by students, and students' comments about the various tasks will also be presented.

Literature Review, Theoretical Framework, and Research Questions

Children model their behaviors on the behaviors of those around them—in particular, their siblings, their parents, the children they meet at home, at school and in the neighbourhood, and their teachers. In their day-to-day lives, they witness others solving problems, and it is natural that sometimes they try to solve problems that they see others solving. They may encounter some evidence of problems being posed—but they are usually on the receiving end of these problems. A significant number of children have the ambition of becoming teachers, and a few of these combine this wish with their love of mathematics, and study to become teachers of mathematics.

Although statements advocating problem posing abound, little problem posing is likely to take place in mathematics classrooms unless teachers have the appropriate knowledge and skills to plan for, introduce, and encourage problem-posing activities. Exposure to ideas about the incorporation of problem posing into the classroom can be via the professional development of practicing teachers through books like *Problem Posing: Reflections and Applications* (Brown & Walter, 1993), or articles such as *Promoting a Problem-Posing Classroom* (English, 1997). But reading *about* problem posing is no substitute for actively *doing* it.

Teacher education can provide fertile contexts in which both practicing teachers and preservice teachers can grow within a culture rich in problem posing. It is not enough to add a segment on problem posing to fill a gap in the curriculum. Ellerton (2013) put forward a framework for locating problem posing in mathematics classrooms. Central to this framework is the *active* involvement of students in posing problems that not only demonstrates their understanding of the structure of the mathematical concepts they have been learning, but also gives students the opportunity to solve and critique the problems of others, and to reflect on and improve their own problems. Incorporating such a framework necessarily involves a shift in the nature of discourse in mathematics classrooms—from teacher-centered to student-centered, with the teacher becoming a facilitator rather than the only authority figure. This framework has been reproduced here as Figure 25.1.

Research on preservice students' use of and experiences with problem posing has generally focused on two important areas: first, as a way of assessing students' understanding of mathematical concepts, based on the problems they posed (Lin, 2004; Silver & Burkett, 1994); and second, as a strategy to help teachers design appropriate problem-posing tasks that might be used to assess particular curricular approaches (Cai et al., 2013). Designing or redesigning of mathematics curricula to incorporate problem posing has received less attention in the research literature (Ellerton, 2013; Staebler-Wiseman, 2011). In fact, very little attention has been given to studying the effectiveness of a more holistic approach to mathematical problem posing—one that aims to involve preservice and practicing teachers in active problem posing—planning, designing, and reflecting about posed problems, critiquing and assessing the problems themselves, and sharing and discussing solutions to the posed problems with their peers. But active problem posing alone is not an end in itself—the activities are part of a holistic approach that make these teachers

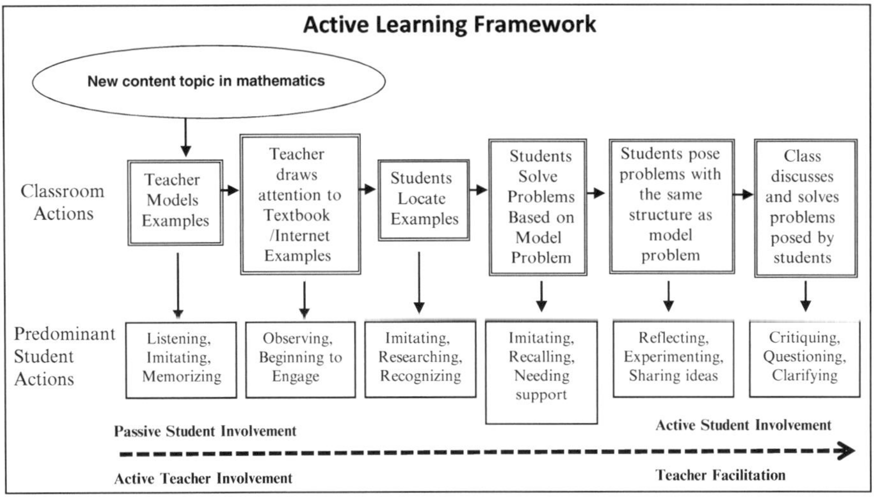

Figure 25.1. Framework for locating problem posing in mathematics classrooms (from Ellerton (2013, p. 99).

(both prospective and practicing) aware of the links between problem solving and problem posing, not only for their own learning of mathematics, but also for the pedagogical strategies they will use with their own (or their future) students. This chapter proposes that a new term—Pedagogy for Problem Posing (PPP)—be used to describe this concept. PPP has been much neglected in discussions about and planning for problem posing in mathematics classrooms.

The following three research questions guided the study:

1. What key features underpin (or work against) the integration of problem posing into mathematics curricula in courses for preservice and practicing teachers?
2. In what ways does a collaborative approach help or hinder the integration of problem posing in such courses?
3. What characterizes Pedagogy for Problem Posing (PPP)?

Design of the Study

A total of 11 students took part in a course on mathematical modelling offered as a capstone undergraduate course for middle-school teachers in a mid-western university in the United States. Seven of these students were preservice middle-school teachers, and four were practicing middle-school teachers who were studying for their Masters degree in mathematics (a limited number of specific undergraduate courses can be taken for credit by graduate students).

The course is intended to expose students to a broad range of mathematical ideas, to introduce concepts of mathematical modelling, and to have students apply mathematical modelling to a range of real-world contexts. The emphasis in the course is on the development of the students' own mathematical conceptions, yet all the time bearing in mind possible pedagogical strategies and assumptions that would be appropriate for the middle school.

Many books (e.g., *Focus in High School Mathematics: Reasoning and Sense Making*, published by the National Council of Teachers of Mathematics in 2009) stress the importance of encouraging students to develop mathematical models that can be used to summarize and make predictions related to real-world situations. Typically, they point out that mathematical modelling begins with an appropriate "mathematization" of a real-world situation or problem. Mathematization involves the students in interpreting a given problem in mathematical terms and posing a suitable mathematical structure for the solution of the problem. This structure and the implied solution needs to be checked against the real-world context, and if necessary, the cycle of posing modifications to the structure is continued until a satisfactory interpretation and solution is obtained. Thus, students are continually involved in problem posing, even though the instructions for writing their report do not explicitly say "problem posing."

The emphasis throughout the course is on the mathematics, and on developing more generalized or generalizable solutions, rather than focusing only on specific models or problems. In this way, the aim is to have students feel that they are drawing on all of their mathematical skills and applying these to different problem contexts. In the study described in this chapter, the students were involved in many cycles of problem posing and problem solving within such mathematizations—yet the labels of problem solving and problem posing were irrelevant. What was important was the smooth running of students' mathematical "vehicles," with little awareness that they were changing gears at different times and places as the terrain demanded. Their focus was on the horizon—the bigger picture—where mathematical elegance rather than mechanics held sway.

Each class session was for 2 hours, two sessions being held for each of the 16 weeks in during the semester. Students were presented with at least one new mathematical modelling problem each week, and were expected to work, in pairs, on writing detailed reports about the problem presented. In these reports, students were asked to consider using or adapting the following headings:

- Interpreting the Problem and Adopting an Approach to the Problem
- Making Assumptions
- Identifying the Variables and Priorities Adopted
- Developing the Mathematics Needed for the Problem
- Making Predictions and Developing Possible Generalizations
- Comparing *your* Approach to Generalizing with Approaches Found in the Literature
- Reflecting on the Problem and on the Process of Mathematical Modelling

Although the initial selection of partners was left to the students, purposeful rotation of student pairs took place every 2 weeks. The aim was for every student to have the opportunity to work with every other student in the class in some way—either as a partner, or as part of larger table groups. Students were free to discuss their possible solution strategies and reflections with any other student in the class.

The problems were created by the instructor specifically for the course, bearing in mind the needs, knowledge, and interests of the students in the course. Problems were designed to challenge the students both mathematically and pedagogically—yet were intended to be within reach of their experience and abilities.

The fact that the instructor herself had created the problems *for* the class was made explicit to the students—in a sense, modelling problem posing in a mathematical modelling class. Emphasis was placed on sense making, explaining the mathematics associated with the problem, and developing conceptual understanding. A brief context was always established for each problem, and some imperative or incentive for the students' expertise to be applied to the problem was usually included.

Problem-Posing Project

The course culminated in the development and presentation of a major group problem-posing project which involved the students in posing a mathematical modelling problem of their own choice. Details of the Project are presented in Figure 25.2. Inherent in the design of the Project were the four perspectives

This set of tasks is designed to give you more experience in both solving and presenting mathematical modeling problems in general, as well as in posing such problems.

Your Task
Each student is to create one mathematical modeling task. The task should be of similar difficulty to those we are tackling in class.
Remember that this class is tailored for middle-school teachers—so the tasks should also be suitable (or easily adaptable) for middle-school students.

1. Exchange this problem with another student. If you have already exchanged your problem with someone, and another student asks you to have a look at their problem then that's fine, but ask them to have a look at yours in return! Two opinions can do no harm! The aim is more to check the basic idea of the problem—no solution is needed at this stage—focus onthe concept, level of difficulty, etc.
2. Refine your problem, based on this initial, informal feedback. Prepare a solution. Problems could still be in draft form at this stage.
3. Share your draft problem (and its solution) with those in your table group. Talk about each of the problems in your group. Give each group member time to explain their problem, and give the group members time to work out solutions or approachesto a solution before handing out your solution(s). Give constructive feedback.
4. Refine your problem again, and its solution, and share it with students in *another* group. Receive and take account of this new round of feedback on your problem.
5. Write up an individual report (about 3 pages, but could be longer) on the process of creating, refining, and producing what you should now believe is a good mathematical modeling problem. Include details of the problem as well as a full well-laid-out solution to your own problem.
6. Write a 2-page individual reflection that states your own views about the project exercise. In addition, this reflection should provide commentary on your experiences throughout the semester as you worked through the different problems.

Figure 25.2. The problem-posing project task.

mentioned at the beginning of the Chapter—specifically, curriculum planning, modes of communication, assessment, and cooperative group work.

The introduction of the course to the students was presented in much the same way as other courses at this university are introduced:

- Students received a brief outline of the course (evidence of curriculum planning).
- The instructor explained that students were expected to talk about their work with their peers, that they would be asked to present their work to the class during the semester, and that the instructor would not be "telling" them how to find solutions to any of the problems presented (modes of communication were clarified, with students having major roles in the classroom communication).
- Assessment of problem-posing tasks would be based on a range of factors involved in posing problems—including planning, mathematical content, appropriate cognitive challenge of the problems posed, and elegance (assessment of mathematics is often associated with right-or-wrong answers only).
- Collaborative group work would be the main mode of operation in the class (although *cooperative* group work was a familiar format for these students, the expected modes of communication in *collaborating* with peers placed greater emphasis on student communication).

Results: Project Problems and Students' Reflections

Two examples of students' created project problems have been presented in Figures 25.3 and 25.4. As part of each figure, quotations from the reflections of the student who created the problem have been given. These quotations are listed under five headings: Inspiration for Problem, Challenges in Creating Problem, Group Feedback, Relevance of Creating Problem to Own Learning, and Additional Comments by Student. It should be noted that students were not asked to respond under these headings; students' reflections were made in response to the task given in Figure 25.2. The responses were selected because these were major points being made by all students in their reflections. Although some of the points made are of a more general nature, there is clearly a sense of spontaneity and openness in the responses that would have been difficult to achieve had the task called for more structured reflections.

Project problems created by the other students, together with excerpts from their reflections, are given in Appendix A (for undergraduate students in the class), and Appendix B (for graduate students in the class). Some students' problems were longer than others, and space limitations for the chapter warranted some abbreviation of their problems. Where this has occurred, a note to that effect has been included in the left-hand column. Preservice middle-school teachers have been called P1, P2, etc., and practicing middle-school teachers taking the course as part of their graduate studies have been called T1, T2, etc.

	Student P7
Problem Created by Student for Project	**Competitive Brothers:** Justin and his brother Matthew are very competitive. Matthew just got his license and wants to challenge Justin to a race. Matthew sets up the rules and says that he will be going 40 mph in his car and Justin will be running 4 mph on the same course. Matthew has done his research and found out that if a runner is going 4 mph, it will take him 45 more minutes than a car going 40 mph to finish the course. Based on these rules, how long will it take Justin to complete the course? How long is the course? How can you change this problem so Justin could win the race? Is it possible for him to win? The next day, Justin challenges his friend to a race. Justin is still running at a speed of 4 mph. His friend, however, is running twice the speed as Justin. Justin is starting from point A and heading to point B, and his friend is starting at point B and heading to point A. If this is the same course as the one that Justin and Matthew took, when will Justin and his friend meet?
Inspiration for Problem	"When I was thinking of a problem, I looked online and used my past math materials to give me some rough ideas. I liked the idea of using rate, time, and distance because I always liked that concept as a student and I think that middle-school students should have some exposure to this formula."
Challenges in Creating Problem	"Creating problems takes a lot of thought, logic, and time in order for it to be a successful problem. When making mathematical modeling problems, you want to allow students to dig deeper and think outside the box to solve the problem. This was one of my main goals in addition to exploring rate, time, and distance."
Group Feedback	"My peers gave me some ideas of how to make the problem better and more beneficial for students. In addition, they suggested giving the characters names. I didn't realize this before, but just having names in a problem makes it better and easier to work through."
Relevance of Creating Problem to Own Learning	"The project helped me and further prepared me to teach middle-school math. . . . It was extremely beneficial to be in groups. . . . Once everyone finished their problems, we were able to present them to the class. I liked this because I got to see each problem and I got to present my own. This was beneficial to me because it was kind of practice for presenting it to my future class. In addition, everyone got to see each other's problems and the creativity and thought that went into each problem."
Additional Comments by Student	"It was really nice to work with the graduate students because I often talked to them about their current classrooms and students. They were able to give me good advice and informed me about how their math classrooms are day-to-day."

Figure 25.3. Project problem created by Student P7, and comments/reflections by the student.

General Characteristics of Problems Created by Students and the Process Involved

All of the problems created by the students relied on different scenarios and involved different mathematics. Furthermore, there was no overlap between the problems posed by the instructor and the problems posed by students. Each student wanted to make his or her problem unique, interesting, and relevant to those who would be engaging with the problems. There was an element of competition involved, but it was also a question of competing with oneself—taking on the challenge to produce a problem that was worthy to be shared with other students in the class.

The students were given details of the project several weeks before their first drafts were to be shared with others in their groups. This gave them plenty of time to find a context and develop a problem with which they would be comfortable. In their reflections, students mentioned different resources that they explored to find

	Student T4
Problem Created by Student for Project (several possible hints for problem were given, mostly to help students organize their ideas)	**Gardening:** In early June you plant an herbal garden, including a small cutting of mint, which is notorious for spreading rapidly. You watch it carefully and notice that at the beginning of the second week there is still just one plant, but by the end of week 2 there are now 2 plants. You read an online article about growing mint and find out that the newest plants won't reproduce the first week, but after that they reproduce every week. According to the article, you can expect to have 3 plants at the end of the third week, and by the end of the fourth, there will be five. You realize that the bed will fill too quickly with mint, and want to introduce another plant. You choose a primrose, for the added color. You know from experience that it also sends shoots, and spreads rapidly. After researching, you find that a primrose will double in size every week. When would you recommend planting the primrose so that the mint will be limited to about ½ of the garden bed? (You guess that your garden could hold a little more than 100 plants.) Your friend, who owns a garden shop, notices your healthy garden. When you tell her about the fast-producing plants, she has an idea to cultivate and sell both the mint and primrose plants. She asks you to tell her how long it will take her to grow 500 plants of each type. What do you tell her? Can you give her a general formula to figure out how many plants she can expect at a given future time? Can you give her any recommendations?
Inspiration for Problem	"I realize that not everyone enjoys gardening, as I do, but I do think that most people have had the experience of plants 'taking over' a garden. I think that the mathematical modelling structure is a perfect fit for 'discovering' a sequence. I would introduce this after the students have spent some time studying some simple sequencing problems, and deriving some simple formulas for given sequences."
Challenges in Creating Problem	"Creating a problem that encourages students to think creatively, use reasoning to solve a problem that they probably have not seen before, as well as be interesting, can be daunting. The problem can't be too difficult, which causes students to give up; but must be challenging enough not to be trivial. . . . Choosing a problem was a challenge because it needed to be a problem that fitted well with the criteria of mathematical modelling: an open-ended, engaging, real-life problem that small groups of students could work on together."
Group Feedback	"Feedback from my classmates, who attempted to solve the problem, helped me refine the writing so that the problem was more understandable. I was surprised when their interpretation was other than what I had intended."
Relevance of Creating Problem to Own Learning	"In teaching mathematics, it is essential to have students engage in experiences in which reasoning and sense making are utilized to understand mathematics in new and meaningful ways. The process of writing a mathematical modeling problem has been a good learning experience."
Additional Comments by Student	"Creating a mathematical modeling problem seemed initially to be a straightforward task. However, it involved a process of having an idea for a problem that is not only creative, but involves thinking of a situation that would encourage students to use critical thinking, applying math to a real-world situation, and finding a workable solution (or solutions)."

Figure 25.4. Project problem created by Student T4, and comments/reflections by the student.

ideas. A few of the problems students posed drew on their experiences in previous mathematics classes. However, all of the posed problems were "owned" by the students in the sense that they were able to pose new problems along different lines while at the same time looking back for ideas and mathematical understanding gained in previous classes.

The four practicing teachers in the class tended to have their particular middle-school students in mind when they prepared their problems—for that was the context in which they found themselves on a day-to-day basis. Their problems tended to focus on concepts that they particularly wanted to give their students experience in applying to real-life situations. Undergraduate students tended to choose contexts to which they themselves (or their peers) would relate.

In their reflections, students acknowledged that creating problems took thought, logic, and time. Several students noted the challenge of finding just the right wording for their problems, and recognized the importance of listening to the feedback from others in the class. Students learned both to give and to take constructive feedback. This proved to be very important for improving the problems they had created. Several commented that the process was more complex than they had expected.

These results will be discussed in relation to the research questions that guided the study.

Reflections of the Teacher as Researcher

As the teacher of this class, my focus was on finding appropriate ways to facilitate students' learning. As a researcher, I wanted to learn more about the pedagogy I was adopting, and why. How did my interactions with students change my approach to a given session or assignment? How did I respond to students' questions? What questions did I ask of students? Was I presenting suitable problem-posing models for the students—both preservice and inservice teachers? Could I achieve my aim of setting challenges for each student, at the same time creating an environment that engendered cooperation and healthy, rather than destructive, competition?

It was also important that I modelled a flexible teaching approach—when extra time was needed, for example, we continued working on the same problem for longer; when additional mathematical skills and knowledge were needed, students felt free to ask for advice, or seek help from other students, or check resources on the internet or in the library. We celebrated student success and welcomed constructive comments on posed problems and solutions. Students began to recognize that problem posing is a basic tenet to student learning. They also realized that posing what others would describe as "good" mathematics problems did not necessarily come without a struggle or without effort. They appreciated being in the shoes of a problem poser rather than being only on the receiving end of a problem that needed to be solved.

As their teacher, I empathized with the students' struggles, but encouraged perseverance rather than accepting second best. But above all, I listened. I heard students' comments and acted upon them. I watched students working together but sometimes hesitating to work together. So I encouraged them to consult with any of their classmates. The table rotations were planned to facilitate students getting to know different people and different approaches.

Although I had worked out a set of tasks and approaches for the semester, I did not have detailed lesson plans for each session. As a teacher, I had an outline of my plan, but as a reflective researcher, I knew that my plans were likely to change during each session. And I wanted students to witness how this evolved during the session. Some students commented on my approach as they felt it gave them confidence to try similar strategies in their own classrooms.

As a teacher-researcher, I believe that PPP can be described as a holistic approach to the teaching and learning of mathematics. In fact, PPP embodies all aspects of the mathematical knowledge for teaching described by Ball et al. (2008)—from both pedagogical and content perspectives. PPP incorporates the aspects of high-quality mathematics instruction summarized by Munter (2014), and is consistent with the development and use of mathematically rich learning environments, as described by Wilhelm (2014).

What distinguishes PPP from other models of teaching are the intertwined connections between problem posing and problem solving, and the symbiotic relationships established between teacher and students. As a holistic approach, PPP embraces rather than shuns constructive complexity, and encourages teachers to find distinctive problem-posing niches that form challenging yet comfortable zones in which to establish powerful learning environments—for themselves as well as for their students.

Answering the Research Questions

Research Question 1: What key features underpin (or work against) the integration of problem posing into mathematics curricula in courses for preservice and practicing teachers?

Three key features have been identified through the data presented in Figures 25.3 and 25.4, Figures 25.A1 through 25.A6, and Figures 25.B1 through 25.B3.

Key feature 1: Establishing goals beyond mere acquisition of problem-posing skills. In their reflections, students did not separate their comments on the two aspects of the course—the project and their work on different problems during the semester. Problem posing and problem solving were not specifically differentiated, even though the instructions for the reflection differentiated between "project exercise" and "experiences" … as they worked through different problems. Integration in this context means just that—moving seamlessly between problem solving and problem posing. The goal to achieve sound conceptual understanding and application of mathematics, with skills in problem posing and problem solving, is a means to an end rather than an end in itself. Thus, the first key feature to underpin this integration is that of establishing goals beyond merely acquiring "problem-posing" skills—consistent with looking towards the right-hand side of the continuum in the *Active Learning Framework* (Figure 25.1).

Key feature 2: Provision of scaffolding as students build confidence with problem posing. Students' confidence can be fragile in any new environment, and students' reflections showed evidence of that fragility. Some needed time to come to terms with the challenge of creating a suitable problem. Others looked to their peers for reassurance and felt disappointed when they seemed to get less feedback than they had hoped for. A second key feature to underpin the integration of problem posing into mathematics curricula is the importance of recognizing how fragile students' confidence can be when they face the instability of trying something they have never tried before—creating a significant mathematics problem. Scaffolding was provided through group work, through the structure of the project process, and through the initial environment that was created within the course itself. Students' reflections are replete with references to such scaffolding.

Key feature 3: A relevant and tangible purpose for problem posing. Having a relevant and tangible purpose to problem posing is a third key feature that underpins the integration of problem posing into mathematics curricula. Undergraduate students were repeatedly reminded by their practicing-teacher peers of how they (the practicing teachers) were excited to prepare problems they could use in their own classrooms. The project was seen to have immediate relevance. So also was the additional work assigned to the practicing teachers in which mathematical problems provided by the instructor for class work were modified by the practicing teachers for use in their classrooms. In particular, both the undergraduate students and their practicing-teacher peers commented positively on this task in their reflections.

Clearly, the lack of any of these three key features would work against the successful integration of problem posing into mathematics curricula.

Research Question 2: In what ways does a collaborative approach help or hinder the integration of problem posing in such courses?

All students commented that the group work had been helpful as they developed and refined their project problems. Initially, some students were either unsure of how groups might function, and others noted that they needed to learn how to give and receive feedback. Many students expressed enthusiasm about the operation of the collaborative groups in this class. Student P6 expressed it this way (see Figure 25.A6): "To do this project you needed to be able to take criticism. It was necessary to understand that when people were critiquing the problem and making suggestions they were only trying to help you and not be mean."

There is a delicate balance between genuine collaboration and groups that hinder rather than support the work of individuals. The instructor needs to be aware of the dynamics of all of the groups. For this class, the instructor intentionally rotated the members of the class to different groups every 2 weeks. This was done openly, and in a planned and coordinated way—the intent being to have every student work (in the context of a group) with every other student in the class. Had this been done in a random way, the potential for disruption could have been a negative influence on the productivity and attitudes of the class.

As already noted, one student (P2, see Figure 25.A2) stated: "However, when I swapped my problem with my peers, I felt as though I did not get as much feedback as I would have liked. This then made it difficult to refine my problem." In shifting the focus from the instructor to the individuals in the class, both the instructor and the students need to be aware of the sensitivities of all individuals. In a supportive and positive classroom environment, this shift in focus can be reflected in the activities in which students are involved. The positioning of the project at the end of the course was intentional, with students being given control of how and what they created.

Research Question 3: What characterizes Pedagogy for Problem Posing (PPP)?

The term pedagogy is associated with the science and art of teaching. The term is also associated with preparatory instruction and training. It was Paulo Freire who coined the term "problem-posing education" (Freire, 1970). I am now introducing the term "pedagogy for problem posing" because I believe that unless there is serious dialogue between stakeholders about how problem posing can become integrated in routine mathematics classroom activities, problem posing will always remain relegated to the fringes. It is in danger of being given only lip service—receiving only a mention in curriculum documents—never having a firm base on which to build, or for which teachers are given support.

Pedagogy for problem posing can be defined as "teaching that integrates problem posing." Informal problem posing in the classroom may not be uncommon, but it is the teacher who transforms a classroom into a constructive and productive learning environment for students. The book *The Art of Problem Posing* (Brown & Walter, 1983) moved the field to look seriously at ways to introduce problem posing in mathematics classrooms. But the art itself has taken many years to mature to the stage that one can look seriously at the underlying pedagogy.

The following characteristics of PPP have been derived from the data presented in this study, including the results summarized under the first two research question headings:

- Students' first experiences with posing mathematical problems need to be scaffolded by the teacher so that the experiences build on what students already know but are nevertheless challenging and interesting for the students.
- Through the classroom activities introduced by the teacher, students need to feel free to share drafts of their problems with their peers, and with the teacher, recognizing that their drafts are likely to be imperfect.
- Teachers need to establish an atmosphere in which students learn both to take and to give critical feedback on posed problems.
- Teachers should encourage students to take on increasing responsibility for creating mathematics problems.

- The act of problem creation needs to be such that the students' knowledge of mathematics develops significantly as a result of their reflection on relationships between mathematical concepts and skills.
- The tasks set for problem posing should be within reach of all of the students.
- Time spent on posing mathematics problems should not be distinguished from time spent on mathematics. Rather, it should be seen by all stakeholders as time well spent on learning mathematics, and should not be seen as an imposition or an extra that somehow needs to be included in an already-busy curriculum.
- Teachers should model problem posing for their students by making it explicit that he/she is actively involved in posing and refining problems.

Although most of these characteristics are also appropriate for mathematics teaching and learning in general, nevertheless the expectation of problem creation adds an edge to the exercise. This observation lends credence to the characteristics identified for PPP.

Conclusions and Implications for Further Research

Establishing, refining, and implementing a pedagogy for problem posing represents a major step forward for the field. Just as art needs to be seen, experienced, and appreciated, so too does pedagogy—especially when there are elements within the pedagogy that are less familiar to the practitioner. The posed problems together with their creators' comments have been included in the Appendices to provide tangible examples of the art of classroom problem posing, as lived by the students who participated.

PPP needs to be modelled for teachers, and for prospective teachers, so that they, too, can model problem posing for their students.

Other stakeholders—including parents, principals, mathematics consultants and supervisors, textbook authors, and curriculum planners—also need to be introduced to the key ideas that underlie pedagogy for problem posing. Constructive dialogue between stakeholders, researchers, and educators needs to continue.

Further research on the eight aspects identified as characteristics of PPP needs to be undertaken. Teachers and prospective teachers who have themselves seen, experienced, and appreciated a pedagogy for problem posing should be followed into their own classrooms so that their implementation of this pedagogy can be monitored, supported, and reported. Lesson Study may prove to be an ideal way of sharing pedagogical ideas about problem posing with colleagues.

Above all, when problem posing is an integral part of mathematics teaching and learning, and when PPP is embraced as a holistic approach to the teaching and learning of mathematics, problem posing becomes both the culmination of one's learning of a mathematical concept, and a natural means for applying one's newly acquired mathematical knowledge and skills. The pedagogy of problem posing represents an untapped opportunity to transform routine tasks into exciting and refreshing discoveries for students and teachers alike.

Acknowledgment The author gratefully acknowledges the willing participation of all students in this study.

Appendix A

Project Problems Created by Undergraduate Students

Six project problems created by undergraduate students have been presented in Figures 25.A1 through 25.A6. As part of each figure, quotations from the reflections of the student who created the problem have been given. These quotations are listed under five headings: Inspiration for Problem, Challenges in Creating Problem, Group Feedback, Relevance of Creating Problem to Own Learning, and Additional Comments by Student.

	Student P1
Problem Created by Student for Project	**Joan's Bakery:** Joan has opened her first bakery and is looking at local sugar companies to see which company she should order her sugar from. Joan needs 75 lb of sugar every other week. There are four different sugar companies in town. Here are what each of the businesses are offering Joan: 1. Sugar U Up: They have 15 lb bags of sugar that are $7.50 each. They charge $7 per block that they have to travel to get to a bakery. 2. Sweet Tooth: They have 5 lb bags that are $2 each. They have a flat rate of $60 for delivery. 3. Sugar Cane: They have 15 lb of sugar that are $13 each and have a minimum order requirement of 7 bags. They do not charge for delivery. 4. Sugar 'N' Spice: They have 3 lb bags of sugar that are $3 each. They charge $1 per block they travel to get to a bakery. (a) If Joan wants 75 lb, what company should Joan buy sugar from? Assume that Joan is only looking at the first order. (b) How much money will Joan spend on sugar for the first year of her bakery being open? (Look at each company) Assume that Joan opened her bakery on the first day of the year. (c) Write a proposal for Joan on what you think would be the best company to buy her sugar from. (Consider: If she buys from Sugar Cane will she have to purchase every week or will she have extra bags of sugar?) (d) Would your response be different if Sugar Cane offered $0.25 off the total price (the sugar and the delivery combined) for every order she places after her first order? (example: First order = $91, second order = $90.75)
Inspiration for Problem	"Joan's Bakery was based off a problem I had found while researching middle-level math topics. I wanted to create a problem that would involve simple addition together with a visual aid, but more importantly, I wanted to create a problem where students had to think outside the box with their mathematics. My mother's name is Joan, and the two of us have always talked about opening a little bakery. Bakeries need many different ingredients and resources, so I chose to have sugar as the product to buy."

Figure 25.A1. Project problem created by Student P1, and comments/reflections by the student.

Challenges in Creating Problem	"Once I started the problem I had a hard time wording different things. I wanted to create four different companies that all had a different plan for selling their sugar to Joan. I also wanted to use whole numbers that were easily divisible by 75 so that the problem would start off very concretely. Each company was then assigned different plans and prices."
Group Feedback	"During the course of this project, I exchanged my problem with my classmates, my instructor, and a teacher at the school where I volunteer. This was important to me as I wanted to see as many ways to solve this problem as I possibly could. Each person who went through this project figured it out differently." "The input that my peers had throughout the process of this project helped so much, so finalizing the project by sharing the final problem was so rewarding."
Relevance of Creating Problem to Own Learning	"I really enjoyed presenting my problem to my classmates. . . . It was great to hear all of my other classmates' final projects as well as the graduate students' modifications [to the modeling problems]. I am excited to look over all of the problems so I can personally modify the problems so that they work in my classroom curriculum."
Additional Comments by Student	"It was so important to work with the people at your table, and to share with the class because you always learned other ways the problems were interpreted and solved. This was especially useful when we were sharing our project problem. We shared our problems with a minimum of six people—not only did I share my problem with my table, but we then rotated our problems to other tables. . . . I was confused on what my final answer was, but I received great feedback from my peers and I learned how to give feedback to others."

Figure 25.A1. (continued)

	Student P2
Problem Created by Student for Project (Map not included for the sake of brevity here)	**Family Reunion:** Your extended family is having a family reunion this summer! Your family has decided to make the journey from Elkhart to the reunion. It is being held where your grandparents live, in Hammond, IN. Your parents have left the travel plans up to you to figure out. Your father is not good with directions so he would like for you to make them easy to understand and follow. Your mother is very strict about spending money so she would like to spend the least possible amount of money on the trip. However, your parents also decided that since the family will already be making the trip, you should stop to see your other cousins in Grovertown before heading off to the reunion. So you must be sure to include this detour into your travel plans,. Then your Aunt Sue calls and asks you to pick her up and take her to the reunion as well. Your Aunt Sue is very persistent and your mother can't say no, so you agree to pick her up in Chesterton on your way. Your first goal is to get to the family reunion in the shortest amount of time with the easiest of directions. Use the included road map and its given distances to find what you feel would be the best possible route for your family to travel. Calculate the total distance you will cover on the trip. Also, use the speed limits posted for each road to find the length of time the entire trip will take. You will need to stop once for gas on the trip, about a 5-minute stop, so this must be included in your plans. The family van gets about 20 miles to every gallon of gas. The tank holds 16 gallons of gas and has only 3 gallons when you begin the trip. You will need to choose a station on the way to fill up. Your second goal is to examine the cost of the trip. Remember, your mom would like to spend the least amount of money possible so make sure that the route chosen achieves this goal.

Figure 25.A2. Project problem created by Student P2, and comments/reflections by the student.

Inspiration for Problem	"At first I had no idea how to even begin creating a good mathematical modeling problem. I started by looking at all of the problems we had done in class to get an idea of what a good problem looks like, I then went back to my high school math classes and looked at those word problems to get ideas as to what types of questions I could create. Once I felt like I had a good idea as to how to write a good mathematical modeling problem I then started thinking about what my actual problem could be. I came up with a road trip with the family because it is something that my family enjoys doing. My dad is very strict about how much time we spend in the car and on the road, and my mom is very strict about how much money we spend, so it was fun to put them in the problem."
Challenges in Creating Problem	"The process to create a good mathematical modeling problem was a lot more difficult and complex than I had originally thought it was going to be. . . . The problem has to be complex in the sense that the students should have to think on a higher cognitive level in order to complete it correctly. Students should have to use their prior math knowledge and expand on it. . . . A good mathematical modeling problem requires students to explore the problem and explain their reasoning for why and how they solved the problem. . . . Also, I believe that a good mathematical modeling problem makes the students interested and want to figure out the answer."
Group Feedback	"The process to refine the problem to make sure it was a solid mathematical problem was very in-depth yet very useful and beneficial. It is very important always to go back and re-check work, and to have an outside opinion makes it even better. By having another person look at your work allows for different opinions and different viewpoints . . . However this process can only be beneficial if the corrections given are applied. When writing a paper it takes many drafts, many revisions, and many different opinions before it is complete. The same goes for writing math problems."
Relevance of Creating Problem to Own Learning	"We had multiple opportunities in class to swap our problem with our peers, have them look at it, solve it, and give us feedback on what they liked and thought we should change. However, when I swapped my problem with my peers, I felt as though I did not get as much feedback as I would have liked. This then made it difficult to refine my problem. . . . I took their suggestions of making the problem more difficult and added more steps in it. By adding more restrictions and guidelines to the problem, it made it more challenging for the students."
Additional Comments by Student	"This project was definitely a lot of work but looking back on it, the process was a good experience. When I am a teacher I am going to have to create problems for my students to complete. I want these problems that I give my students to challenge them and really make them think because it will only benefit them in the end. By going through this project and the process that came along with it, I feel as though I am much more prepared to create good mathematical modeling problems." "With this problem there is no set correct answer—students can have different opinions as to which route they should take. This freedom allows students to use their knowledge, be creative, come up with an answer, and then defend and explain their answer in their written report. Students enjoy being given freedom and the opportunity to be creative; therefore this problem will engage them and make them want to work it out."

Figure 25.A2. (continued)

	Student P3
Problem Created by Student for Project (possible hint for problem was a sketch for placing the first noodle)	**Fear of Snow:** Luna, Terry's dog, absolutely loves to be outside at all hours of the day. Strangely enough Luna is terrified of snow and refuses to step "paw" onto the snow. Last weekend there was a big snowstorm. With such strong winds throughout the snow fall, it created a 4-yeard wide snow bank around the entire house. Note that this family lives in a circular house. Terry looked in her utility closet and noticed that they had two old pool noodles/toys that were 3.5 yards long each, which Luna could walk across to avoid the snow. Unfortunately because of the slippery material, there is no way to fasten the two noodles together using tape or anything else. Terry can be the one to place the noodles down for Luna to walk on. One noodle is lying parallel to the back door, on top of the snow bank. Is it possible that Terry could place the second noodle somewhere else so that Luna could get over the snow bank and to the house successfully? You can assume that from the center of the house to the back door is 7 yards.
Inspiration for Problem	"I reviewed all of the problems I had done in my previous class, and chose one that I found most interesting and engaging. I also chose this specific problem because the mathematics a student would use is the Pythagorean Theorem." "... I chose my dog because in families today dogs are a huge part of the family."
Challenges in Creating Problem	"When I first heard that we would be constructing a problem for our project I began to worry. I did not feel confident that I would be able to come up with a solid problem and solution." "After I knew what exactly I wanted to do I had to start drawing out the plan as well as substituting in the values for each measurement. That part of the process was difficult because I was trying to be as realistic as possible but still have the answer be what I wanted it to be."
Group Feedback	"One girl [in the group] had trouble understanding the problem. She made a few comment on my wording and told me to clarify things more .. . . Her comments and assistance helped me make my problem clearer. Some students struggled with how to draw this picture and how to interpret it as well. . . . After all of these comments, I produced an edited problem and gave it to a student who had never seen it. She said that she really enjoyed the problem and that it was creative."
Relevance of Creating Problem to Own Learning	"Out of all the math classes I have taken thus far in my collegiate career, I believe that this is my favorite one. I did not learn too many different mathematical skills, but developed a way to explain my mathematics. That was one thing I always struggled with: I understood the problem but was not sure how to explain what I was doing. This class has prepared me for many more and I am glad I took it!"
Additional Comments by Student	"Not only did I enjoy presenting my own problem, but I enjoyed listening to everyone else present theirs. We all came up with different ideas which were interesting to hear. My favorite part of the presentations was when the graduate students presented their modifications. This made me realize I really could use these problems in a middle-school classroom. Minor modifications may need to be made depending on the age of the students."

Figure 25.A3. Project problem created by Student P3, and comments/reflections by the student.

<table>
<tr><th></th><th>Student P4</th></tr>
<tr><td>Problem Created by Student for Project (Illustrations for each type of display were provided with the original problem. A rectangular sketch of the empty wall was also provided.)</td><td>**Designing a Shop Display:** Designing a wall for clothes in the retail world is a task that is often overlooked. There is not one set way to design a wall, but there are ways to make the wall more visually appealing to a customer and more accessible to a customer. There are also ways to make a wall easily accessible for employees. There are numerous ways to set a wall, and all ways are justifiable. For example, a shirt can be folded into a pile and placed on a shelf, hung on a face-out bar, or hung on an across bar.
Shelves. If you decide to fold shirts, you will need a shelf which is 4 ft x 2 ft. Shelves can hold a maximum of 20 sweatshirts, with 4 piles consisting of 5 sweatshirts each. Shelves should be a minimum of 1 ft apart to leave room for folded piles.
Hanging on the Face-out Bar. If you decide to hang up your sweatshirts, you can hang them with a bar that can hold 25 sweatshirts. This bar is perpendicular to the wall. Note that hung shirts take up a space of 2 ft x 2 ft. ***Across Bar.***You also can hang the shirts on an across bar, which hold more shirts than a face-out bar. The across bar holds up to 50 shirts. It takes up a space of 4 ft x 2 ft, leaving less shopping room for customers.
Your Task. It is 24 hours before Black Friday, the busiest day in the retail world in the United States. Your store manager, Sandy, just asked you to design a wall for the sweatshirts that will be on sale tomorrow. Your task is to put as many sweatshirts out as possible to ensure that they will sell. The more sweatshirts you place on display, the less your employees will need to restock! Sweatshirts can be folded on a shelf or hung on bars that are listed above. You have unlimited access to hangers, shelves, and any bars to help you design the wall which is 20 ft long and 10 ft high.

1. How would you design the wall for Black Friday? How many sweatshirts will your design hold? What could you do to display additional sweatshirts?
2. Who is your display made for, customers or employees? Justify your answer.
3. After showing your design to Sandy, she realized that your design needs 4 signs that mark the price of the sweatshirts. These signs help the customers know the price of the item. The sign can be placed on a shelf, or at the end of a face-out bar, but the signs limit the amount of stock on that shelf or bar by ¼. Where would you place the signs? How many sweatshirts can your wall hold after placing signs on the wall?

Limitations.
1. You cannot have more than 22 shelves in the wall.
2. Shelves need to be at least one foot off the ground, and no higher than 8 ft so customers can easily access the merchandise.
3. Hanging shirts need to be at least 2 feet off the floor so they will hang properly.
4. Regardless of where hanging shirts are located, if you plan to place a shelf below hanging shirts, you need to allow a minimum of 1 foot of space below the bottom of the shirts. This will allow you to place your pile of shirts so that the hanging shirts and folded shirts do not overlap.
The aim of this problem is not to find a right or wrong answer. This is a complex task that has a wide range of solutions that puts students in a place where they need to justify their answer and think through their mathematics. There is no single "correct" answer, but a variety of possibilities that could work for the design. Think through your mathematics, and experiment with different possibilities.</td></tr>
<tr><td>Inspiration for Problem</td><td>"I wanted to create a problem that would be a real-life situation that students would connect with. More importantly, I did not want my problem to have one 'correct' answer because students are always told that math has a right or wrong answer.</td></tr>
</table>

Figure 25.A4. Project problem created by Student P4, and comments/reflections by the student.

Challenges in Creating Problem	"I created my problem by thinking of a real-world challenge that I have faced in my summer career in a retail store. Designing the wall and considering various fixtures for a clothing store is a task that is overlooked. . . . Approaching this problem may be difficult due to different sales, limitations or restrictions that retail managers put in place. More importantly, there are many different answers to the task, but *justifying* your approach is what leads to a correct answer. And one must be able to justify one's ideas in order to succeed in the retail world. . . . Refining the problem was a challenge for me because the problem is open ended. . . . Another challenge . . . was defining words and terms used.
Group Feedback	"My peers' comments led me to believe that many were confused by my problem. I grew overwhelmed at their questions . . . I realized that my future students will also have questions, and it is important to clarify any questions and misunderstandings they have. For this open-ended problem, there will be questions because there are numerous ways to solve the problem."
Relevance of Problem to Own Learning	"Through the semester, I have grown as a student, but more importantly as an educator. I have developed skills and ideas that will allow me to help my students in the future. The course has helped me to develop various math problems, questions and tasks that relate to real-life experiences and problems."
Additional Comments by Student	"I enjoyed writing and editing the math project for this class. I felt that the possibilities for writing and making mathematical tasks for students is truly endless, and finding various modeling problems in everyday life was a fun challenge. There are numerous adaptations and limitations that one can put on mathematics problems, but I feel as though that is reality."

Figure 25.A4. (continued)

	Student P5
Problem Created by Student for Project	**Track-and-Field:** Aiden is the Athletics Director at Red Hill College. He has just found out that Red Hill College will be playing host to the Central Valley Conference championships for Track-and-Field; the only problem is that all of the stagger lines on the track have faded away. Aiden is also involved with the Math Department, and hears from Dr Ross that you are quite the talented mathematician. He asks for your recommendation on where to place the staggers for the start of the 400-meter dash. You know that the track's and the track is an oval with each turn and straight away measuring 100 meters each. Assume that the width of the lane markers and stagger lines make no difference in the distance. Also, Aiden wants the starting stagger for the first lane and the finish line to be in the same spot. What would your recommendation be to Aiden in order to paint the stagger start lines so that no runner is given an advantage? If he also needed to paint the lines at the 200 meter mark, where would you place those? What would change if Aden decided he wanted six lanes? What would be your recommendation if Red Hill University employed a circular track? Can you create a generalized solution so that Aiden can do the mathematics in the future?
Inspiration for Problem	"The first step taken was the creation of the problem. I had no idea where to go, so I went to the instructor and asked for advice. The instructor asked me what I had done in high school, in terms of extracurricular activities—the reasoning being to get me to think about mathematics in areas that I may not have initially been thinking about. I told the instructor that I had been involved in track-and-field—one of the ideas from this was the use of staggered start lines for the 400m and 200m dashes. This immediately grasped my attention as it sounded as if it would be using both algebraic and geometric thought in order to solve it."

Figure 25.A5. Project problem created by Student P5, and comments/reflections by the student.

Challenges in Creating Problem	"I knew that the problem had to be something that I found interesting, yet could easily translate to the mathematics that was being done in the course. The problem I created was one that could easily be changed to fit the needs of middle-level classrooms, but at the same time would present enough of a challenge for my classmates."
Group Feedback	"Following the creation of my rough draft, I took my work in to my tablemates for review. The initial response was that it was a well-written problem that allowed for expansion, along with open-ended thought processes. Both of these were keys in the process of creating the problem. My tablemates found a few grammatical and syntax errors . . . I made some quick changes to create a better-rounded problem." "The task of creating a solution that I believed could be replicated was both challenging and rewarding. . . . The day came for formal review with the class [problems were exchanged with other groups]. P2's first attempt to solve the question posed was much closer than what I had expected anyone to get. P2's idea was to create some basic hints that alluded to, but did not give away, the answer. She also helped to make the problem more accessible, stating that some parts were poorly worded or were not necessary."
Relevance of Creating Problem to Own Learning	"The assignments [that I completed throughout this semester-long course] truly helped me grow not only as a student, but also as a future educator. Initially, working with mathematical modeling felt clunky and unwieldly, Modeling felt as if I was working so hard on something, but never coming up with that one true answer. Throughout my collegiate and high school career, I had been told to do this, this and this. . . . The idea that I had freedom in a subject that for so long held rigidly in my academics made the transition difficult." "My views on mathematics have changed drastically."
Additional Comments by Student	"As I have worked more in this powerful tool [mathematical modeling], I have come to realize just how much more a student would learn from using this process. This method of learning creates a sense of ownership: an ownership based in the fact that a group of individuals devised their own solution to problems. They were not told step by step what to do, rather they made their path." "The creation of a problem differs from only solving one greatly. In creation, one must think of the potential audience for the problem. The eventual solver must be able to solve the problem, yet it should pose enough of a difficulty that they do not know the answer right away. Also, the problem needs to be worded in a manner that is simple enough to be understood easily, but still passes along the necessary information to the reader. As I worked on my own problem, I strove to create something that would fulfil the requirement set not only by the instructor but also by myself. This proved difficult, as I had never truly created a problem. My rough draft was just that—rough."

Figure 25.A5. (continued)

	Student P6
Problem Created by Student for Project	**Comparing Insurance Policies:** An insurance company sells insurance policies tp people to protect some of their belongings or to help with some of the financial burdens that might be due to a medical reason or a car accident. People who buy a policy from an insurance company are known as policyholders. People who buy an insurance policy will have a contract with the insurance company, and they will have to pay the insurance company for a period of time. If the policyholder does not experience an accident that is covered, they will not get their money back for that policy, and it will be a profit for the insurance company. In this problem, we are going to see how much profit an insurance company can make for each policyholder that agrees to buy a policy from that insurance company. **Part 1:** The insurance company, A, has a policy available for customers that is \$100 per year. If customers get this policy, the insurance company will pay their policyholders \$10,000 if they suffer a major injury, A major injury is classified as having to be hospitalized. The insurance company will also pay their policyholders \$3,000 if they suffer a minor injury. A minor injury is classified as having to take off work. The company estimates that each year 1 in every 2,000 policyholders will have a major injury, and 4 in every 2,000 policyholders will have a minor injury. Therefore, the company estimates that there will be 1,995 people out of every 2,000 who will not have any kind of injury. What is the expected profit for the insurance company per policyholder? **Part 2:** There is another insurance company, B, in town that provides a similar plan to State Farm's injury policy. This insurance company has a policy available for customers for \$150 per year. If customers get this policy, the insurance company will pay their policyholders \$20,000 if they suffer a major injury. A major injury is classified as having to be hospitalized. The insurance company will also pay their policyholders \$8000 if they suffer a minor injury. . A minor injury is classified as having to take off work. The company estimates that each year 1 in every 1,000 policyholders will have a major injury, and 10 in every 1,000 policyholders will have a minor injury. Therefore, the company estimates that there will be 989 people out of every 1,000 who will not have any kind of injury. What is the expected profit for the insurance company per policyholder? **Part 3:** Write a report which sets out your answers to the following questions: 1. Which of the two insurance companies, A or B, has a better insurance policy? Use your answers to Part 1 and Part 2 to find (a) Which policy is more beneficial for the customer? and (b) Which policy is more beneficial for the insurance companies? 2. Which of the two insurance policies would you want to have in case you had an injury? 3. What would you do if you opened an insurance company? Would you want your policies to be fair for the insurance company or for your customers?
Inspiration for Problem	"I wanted my problem to deal with money because I think that it is important for middle-school students to understand the value of money. I want them to think about what some things in life cost and if they should spend their money on these things. I chose to work my problem around an insurance company because insurance is a service that people should have."
Challenges in Creating Problem	"I looked back at my old papers from my previous math class to try and find some inspiration for some problems and to remember some of the mathematics that I learned from class that I might have forgotten."

Figure 25.A6. Project problem created by Student P6, and comments/reflections by the student.

Group Feedback	"My group had some great suggestions to improve my problem. . . . My group was confused on what I was expecting a student to do to solve this problem. They were confused on what a policyholder was and how you were supposed to find the expected profit for the insurance company and for each policyholder. . . . They suggested that I change the wording of my question at the end, and they thought it would be good to add another insurance company. . . . After I made these changes, I gave the problem to another group to solve. They suggested that I define what a policyholder is, and suggested an additional question in Part C."
Relevance of Creating Problem to Own Learning	"This project at the beginning did not seem to be a big deal. I thought it would be easy to create a mathematical problem that was appropriate for middle-school students. I was proved wrong. Creating a math problem is very difficult. It is all about how you word the problem. . . . you do not want the student to be confused on what they are supposed to solve for. . . . it is important to have the problem relate to the real world. It has to be creative. . . . The most beneficial part of the project was to allow our classmates to help us refine our problem. It was great seeing their opinions on the problem and how they thought the problem could be improved."
Additional Comments by Student	"To do this project you needed to be able to take criticism. It was necessary to understand that when people were critiquing the problem and making suggestions they were only trying to help you and not be mean. By the end of this project, I got to see other great mathematical modeling problems that my classmates created, and I believe I created a good problem. I got to experience that problem making is harder than it looks."

Figure 25.A6. (continued)

Appendix B

Project Problems Created by Practicing Teachers

Three project problems created by practicing teachers have been presented in Figures 25.B1 through 25.B3. As part of each figure, quotations from the reflections of the student who created the problem have been given. These quotations are listed under five headings: Inspiration for Problem, Challenges in Creating Problem, Group Feedback, Relevance of Creating Problem to Own Learning, and Additional Comments by Student.

<table>
<tr><th></th><th>Student T1</th></tr>
<tr><td>Problem Created by Student for Project</td><td>
T-Shirt Design: The student council at Mytown Grade School has been assigned the task of presenting a proposal for buying T-shirts for the fund-raiser for the school. There are criteria that must be considered:
1. The school board will only allow an account with one vendor.
2. The design for each type of shirt is different and all types of shirts will continue to be sold.
3. Each design is considered a separate order.
4. Data collected from last year's shirt sales need to be considered for the purchase of the new inventory.
<table>
<tr><td>Type of Shirt</td><td>Number of shirts sold</td></tr>
<tr><td>Spirit shirt</td><td>300</td></tr>
<tr><td>Redlife pride shirt</td><td>120</td></tr>
<tr><td>Sport shirt</td><td>100</td></tr>
<tr><td>Student council shirt</td><td>50</td></tr>
</table>
Student council members have the following information from manufacturers around the immediate area:
Designwear, a local business, requires a minimum purchase of 12 shirts. The design charge for the logo is $20 and the charge before the set up of screen printing is $25. The charge per shirt up to 100 is $10-.50. For a purchase of over 100 shirts, cost per shirt is $7,50 and the design charge is waived. With a purchase of over 200 shirts, cost per shirt is $6.00. Children from the family which runs Designwear go to Mytown Grade School.
Boyds Sporting Goods in a nearby town charges $6.95 for a colored shirt and $7.45 for a white shirt. All design fees and screen-printing set up are included in the prices. With an order over 300 T-shirts, the cost is reduced by $0.35 per shirt. They require a minimum order of 12 T-shirts.
T-works is another company in a nearby town. They charge $7.50 per shirt with all costs included. Please recommend to the school board which company you think is best. Explain why it is superior by comparing to other companies. What selling price would you recommend for each shirt, in order to make a profit of at least $1000?
</td></tr>
</table>

Figure 25.B1. Project problem created by Student T1, and comments/reflections by the student.

Inspiration for Problem	"I wanted my sixth-grade math students to be presented with options to make a business decision and report their findings to the school administration. The businesses I chose for the problem are actual businesses in a ten-mile radius of Mytown. I contacted each of the businesses to gather the information needed for the modelling problem. . . . I wanted this problem to be somewhat open-ended and have the students support their decisions with reasons for their particular choice of a vendor."
Challenges in Creating Problem	"As you can see, some of the criteria involved in the design and production of shirts is very detailed. . . . Some of the business I contacted to get realistic details for pricing, etc., were vague about exact costs. One kept saying it depended on the design of the shirt. Finally, he gave me an average cost of a typical T-shirt with a design."
Group Feedback	"With the help of my classmates, I refined my modelling problem to a version that will work well with my sixth-grade students. First of all, I needed to make my concluding questions clearer and more concise. Second, I needed to make clear the range . . . next, I restated in the details . . . Finally, after a fellow classmate worked on a solution for my problem, she was not sure if this amount of T-shirts was totalled for a complete order or if these shirts with a different design were considered individual orders. The next day, I called the three businesses and had them clarify what they meant by a total order. They made it clear that each different design for a T-shirt was a separate order. Therefore, I added that fact to my criteria for correct cost estimates."
Relevance of Creating Problem to Own Learning	"I loved working in groups for this course. The brainstorming was amazing. With some of the problems, I did not know where to begin. Instead of getting frustrated and feeling defeated, my group members explained what I did not understand. We all brought a certain relevance to the problem. Some students were stronger in algebra while other students were stronger in geometry. The opportunity to come up with our own solution that was supported with detail, made it comfortable for groups to think outside-of-the-box." "I appreciated the presentations [by the other students] of different processes to obtain solutions to the mathematical modelling problems they had created."
Additional Comments by Student	"I am excited to use the mathematical modeling problems from the course that I modified for my sixth-graders. Classroom group work and discussion is a major part of learning in math. Students help each other understand the depths of different concepts in math. They bring ideas, concerns, and opinions that are relevant to different solutions. A major goal of the new core math standards is to create students who will be responsive and productive citizens. The type of mathematical modelling problems presented throughout the class make connections to real-life situations for students. They will come to realize that math is utilized constantly to make decisions in their everyday lives."

Figure 25.B1. (continued)

	Student T2
Problem Created by Student for Project	**Kayla and Lexi's Rumor:** Kayla, a sixth grader, and Lexi, a seventh grader, are known for their love of the latest boy band, One Direction. Kayla and Lexi love to pull pranks on their classmates and have decided that they would start the rumor that One Direction will be holding a special concert on May 15. Kayla decided that she will be spreading the rumor to the sixth graders on even-numbered days, while Lexi spreads the rumor to the seventh graders on odd numbered days. Kayla begins by sharing her rumor with three sixth graders on May 2, and each of those three sixth graders tell three more sixth graders on May 4, who each tells three more sixth graders on May 6, and so on until the day of the concert. On the other hand, Lexi begins by sharing her rumor with two seventh graders on May 1 who then each share the rumor with two more seventh graders on May 3, who then each tell the rumor to two more seventh graders on May 5, and so on, until the day of the concert. The theater at the Civic Center seats 2,244 people. Kayla believes that their plan to spread the rumor will tell enough sixth and seventh graders about the rumored concert that they could fill the theater. Lexi believes that there will not be enough people told about the rumored concert by May 15 using their plan. Who do you believe is correct? If Kayla has spread her rumor every five days, and Lexi spreads her rumor every three days, would enough students hear the rumor to fill the theater at the Civic Center? Develop a plan for Kayla and Lexi to spread their rumor about One Direction playing a concert at the Coliseum in the neighboring city, which holds a maximum of 8,000 people. Make a mathematical model to help predict how many sixth and seventh graders hear the rumor that Kayla and Lexi have started, given any number of days the rumor is told.
Inspiration for Problem	"To start the process of creating the problem, I started with a problem similar to an investigation problem that I give my 8th grade algebra class when investigating exponential growth. My goal for this problem was to give students a problem that they can relate to while looking at the mathematics associated with exponential growth and geometric expressions."
Challenges in Creating Problem	"One student felt that parts were too easy, and another felt that it was not open-ended enough for what was intended. I modified the problem and my group seemed to like the changes. After giving the problem to a different group, I realized that I needed to make adjustments with the first part to make the number of students who heard the rumor to be closer to the maximum number of seats at the civic center, so I changed Kayla's rumor to be told on even days and Lexi's rumor to be told on odd days."
Group Feedback	"The feedback from peers was invaluable for me as I am trying to improve on writing my own problems for my students. One of my worries is with semantics since having wording that all students understand will help the student feel more comfortable with open-ended problems. I also wanted to make sure that the problem was solvable and that students would come up with similar solutions while using multiple approaches."
Relevance of Creating Problem to Own Learning	Much of the course, either working on problems or creating problems, has utilized communication with peers. I felt that my mind could be jump started and helped me with thinking differently when solving problems. . . . through the different types of mathematics needed to for the problems, I could honestly see my strengths and weaknesses. With having many people working together, I have regained the appreciation for everyone having strengths and weaknesses and helping each other."
Additional Comments by Student	"Through this class, I feel that I will be more confident with writing problems, utilizing problems, and facilitating students while working with mathematical modeling to make my own classroom more mathematically rich." "Through the reading and aiding other students with revising their own problems they wrote, I think that I will be able to check for the semantics of a problem, the richness of the task, as well as extension questions I can ask."

Figure 25.B2. Project problem created by Student T2, and comments/reflections by the student.

	Student T3
Problem Created by Student for Project (A sketch of a dog house was included with the problem)	**Dog House:** Mackenzie wants a dog more than anything in the world, but first, she needs to build a dog house with her dad. The plan is a rectangular design with an A-shaped peaked roof, and a floor. The dog house has the following dimensions: • base (floor) of the dog house is 59 in. by 29.5 in. • The height from the floor panel to the peak of the roof is 47 in. • The end pieces from the floor to the bottom of the roof are 29.5 in. by 29.5 in. • The slanted edge of the roof is 23.5 in. in length **Part 1** 1. Add the given dimensions to the diagram. Sketch the nine pieces of wood (roof, sides, floor, ends, roof ends) required to construct the dog house, and find the area of each piece. 2. What is the total area (in square inches) of all nine pieces of wood? **Part 2** Mackenzie and her dad decide that they have enough nails and tools for the dog house, and will only need to buy the plywood. When Mackenzie and her Dad go to the home improvement store to buy the wood for the dog house, they find out that they have two options. They can buy the plywood in sheets which are 4 ft. by 8 ft., and cost \$16.50 each, or they can buy the plywood in sheets which are 4 ft. by 6 ft., and cost \$14.00 each. They have to pick one size of plywood to buy- either the 4 ft. by 8 ft., or the 4 ft. by 6 ft. 3. If each piece must be cut whole from a sheet of plywood, what is the minimum number of sheets they will need for each size (4 ft. by 8 ft., and 4 ft. by 6 ft.)? Make a diagram showing the layout of pieces on each sheet of plywood. 4. How much will the dog house cost for each size option? 5. How much plywood will be wasted for each size option? 6. What size of plywood would you recommend to Mackenzie and her dad to buy, and why? Support your answer with reasons and the mathematics to support your answer.
Inspiration for Problem	"I think this is a strong mathematical modeling problem for middle schools students because it starts out being pretty straight-forward, and then becomes more open-ended. I think that this is beneficial, especially for younger middle school students so that they can get started right away, and build some confidence. I have found that when problems start out being very open-ended students sometimes have a hard time getting started. . . . This problem involves area of rectangles, areas of triangles, surface area of solids, unit conversion, and figuring out cost, which are all concepts that are accessible for middle school students. Furthermore, this problem is one that middle school students can relate to."
Challenges in Creating Problem	"At first, it was challenging to decide what to do for my problem. I had all these ideas, and concepts in my head and it was hard to figure out what concepts to focus on. I decided that I would want to use the problem with my own students, and I knew that I would be covering geometry concepts in both of my 6th grade classes, and my advanced 6th grade class. This problem is one that I could easily adapt to use in all three of my math classes and I am planning on using it soon."

Figure 25.B3. Project problem created by Student T3, and comments/reflections by the student.

Group Feedback	"I found that the process of creating the problem and refining it after getting feedback from my classmates was very helpful. My classmates were able to point out things in my problem that I had missed (such as confusing parts), and made suggestions to improve my problem. I also like having both undergraduate and graduate students in the class, because each group of students saw things from a different perspective."
Relevance of Creating Problem to Own Learning	"Furthermore, this process made me reflect on how I do things at school. I often share problems/activities that I have created with my peers. However, the process that we used in class made me think about getting more feedback from my colleagues before giving the problem to my students. . . . Another aspect of this class that I enjoyed and plan on implementing in my classes is providing the students with high-level problems/tasks that are sometimes open-ended. The problems that we worked on were real-world problems that I could relate to and could be easily adapted to be real-world for my students. As I reflect on this semester I have found that I left almost every class still thinking about the problem/task we were working on and I often came home to discuss the problem with my husband."
Additional Comments by Student	"Group work was a huge aspect of this class. I have my students work in groups a lot, but the environment that we created in in the course was different than the environment I have created in my classroom. In the course we worked in groups for every problem, which I found crucial for the tasks we were given. These problems were sometimes too difficult to do on my own, but with the contributions and ideas from my classmates, I was able to complete each problem/task given to us successfully. Although, I often have my students work in groups there usually is not the openness to talk to other groups and see what they think. Part of the reason is because I like to be in control in my classroom, and I do try to avoid chaos if at all possible. Before this class, I would have thought giving multiple groups the ability to work together would create chaos. Now, I think there is a lot of power in that, and that it can be done without creating chaos."

Figure 25.B3. (continued)

References

Ball, D. L., Bass, H., Lewis, J., Sleep, L., Suzuka, K., Babson, A., …, Zopf, D. (2008, April). *Designing and using problems to teach mathematical knowledge for teaching*. Paper presented at the National Council of Teachers of Mathematics Research Pre-session, Salt Lake City, UT.

Brown, S. I., & Walter, M. I. (1983). *The art of problem posing*. Philadelphia, PA: Franklin Institute Press.

Brown, S. I., & Walter, M. I. (1993). *Problem posing: Reflections and applications*. Hillsdale, NJ: Lawrence Erlbaum.

Cai, J., Moyer, J. C., Wang, N., Hwang, S., Nie, B., & Garbern, T. (2013). Mathematical problem posing as a measure of curricular effect on students' learning. *Educational Studies in Mathematics, 83*(1), 57–69.

Clements, M. A., & Ellerton, N. F. (2006). Historical perspectives on mathematics elegance: To what extent is mathematical beauty in the eye of the beholder? In P. Grootenboer, R. Zevenbergen, & M. Chinnappan (Eds.), *Identities, cultures, and learning spaces* (pp. 147–154). Adelaide, Australia: Mathematics Education Research Group of Australasia.

Del Campo, G., & Clements, M. A. (1987). *A manual for the professional development of teachers of beginning mathematicians*. Melbourne, Australia: Association of Independent Schools of Victoria, and the Catholic Education Office of Victoria.

Ellerton, N. F. (1986). Children's made-up mathematics problems: A new perspective on talented mathematicians. *Educational Studies in Mathematics, 17*(3), 261–271.

Ellerton, N. F. (2013). Engaging preservice middle-school teacher-education students in mathematical problem posing: Development of an active learning framework. *Educational Studies in Mathematics, 83*(1), 87–101.

English, L. (1997). Promoting a problem-posing classroom. *Teaching Children Mathematics, 4*(3), 172–179.

Freire, P. (1970). *Pedagogy of the oppressed*. New York, NY: Continuum Press.

Lin, P.-J. (2004). Supporting teachers on designing problem-posing tasks as a tool of assessment to understand students' mathematical learning. In M. J. Hoines (Ed.), *Proceedings of the 28th Conference of the International Group for the Psychology of Mathematics Education* (Vol. 3, pp. 257–264). Bergen, Norway: PME.

Munter, C. (2014). Developing visions of high-quality mathematics instruction. *Journal for Research in Mathematics Education, 45*(5), 584–635.

National Governors Association Center for Best Practices, Council of Chief State School Officers. (2010). *Common core state standards for mathematics*. Washington, DC: Author.

National Council of Teachers of Mathematics. (2000). *Principles and standards for school mathematics*. Reston, VA: Author.

Silver, E. A., & Burkett, M. L. (1994, April). *The posing of division problems by preservice elementary school teachers: Conceptual knowledge and contextual connections*. Paper presented at the Annual Meeting of the American Educational Research Association, New Orleans, LA.

Staebler-Wiseman, H. A. (2011). *Problem posing as a pedagogical strategy: A teacher's perspective* (PhD dissertation). Illinois State University, Normal, IL.

Stoyanova, E., & Ellerton, N. F. (1996). A framework for research into students' problem posing in school mathematics. In P. Clarkson (Ed.), *Technology in mathematics education* (pp. 518–525). Melbourne, Australia: Mathematics Education Research Group of Australasia.

Tarr, J. E., Grouws, D. A., Chávez, Ó., & Soria, M. (2013). The effects of content organization and curriculum implementation on students' mathematics learning in second-year high school courses. *Journal for Research in Mathematics Education, 44*(4), 683–729.

Victorian Curriculum and Assessment Authority. (2008). *Victorian essential learning standards*. Melbourne, Australia: Author.

Westbury, I. (1980). Change and stability in the curriculum: An overview of the questions. In H. G. Steiner (Ed.), *Comparative studies of mathematics curricula: Change and stability 1960–1980* (pp. 12–36). Bielefeld, Germany: Institut für Didaktik der Mathematik-Universität Bielefeld.

Wilhelm, A. G. (2014). Mathematics teachers' enactment of cognitively demanding tasks: Investigating links to teachers' knowledge and conceptions. *Journal for Research in Mathematics Education, 45*(5), 636–674.

Part IV
Mathematics Problem Posing: Some Concluding Comments

Chapter 26
Problem Posing in Mathematics: Reflecting on the Past, Energizing the Present, and Foreshadowing the Future

Nerida F. Ellerton, Florence Mihaela Singer, and Jinfa Cai

Abstract Five themes from the first 25 chapters of this book are identified: (a) the object of mathematical investigation as the construction of the problem itself and not just as finding the solution to a problem; (b) problem posing as an agent of change in the mathematics classroom; (c) integrating problem posing into mathematics classrooms; (d) problem posing as a conduit between formal mathematics instruction, problem solving, and the world outside the classroom; and (e) the need for appropriate theoretical frameworks for reflecting on problem posing. The fact that the chapters were prepared by a total of 52 authors from 16 countries is used to justify the claims that problem posing is not merely a local phenomenon, and that its place in school mathematics is gaining increasing recognition. Several imperatives for the field are set out, with mathematics educators urged to find ways and means of translating the obvious authenticity and enthusiasm displayed in this book into active research and practice in mathematics classrooms around the world.

Contents

N.F. Ellerton (✉)
Department of Mathematics, Illinois State University,
Normal, IL, Campus Box 4520, 61790-4520, USA
e-mail: nellert@ilstu.edu

F.M. Singer
Department of Educational Sciences, University of Ploiesti,
B-dul Bucuresti 39, Bucharest 100680, Romania

J. Cai
Department of Mathematical Sciences, University of Delaware, Newark, DE, USA

F.M. Singer et al. (eds.), *Mathematical Problem Posing*,
Research in Mathematics Education, DOI 10.1007/978-1-4614-6258-3_26

Origins and Emerging Themes

The origin of this book can be traced to appeals by mathematics educators committed to problem posing for a line of research to express their ideas, share their findings, and put forward their visions about their work. Within its chapters, there are contributions from 52 authors from 16 countries. Although some of the authors have met each other at different conferences and shared, first-hand, their work on problem posing, there has never been an opportunity for all of the authors to be together at any one place and time—except through the written words in this book.

Given this history, it should come as no surprise that the chapters present very different perspectives on problem posing at all levels of education. Some chapters discuss possible applications of problem posing at the elementary school level, while others set out to demonstrate the potential of problem posing with students at the college level. What the editors have found particularly striking, though, is the commonality of many of the emerging themes, and the overall authenticity of the issues addressed.

Reflecting on the Past

Traditional wisdom emerging from the experiences and practices of several centuries of school mathematics point to two fundamental assumptions held by many currently engaged in the teaching and learning of mathematics—that mathematics problems must be provided either by the teacher or by a textbook, and that the task for students is to solve these problems. Although stark contrasts can be drawn between the teaching of school mathematics in the 18th and 21st centuries (see, e.g., Ellerton & Clements, 2012; Siu, 2004), any attempt to challenge the first of these assumptions is likely to face direct resistance. As a consequence, problem posing is generally *not* placed at center stage in important curriculum documents.

Calls to engage students in posing mathematical problems are not new. In 1887, for example, in his book *Revised Model Elementary Arithmetic*, Henry Belfield (1887) wrote under the heading *Suggestions for Teachers*, "Children become interested in making their own problems. Let some abstract examples be assigned for the children to change to concrete problems" (p. 4). Some 50 years later, Albert Einstein and Leopold Infeld (1938) noted: "The formulation of a problem is often more essential than its solution, which may be merely a matter of mathematical or experimental skills" (p. 92). Karl Duncker (1945) observed that successive reformulations of a problem are part of the process of problem solving. Duncker talked about problem reformulation "as sharpening the original setting of the problem" (p. 9), noting that:

> *It is therefore meaningful to say that what is really done in any solution of problems consists in formulating the problem more productively.* To sum up: *The final form of a solution is typically attained by way of mediating phases of the process, of which each one, in retrospect, possesses the character of a solution, and, in prospect, that of a problem.* (p. 9, original emphasis)

Duncker was clearly making reference to what he regarded as an essential part of the process of solving a problem—that of finding ways to reformulate a problem until a solution path could be identified.

Two decades ago, in his reflections about problem-posing research, Edward Silver (1994) concluded that:

> Three major conclusions seem warranted from this review. First, it is clear that problem-posing tasks can provide researchers with both a window through which to view students' mathematical thinking and a mirror in which to see a reflection of students' mathematical experiences. Second, problem-posing experiences provide a potentially rich arena in which to explore the interplay between the cognitive and the affective dimensions of students' mathematical learning. Finally, much more systematic research is needed on the impact of problem-posing experiences on students' problem posing, problem solving, mathematical understanding and disposition towards mathematics. (p. 25)

Much more recently, Silver (2013), in his commentary for a special issue on problem posing published by *Educational Studies on Mathematics*, pondered over the diversity of form, context, and participants involved in contemporary problem-posing research. He asked:

> Is the time ripe for the field to make sharper distinctions among the several manifestations of problem posing as a phenomenon? To what extent is the activity of problem posing that a teacher does in order to provide an appropriate task to her students similar to or different from the activity of problem posing that a teacher does when posing problems for herself, or the activity of problem posing when her students pose problems for their classmates? (p. 159).

Silver then expressed his belief that the task of making sense of the diverse studies on problem posing would be greatly assisted through the development of theoretical frameworks which could help guide analysis and interpretation of the different approaches. Indeed, in the first chapter of this book, Cai, Hwang, Jiang, and Silber have made an attempt to review problem-posing research systematically and provide a comprehensive survey of the answered and unanswered questions in the field. Based on this survey, they, too, conclude that the field is now ripe for the development of cohesive theoretical frameworks that can organize the empirical results and our understanding of problem posing as a phenomenon.

Five Themes

The authors in this book have, collectively, addressed major themes in the contemporary problem-posing literature. In writing this final chapter, we have chosen to highlight the following five themes, based on the challenges laid down by Silver (1994, 2013), and in response to the questions raised in Chapter 1 by Cai et al.:

- Problem posing can transform attitudes towards mathematics so that the object of mathematics is the problem not just the solution to a problem.
- Problem posing can be an agent of change in the mathematics classroom.
- Through purposeful planning, problem posing can be integrated into school mathematics curricula.
- Since problem posing can be inextricably linked with real-life situations, it can be seen as a natural link between formal mathematics instruction, problem solving, and the world outside the classroom.
- Appropriate theoretical frameworks need to be developed to help analyze, interpret, and ultimately adapt and apply different models of problem posing to a wide range of contexts.

Rather than offer literal chapter-by-chapter breakdowns of the research reported in this book, these themes have been chosen by the editors as ways to begin to uncover emergent ideas on problem posing, and to identify specific questions that will form bases for further investigations. We do not consider chapters as falling exclusively under only one of these five themes; some chapters clearly present research that is relevant to several of them. In the discussion which follows, therefore, reference to a chapter as an example under one theme should not be taken to imply that that chapter addresses research covered only under that theme.

Theme 1: The Object of Mathematics Is the Problem Itself, Not Just the Solution to a Problem

Problem posing has often been seen as a means to an end—the end being the solution to the problem that was posed. Several authors in this book have turned this around, making the problem itself the object of the activity, either to gain different insights into the act of problem posing, or to use the activity as a means of assessing student understanding of the mathematical concepts involved. In mathematical modelling, for example, the formulation and reformulation of appropriate problems is central to the interpretation and application of the model; thus by the predictive nature of the problems posed, the object of the activity is to generate problems rather than to find solutions (e.g., Hansen & Hana, Chapter 2). Investigations that look at the teacher's role in posing problems provide another approach to focusing on mathematics problems and their development, rather than only on problem solutions (e.g., Milinković, Chapter 3; Klinshtern, Koichu and Berman, Chapter 22).

It goes without saying that unless teachers feel comfortable posing problems themselves, they are unlikely to involve students directly in posing problems. And those who struggle to pose problems will be less likely to find ways of linking problem posing with real-life situations. If teachers have little confidence in their own ability to pose problems, their attempts to implement problem-posing activities and tasks called for in any intended curriculum produced by the school, district, state, or nation are likely to be clumsy and uninspiring.

Logic would suggest that the earlier the teachers gain experience with posing problems, or with seeing ways in which they can use problem posing with their students, the more likely they will be to feel that problem posing is a natural and fundamental aspect of the whole process of teaching and learning mathematics. From that perspective, it would seem to be important that problem posing be a central aspect of mathematics teacher education. Along those lines, Grundmeier (Chapter 20); Tichá and Hošpesová (Chapter 21); Osana and Pelczer (Chapter 23), Crespo (Chapter 24), and Ellerton (Chapter 25) all focus on problem-posing research involving preservice mathematics teachers. They provide an informative overview that can be used as a resource of problem-posing situations and ways to conceptualize teaching strategies.

Theme 2: Problem Posing Can Be an Agent of Change in the Mathematics Classroom

An agent of change generates changes in attitudes on the part of all stakeholders. In the context of mathematical problem posing, changes in teaching and learning sequences and activities need to occur in mathematics classrooms. As noted in Chapter 1, a largely unexplored question is how students and teachers can renegotiate traditional roles and norms of the mathematics classroom so that problem posing becomes an accepted practice. Mathematics is viewed by some as difficult or uninteresting or mechanical—yet to many it can be a source of fascination, excitement, and beauty. If, through problem posing, mathematics can be seen by learners and teachers as a creative endeavor, then problem posing can become an agent of change in classrooms. Several authors (e.g., Bonotto and Dal Santo, Chapter 5; Matsko amd Thomas, Chapter 6; Singer and Voica, Chapter 7; Van Harpen and Presmeg, Chapter 14) discuss different aspects of the use of problem posing for encouraging or investigating creativity in the mathematics classroom. These authors offer promising leads worthy of further investigation.

Theme 3: Integrating Problem Posing into School Mathematics Curricula

Curriculum issues are a central concern to schools (the intended curriculum), to teachers (the implemented curriculum), and to students (the attained curriculum). As indicated in Chapter 1, at present, little attention has been paid to the issue of including problem posing in the school mathematics curriculum. More effort is needed to study how best to achieve this integration. The authors of several chapters in this book propose innovative ways of including problem posing in intended and implemented mathematics curricula and suggest different approaches for facilitating that inclusion. Technology surfaces as an important mediator in problem-posing activities (e.g., Abramovich and Cho, Chapter 4; Imaoka, Shimomura and Kanno, Chapter 12; Leikin, Chapter 18; Lavy, Chapter 19), and the use of everyday problem-posing contexts (such as the school playground, or sociocultural language-mediated activities) has been found to provide fertile grounds for incorporating problem posing into curricula (e.g., Gade and Blomqvist, Chapter 9; Klaassen and Doorman, Chapter 10; English and Watson, Chapter 11; Crespo, Chapter 24). Moreover, with any discussion of the integration of problem posing into curricula, attention should be given to the potential of using problem posing to assess mathematical understanding. This issue is addressed in a number of chapters and is the main focus used by Kwek (Chapter 13). A framework for assessing mathematical understanding is also developed by Singer and Voica (Chapter 7) who show that problem posing can be used in informal contexts with effective results for the development of students' mathematical thinking.

As the experimental approaches developed in this book show, the inclusion of students' problem-posing sessions into the implemented curriculum can cover various time frequency ranges—from a systematic base (for example, 10 minutes during each lesson) to just ad-hoc insertions that fit particular learning situations. Various ranges of contexts—from open-ended proposals, to reformulations of given problems, to generation of problems based on one or more given conditions, and different phases of the lesson, from facilitating the introduction of a new concept or procedure to the assessment of learning—are included in the book.

Theme 4: Problem Posing Can Provide Natural Links Between Formal Mathematics Instruction, Problem Solving, and the World Outside the Classroom

Strategies to integrate problem posing into intended and implemented mathematics curricula can often be seen as natural links between formal mathematics instruction, problem solving, and the real world.

Problem solving has frequently been viewed as both a means to an end and an end in itself—in relative isolation from problem posing. Although some relationships between problem solving and problem posing have been explored by a number of researchers (see the review in Chapter 1 and also, for example, Kar, Özdemir, İpek, & Albayrak, 2010; Showalter, 1994; Singer & Voica, 2013), it is only recently that researchers have begun to make direct links between the theories, practices, and challenges of problem solving and problem posing. Some of these attempts are successfully reflected in the present book (e.g., Cifarelli and Sevim, Chapter 8; Chen, Van Dooren and Verschaffel, Chapter 15; Rosli et al., Chapter 16; Prabhu and Czarnocha, Chapter 17). By balancing the two facets of mathematical activity, these authors bring evidence for the importance of linking problem solving and problem posing within the course of students' conceptual development.

More and more researchers are becoming aware that in the dynamic world of today, schools are at risk if they function in isolation from society. A big question is how to address this issue. This book brings a plea that problem-posing activities incorporated into intended and implemented curricula can facilitate connections between classroom mathematics, the worlds of work, and social activity outside the classroom (e.g., Singer and Voica, Chapter 7; Gade and Blomqvist, Chapter 9; Klaassen and Doorman, Chapter 10; English and Watson, Chapter 11). Real world artefacts (Bonotto and Dal Santo, Chapter 5), real-life situations (Singer and Voica, Chapter 7), and recent technology (Abramovich and Cho, Chapter 4; Lavy, Chapter 19) may become access points for developing mathematical thinking that is embedded into contemporary reality.

Theme 5: Theoretical Frameworks and Questions: Reflections on the Future of Problem Posing

Although problem posing has often been recognized as central to mathematics teaching and learning, it has nevertheless remained on the periphery in curriculum documents. Problem posing is at once fully embraced as an important element of mathematics, yet all but rejected when it comes to discussions about ways of incorporating it into regular classroom activities. Research on more traditional aspects of the teaching and learning of mathematics—including student understanding of content, classroom discourse, and approaches to teaching and teacher education—seem to be more attractive than problem posing to mathematics education researchers, and efforts to incorporate problem posing into the mainstream of mathematics education research have often been marginalized.

Collectively, the research presented in this book boasts two key features that are of particular note. First and foremost, the chapters all reflect *authenticity*—the problem-posing research reported in each chapter is grounded in the reality of classrooms. This authenticity is carried forward, in part, by the obvious enthusiasm of the authors for involving students in problem posing in the classroom. Second, the chapters present a diverse range of creative approaches to problem posing in a wide range of classroom contexts. It is significant that this book includes research on problem posing which was carried out in 16 different nations; clearly, the importance of problem posing is recognized in many parts of the world. That said, however, it would be an exaggeration to say that problem posing is being widely used in the world's classrooms. Gaps between research and practice, and between advocacy and actuality, need to be recognized and addressed.

Various frameworks are conceptualized in this book for facilitating students' problem posing, namely, the importance of children recognizing and utilizing problem structures and problem qualities (such as coherence and consistency), the need to consider students' perceptions of and preferences for different problem types, and a focus on developing their diverse mathematical thinking and transfer capabilities (e.g., Bonotto and Dal Santo, Chapter 5; English and Watson, Chapter 11; Singer and Voica, Chapter 7). Other frameworks focus on professional development aimed at enhancing teachers' problem-posing skills so that they will be able to pose appropriately challenging problems for their students (Ellerton, Chapter 25), engage in problem posing with their students, and take account of sociocultural factors (Crespo, Chapter 24), and as they pose creative new problems in what some have described as a discovery experience (Abramovich and Cho, Chapter 4).

The production of this book carries with it a certain momentum. It represents the culmination of several years' work aimed at showing how problem posing can provide a powerful approach for the teaching and learning of mathematics around the world and builds on the research presented in a special issue of *Educational Studies in Mathematics* (see Singer, Ellerton, & Cai, 2013). But the excitement with which one can view this authentic, diverse, and international collection of research studies must be tempered by considering the following ongoing challenges. How can these individual groups of researchers influence school mathematics in their respective countries to such an extent that genuine problem posing will become an integral and valued part of the mathematics curriculum? How can teachers see problem posing as being a natural part of the teaching and learning of mathematics? We hope that this book will, in fact, encourage mathematics teachers at all levels to adopt and adapt the strategies explored in its chapters. We also hope that this book will make an impact on the field, and that the theoretical frameworks put forward will help support the efforts of mathematics education researchers investigating problem posing. Many questions about problem posing remain—but the book provides a sense of direction for those willing and able to take up the mantle. Indeed, it is in this sense that we, the editors, believe that the book can foreshadow the future.

Final Comments

Although we feel privileged to have been editors for this book, we recognize that the role has carried with it great responsibility to the field. Through working closely with the authors and reviewers of the chapters, we have had the opportunity to reflect on the questions raised in Chapter 1 and to examine more closely the *Active Learning Framework* and a *Pedagogy for Problem Posing* put forward in Chapter 25.

Rather than repeat details here, or reproduce the *Framework*, we draw attention to the following points as imperatives for the field:

- If we accept the assumption that problem posing is important at all levels of school mathematics, then ways must be found to convince all stakeholders of its key role in the teaching and learning of school mathematics.
- It follows that appropriate support for both the appreciation of the role of problem posing, and the development of problem-posing skills for practicing teachers and for preservice teacher education students, need to be provided.
- Research which underpins our understanding of the cognitive processes associated with problem posing needs to be extended. In particular, further research is needed on the links between problem posing and mathematical creativity on the one hand, and between problem posing and problem solving on the other.
- We need to know more about the kinds of classroom environments that have the potential to support students' problem-posing activities, and how technology can be used to enhance and enrich opportunities for, and attitudes towards, problem posing.
- Ways in which problem posing can be used both to enhance and to assess students' learning need to be explored.
- Research is needed to understand more about a *Pedagogy for Problem Posing*.
- Problem posing is rarely seen as the pinnacle of mathematics achievement; rather, it is often viewed as a poor cousin of problem solving. Strategies for achieving changes in attitudes and values among mathematics educators are needed. Frameworks such as the *Active Learning Framework* (Ellerton, 2013) may provide a means to explore how traditional views can be turned around, so that problem posing can become the crown jewel rather than an optional extra of mathematics education.

As this book was in the final stages of preparation, Nicholas Wasserman's (2014) article *A Rationale for irrationals*: *An unintended exploration of e* appeared. An introductory sentence placed above the title reads: "The practice of problem posing is as important to develop as problem solving. The resulting explorations can be mathematically rich" (p. 500). Students adopted a "What-If-Not" approach (Brown & Walter, 2005), and followed one of Polya's (1945) strategies for solving problems: "Solve a simpler problem." Although it was encouraging to see this example of problem posing being incorporated into a mathematics classroom, at the same time it was discouraging to realize that the use of problem posing for the example had been unintended. For teachers and students alike, an *awareness* of the challenges and learning opportunities presented by the posing of mathematical problems is fundamental.

It is our hope that the chapters in this book can act as a springboard for future research, and as a source of problem-posing ideas and strategies for a range of classrooms and contexts. In particular, it is our hope that those who read this book will take steps to incorporate problem posing more fully into their own practices, as well as inspire others to undertake the research and reflection on problem posing that is needed to carry the field forward.

References

Belfield, H. H. (1887). *Revised model elementary arithmetic*. Chicago, IL: Geo. Sherwood.

Brown, S., & Walter, M. I. (2005). *The art of problem posing* (3rd ed.). Mahwah, NJ: Erlbaum.

Duncker, K. (1945). On problem solving. *Psychological Monographs* (Vol. 58). New York, NY: American Psychological Association.

Einstein, A., & Infeld, L. (1938). *The evolution of physics from early concepts to relativity and quanta*. New York, NY: Simon & Schuster.

Ellerton, N. F. (2013). Engaging preservice middle-school teacher-education students in mathematics problem posing: Development of an active learning framework. *Educational Studies in Mathematics, 83*(1), 87–101. doi:10.1007/s10649-012-9449-z.

Ellerton, N. F., & Clements, M. A. (2012). *Rewriting the history of school mathematics*. New York, NY: Springer.

Kar, T., Özdemir, E., İpek, A. S., & Albayrak, M. (2010). The relation between the problem-posing and problem-solving skills of prospective elementary mathematics teachers. *Procedia Social and Behavioral Sciences, 2*(2), 1577–1573.

Polya, G. (1945). *How to solve it*. Princeton, NJ: Princeton University Press.

Showalter, M. E. (1994). Using problems to implement the NCTM's professional teaching standards. *Mathematics Teacher, 87*(1), 5–7.

Silver, E. A. (1994). On mathematical problem posing. *For the Learning of Mathematics, 14*(1), 19–28.

Silver, E. A. (2013). Problem posing research in mathematics education: Looking back, looking around, and looking ahead. *Educational Studies in Mathematics, 83*(1), 157–162. doi:10.1007/s10649-013-9477-3.

Singer, F. M., Ellerton, N. F., & Cai, J. (2013). Problem-posing research in mathematics education: New questions and directions. *Educational Studies in Mathematics, 83*(1), 1–7. doi:10.1007/s10649-013-9478-2.

Singer, F. M., & Voica, C. (2013). A problem-solving conceptual framework and its implications in designing problem-posing tasks. *Educational Studies in Mathematics, 83*(1), 9–26. doi:10.1007/s10649-012-9422-x.

Siu, M. K. (2004). Official curriculum in mathematics in ancient China: How did candidates study for the examination. In L. Fan, N.-Y. Wong, J. Cai, & S. Li (Eds.), *How Chinese learn mathematics: Perspectives from insiders*. Singapore: World Scientific.

Wasserman, N. H. (2014). A rationale for irrationals: An unintended exploration of e. *Mathematics Teacher, 107*(7), 500–507.

Author Index

F.M. Singer et al. (eds.), *Mathematical Problem Posing*,
Research in Mathematics Education, DOI 10.1007/978-1-4614-6258-3

Subject Index

[1] Note: "Some entries in the index such as "context," "creativity," "mathematics," "teaching," or "problem solving," include page references only to the first instance of that entry in a given chapter,and if applicable, where the term is explained in detail."

F.M. Singer et al. (eds.), *Mathematical Problem Posing*, Research in Mathematics Education, DOI 10.1007/978-1-4614-6258-3

Q

R

S

Printed by Printforce, the Netherlands